LASER PHYSICS

MURRAY SARGENT III

MARLAN O. SCULLY

WILLIS E. LAMB, JR.

Optical Sciences Center, University of Arizona

1974

Addison-Wesley Publishing Company
Advanced Book Program
Reading, Massachusetts

London · Amsterdam · Don Mills, Ontario · Sydney · Tokyo

Laser Physics

Murray Sargent III, Marlan O. Scully, Willis E. Lamb, Jr.

First printing, 1974
Second printing, with corrections, March 1977
Third printing, with additional corrections, November 1977
ISBN 0-201-06904-0
ISBN 0-201-06903-2 (pbk.)
Fourth printing, 1982

Library of Congress Cataloging in Publication Data

Sargent, Murray.
Laser physics.

1. Quantum optics. 2. Lasers. I. Scully, Marlan O., joint author. II. Lamb, Willis Eugene, 1913– joint author. III. Title.
QC476.2.S27 535.5′8 74–5049

ISBN 0–201–06718–8
ISBN 0–201–06719–6 (pbk.)

Printed in the United States of America

DEFGHIJKLM-MA-898765432

To Helga

Judith

Ursula

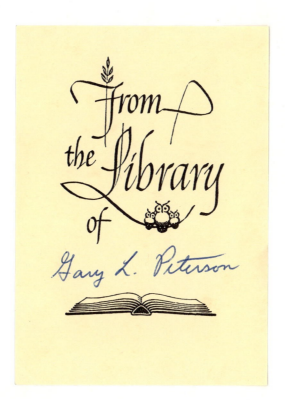

Laser Physics

CONTENTS

PREFACE

This book treats the interaction of radiation with matter, particular attention being paid to the laser. Knowledge is assumed of the usual half-year introduction to quantum mechanics found in undergraduate physics curricula. The material can be covered in two semesters, or, alternatively, the first part (Chaps. 1–13) can be used as a one-semester course in which quantum-mechanical aspects of the electromagnetic field are ignored. Each chapter is accompanied by problems that illustrate the text and give useful (occasionally new) results.

Existing laser media are intrinsically quantum mechanical and are most easily studied with the quantum theory. Understanding the laser along these lines enlivens one's understanding of quantum mechanics itself. In fact, the material constitutes a viable, applied alternative for the usual second and third semesters of quantum mechanics.

The text format lends itself quite well to reference use, particularly with regard to the fundamental concepts of laser physics. There is leeway for some detailed extensions, notably in the problems, but we have deliberately sacrificed generality for the sake of clarity. An understanding of the simpler theories enables one to work with the more general extensions available in the original literature or to treat new problems oneself.

With the exception of Chaps. 19 and 20, the laser theory discussed in this book tends to follow the approaches of the Lamb school. Parallel work of the Bell Telephone Laboratories group has been presented in the book by W. H. Louisell, *Quantum Statistical Properties of Radiation,* (John Wiley & Sons, New York, 1973), while H. Haken's contribution, "Laser Theory," in *Encyclopedia of Physics,* Vol. XXV/2c, edited by S. Flügge, Springer-Verlag, Berlin, 1970; also Chap. 23 in *Laser Handbook,* gives a very complete account of the Stuttgart work.

In keeping with the text format, no uniform attempt is made to assign credit to the original papers. Rather, we have referred the reader to review articles and books which he might find useful for further study. Even in this capacity, we do not give all possible references; A. E. Siegman's list of laser books (*Appl. Opt.,* **18,** A38, 1971) includes 135 entries alone and is 3 years older than this book.

The first 13 chapters use the semiclassical theory of the interaction of radiation with matter. For this, the atoms are assumed to obey the laws of quantum mechanics and the field is governed by classical Maxwell equations. Chapters 14 through 20 treat both field and atoms quantum mechanically. More specifically, Chaps. 1 and 2 present basic quantum preliminaries. Chapter 3 discusses classical and quantum dipole moments and their interaction with electric fields. Chapter 4 traces historical analogs of some laser phenomena from Huygens to Van der Pol. Chapter 5 applies the theory developed to the first maser, introducing important laser concepts like cavity Q, saturated gain, and cavity tuning. Chapter 6 reviews quantum formalism especially useful for Chaps. 14 through 20 (Dirac notation, Schrödinger, interaction, and Heisenberg pictures). Chapter 7 introduces the density matrix, which is used to considerable analytical and pedagogical advantage in subsequent chapters.

Chapters 8 through 11 develop the theory of the laser with single- and multimode electromagnetic fields, with homogeneously and Doppler broadened media, and with two mirror and ring cavities. The atoms are approximated by two level systems, and the field is taken to be scalar. In Chap. 12 these restrictions are removed, and an applied DC magnetic field is considered. Chapter 13 treats pulse propagation, self-induced transparency, and photon echo in terms of the semiclassical theory.

Chapter 14 presents the quantum theory of radiation in a form suitable for treatment of the laser. The electromagnetic field is shown to consist of simple harmonic oscillators with straightforward quantization rules. The Weisskopf-Wigner theory of spontaneous emission is given as an important example. An accompanying appendix deals with the phenomenon of superradiance. Chapter 15 develops the coherent state—the state most closely approximating the classical electromagnetic field. The state provides not only a quantum-classical bridge, but also an instructive formalism for the quantum laser. Chapter 16 defines the system-reservoir problem with the use of the reduced density operator. This problem is fundamental to all physical situations involving the interaction of a system of interest with an environment having many degrees of freedom. The techniques of Chaps. 14 through 16 are used in Chap. 17 to treat the laser from a fully quantum-mechanical point of view, yielding the semiclassical theory in the appropriate limit and providing information about photon statistics, laser linewidth, and buildup from noise. Chapter 18 discusses the theory of measurement as applied to laser problems. Chapter 19 revisits the system-reservoir problem from a Brownian motion viewpoint, which is based on the Heisenberg picture of quantum mechanics. This approach is employed in Chap. 20 to treat the laser. These two chapters are patterned after the work of Lax with suitable pedagogical modifications. Chapter 21 provides an overview of laser physics, relating it to other subjects in the field loosely called "quantum optics," including Josephson radiation. In this connection, the approaches developed in the book are eminently applicable to laser problems; they are, furthermore, general and hence can often be applied elsewhere. It is hoped that the reader can obtain from the book not only an increased understanding of the laser, but also a more profound comprehension of many-system physical phenomena in general.

Acknowledgments

We are indebted to numerous students for corrections and suggestions improving the clarity of the material. In particular we would like to thank C. L. O'Bryan, W. W. Chow, D. Jones, J. A. Gorrell, E. L. Giezlemann, J. B. Hambenne, D. R. Hanson, J. C. Matter, A. L. Smirl, E. W. Van Stryland, and H. Weichel. We are grateful also to many colleagues for helpful suggestions and encouragement, and thank especially P. R. Berman, H. A. Haus, A. G. Hill, F. A. Hopf, S. F. Jacobs, W. H. Louisell, J. Moore, T. Oka, D. N. Rogovin, R. F. Shea, H. Shih, and R. L. Shoemaker.

MURRAY SARGENT III
MARLAN O. SCULLY
WILLIS E. LAMB, JR.

LASER THEORY NOTATION

a	annihilation operator (boson)	(14.11)
$a^\dagger$	creation operator (boson)	(14.12)
a	as subscript, refers to the upper level a of a two-level atom	(2.14)
a', a''	magnetic quantum numbers for level a	(Chap. 12)
a_0	Bohr radius (0.53 Å)	(1.27)
a_s	annihilation operator for mode s (boson)	(14.41)
$a_\mu(t)$	general Heisenberg picture operator	(19.52)
$\{a\}$	set of quantum operators $a_1, \ldots, a_\mu, \ldots$ (not necessarily annihilation operators)	(19.52)
$\|a\rangle$	energy eigenstate for level a	(6.46)
A	Einstein A coefficient	(2.38)
$\mathbf{A}$	vector potential	(2.7)
$A(t)$	slowly varying annihilation operator $[a(t) = A(t)\exp(i\Omega t)]$	(19.33)
$A_\mu(t)$	slowly varying operator for a_μ $[a_\mu(t) = A_\mu(t)\exp(-i\Omega_\mu t)]$	(19.58)
$\mathscr{A}$	linear gain constant in quantum theory of laser	(17.13)
$\mathscr{A}_b$	active medium linear absorption constant	(17.58)
$\mathscr{A}_c$	complex linear gain coefficient in quantum theory of laser	(20.49)
b	as subscript, refers to the lower level b of two levels	(2.14)
$b_k(t)$	annihilation operator for mode k	(19.24)
$\|b\rangle$	energy eigenstate of lower of two levels	(6.46)
B	Einstein B coefficient	(2.38)
$\mathbf{B}$	magnetic induction	(8.2)
$\mathscr{B}$	lowest-order saturation coefficient in quantum theory of laser	(17.14)
$\mathscr{B}_c$	complex saturation coefficient for $\mathscr{B}$	(20.50)
$\boldsymbol{\mathscr{B}}$	effective magnetic induction	(7.77)
c	speed of light (2.99793×10^8 m/sec)	
$c_a(t)$	probability amplitude for upper level a (Schrödinger picture)	(6.71)
$c_b(t)$	probability amplitude for lower level b (Schrödinger picture)	(6.72)

$c_n(t)$	probability amplitude for evergy level n (Schrödinger picture)	(6.67)
$C_a(t)$, $C_b(t), C_n(t)$	as above, but in interaction picture [$C_x = c_x \exp[(i\omega_x t)]$, $x = a, b, n$	(2.14), (1.13)
$C_{an}(t)$	probability amplitude for atom in upper level, field with n photons (interaction picture)	(14.68)
$c_{n_1, n_2, \cdots n_r, \cdots} \equiv c_{\{n_r\}}$	probability amplitude for multimode field with n_1 photons in mode 1, n_r photons in mode r.	(14.48)
$C_n(t)$	in Sec. 10-3 and Prob. 8-1 only, cosine (in-phase) component of polarization	(8.61)
$\mathscr{C}_{an}(t)$	probability amplitude for atom in upper state, an n-photon laser field and vacuum otherwise (interaction picture)	(I.18)
c.c.	abbreviation for "complex conjugate"	
d	constant term in mode-locking equation	(9.54)
$\mathbf{D}$	displacement vector in Maxwell's equations	(8.2)
D_{vv}	diffusion coefficient for Brownian motion	(19.6)
$D(\varphi)$	phase diffusion coefficient in quantum theory laser	(20.74)
$D(\alpha)$	displacement operator for coherent states	(15.61)
$D(z, v, t)$	population difference for two-level systems	(10.58)
D_μ	drift coefficient for operator A_μ	(19.60)
$\langle D_{\mu\nu}\rangle$	diffusion coefficient for operators A_μ and A_ν	(19.63)
$\mathscr{D}_x(\Delta\omega)$	complex denominator $= 1/(\gamma_x + i\,\Delta\omega)$	(9.6)
$\mathscr{D}_n$	sums of complex denominators	(E. 12)
$\mathfrak{D}(\omega)$	density of free-field radiation states	(14.101)
e	electron charge $= -1.6022 \times 10^{-19}$ coulombs	(1.24)
e^n	$(e)^n$, n an integer	
$\hat{e}_1$, $\hat{e}_2$, $\hat{e}_3$	unit vectors in abstract Cartesian coordinate system, used for the pictorial representation of the density matrix	(7.70)
E_0	scalar electric field amplitude	(2.13)
$E(z, t)$	plane wave electric field	(8.8)
$\mathbf{E}(\mathbf{R}, t)$	electric field vector	(2.6)
$E_n(t)$	slowly varying Fourier amplitude for $E(z, t)$	(8.8)
$\mathscr{E}(z, t)$	complex electric field envelope in pulse propagation	(13.1)
$\mathscr{E}(t)$	Positive frequency part of electric field E(t)	(16.36)
$\mathscr{E}$	electric field "per photon"	(14.17)
$\bar{E}$	dimensionless electric field amplitude $= \sqrt{I}$	(E. 11)

$f(x)$	arbitrary function of x	
$f(t)$	rapidly varying noise operator for annihilation operator $a(t)$	(19.32)
$F(t)$	slowly varying noise operator for $A(t)$ $[F(t) = f(t) \exp(i\Omega t)]$	(19.34)
$F_\mu(t)$	slowly varying noise operator for $A_\mu(t)$	(19.60)
$F_\nu(t)$	random classical force on particle in Brownian motion	(19.2)
F_1	first-order factor in laser coefficients	(Tables 8-1, 10-1)
F_3	third-order factor in laser coefficients	(Tables 8-1, 10-1)
$\mathscr{F}\{f(x)\}$	Fourier transform of $f(x)$	(15.48)
$\mathfrak{F}$	real part of continued fraction in strong signal theory	(E. 32)
g	coupling constant in quantum theory of radiation	(14.60)
$g(\omega)$	coupling constant for frequency ω	(14.101)
g_s	coupling constant for sth mode of field	(14.88)
g_a	Landé g factor for upper level of two-level atom	(12.18)
g_b	Landé g factor for lower level of two-level atom	(12.18)
g_e	electron g factor	(1.37)
g_{ij}	$(i, j = \pm)$ conductivity matrix elements	(12.7)
G	conductivity matrix (in ring and Zeeman laser theory)	(12.7)
$G(x, x_0, t)$	Green's function	(H. 17)
$G^{(n)}(x_1, \ldots, x_{2n})$	n-th order correlation function	(15.59)
h	Planck's constant $= 6.626196 \times 10^{-34}$ joule-sec	
$\hbar$	Planck's constant$/2\pi = 1.05459 \times 10^{-34}$ joule-sec	(1.6)
$\mathbf{H}$	magnetic field	(1.34)
H	magnitude of magnetic field $\mathbf{H}$	(1.34)
H_y	y component of magnetic field	(14.7)
$H_n(\xi)$	Hermite polynomial	(1.22)
$\mathscr{H}$	total Hamiltonian	(2.1)
$\mathscr{H}_0$	unperturbed Hamiltonian	(2.1)
$\mathscr{H}_C$	classical energy	(1.24)
i	$\sqrt{-1}$	
I_n	dimensionless intensity for nth mode $= \frac{1}{2}(\wp E_n/\hbar)^2 (\gamma_a \gamma_b)^{-1}$	(8.45)
$\mathscr{I}$	identity operator for matrix	(6.13)

$\mathrm{Im}(\mathscr{P})$	imaginary part of $\mathscr{P}$	
$I(z, t)$	intensity envelope in pulse propagation	(13.28)
$\mathscr{I}(z,t)$	partial energy integral	(13.32)
$\mathbf{J}$	current density in Maxwell's equations	(8.2)
$J_n(\Gamma)$	nth-order Bessel function	(9.101)
J_a, J_b	angular momenta for upper and lower levels	(Fig.12-2)
k_B	Boltzmann's constant $= 1.38062 \times 10^{-23}$ J/K	
K	wave number $\approx K_n$ (when difference between K_n's does not matter)	(3.38)
K_n	wave number for nth mode of electric field	(8.5)
Ku	$= K$ times u: Doppler broadening constant	(10.2)
l, l_c, l_s	locking parameters	(9.59), (9.54)
l_x, l_y, l_z	lengths of x, y, and z dimensions in maser cavity	(5.23)
L	mirror separation in laser	(Fig. 8-2a)
$\mathbf{L}$	orbital angular momentum vector	(after (6.41))
L_z	z component of orbital angular momentum $\mathbf{L}$	(after (6.41))
$\mathscr{L}_x(\omega - \nu)$	dimensionless Lorentzian $= \gamma_x^2/[\gamma_x^2 + (\omega - \nu)^2]$ (x can be blank)	(8.36)
m	particle mass	(1.8)
M	mass constant for radiation field oscillator	(14.6)
$M(z, v, t)$	population sum in strong-signal laser theory	(10.59)
M_n	nth-order moment in Langevin process	(19.8)
$\|n\rangle$	energy eigenstate with eigenvalue $\hbar\omega_n$ (usually photon number state)	(6.28)
$\langle n(t)\rangle$	average number of photons	(16.31)
n_a	number of atoms in upper level (Einstein theory)	(2.38)
n_b	number of atoms in lower level (Einstein theory)	(2.38)
$\bar{n}_{ss}$	steady-state average number of photons in laser field	(17.33)
$\|n_1 n_2 \ldots n_r\rangle$	$\equiv \|\{n_s\}\rangle$, multimode field eigenstate	(14.46)
N	number of modes in multimode laser operation [or $N(z, t)$, see below]	(sec. 9-4)
N	number of atoms in Langevin theory	(20.10)
$\bar{N}$	average population inversion density	(8.42)
N_{2l}	$2l$th component of the population inversion density	(9.16)
$N(z, t)$	population inversion density	(8.39)
N_a, N_b	number of atoms in upper and lower states	(20.39)

$\mathfrak{N}$	relative excitation	(8.54)
$\mathcal{N}_{nm}, \mathcal{N}_{nm}'$	number factors in quantum theory of the laser	(17.15)
$\mathcal{N}$	population inversion operator	(20.45)
$\mathcal{N}$	normalization constant	(8.22)
$\mathfrak{N}$	set of index values whose corresponding field amplitudes are nonzero	(9.72)
$\mathcal{O}$	arbitrary quantum-mechanical operator	(1.4)
P_n	probability of n photons	(15.13)
$\mathbf{p}$	particle momentum	(1.7)
$\mathbf{P}$	polarization vector in Maxwell's equations	(8.2)
$P(z, t)$	scalar polarization of medium	(8.9)
$P(\alpha), P(\alpha,t)$	diagonal coherent state representation of density operator	(15.2)
$P(x, t)$	probability density for a particle at x (time t)	(16.66)
$P(x, y), P(r, \theta)$	two-dimensional probability densities	(16.119), (16.86)
P_a	transition probability to state $\|a\rangle$	(2.31)
P_s	probability of stimulated emission	(2.50)
P_ψ	probability of the state vector $\|\psi\rangle$ in a mixture	(7.17)
$\mathcal{P}_n(t)$	slowly varying complex polarization for mode n	(8.9)
$\mathcal{P}(z, t)$	slowly varying complex polarization in pulse propagation	(13.2)
$\wp$	electric-dipole matrix element between levels a and b (taken real)	(2.16)
$\wp_{a'b'}$	electric-dipole matrix element between levels a' and b'	(12.23)
q	position coordinate	(14.6)
q_n	Fourier coefficient of population difference $D(z, v, t)$ and in quadrature coefficients $S(z, v, t)$	(10.60)
Q	cavity quality factor	(5.19)
Q_n	cavity quality factor for nth mode	(8.10)
Q_x, Q_y	Q's for x and y polarizations of electric field	(12.16)
r	radial coordinate in polar coordinate system	(Fig. 1-4)
$\mathbf{r}$	position vector in polar coordinates	(Fig. 1-4)
r_a, r_b	excitation rates to states $\|a\rangle$ and $\|b\rangle$	(16.1)
r_n	complex ratios of atomic Fourier coefficients (q_n/q_{n-1})	(E.14)
r_0	classical electron radius $\approx 2.8 \times 10^{-15}$ m	(3.22)

R	rate constant	(8.35)
$\mathbf{R}$	vector in pictorial representation of density matrix	(7.70)
R_1, R_2, R_3	components of $\mathbf{R}$ in abstract Cartesian space	(7.66), (7.67), (7.69)
$R_{nl}(r)$	associated Laguerre polynomials	(1.26)
R_s	saturation parameter $= 1/(\gamma_a^{-1} + \gamma_b^{-1})$	(8.38)
$\mathscr{R}_a, \mathscr{R}_b$	single-mode rate coefficients in quantum theory of reservoirs	(16.25) (16.22)
$R(\alpha^*, \beta)$	coherent state representation of density operator	(15.36)
$\text{Re}\{\mathscr{P}\}$	real part of complex quantity $\mathscr{P}$	
R_∞	Rydberg constant (109737 cm^{-1})	
s_n	constants in third-order complex polarization integrals T_{lw}	(D.6)
S	similarity transformation (Zeeman theory)	(12.14)
$S(z, v, t)$	in-quadrature component of polarization (strong-signal theory)	(10.56)
$S_n(t)$	in-quadrature component of nth-mode polarization	(8.60)
$\mathscr{S}$	stabilization factor (laser theory)	(8.57)
$t, t', t'', t''', t_1, t_2, t_0, t_n$	various times	
T	temperature in degrees Kelvin	(16.1)
T_1	level lifetime of atomic system $[= \frac{1}{2}(\gamma_a^{-1} + \gamma_b^{-1})]$	(7.86)
T_2	dephasing time of atomic dipoles $(= 1/\gamma)$	(13.26)
$T_2{}^*$	inhomogeneous dephasing time of atomic dipoles	(13.56)
T_{lw}	third-order complex polarization integrals	(D.7)
$\mathscr{T}(z)$	total energy crossing x-y plane at position z (pulse propagation)	(13.35)
u	average atomic speed in a gas	(10.2)
$u_a(\mathbf{r}), u_b(\mathbf{r})$	energy eigenfunctions for upper and lower levels	(2.14)
$u_n(x), u_n(\mathbf{r})$	energy eigenfunctions for eigenvalue $\hbar\omega_n$	(1.10)
$U_n(z)$	nth-mode function of a cavity (has wave number K_n)	(8.8)
$\mathscr{U}(\omega)$	energy distribution, e.g., for blackbody radiation	(2.31)
v	z component of velocity in gas laser (and for Brownian motion)	(10.2)
V	volume of a cavity	(5.20)
$V(r)$	atomic potential energy	(2.7)

$\mathscr{V}(t)$	atom-field interaction energy (usually electric dipole)	(2.1)
$W(v)$	dimensionless velocity distribution (usually Maxwellian)	(10.2)
$W(\omega)$	dimensionless frequency distribution for inhomogeneously broadened line	(10.2)
$\tilde{W}(T)$	Fourier transform of $W(\omega)$	(13.25)
x, y, z	x, y, and z Cartesian coordinates	(Fig. 1-4)
$\hat{x}, \hat{y}, \hat{z}$	unit vectors for x, y, and z Cartesian axes	(Fig. 1-4)
$X(t)$	slowly varying complex displacement $[x(t) = \text{Re } X(t) \exp(i\omega t)]$	(3.29)
$Y_{lm}(\theta, \varphi)$	spherical harmonics	(1.26)
$Z(v)$	plasma dispersion function	(10.29)
$\mathfrak{Z}$	set of index values whose corresponding field amplitudes vanish	(9.72)
α	complex dimensionless field amplitude of coherent state $\|\alpha\rangle$ (eigenvalue of annihilation operator $\{a\|\alpha\rangle = \alpha\|\alpha\rangle\}$)	(15.12)
α	energy gain coefficient in pulse propagation	(13.30)
α	as subscript, means a or b for energy levels of two-level atom	(8.20)
α	net gain coefficient in classical sustained oscillator	(4.1)
α'	gain parameter	(13.18)
$\|\alpha\rangle$	coherent state	(15.12)
α_n	net gain coefficient for nth laser mode	(8.50)
β	as subscript, means a or b for energy levels of two-level atom	(14.90)
β	self-saturation coefficient in classical sustained oscillator	(4.1)
$\|\beta\rangle$	auxiliary coherent state	(15.35)
β_n	self-saturation coefficient for the nth laser mode	(8.50)
γ	atomic dipole decay constant $(=1/T_2 = \gamma_{ab} + \gamma_{ph})$	(7.48)
γ_a, γ_b	upper and lower-level decay constants	(2.46) (2.47)

γ_{ab}	$\frac{1}{2}(\gamma_a + \gamma_b)$, spontaneous emission and inelastic collision contribution to decay of atomic dipole	(7.37)
γ_{ph}	elastic collision contribution to dipole decay	(7.43)
Γ	modulation depth	(9.104)
Γ	damping constant in Brownian motion	(19.2)
$\delta(x - x')$	one-dimensional Dirac delta function	
$\delta(a - a_0)$	two-dimensional Dirac delta function	
$\delta(\mathbf{r} - \mathbf{r}')$	three-dimensional Dirac delta function	
δ_{ij}	Kronecker delta function $= \begin{Bmatrix} 1 & i = j \\ 0 & i \neq j \end{Bmatrix}$	
$\delta\omega$	small frequency shift	(7.40)
δt	small time interval	
Δ	intermode beat frequency ($= \nu_n - \nu_{n-1}$)	(9.79)
$\overline{\Delta\nu}$	low-frequency beat note	(9.61)
Δt	time interval	(19.8)
ε	permittivity of medium	(13.3)
ε_n	small displacement from stationary intensity I_n	(9.73)
ε_0	permittivity of vacuum $= (4\pi c^2)^{-1}\, 10^7 \simeq 8.8542 \times 10^{-12}$ F/m	(8.2)
$\hat{\varepsilon}_+, \hat{\varepsilon}_-$	complex unit vectors for circular polarization of electric field	(12.1)
$\lvert\zeta\rangle, \lvert\xi\rangle$	arbitrary vectors in quantum-mechanical space	(6.16)
η	index of refraction	(8.18)
θ	polar angle in polar coordinates	(Fig. 1-4)
θ_{nm}	real, cross-saturation coefficients (by mode E_m on E_n)	(9.71)
$\theta(z)$	area under pulse envelop at position z	(13.40)
$\vartheta(z, t)$	partial area under pulse envelop (up to time t)	after (13.53)
$\vartheta_{n\mu\rho\sigma}$	complex, third-order general saturation coefficient	(9.18)
Θ	Multimode stability matrix	(9.41)
κ	loss coefficient in amplifier theory	(13.3)
λ	wavelength of light	
λ_n	wavelength of mode n ($=2\pi/K_n$)	
λ_a, λ_b	numbers of atoms per second-per volume excited to upper and lower levels	(8.25)

Λ_a, Λ_b	excitation operators for upper and lower states	(20.7)
μ	power-broadened frequency in Rabi flopping	(2.61)
μ	as subscript, indexes field amplitude and phase: $E_\mu(t)$, $\phi_\mu(t)$	(9.18)
$\boldsymbol{\mu}$	magnetic-dipole moment	(1.37)
μ_B	Bohr magneton $= 9.2741 \times 10^{-24}$ J/T	(1.37)
μ_n	eigenvalue in Rabi flopping	(2.58)
μ_0	permeability of vacuum $= 4\pi \times 10^{-7}$ H/m	(8.2)
ν	laser (optical) oscillation frequency in radians/sec not Hertz (circular frequency)	(8.10)
ν_n	laser frequency for mode n	(8.8)
ν_M	modulation frequency	(9.84)
ν_0	average frequency $= \frac{1}{2}(\nu_+ + \nu_-)$	(Table 12-1)
$\nu_\pm$	frequency of $\pm$ circular polarizations in Zeeman laser or right- and left-traveling waves in ring laser	(11.2), (12.1)
ξ	dimensionless coordinate for simple harmonic oscillator $= (m\omega/\hbar)^{1/2}x$	(1.22)
π	3.1415926535897. . . .	
$\rho, \rho(t)$	density matrix or operator	(7.5), (7.17)
ρ	mode subscript in laser theory: for example, $E_\rho(t)$	(9.18)
$\rho(z, t)$	population matrix for laser medium	(8.23)
$\rho(z, v, t)$	population matrix for ensemble moving with z component of velocity, v	(10.6)
$\rho_{\alpha\alpha}(z, t)$	$\alpha = a, b$, number of atoms in the α th level	(8.23)
$\rho_{ab}(z, t)$	population matrix element proportional to the complex polarization	(8.29)
$\rho(\alpha, z_0, t_0, v, t)$	pure case density matrix for atom excited to level $\alpha = a$ or b, at time t_0, with z component of velocity v, located at z at time t	(10.4)
$\rho_{nm}(t)$	number representation of the density operator $\rho(t)$	(7.19)
$\rho_A(t), \rho_B(t)$	reduced density operators for the system and reservoir	(16.87), (16.88)
$\rho_c(t)$	correlation part of system-reservoir density operator	(16.91)

σ	spin-flip operator $= \frac{1}{2}(\sigma_x - i\sigma_y)$	(1.41)
σ	as subscript, indexes field amplitudes and phases, for example, $E_\sigma(t)$	(9.18)
$\overleftrightarrow{\sigma}$	conductivity tensor in Maxwell's equations	(12.3)
σ	scalar conductivity in Maxwell's equations	(8.2)
$\boldsymbol{\sigma}$	Pauli spin vector	(1.38)
$\sigma_x, \sigma_y, \sigma_z$	components of Pauli spin vector	(1.38)
$\sigma_a(t), \sigma_b(t)$	projection operators for upper and lower states	(20.11)
$\sigma_a{}^i(t)$	projection operator for *i*th atom	(20.1)
$\sigma_{k,k}$	density matrix in Josephson radiation theory	(21.24)
σ_n	linear mode pulling for laser mode $E_n(t)$	(8.52)
$\Sigma(t)$	slowly varying spin-flip operator in Langevin theory	(20.12)
$\tau, \tau', \tau'', \tau'''$	time intervals	(10.47)
τ_{nm}	mode cross pushing coefficients in laser theory	(9.22)
τ_c	correlation time of random functions	after (19.5)
τ_p	pulse duration	(13.26)
τ_s	hyperbolic secant time parameter	(13.46)
υ	complex frequency, for example, $= \gamma + i(\omega - \nu)$	(10.36)
υ_{lw}	complex frequency	(Table D-2)
ϕ	azimuthal angle in polar coordinates	(Fig. 1-4)
$\phi_n(x)$	Hermite-Gaussian functions, eigenfunctions of simple harmonic oscillator	(1.22)
$\phi_n(t)$	slowly varying phase of the *n*th laser mode	(8.8)
χ_n	complex susceptibility for *n*th mode	(8.13)
$\chi(z, T, t)$	complex susceptibility in pulse propagation	(13.21)
$\psi(\mathbf{r}, t)$	Schrödinger wave function	(1.1), (6.41)
$\|\psi(t)\rangle$	state vector	(6.26)
$\Psi_{n\mu\rho\sigma}$	third-order relative phase angle	(9.15)
Ψ	relative phase angle	(9.44)
Ψ_1, Ψ_2	stationary-state values of ψ	(9.62), (9.63)
Ψ_0	a relative phase angle value	(9.60)
ω	atomic line center frequency in laser media $= \omega_a - \omega_b$	(2.19)
ω_0	atomic line center frequency	(13.56)
ω_n	eigenfrequency of an unperturbed Hamiltonian	(1.9)

ω_a, ω_b	eigenfrequencies of upper and lower levels (a and b)	(2.14)
ω_{nm}	frequency difference $= \omega_n - \omega_m$	
$\omega_{a'b'}$	$= \omega_{a'} - \omega_{b'}$	(Fig. 12-2)
Ω	frequency of single-mode radiation	(14.6)
Ω_n	frequency of nth mode (passive cavity or free space)	(8.5)

Laser Physics

I
WAVE MECHANICS

In this book we treat the interaction of radiation with matter with special application to gain (amplifying) media typically existing in lasers. Traditionally the interaction of radiation with matter is treated in terms of the index of refraction and the absorption of light, utilizing the model of electrons on springs. In Chap. 3 we show that this model is reasonable for some linear absorbing systems but is very awkward, if not outright unworkable, for a microscopic description of gain media. The essential failure of the electron-spring model is its difficulty in yielding net stimulated emission (and therefore gain) for an ensemble of systems.† In contrast, a more physically sound, quantum treatment of matter provides for the concept of population inversion and explains stimulated emission in a natural way. Hence from the very start we are compelled to base our understanding of the interaction of radiation with matter on a quantum-mechanical description of atoms. In Chaps. 1–13 a classical field interacts with quantum-mechanical atoms; in the remainder of the book, both field and atoms are treated quantum mechanically.

In the present chapter, we review some highlights of wave mechanics which have particular interest in the interaction of radiation with matter. Our discussion defines notation required later and guides the reader to more complete references on background material. In Sec. 1–1, we introduce the central concept of quantum mechanics, namely, the wave function, and give an interpretation in terms of a probability density. The wave function is then used in the expectation values of operators to provide connection with sets of physical measurements. The Schrödinger equation of motion for the wave function is given, along with a formal solution written as a superposition of energy eigenfunctions. In Sec. 1–2 the theory is illustrated by a particle in a one-

†But see Fig. 3-7 and associated text for description of a classical maser.

dimensional potential well. This problem has formal similarity with the mode problem of a laser resonator (Chap. 8). In Sec. 1–3 we summarize the results of the quantum-mechanical simple harmonic oscillator. This system has particular relevance in the quantum theory of radiation, for which modes of the radiation field are shown (Chap. 14) to act like simple harmonic oscillators. In Sec. 1–4 we review the hydrogen atom. This constitutes a simple, concrete example of a quantum-mechanical system, which we later use to illustrate the interaction of radiation with matter. Finally, in Sec. 1–5 we consider a particle with spin $\frac{1}{2}$ and a magnetic moment. This problem and the Stern-Gerlach experiment illustrate major differences between classical and quantum-mechanical systems. The analysis is formally very similar to that of the two-level atomic systems treated later and provides a useful notation for many related problems.

1-1. The Wave Function

In our analysis of systems encountered in quantum optics, we shall initially make use of quantum-mechanical wave functions like $\psi(\mathbf{r}, t)$. In later work, we use related quantities, namely, the state vector $|\psi\rangle$ and density matrix ρ. These entities contain all (possible) knowledge concerning a given system, such as a gas, a bound electron, or a radiation field. To introduce the wave function, let us consider, in particular, the bound electron. The most direct physical interpretation of the electron's wave function is in terms of the absolute value squared:

$$\psi^*(\mathbf{r}, t)\psi(\mathbf{r}, t) \tag{1}$$

which is the probability density of finding the electron at point $\mathbf{r}$ at time t. Consequently, the probability of finding the electron in volume d^3r about point $\mathbf{r}$ at time t is given by

$$\psi^*(\mathbf{r}, t)\psi(\mathbf{r}, t)\, d^3r. \tag{2}$$

Inasmuch as the probability of finding the particle somewhere is unity, the wave function is normalized, that is,

$$\int \psi^*(\mathbf{r}, t)\psi(\mathbf{r}, t)\, d^3r = 1. \tag{3}$$

In view of result (1), it is clear that the charge density is given by $e\psi^*\psi$, where e is the (negative) charge of the electron. We see in Sec. 3–1 that the density can oscillate back and forth, giving rise to an oscillating dipole moment.

In quantum mechanics, observables, such as the dipole moment or energy, are represented by operators which act on the wave function. We assume that it is possible to measure these observables. The expectation value of the operator $\mathcal{O}$,

$$\langle \mathcal{O} \rangle = \int d^3r \psi^*(\mathbf{r}, t)\, \mathcal{O}(\mathbf{r})\psi(\mathbf{r}, t), \tag{4}$$

gives the average value of the measurements for an ensemble of systems each member of which is described by $\psi(\mathbf{r}, t)$.

A particularly important quantity for laser theory is the electric-dipole moment. Classically this is defined by $\Sigma_i q_i \mathbf{d}_i$, where q_i is the ith charge and $\mathbf{d}_i$ is its position vector with respect to a conveniently chosen origin. For a single-electron atom, this reduces to $e\mathbf{r}$, where e is the (negative) electron charge and $\mathbf{r}$ is the position vector of the electron with respect to the nucleus.

The corresponding quantum-mechanical operator is then $e\mathbf{r}$, and the dipole moment is given by the expection value,

$$\langle e\mathbf{r} \rangle = \int d^3r\ \psi^*(\mathbf{r}, t) e\mathbf{r}\ \psi(\mathbf{r}, t) = \int d^3r\ (e\psi^*\psi)\mathbf{r}. \tag{5}$$

Inasmuch as $e\psi^*\psi$ is the charge density, the expectation value for $e\mathbf{r}$ is literally the average dipole moment in space, as indicated in Fig. 1–1. This formula is used, for example, in the calculation of the atomic polarization of a laser medium.

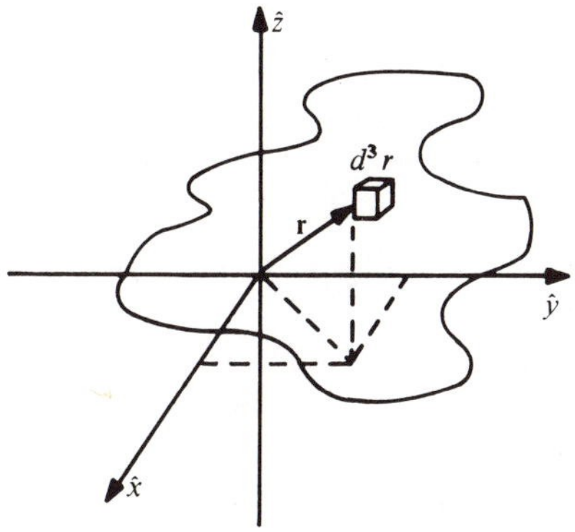

Figure 1-1. Schematic diagram of probability density $\psi^*\psi$ for electron in atom with representative position vector **r** *and corresponding volume element d^3r. The expectation value of the electric-dipole moment* $\langle e\mathbf{r}\rangle$ *is given by the average value of* e**r** *with respect to this probability density* [*see Eq. (5)*].

The time development of the wave function is determined by the Schrödinger equation:

$$i\hbar\dot{\psi}(\mathbf{r}, t) = \mathcal{H}(\mathbf{r}, \mathbf{p})\psi(\mathbf{r}, t), \tag{6}$$

in which $\mathcal{H}(\mathbf{r}, \mathbf{p})$ is the Hamiltonian for the system described by $\psi(\mathbf{r}, t)$, and $\mathbf{p}$

is the momentum. The Hamiltonian is usually given by the classical energy, in which the measurable quantities such as position and momentum are replaced by appropriate operators. In the absence of an electromagnetic field,

$$\mathbf{p} = -i\hbar\nabla. \tag{7}$$

For example, a free particle has only kinetic energy and hence has the Hamiltonian

$$\mathscr{H}(\mathbf{r}, \mathbf{p}) = \mathscr{H}(\mathbf{p}) = \frac{p^2}{2m} = -\frac{\hbar^2}{2m}\nabla^2. \tag{8}$$

The stationary solutions $\psi_n(\mathbf{r}, t)$ of the Schrödinger equation are those for which the time dependence can be separated from the space dependence, that is, for which

$$\psi_n(\mathbf{r}, t) = u_n(\mathbf{r}) \exp(-i\omega_n t). \tag{9}$$

Here ω_n is a circular frequency, that is, it is given in *radians* per second, not in hertz. Substituting Eq. (9) into (6), we find the time-independent equation:

$$\mathscr{H}(\mathbf{r}, \mathbf{p})\, u_n(\mathbf{r}) = \hbar\omega_n u_n(\mathbf{r}). \tag{10}$$

This is an eigenvalue equation for the Hamiltonian $\mathscr{H}(\mathbf{r}, \mathbf{p})$ with eigenfunctions $u_n(\mathbf{r})$ and eigenvalues $\hbar\omega_n$. The functions $u_n(\mathbf{r})$ can generally be shown to be orthonormal;

$$\int u_n^*(\mathbf{r}) u_m(\mathbf{r})\, d^3r = \delta_{n,m} = \begin{cases} 1, & n = m \\ 0, & n \neq m \end{cases} \tag{11}$$

and complete: †

$$\sum_n u_n^*(\mathbf{r}) u_n(\mathbf{r}') = \delta(\mathbf{r} - \mathbf{r}'), \tag{12}$$

where $\delta_{n,m}$ and $\delta(\mathbf{r} - \mathbf{r}')$ are the Kronecker and Dirac delta functions, respectively. We have occasion to use relations (11) and (12) frequently and return to them shortly.

In this book we assume, for the most part, that the eigenfunctions $u_n(\mathbf{r})$ and values $\hbar\omega_n$ are known from other treatments (e.g., Schiff, 1968), and we are concerned primarily with mechanisms causing transitions between the different states of the system. The general solution of the Schrödinger equation (6) is a linear superposition of these simple oscillatory solutions:

$$\psi(\mathbf{r}, t) = \sum_n C_n u_n(\mathbf{r}) \exp(-i\omega_n t). \tag{13}$$

Because both the wave function $\psi(\mathbf{r}, t)$ and the eigenfunctions are normalized [see Eqs. (3) and (11)], the expansion coefficients satisfy the condition

$$\sum_n |C_n|^2 = 1. \tag{14}$$

†See Prob. 1–1 for an alternative definition.

As such, the quantities $|C_n|^2$ have physical interpretation as the probabilities that the system is in the nth state.

We note that the superposition of states in Eq. (13) leads to quantum-mechanical interference between the probability amplitudes, just as there is electric field amplitude interference in a Michelson interferometer. This phenomenon is implicit in much of our work and is discussed, for example, in Sec. 3–1 on the oscillating dipole moment. In the rest of this chapter, we illustrate these general principles in terms of some useful special cases.

1–2. Particle in One-Dimensional Well

In the one-dimensional well (Fig. 1–2), the Hamiltonian is that for a free particle $p^2/2m$ within the well and infinite elsewhere:

$$\mathscr{H} = \begin{cases} \dfrac{p^2}{2m} = -\dfrac{\hbar^2}{2m}\dfrac{d^2}{dz^2}, & 0 \leq z \leq L, \\ \infty & , z < 0, z > L. \end{cases} \tag{15}$$

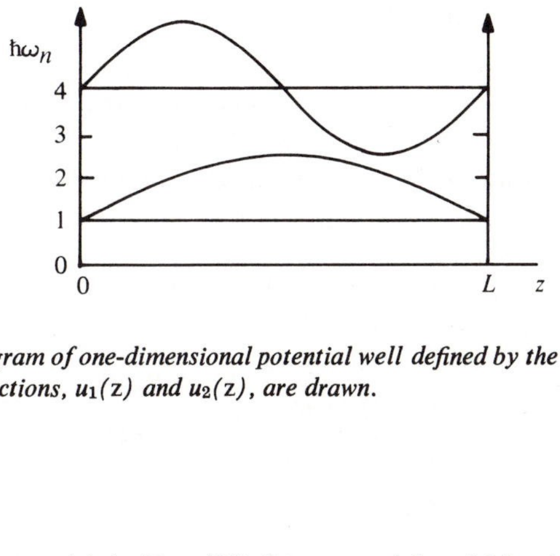

Figure 1-2. Diagram of one-dimensional potential well defined by the Hamiltonian (15). Lowest two eigenfunctions, $u_1(z)$ and $u_2(z)$, are drawn.

The eigenfunctions $u_n(z)$ in Eq. (10) for a particle within the well are those satisfying the boundary conditions $u_n(0) = u\ (L) = 0$:

$$u_n(z) = \left(\frac{2}{L}\right)^{1/2} \sin K_n z, \quad 0 \leqslant z \leqslant L, \tag{16}$$
$$= 0, \qquad \text{elsewhere,}$$

where the wave number

$$K_n = \frac{n\pi}{L}, \quad n = 1, 2, \ldots, \tag{17}$$

and the factor $\left(\frac{2}{L}\right)^{1/2}$ is included for normalization.

Combining (10), (15), and (16), we see that the corresponding energy eigenvalues are

$$\hbar\omega_n = \frac{1}{2}\frac{\hbar^2 K_n{}^2}{m} = \frac{1}{2}\frac{\hbar^2\pi^2}{L^2 m}n^2. \tag{18}$$

Here we notice that the energies allowed a particle in the well are discrete or quantized, in distinct contrast to the continuum allowed classically. However, an ensemble average measurement of the energy can take on any value (not necessarily proportional to n^2), for the wave function is in general an arbitrary superposition of eigenstates (see Prob. 1–4). Specifically, Eq. (13) becomes

$$\psi(z, t) = \left(\frac{2}{L}\right)^{1/2} \sum_n C_n \sin K_n z \exp(-i\omega_n t). \tag{19}$$

We see later that the form for this wave function (probability amplitude) is the same as that for the complex electric field amplitude in a laser [compare with Eq. (8.8)†].

1-3. The Simple Harmonic Oscillator

A problem which is of fundamental importance is the quantum-mechanical treatment of the simple harmonic oscillator. The classical energy for a particle of mass m in a simple harmonic potential characterized by oscillation frequency Ω is

$$\mathscr{H}_C = \frac{p^2}{2m} + \tfrac{1}{2}m\Omega^2 x^2, \tag{20}$$

and hence the Hamiltonian operator is

$$\mathscr{H}(x, p) = -\frac{\hbar^2}{2m}\frac{d^2}{dx^2} + \tfrac{1}{2}m\Omega^2 x^2. \tag{21}$$

This Hamiltonian has eigenfunctions

$$u_n(x) = \phi_n(\xi) = \left(\frac{m\Omega}{\hbar\pi}\right)^{1/4}\left(\frac{1}{2^n n!}\right)^{1/2} H_n(\xi)\exp(-\tfrac{1}{2}\xi^2), \tag{22}$$

†Citations of this type refer to equations in other chapters here, specifically, to Eq. (8) of Chap. 8.

where the $H_n(\xi)$ are Hermite polynomials, and the dimensionless coordinate $\xi = (m\Omega/\hbar)^{1/2}x$. The corresponding energies (see Fig. 1-3) are given by

$$\hbar\omega_n = (n + \tfrac{1}{2})\hbar\Omega, \quad n = 0, 1, 2, \ldots. \tag{23}$$

In Chap. 14 we derive these eigenfunctions and values which are equally spaced (by $\hbar\Omega$), in contrast to those (18) for the infinite potential well, by a purely algebraic means. The simple harmonic oscillator is particularly useful both in the quantum theory of radiation (Chap. 14) and in the description of vibrational states of molecules.

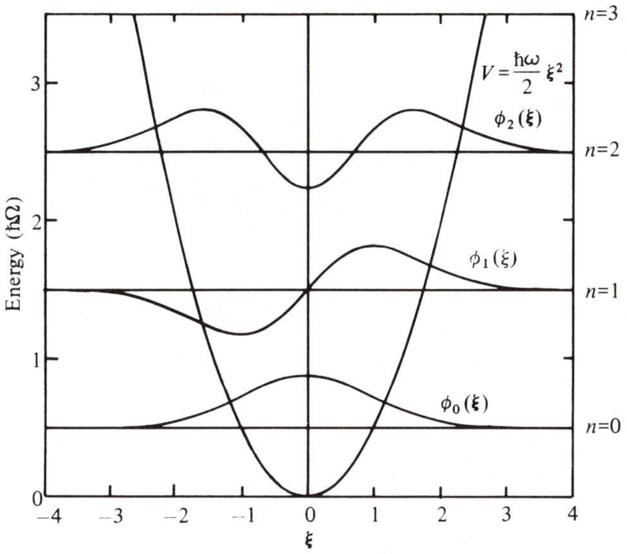

Figure 1-3. Energy-level diagram for simple harmonic oscillator, including the first three eigenfunctions of Eq. (22). The energy axis is marked in units of $\hbar\Omega$.

1-4. Hydrogen Atom

A more difficult (but very useful) example is that of the hydrogen atom. In the absence of external forces, the classical energy of the electron bound by its negative charge to a fixed, positively charged nucleus is

$$\mathscr{H}_C = \frac{p^2}{2m} - \frac{e^2}{r}, \tag{24}$$

where $\mathbf{p}$, m, and e are the momentum, mass, and charge of the electron, respectively, and r is the distance between the electron and the nucleus. The eigenvalue equation (10) that must be solved for hydrogen is, then,

$$\left(-\frac{\hbar^2\nabla^2}{2m}-\frac{e^2}{r}\right)u(\mathbf{r})=\hbar\omega u(\mathbf{r}), \tag{25}$$

where it is advantageous to express ∇^2 in spherical coordinates r, θ, ϕ (see Fig. 1-4) because of the spherical symmetry of the potential energy $-e^2/r$.

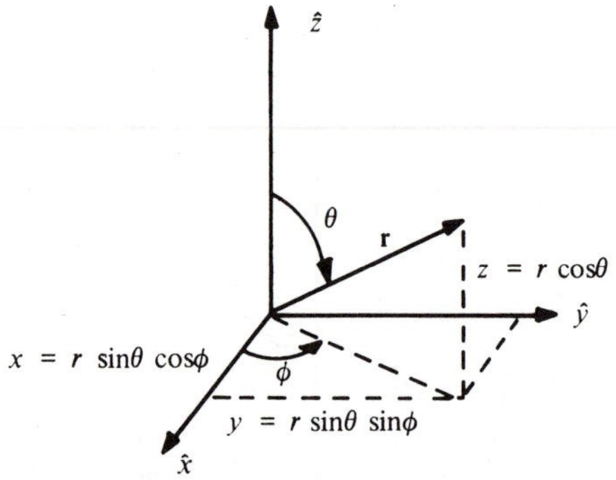

Figure 1-4. Polar coordinates describing atomic radius vector **r** *in three dimensions. Projections on the Cartesian axes are given for which $x = r \sin\theta \cos\phi$, $y = r \sin\theta \sin\phi$, and $z = r \cos\theta$, A differential volume element d^3r is given by $dx\, dy\, dz$ or by $r\, d\theta\, r \sin\theta\, d\phi dr$.*

This equation can be separated into three equations, each containing a single spherical coordinate. The solutions of the equations correspond to discrete energies, as for all bound problems, and have the values

$$u_{nlm}(r, \theta, \phi) = R_{nl}(r) Y_{lm}(\theta, \phi), \tag{26}$$

where three indices, n, l, and m, are used in place of the single index of (13).

Here the $R_{nl}(r)$ are Laguerre polonomials multiplied by the exponential factor $\exp(-r/na_0)$, $a_0 = \hbar^2/me^2 = 0.53$ Å is the Bohr radius, and the $Y_{lm}(\theta, \phi)$ are spherical harmonics well known also in classical electrodynamics. In particular, the first few $u_{nlm}(r, \theta, \phi)$ are as follows:

$$u_{100}(r, \theta, \phi) = (\pi a_0^3)^{-1/2} \exp(-r/a_0), \tag{27}$$

$$u_{200}(r, \theta, \phi) = (32\pi a_0^3)^{-1/2} (2 - r/a_0) \exp(-r/2a_0), \tag{28}$$

$$u_{210}(r, \theta, \phi) = (32\pi a_0^3)^{-1/2} (r/a_0) \cos\theta \exp(-r/2a_0), \tag{29}$$

$$u_{21,\pm1}(r, \theta, \phi) = (64\pi a_0^3)^{-1/2} (r/a_0) \sin\theta \exp(\pm i\phi) \exp(-r/2a_0). \tag{30}$$

The corresponding energies are

$$\hbar\omega_{nlm} = \hbar\omega_n = -\frac{e^2}{2a_0n^2} = -\frac{R_\infty}{n^2}, \tag{31}$$

which are, once again, discrete (see Fig. 1-5). Notice that the states (28), (29), and (30) are degenerate, that is, they have the same energy according to

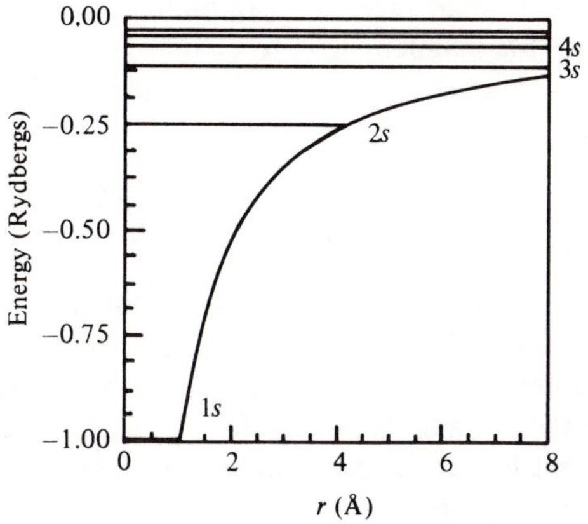

Figure 1-5. Energy-level diagram for hydrogen atom. Curved line is potential well, $V(r) = -e^2/r$. Horizontal lines indicate energy levels given by Eq. (31) for $n = 1$ ($-1.0\ R_\infty$), $n = 2$ ($-0.25\ R_\infty$), $n = 3$ ($-\frac{1}{9}R_\infty$), etc. ($1\ R_\infty = 13.6$ eV).

(31). This degeneracy may be lifted by application of a perturbation. The stationary solutions are

$$\psi_{nlm}(r, \theta, \phi, t) = \exp(-i\omega_{nlm}t)u_{nlm}(r, \theta, \phi) \tag{32}$$

with the most general solution given by

$$\psi(r, \theta, \phi, t) = \sum_n\sum_l\sum_m C_{nlm} \exp(-i\omega_{nlm}t)u_{nlm}(r, \theta, \phi). \tag{33}$$

1-5. Particle with Spin and Magnetic Moment

We next discuss a problem which does not, in general, require the spatial dependence of the wave function, although it does involve distinct states. This is the problem of a spin magnetic moment placed in a magnetic field. The electron has a spin, and hence it is necessary to generalize its wave function to allow for the extra degree of freedom.

Classically the energy of a magnetic moment $\boldsymbol{\mu}$ subject to a magnetic induction $\mathbf{B}$ is given by

$$\mathscr{H}_C = -\boldsymbol{\mu} \cdot \mathbf{B} = -\mu B \cos\theta, \tag{34}$$

where the minus sign appears because a moment aligned antiparallel to the field has higher energy than one aligned parallel. It is a most remarkable fact that quantum mechanics allows only discrete energy eigenvalues of (34) corresponding to discrete orientations of the magnetic moment.

A particularly striking consequence of this fact occurred in the experiment of Stern and Gerlach. They sent a beam of silver atoms between the poles of a magnet with the pole faces as diagrammed in Fig. 1–6. The sharply pointed face causes the magnetic field to be inhomogeneous, that is, stronger at one pole face than at the other. This variation exerts a force on the magnetic moments given by

$$F_z = -\frac{\partial \mathscr{H}_C}{\partial z} = \mu \cos\theta \frac{\partial B}{\partial z}, \tag{35}$$

where we take the direction perpendicular to the pole faces to be z. In addition, a torque $\boldsymbol{\mu} \times \mathbf{B}$ is exerted on the moments, tending to cause precession of $\boldsymbol{\mu}$ about $\mathbf{B}$. The classical result of these forces is that the moments can arrive at the glass plate (depicted in Fig. 1–6) deflected through all possible angles, and hence the beam should be spread out (proportionately to the magnetic moment $\boldsymbol{\mu}$) in a continuous fashion. Instead, two separate spots were observed when the beam was deposited on the glass plate! It can be concluded that for the silver atom only two orientations and, therefore, only two energy states exist.

In quantum mechanics as first worked out by Pauli, the spin states are represented by the two-dimensional column vectors

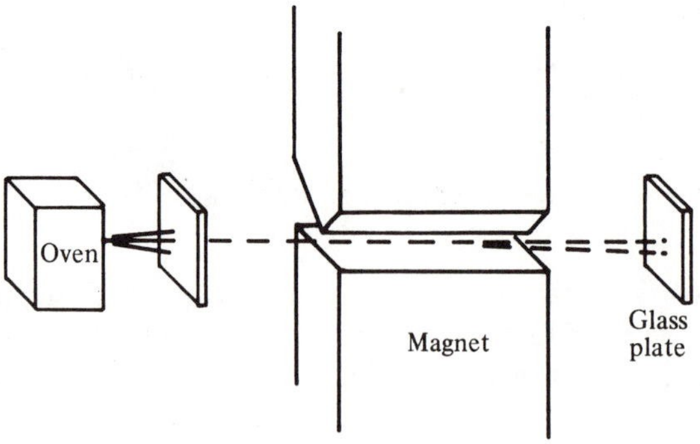

Figure 1-6. Stern-Gerlach apparatus for separating a single beam of silver atoms into two beams. In general a beam of spin s *separates into* 2s + 1 *beams. The silver atoms have spin* $\frac{1}{2}$.

$$\begin{pmatrix} 1 \\ 0 \end{pmatrix} \quad \text{and} \quad \begin{pmatrix} 0 \\ 1 \end{pmatrix} \tag{36}$$

corresponding to alignment antiparallel (higher energy) and parallel to the magnetic field. This notation can, of course, be used for any two-level system.

In general, a beam of particles is split into $2s + 1$ beams, where s is the spin of the particle. The spin is a quantum-mechanical property of particles which gives rise to their magnetic moments. The silver atom and other atoms having one important electron have spin $\frac{1}{2}$, the spin of the electron. The spin of the electron points opposite to its magnetic moment; hence the states in Eq. (36) correspond to "spin up" and "spin down," respectively. The two-dimensional column vectors are often called spinors. We deal with the corresponding forms for the Schrödinger equation in Chap. 5.

In our later work, particularly in the description of coherent pulse propagation, we use the quantitative relationship between the electron magnetic moment operator $\boldsymbol{\mu}$ and the spin operator $\boldsymbol{\sigma}$:

$$\boldsymbol{\mu} = -\tfrac{1}{2}\mu_{\mathrm{B}} g_e \boldsymbol{\sigma}. \tag{37}$$

Here μ_{B} is the Bohr magneton, g_e is the electron g factor, and $\boldsymbol{\sigma}$ is given by the Pauli spin matrices:

$$\boldsymbol{\sigma} = \hat{x}\begin{pmatrix} 0 & 1 \\ 1 & 0 \end{pmatrix} + \hat{y}\begin{pmatrix} 0 & -i \\ i & 0 \end{pmatrix} + \hat{z}\begin{pmatrix} 1 & 0 \\ 0 & -1 \end{pmatrix}. \tag{38}$$

These matrices obey the commutation relations

$$\begin{aligned} [\sigma_x, \sigma_y] &\equiv \sigma_x\sigma_y - \sigma_y\sigma_x = 2i\sigma_z, \\ [\sigma_y, \sigma_z] &= 2i\sigma_x, \ [\sigma_z, \sigma_x] = 2i\sigma_y, \end{aligned} \tag{39}$$

which can be written compactly as

$$\boldsymbol{\sigma} \times \boldsymbol{\sigma} = 2i\boldsymbol{\sigma}. \tag{40}$$

This cross product does not vanish because the vector components do not commute with one another. The linear combinations

$$\sigma = \tfrac{1}{2}(\sigma_x - i\sigma_y) = \begin{pmatrix} 0 & 0 \\ 1 & 0 \end{pmatrix} \tag{41}$$

$$\sigma^\dagger = \tfrac{1}{2}(\sigma_x + i\sigma_y) = \begin{pmatrix} 0 & 1 \\ 0 & 0 \end{pmatrix} \tag{42}$$

lower and raise the states $\begin{pmatrix} 1 \\ 0 \end{pmatrix}$ and $\begin{pmatrix} 0 \\ 1 \end{pmatrix}$, respectively, that is,

$$\sigma\begin{pmatrix} 1 \\ 0 \end{pmatrix} = \begin{pmatrix} 0 \\ 1 \end{pmatrix}, \tag{43}$$

$$\sigma^\dagger \begin{pmatrix} 0 \\ 1 \end{pmatrix} = \begin{pmatrix} 1 \\ 0 \end{pmatrix}. \tag{44}$$

These lowering and raising operators can be used to represent the emission and absorption of electromagnetic radiation, a process discussed in Chap. 2.

The quantum-mechanical Hamiltonian is found by replacing the classical magnetic moment in Eq. (34) by that in (37), that is,

$$\mathscr{H} = -(-\tfrac{1}{2}\mu_{\mathrm{B}} g_e \sigma) \cdot \mathbf{B} = \tfrac{1}{2}\mu_{\mathrm{B}} g_e B \begin{pmatrix} 1 & 0 \\ 0 & -1 \end{pmatrix}. \tag{45}$$

The corresponding energy eigenvalues $\pm\tfrac{1}{2}\mu_{\mathrm{B}} g_e B$ are depicted in Fig. 1–7.

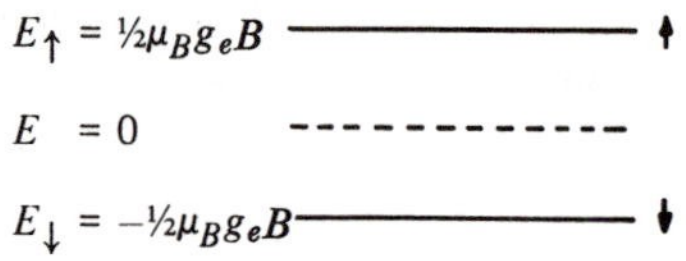

Figure 1-7. Energy-level diagram for electron subject to a magnetic induction **B**. *Here μ_{B} is the Bohr magneton, and g_e is the electron g factor.*

Problems

1-1. Show that with (11) the completeness relation (12) is equivalent to the alternative definition that any function $f(\mathbf{r})$ can be expanded as

$$f(\mathbf{r}) = \sum_n d_n u_n(\mathbf{r}). \tag{46}$$

1-2. Calculate the dipole moment $\langle e\mathbf{r}\rangle$ of an atom with wave function

$$\psi(\mathbf{r}, t) = C_{210}u_{210}\exp(-i\omega_{210}t) + C_{100}u_{100}\exp(-i\omega_{100}t). \tag{47}$$

Hint: Use spherical coordinates as shown in Fig. 1-4 and write the position vector in the form

$$\mathbf{r} = \tfrac{1}{2}r\sin\theta\,[(\hat{x} - i\hat{y})\exp(+i\phi) + (\hat{x} + i\hat{y})\exp(-i\phi)] + r\cos\theta\,\hat{z}. \tag{48}$$

1-3. Verify the commutation relations in Eq. (39).

1-4. What is the expectation value of the energy (15) for the wave function (19)? Is this value ever actually measured?

References

Most books on quantum mechanics discuss the material given in this chapter. Among these are the folloming:

D. Bohm, 1951, *Quantum Theory*, Prentice-Hall, Englewood Cliffs, N. J.

C. Cohen-Tannoudji, B. Diu, and F. Laloë, 1973, *Mécanique Quantique*, Hermann, Paris. English Translation in Press.

R. H. Dicke and J. P. Wittke, 1960, *Introduction to Quantum Mechanics*, Addison-Wesley Publishing Co., Reading, Mass.

R. M. Eisberg, 1967, *Fundamentals of Modern Physics*, John Wiley & Sons, New York.

R. P. Feynman, R. B. Leighton, and M. Sands, 1965, *The Feynman Lectures on Physics*, Vol. III, Addison-Wesley Publishing Co., Reading, Mass., esp. Chaps. 1–5, 9, 11, 16, 18, 19.

R. B. Leighton, 1959, *Principles of Modern Physics*, McGraw-Hill Book Co., New York.

E. Merzbacher, 1970, *Quantum Mechanics*, 2nd ed., John Wiley & Sons, New York.

L. I. Schiff, 1968, *Quantum Mechanics*, 2nd ed., McGraw-Hill Book Co., New York.

H. L. Strauss, 1968, *Quantum Mechanics: An Introduction*, Prentice-Hall, Englewood Cliffs, N. J.

The Stern-Gerlach experiment was originally proposed by O. Stern, 1921, *Z. Physik* **7**, 249 and carried out by him with Gerlach: W. Gerlach and O. Stern, 1922, *Z. Physik* **8**, 110, and **9**, 349.

II

ATOM-FIELD INTERACTION

2. Interaction Between Electromagnetic Field and Atom

In Chap. 1 we introduced the Schrödinger equation and gave solutions for a number of simple cases. The general solution was then obtained by a linear superposition of these solutions. In our work, we are particularly concerned with the way in which atomic systems interact with electromagnetic radiation. These systems are still described by wave functions given by superpositions of simple solutions, but the expansion coefficients C_n (probability amplitudes) become functions of time. In this chapter we derive equations of motion for the expansion coefficients using Schrödinger's equation and introduce the electric-dipole interaction energy. Then, specializing to a two-level system and a monochromatic electromagnetic field, we solve the equations of motion both by time-dependent perturbation theory and by the more exact Rabi method.

If the atomic Hamiltonian $\mathscr{H}$ includes a perturbing term $\mathscr{V}$:

$$\mathscr{H} = \mathscr{H}_0 + \mathscr{V}, \tag{1}$$

where the eigenfunctions $u_k(\mathbf{r})$ for $\mathscr{H}_0$ are known, the wave function $\psi(\mathbf{r}, t)$ is given by Eq. (1.13), provided the C_k are considered to be functions of time. Given initial values for the C_k, we can use the Schrödinger equation (1.6) to determine values for later times. In fact, substituting (1.13) into (1.6), we have

$$i\hbar \sum_k (\dot{C}_k - i\omega_k C_k) \exp(-i\omega_k t) u_k(\mathbf{r}) = \sum_k (\hbar\omega_k + \mathscr{V}) \exp(-i\omega_k t) u_k(\mathbf{r})\, C_k. \tag{2}$$

Multiplying both sides by $u_n^*(\mathbf{r})$ and integrating over spatial coordinates, we find

$$\dot{C}_n = -\frac{i}{\hbar} \sum_k C_k (u_n | \mathscr{V} | u_k) \exp[-i(\omega_k - \omega_n)\, t], \tag{3}$$

where we have used the orthonormality of the u_k:

$$\int d^3r\, u_n^*(\mathbf{r})\, u_k(\mathbf{r}) = \delta_{nk}, \tag{4}$$

and where the matrix element of the perturbing Hamiltonian

$$\mathscr{V}_{nk} = (u_n|\mathscr{V}|u_k) = \int d^3r\, u_n^*(\mathbf{r})\, \mathscr{V} u_k\,(\mathbf{r}). \tag{5}$$

Equation (3) is equivalent to Schrödinger's equation. At any given time, $C_k(t)$ is the probability amplitude ($|C_k|^2$ is the probability) that the system is described by the eigenfunction $u_k(\mathbf{r})$ of the unperturbed Hamiltonian. We see that the effect of adding the perturbing term to the Hamiltonian is not to alter the eigenfunctions used as basis functions for $\psi(\mathbf{r}, t)$, but rather to change the superposition, *i. e.*, the C_n*'s*, in time.

In particular, let us consider the perturbation energy due to an applied electric field:

$$\mathscr{V} = -e\mathbf{E}(\mathbf{R}, t) \cdot \mathbf{r}, \tag{6}$$

where $\mathbf{E}$ is the electric field vector evaluated at the nucleus of the atom (position $\mathbf{R}$ in a laboratory coordinate system). Here the wavelength of the electromagnetic radiation must be long compared to the size of the atom (electric-dipole approximation). This form for the electric field interaction is compatible with the Schrödinger equation for an electron subject to a vector potential $\mathbf{A}(\mathbf{R}, t)$ (in the dipole approximation):

$$i\hbar \frac{\partial}{\partial t} \psi(\mathbf{r}, t) = \left\{\frac{1}{2m}[\mathbf{p} - e\,\mathbf{A}(\mathbf{R}, t)]^2 + V(r)\right\} \psi(\mathbf{r}, t). \tag{7}$$

In fact, substituting

$$\psi(\mathbf{r}, t) = \exp\left[\frac{ie}{\hbar}\mathbf{A}(\mathbf{R}, t) \cdot \mathbf{r}\right] \phi(\mathbf{r}, t) \tag{8}$$

into (7), and noting that

$$[\mathbf{p} - e\,\mathbf{A}(\mathbf{R})]\, \psi(\mathbf{r}, t) = \exp\left[\frac{ie}{\hbar}\mathbf{A} \cdot \mathbf{r}\right] \mathbf{p}\, \phi(\mathbf{r}, t),$$

we find

$$i\hbar\left[\frac{ie}{\hbar}\dot{\mathbf{A}} \cdot \mathbf{r}\phi + \dot{\phi}(\mathbf{r}, t)\right] \exp\left(\frac{ie}{\hbar}\mathbf{A} \cdot \mathbf{r}\right) \tag{9}$$

$$= \exp\left(\frac{ie}{\hbar}\mathbf{A} \cdot \mathbf{r}\right)\left[\frac{p^2}{2m} + V(r)\right] \phi(\mathbf{r}, t).$$

Since $\mathbf{E} = -\dot{\mathbf{A}}$, this is just

$$i\hbar\dot{\phi}(\mathbf{r}, t) = [\mathscr{H}_0 - e\mathbf{E} \cdot \mathbf{r}]\, \phi(\mathbf{r}, t). \tag{10}$$

The new wave function $\phi(\mathbf{r}, t)$ can be used in place of $\psi(\mathbf{r}, t)$ to calculate expectation values for any (gauge-invariant) operator. We note that the interaction Hamiltonian

$$\mathscr{H}_1 = em^{-1}\mathbf{A} \cdot \mathbf{p} \tag{11}$$

is not as accurate as Eq. (6) in the dipole approximation (does not include the A^2 term) and can lead to incorrect results. Inasmuch as the electric field does not depend on the atomic coordinates, the matrix element (5) for $e\mathbf{E}\cdot\mathbf{r}$ reduces to

$$\mathscr{V}_{nk} = -e\mathbf{E} \cdot \int d^3r\, u_n^*(\mathbf{r})\mathbf{r}u_k(\mathbf{r}) = -e\mathbf{E} \cdot \mathbf{r}_{nk}. \tag{12}$$

2-1. Induced Resonant Transitions

Let us turn now to the important problem of induced resonant transitions. Consider in particular a hydrogen atom which is initially in the ground state u_{100}. At time $t = 0$ we "turn on" an oscillating electric field:

$$\mathbf{E}(t) = \hat{x}E_0 \cos \nu t, \tag{13}$$

for which $\nu \cong (E_2 - E_1)/\hbar$, that is, the field is nearly resonant with the transitions from $n = 1$ to $n = 2$. From Fig. 1-5, we see that states u_{nlm} with $n > 2$ are way off resonance with this field and can be neglected. The u_{200} state has the correct energy for interaction but has a vanishing electric dipole matrix element, as can be shown along the lines of Prob. 1-2. Without loss of generality, we can choose the atomic z axis to be parallel to the electric vector $\mathbf{E}$. The matrix elements between the states $u_{21,\pm 1}$ and u_{100} then also vanish, and we are left with a two-level (scalar) problem. This kind of simplification is often used in the interaction of radiation with matter (see also the ammonia molecule in Chap. 5). Hence we can describe the atom by the two-level wave function (see Fig. 2-1)

$$\psi(\mathbf{r}, t) = C_a u_a(\mathbf{r}) \exp(-i\omega_a t) + C_b u_b(\mathbf{r}) \exp(-i\,\omega_b t) \tag{14}$$

and take the eigenfunctions $u_a = u_{210}$ and $u_b = u_{100}$ in numerical examples.

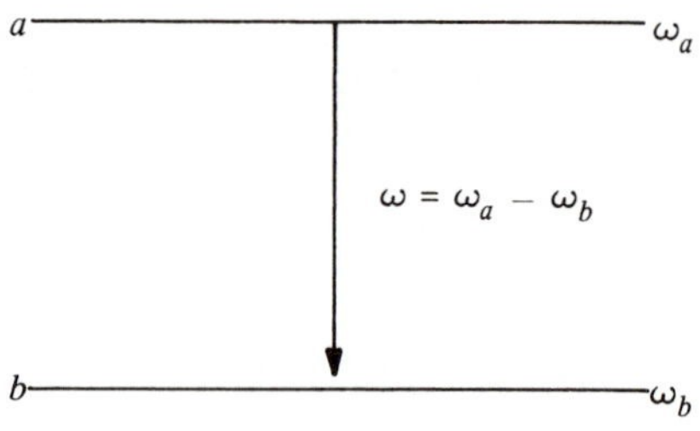

Figure 2-1. Diagram depicting two energy levels a and b of the unperturbed Hamiltonian $\mathscr{H}_0$. The energies are given by $\hbar\omega_a$ and $\hbar\omega_b$ respectively. Maser (laser) action takes place between two such levels in much of our discussion.

Here the coefficients satisfy the simple normalization condition [compare with Eq. (1.14)]

$$|C_a|^2 + |C_b|^2 = 1, \tag{15}$$

which implies that the probability for finding the system in a state other than a or b is zero. Although we bear in mind the hydrogen problem, the analysis that follows applies to any two-level problem for which there exists a nonvanishing electric-dipole matrix element.

The perturbation energy matrix element (12) reduces, in the two-level problem with field (13), to

$$\mathscr{V}_{ab} = -\wp E_0 \cos \nu t, \tag{16}$$

where $\wp = ez_{ab} = \wp^*$ is the magnitude of the electric-dipole matrix element.† Substituting this into the equations of motion (3), we find

$$\dot{C}_a = \tfrac{1}{2} i \wp \frac{E_0}{\hbar} \{\exp[i(\omega - \nu)t] + \exp[i(\omega + \nu)t]\} C_b, \tag{17}$$

$$\dot{C}_b = \tfrac{1}{2} i \wp \frac{E_0}{\hbar} \{\exp[-i(\omega - \nu)t] + \exp[-i(\omega + \nu)t]\} C_a, \tag{18}$$

where the frequency difference

$$\omega = \omega_a - \omega_b. \tag{19}$$

If $\wp E_0 t/\hbar$ is a small quantity, the values of C_a and C_b are not affected very much. This fact lends itself to an approximation method for solving (17) and (18). In particular, suppose that $C_b(0) = 1$ and $C_a(0) = 0$, that is, the system is in the ground state at time $t = 0$. Then the equations of motion for C_a and C_b reduce to

$$\dot{C}_b \simeq 0, \tag{20}$$

$$\dot{C}_a \simeq \tfrac{1}{2} i \wp \frac{E_0}{\hbar} \{\exp[i(\omega - \nu)t] + \exp[i(\omega + \nu)t]\}. \tag{21}$$

This yields

$$C_b(t) \simeq C_b^{(0)}(t) = 1, \quad C_b^{(1)} = 0, \tag{22}$$

$$C_a(t) \simeq C_a^{(1)}(t) = \tfrac{1}{2} \wp \frac{E_0}{\hbar} \left\{ \frac{\exp[i(\omega - \nu)t] - 1}{\omega - \nu} + \frac{\exp[i(\omega + \nu)t] - 1}{\omega + \nu} \right\}, \tag{23}$$

where the superscript (n) indicates that the field has acted n times (here no times or once) in perturbation.

For optical frequencies, the denominator $\omega + \nu$ is very large ($\omega \gg 1$), and the second term in (23) can be neglected with respect to the first since

†In our lectures on this material, we have found it useful to refer to $\wp$ as "squiggle."

$\nu \cong \omega$. This is called the rotating-wave approximation, for only the term in which the atomic and field "waves" [phasors $\exp(-i\omega t)^*$ and $\exp(-i\nu t)$] rotate together is kept. In this approximation, the perturbation matrix element (16) is given by

$$\mathscr{V}_{ab} = -\tfrac{1}{2}\wp E_0 \exp(-i\nu t) = \mathscr{V}_{ba}^*, \tag{24}$$

and the equations of motion (17) and (18) for C_a and C_b reduce to

$$\dot{C}_a = \tfrac{1}{2} i \frac{\wp E_0}{\hbar} \exp[i(\omega - \nu)t] C_b, \tag{25}$$

$$\dot{C}_b = \tfrac{1}{2} i \frac{\wp E_0}{\hbar} \exp[-i(\omega - \nu)t] C_a. \tag{26}$$

Thus the first-order contribution to C_a is

$$C_a^{(1)}(t) = \frac{1}{2}\frac{\wp E_0}{\hbar}\left\{\frac{\exp[i(\omega - \nu)t] - 1}{\omega - \nu}\right\}. \tag{27}$$

Noting that

$$\begin{aligned}\exp(ix) - 1 &= \exp(ix/2)\,[\exp(ix/2) - \exp(-ix/2)]\\ &= 2i \exp(ix/2) \sin(x/2)\end{aligned} \tag{28}$$

we find

$$C_a(t) \cong C_a^{(1)}(t) = \tfrac{1}{2} i \frac{\wp E_0}{\hbar} \exp[i(\omega - \nu)t/2] \times \frac{\sin[(\omega - \nu)t/2]}{(\omega - \nu)/2}, \tag{29}$$

and the probability for being in the upper state

$$|C_a(t)|^2 = |C_a^{(1)}(t)|^2 = \frac{1}{4}\left(\frac{\wp E_0}{\hbar}\right)^2 \left\{\frac{\sin^2[(\omega - \nu)t/2]}{[(\omega - \nu)/2]^2}\right\}. \tag{30}$$

Inasmuch as the atom gains energy, the field should lose the same amount, and $|C_a(t)|^2$ is called the probability for stimulated absorption of radiation. In Chap. 3 we show how the effect of the atom on the classical field can be included; in Chap. 14 the fully quantum-mechanical version is presented: On resonance ($\nu = \omega$), $|C_a(t)|^2 = \frac{1}{4}(\wp E_0/\hbar)^2 t^2$, as discussed in Prob. 2-3. The probability is plotted in Fig. 2-2 for three nonvanishing values of the detuning $\omega - \nu$. It oscillates in time with frequency $\omega - \nu$ [$\sin^2(\omega t/2) = \frac{1}{2}(1 - \cos \omega t)$] and magnitude proportional to $(\omega - \nu)^{-2}$. It is left as an exercise for the reader to show that, for $\frac{1}{2}(\wp E_0 t/\hbar)^2 \ll 1$ or for $(\wp E_0/\hbar)^2/(\omega - \nu)^2 \ll 1$, the first-order solution (30) remains a good approximation.

Some physical insight can be gained by plotting $|C_a^{(1)}(t)|^2$ versus the detuning, as is done in Fig. 2-3. We see that the probability of being in the a state decreases rapidly as the detuning is increased. Furthermore, the approximate width of the central peak is inversely proportional to t, and the height is proportional to t^2. This gives an area under the curve proportional to time.

This fact is manifested in a formula with sufficiently useful application that Fermi called it the golden rule of quantum mechanics. The rule is based on the assumption that the detuning $\omega - \nu$ has a range of values due either to a field with a continuous spectrum, as in blackbody radiation, or to a spread in the energy levels themselves, as applies to the photoionization of an atom in a detector. We consider here the former possibility (see Prob. 2-9 for the latter) and use the result to calculate the Einstein B coefficient for blackbody radiation of energy density $\mathscr{U}(\nu)$.

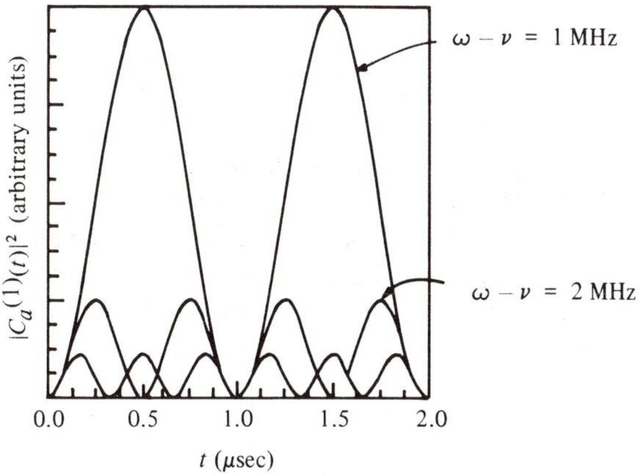

Figure 2-2. The probability (30) of a transition to the upper level under the influence of an applied oscillating electric field. The curves also represent the exact solution (63), provided $\omega - \nu$ is replaced by the "power-broadened" frequency $[(\omega - \nu)^2 + (\wp E_0/\hbar^2)]^{1/2}$. The first order results cannot be trusted if $|C_a^{(1)}(t)|^2$ does not remain much less than unity.

To allow for the continuous spectrum, we replace E_0^2 in (30) by $\varepsilon_0^{-1}\mathscr{U}(\nu)\, d\nu$ and sum over all frequencies. This yields the transition probability

$$P_a = \frac{1}{4}\frac{\wp^2}{\hbar^2\varepsilon_0}\int d\nu\, \mathscr{U}(\nu)\frac{\sin^2[(\omega - \nu)t/2]}{[(\omega - \nu)/2]^2}\,. \tag{31}$$

We suppose that the energy density $\mathscr{U}(\nu)$ varies slowly compared to the $\sin^2 x/x^2$ factor, and we replace it by the resonance value $\mathscr{U}(\omega)$. The remaining integral has the value

$$\int d\nu \frac{\sin^2[(\omega - \nu)t/2]}{[(\omega - \nu)/2]^2} = 2t\int_{-\infty}^{\infty} dx \sin^2(x)/x^2 = 2\pi t.$$

Collecting these results, we find for (31)

$$P_a = \tfrac{1}{2}\pi \frac{\wp^2}{\hbar^2\varepsilon_0}\mathscr{U}(\omega)t, \tag{32}$$

which is proportional to t, as was the area in Fig. 2-3. Equation (32) yields the transition rate:

$$\frac{dP_a}{dt} = \tfrac{1}{2}\pi \frac{\wp^2}{\hbar^2\varepsilon_0}\mathscr{U}(\omega). \tag{33}$$

For blackbody radiation, the field is unpolarized and isotropic, unlike the field (13) used in the calculation of (33). We can generalize (33) for this kind of radiation by setting the energy density

$$\mathscr{U}(\omega) \rightarrow \frac{\mathscr{U}(\omega)}{3}, \tag{34}$$

see Bonifacio et al., (1976). We then find the transition rate for blackbody radiation:

$$\frac{dP_a}{dt} = \tfrac{1}{6}\pi \frac{\wp^2}{\hbar^2\varepsilon_0}\mathscr{U}(\omega) = B\mathscr{U}(\omega), \tag{35}$$

where B is the Einstein B coefficient of Sec. 2-2.

In our discussion of the two-level problem, we have considered initial conditions suitable for (stimulated) absorption. We could have chosen $C_a(0) = 1$ and $C_b(0) = 0$, that is, the atom initially in the upper state. We can then calculate (Prob. 2-1) the probability that the atom makes a transition to the lower state:

$$|C_b(t)|^2 = |C_b^{(1)}(t)|^2 = \frac{1}{4}\left(\frac{\wp E_0}{\hbar}\right)^2 \left\{\frac{\sin^2[(\omega-\nu)t/2]}{[(\omega-\nu)/2]^2}\right\}. \tag{36}$$

This is the probability that stimulated emission has taken place and has the same value as $|C_a(t)|^2$ had in (30) for the absorption problem.

The analysis given for the transition rate (33) applies equally well to (36). Hence we see that the rates for stimulated absorption and emission are equal.

2-2. Blackbody Radiation

The experimentally observed spectral distribution of blackbody radiation is very well fitted by the formula discovered by Planck:

$$\mathscr{U}(\omega) = \frac{\hbar\omega^3/\pi^2c^3}{\exp(\hbar\omega/k_BT)-1}, \tag{37}$$

where ω is the frequency of the radiation, T is the absolute temperature, c is the speed of light in vacuum, and k_B is Boltzmann's constant. Einstein (1917) revealed that this formula can be explained in terms of ensembles of atoms with discrete energy levels of various frequencies provided that (1) spontaneous

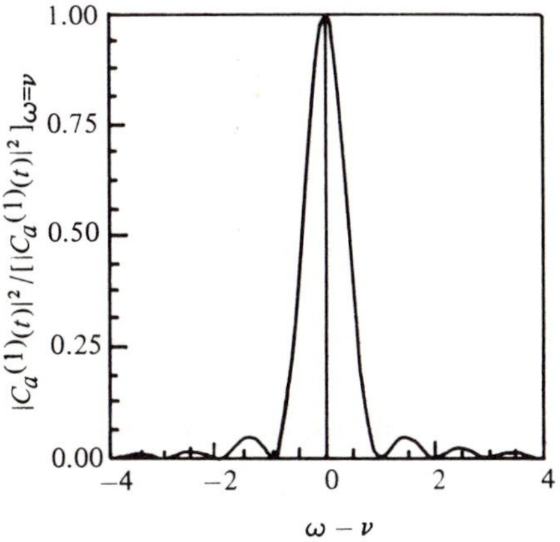

Figure 2-3. Normalized probability [from (30)] of a transition to the upper level versus detuning. This is also the graph for the exact solution, provided the frequency difference $|\omega - \nu|$ *is replaced by the "power-broadened" difference* $\sqrt{(\omega-\nu)^2 + (\wp E_0/\hbar)^2}$. *The width is inversely proportional to t; the height is proportional to* t^2. *Thus the area under the curve is approximately proportional to t. This fact is reflected in Eq. (32).*

emission is included along with the processes of stimulated emission and absorption, and (2) the upper and lower energy levels are populated according to the Boltzmann distribution. In this section we follow this derivation but include information not available to Einstein about the stimulated transition rates obtained by the use of quantum mechanics in the preceding section. We conclude this section with a formula for the spontaneous emission rate constant A based on the stimulated rate in conjunction with the Planck formula (37).

We suppose that the populations (number of atoms) n_a and n_b in the upper and lower levels obey the equations of motion

$$\dot{n}_a = -An_a - B\mathscr{U}(\omega)(n_a - n_b), \tag{38}$$

$$\dot{n}_b = +An_a + B\mathscr{U}(\omega)(n_a - n_b). \tag{39}$$

As depicted in Fig. 2-4, An_a is the rate at which atoms are leaving the upper level because of spontaneous emission, and $B\mathscr{U}(\omega)\,(n_a - n_b)$ is the corresponding rate for stimulated processes. Atoms which leave the upper level enter the lower level. Because (38) and (39) consist entirely of rates, they are called rate equations. We have use for similar rate equations in the theory of the laser.

In equilibrium, the populations should be constant, that is, $\dot{n}_a = \dot{n}_b = 0$. Then both (38) and (39) yield

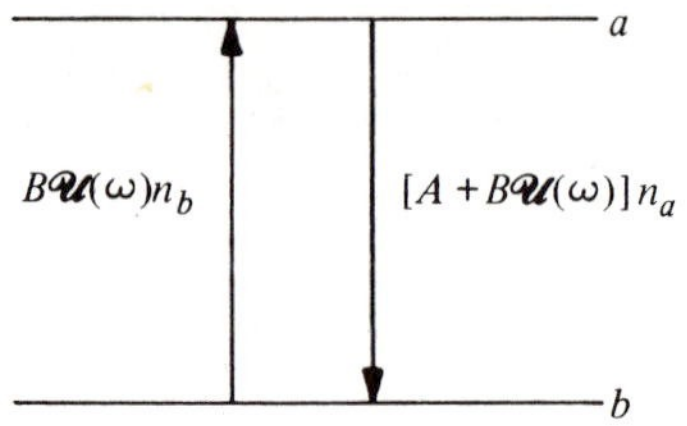

Figure 2-4. Flow of atomic populations between the energy levels a and b.

$$[A + B\,\mathcal{U}(\omega)]n_a = B\,\mathcal{U}(\omega)n_b. \tag{40}$$

Furthermore, in thermal equilibrium, the populations are related by the Boltzmann distribution:

$$n_a = n_b \exp(-\hbar\omega/k_{\mathrm{B}}T). \tag{41}$$

Combining this with (40), we find the energy density

$$\mathcal{U}(\omega) = \frac{A}{B}[\exp(\hbar\omega/k_BT) - 1]^{-1}. \tag{42}$$

This gives the Planck formula (37), provided that the ratio of the spontaneous and stimulated coefficients is given by

$$\frac{A}{B} = \frac{\hbar\omega^3}{\pi^2c^3}\,. \tag{43}$$

This is the energy per photon $\hbar\omega$ multiplied by the density of modes per unit volume between ω and $\omega + d\omega$. If we take (43) as an experimental fact, we can derive the spontaneous rate constant A by using the transition rate (35) for B:

$$A = \frac{\frac{1}{6}\hbar\omega^3\,\wp^2}{\pi\hbar^2c^3\varepsilon_0}\,. \tag{44}$$

Because this factor is proportional to ω^3, spontaneous emission is considerably more important at optical frequencies than at radio frequencies. In Chap. 8, we see how this fact affects the gain in a laser.

It is interesting to note that if we had dropped the stimulated emission term $B\mathcal{U}(\omega)n_a$ in the equilibrium condition (40), we would have obtained the radiation density:

$$\mathcal{U}(\omega) = \frac{A}{B}\exp(-\hbar\omega/k_{\mathrm{B}}T), \tag{45}$$

which, with A/B given by (43), is the Wien law. This agrees quite well with the experimentally observed Planck formula (37) for $\hbar\omega/k_BT > 5$, but differs

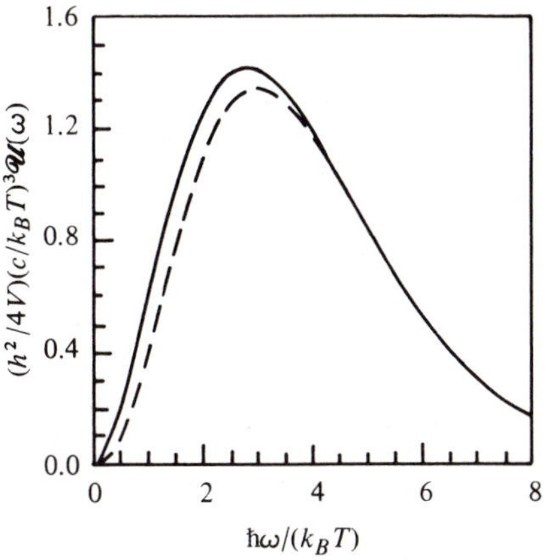

Figure 2-5. Planck (37) (solid line) and Wien (45) (dashed line) radiation laws.

for smaller values, as depicted in Fig. 2-5. Although the distributions seem close in value even for low frequencies (or high temperatures), they produce very different values for the higher-order correlations such as the average $\langle n^2 \rangle$, where n is the number of photons for a given mode. Here the energy density $\mathcal{U}(\omega)$ is proportional to the first-order correlation,the average number of photons of frequency ω, $\langle n(\omega) \rangle$. (See Prob. 16-15.)

2-3. Inclusion of Decay Phenomena

We have seen how excited atomic levels decay in time because of spontaneous emission. They can also decay because of collisions and other phenomena. In Fig. 2-6 we indicate such decay from both the a and the b levels, a situation which occurs in typical laser media. The finite lifetimes can be described very well by adding phenomenological decay terms to the equations of motion (25) and (26). We write

$$\dot{C}_a = -\tfrac{1}{2}\gamma_a C_a + \tfrac{1}{2} i \frac{\wp E_0}{\hbar} \exp[i(\omega - \nu)t] C_b, \tag{46}$$

$$\dot{C}_b = -\tfrac{1}{2}\gamma_b C_b + \tfrac{1}{2} i \frac{\wp E_0}{\hbar} \exp[-i(\omega - \nu)t] C_a. \tag{47}$$

The factors of $\frac{1}{2}$ are included so that, for example, the probability $|C_a|^2$ decays as $\exp(-\gamma_a t)$ in the absence of E_0. The lifetimes are defined as the times at which the probabilities have decayed to $1/e$ of their original values. Hence they are given by the reciprocals of the decay constants γ_a and γ_b.

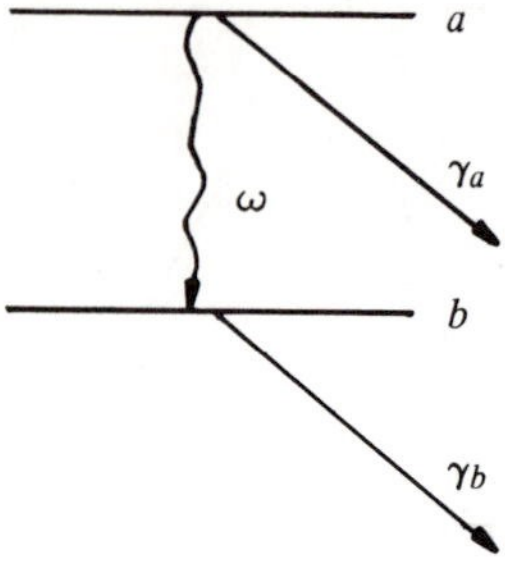

Figure 2-6. Energy-level diagram for two-level atom, showing decay rates γ_a and γ_b for the probabilities $|C_a|^2$ and $|C_b|^2$.

We can solve these equations by first-order perturbation theory, as in Sec. 2-1, or by the more exact formulations in Sec. 2-4. We note here that for the simple case of equal decay constants the substitutions

$$\begin{aligned} C_b' &= C_b \exp(\gamma_b t/2), \\ C_a' &= C_a \exp(\gamma_b t/2), \end{aligned} \tag{48}$$

reduce Eqs. (46) and (47) to the undamped versions (25) and (26). The solutions for this damped case are, then, just those for the undamped case multiplied by the exponential decay factor $\exp(-\frac{1}{2}\gamma_b t)$. In particular the probability that stimulated absorption takes place changes from (30) to the damped formula

$$|C_a(t)|^2 = |C_a{}^{(1)}(t)|^2 = \frac{1}{4}\left(\frac{\wp E_0}{\hbar}\right)^2 \exp(-\gamma_b t) \left\{\frac{\sin[(\omega-\nu)t/2]}{(\omega-\nu)/2}\right\}^2. \tag{49}$$

This is illustrated in Fig. 2-7 along with the undamped case.

We can find the spectral distribution for stimulated emission by starting in the upper level and calculating the total probability that the atom decays by spontaneous emission from the lower level. This is given by

$$P_s = \gamma_b \int_0^\infty dt\, |C_b(t)|^2, \tag{50}$$

for $\gamma_b |C_b(t)|^2$ is the probability per unit time that the atom decays from the b level, and $|C_b(t)|^2$ is the probability that the atom is in the b level because of stimulated emission. Multiplying (36) by the damping factor $\exp(-\gamma_b t)$, we find

$$|C_b(t)|^2 = \frac{1}{4}\left(\frac{\wp E_0}{\hbar}\right)^2 \exp(-\gamma_b t) \left\{\frac{\sin[(\omega-\nu)t/2)]}{(\omega-\nu)/2}\right\}^2. \tag{51}$$

Substituting this into (50), we find that the profile is Lorentzian with width γ_b:

$$P_s = \frac{1}{2}\frac{(\wp E_0/\hbar)^2}{(\omega - \nu)^2 + \gamma_b{}^2}. \tag{52}$$

Of course, if the transition probability $|C_b(t)|^2$ fails to remain always much less than unity, Eq. (51) cannot be trusted. It is still possible, however, to calculate P_s by using a strong-signal theory developed in the next section [see Eq. (65)] and Probs. 2-7 and 2-8.

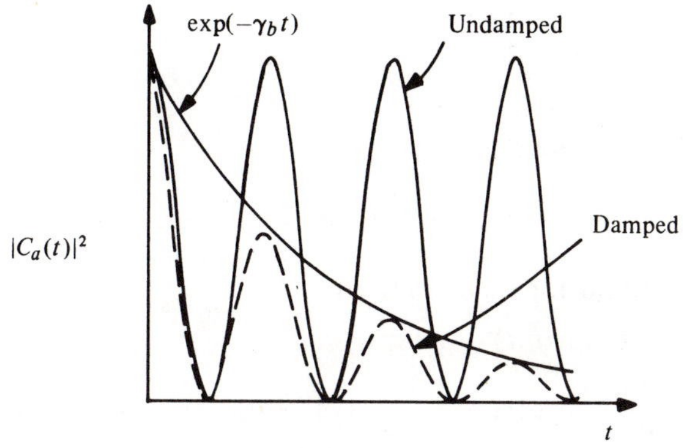

Figure 2-7. Transition probability (49) with nonzero γ_b (dashed line) and $\gamma_b = 0$ (solid line). The latter corresponds to an infinite lifetime. Note that we have taken $\gamma_a = \gamma_b$.

2-4. Exact Rabi Solution

A more exact solution for the probability amplitudes can be obtained by substitution of improved values for C_a and C_b back into the equations of motion (25) and (26) [or the damped versions of (46) and (47)]. Note that, for the initial conditions $C_a(0) = 0$, $C_b(0) = 1$, only even powers of the electric field interaction $-\wp E_0$ contribute to $C_b(t)$, that is,

$$C_b(t) = C_b{}^{(0)} + C_b{}^{(2)}(t) + C_b{}^{(4)}(t) + \cdots, \tag{53}$$

and only odd powers contribute to $C_a(t)$:

$$C_a(t) = C_a{}^{(1)}(t) + C_a{}^{(3)}(t) + C_a{}^{(5)}(t) + \cdots. \tag{54}$$

This iteration technique is particularly useful when complicated fields (e.g., multimode) are involved or when extensions are desired to treat three or more atomic levels.

In the present case of two levels and a monochromatic field, Eqs. (25) and (26) can be solved exactly (in the rotating-wave approximation). Assuming a solution†

†Simpler methods of solution are possible for resonance (see Prob. 2-4).

$$C_b = \exp(i\mu t), \tag{55}$$

and using (26), we find

$$C_a(t) = \frac{2\hbar\mu}{\wp E_0} \exp[i(\omega - \nu + \mu)t]. \tag{56}$$

Substituting this, in turn, into (25), we find

$$i\frac{2\hbar}{\wp E_0}\mu(\mu + \omega - \nu) = \tfrac{1}{2}i\frac{\wp E_0}{\hbar}, \tag{57}$$

or

$$\mu(\mu + \omega - \nu) - \frac{1}{4}\left(\frac{\wp E_0}{\hbar}\right)^2 = 0.$$

This has the two solutions

$$\mu_{1,2} = -\tfrac{1}{2}(\omega - \nu) \pm \frac{1}{2}\left[(\omega - \nu)^2 + \left(\frac{\wp E_0}{\hbar}\right)^2\right]^{1/2}. \tag{58}$$

The general solutions for C_b and C_a are

$$C_b(t) = A\exp(i\mu_1 t) + B\exp(i\mu_2 t), \tag{59}$$

$$C_a(t) = 2\left(\frac{\wp E_0}{\hbar}\right)^{-1}\exp[i(\omega - \nu)t][A\mu_1\exp(i\mu_1 t) + \mu_2 B\exp(i\mu_2 t)]. \tag{60}$$

In particular, suppose that $C_b(0) = 1$ and $C_a(0) = 0$ (absorption). From (60),

$$A = -\frac{B\mu_2}{\mu_1}.$$

Combining this with (59), we find

$$1 = B + A = B\left(1 - \frac{\mu_2}{\mu_1}\right) = \frac{B\mu}{\mu_1},$$

where we define the Rabi "flopping frequency":

$$\mu \equiv \mu_1 - \mu_2 = \sqrt{(\omega - \nu)^2 + (\wp E_0/\hbar)^2}. \tag{61}$$

This gives coefficients $A = -\mu_2/\mu$ and $B = \mu_1/\mu$. Combining these with (60) and noting that $\mu_1\mu_2 = -\frac{1}{4}(\wp E_0/\hbar)^2$, we have

$$C_a(t) = i\,\wp E_0(\hbar\mu)^{-1}\exp[i(\omega - \nu)t/2]\sin(\mu t/2). \tag{62}$$

Comparing this to the first-order result (29), we see that the only difference is that the frequency difference $\omega - \nu$ in $\sin[(\omega - \nu)t/2]/[(\omega - \nu)/2]$ has been replaced by the "power-broadened" version $\mu = [(\omega - \nu)^2 + (\wp E_0/\hbar)^2]^{1/2}$. Hence Figs. 2-2, 2-3, and 2-7 apply here, provided this replacement is made.

Taking the absolute-value squares of (62), we find the strong-signal transition probability:

$$|C_a(t)|^2 = \left[\frac{\sin^2(\mu t/2)}{(\mu/2)^2}\right]\left(\frac{\wp E_0}{2\hbar}\right)^2. \tag{63}$$

The same value is obtained for $|C_b(t)|^2$ for an initially excited system. This is the Rabi "flopping formula," originally derived by I.I. Rabi (1937) to express the probability that a spin-$\frac{1}{2}$ atom incident on a Stern-Gerlach apparatus could be flipped from the $\begin{pmatrix}1\\0\end{pmatrix}$ or $\begin{pmatrix}0\\1\end{pmatrix}$ state to the $\begin{pmatrix}0\\1\end{pmatrix}$ or $\begin{pmatrix}1\\0\end{pmatrix}$ state by an applied radio-frequency magnetic field. Rabi used this calculation in combination with a special apparatus to measure the magnetic moments of atoms much more accurately than is possible with the original Stern-Gerlach apparatus.

On resonance, (63) reduces to

$$[|C_a(t)|^2]_{\nu=\omega} = \sin^2\left[\frac{(\wp E_0/\hbar)t}{2}\right], \tag{64}$$

which shows that at time $t = \pi\hbar/(\wp E_0)$ the probability of a transition to the upper state is unity. A "pulse" characterized by this value for $\wp E_0$ and with duration t is sometimes called a π pulse, for $\wp E_0 t/\hbar = \pi$. In Sec. 7-5 we show that for the spin-$\frac{1}{2}$ system this corresponds to a rotation of π radians in space.

It is possible to integrate the damped equations of motion (46) and (47) for C_a and C_b (see Probs. 2-7 and 2-8). The resulting formula for $|C_b(t)|^2$ can then be substituted into Eq. (50) to find P_s, the net probability that the atom decays by stimulated emission from the upper to lower levels. The result is

$$P_s = \frac{\gamma_{ab}{}^2\, I_0(\gamma_b/\gamma_{ab})}{(\omega - \nu)^2 + \gamma_{ab}{}^2(1 + 2I_0)}, \tag{65}$$

where the average decay constant

$$\gamma_{ab} = \tfrac{1}{2}(\gamma_a + \gamma_b), \tag{66}$$

and the dimensionless intensity

$$I_0 = \frac{\frac{1}{2}(\wp E_0)^2}{\hbar^2\gamma_a\gamma_b}. \tag{67}$$

We see that the frequency response of the atom to an applied field is broadened not only because of decay, but also because of saturation. This second effect is called power broadening.

Problems

2-1. Starting with the initial conditions $C_a(0) = 1$, $C_b(0) = 0$, solve the Schrödinger equations of motion (17) and (18) in the rotating-wave approximation to first order. Here $|C_b^{(1)}(t)|^2$ is the probability at time t that stimulated emission has taken place.

2-2. Starting with the initial conditions $C_a(0) = 1$ and $C_b(0) = 0$, solve the equations of motion (17) and (18) in the rotating-wave approximation to third order in the electric field interaction energy.

2-3. Show that, for $\frac{1}{4}(\wp E_0 t/\hbar)^2 \ll 1$ or for $(\wp E_0/\hbar)^2/(\omega - \nu)^2 \ll 1$, the first-order solution (27) is a good approximation.
2-4. Derive resonant ($\omega = \nu$) solutions to the equations of motion (25) and (26) for the probability amplitudes C_a and C_b in two ways: (*a*) solve first for $\ddot{C}_b$ in terms of C_b; (*b*) solve first for the difference $\dot{C}_a - \dot{C}_b$ and the sum $\dot{C}_a + \dot{C}_b$.

These methods are considerably shorter than the one used for the general case in the text. The second method is particularly useful when E_0 is replaced by a function of time, as in coherent pulse propagation (Chap. 13).
2-5. Show that, in effect, $\wp^2$ is replaced by $\wp^2/3$ in Eq. (34). *Hint:* $\mathbf{r} = x\hat{x} + y\hat{y} + z\hat{z}$, and for random systems $|\langle ex\rangle|^2 = |\langle ey\rangle|^2 = |\langle ez\rangle|^2$.
2-6. Show that, for an initially excited system, the strong-signal formula (63) is replaced by

$$|C_b(t)|^2 = \left(\frac{\wp E_0}{2\hbar}\right)^2 \frac{\sin^2(\mu t/2)}{(\mu/2)^2} \,. \tag{68}$$

2-7. Solve the damped equations of motion (46) and (47), using the Rabi method of Sec. 2-4. Specifically, find $|C_b(t)|^2$ for the initial conditions $C_a(0) = 1$, $C_b(0) = 0$.

Ans.: The eigenvalues are [setting $C_b(t) = \exp(i\mu t)$]

$$\mu_{1,2} = -\tfrac{1}{2}(\omega - \nu - i\gamma_{ab}) \pm \frac{1}{2}\left\{[\omega - \nu - \tfrac{1}{2}i(\gamma_a - \gamma_b)]^2 + \left(\frac{\wp E_0}{\hbar}\right)^2\right\}^{1/2} \tag{69}$$

and

$$|C_b(t)|^2 = (\wp E_0)^2 (\hbar^2 \mu\mu^*)^{-1} \exp(-\gamma_{ab} t) \sin(\tfrac{1}{2}\mu t) \sin(\tfrac{1}{2}\mu^* t), \tag{70}$$

where the complex Rabi flopping frequency

$$\mu = \left\{[\omega - \nu - \tfrac{1}{2}i(\gamma_a - \gamma_b)]^2 + \left(\frac{\wp E_0}{\hbar}\right)^2\right\}^{1/2} . \tag{71}$$

2-8. Find the net probability for stimulated emission P_s of (50). *Hint:* Do time integration and simplify *before* using (71) for μ. The answer is Eq. (65) of the text.
2-9. Consider a monochromatic field of frequency ν incident on a collection of atoms with a spread of resonant frequencies given by $\rho(\omega)$. For convenience let $\rho(\omega)$ be symmetrical about ν. Define $\langle P_a \rangle$ to be the average probability for atoms in the upper state:

$$\langle P_a \rangle = \int d\omega\, \rho(\omega) |C_a(t)_\omega|^2.$$

(*a*) Derive Fermi's golden rule for this case, assuming that first-order theory is valid.

(*b*) How does the finite width of $\rho(\omega)$ limit the validity of Fermi's golden rule?

(*c*) Assuming that the formula in part (*a*) is valid, and that the atoms are distributed uniformly in a volume with a density $N =$ number of atoms per unit volume, show that the intensity $I = E_0^2$ of the light is attenuated in the volume according to the formula

$$\frac{dI}{dz} = -\ aI,$$

where z is the direction of the light, and give an expression for α in terms of N, $\rho(\omega)$, $\wp$, and other parameters.

(*d*) Let the incident beam have a finite leading edge in the form of a step function. How do the considerations in part (b) affect the output? Do the limitations mean that more or less light is absorbed, when the golden rule breaks down, than you expect from the answer in (*c*)? (Explain why.)

(*e*) Sketch the light (as a function of time) as it exits from the volume.

2-10. Using Fermi's golden rule, show the following properties of the photoelectric effect: [note: *classical* field]

(*a*) When light shines on a photoemissive surface, electrons are ejected with a kinetic energy equal to Planck's constant times the frequency ν of the incident light less some work function ϕ, usually written as

$$\hbar\nu = \frac{mv^2}{2} + \phi.$$

(*b*) The rate of electron ejection is proportional to the square of the electric field of the incident light (ejection rate $\propto E_0^2$).

(*c*) There is not necessarily a time delay between the instant the field is turned on and the ejection of photoelectrons.

2-11. Solve the three-level Rabi flopping problem for which two of the three possible transitions are allowed and subjected to resonant radiation. Take initial probability of lowest level equal to unity. Ans: M. Sargent III and P. Horwitz, 1976, Phys. Rev. *A*13, 1962, Eqs. (16)–(18).

References

See also the references given at the end of Chap. 1.

R. Bonifacio, F. Hopf, P. Meystre, M. Scully, 1976, Chap. 9 in: *Laser Induced Fusion and X-Ray Laser Studies*, ed. by S. F. Jacobs, M. O. Scully, M. Sargent III, C. D. Cantrell III, Addison-Wesley Pub. Co., Reading, Mass.

A. Einstein, 1917, *Phys. Z.* **18**, 121.

J. Fiutak, 1963, *Canadian J. Phys.* **41**, 12.

W. E. Lamb, Jr., 1950, *Phys. Rev.* **85**, 268.

W. E. Lamb, Jr. and M. O. Scully, 1969, in *Polarization, matter and radiation. Jubilee volume in honor of A. Kastler,* Presses Universitaires de France, Paris. See also M. O. Scully and M. Sargent III, 1972, Phys. Today **25**, 38.

E. Power and S. Zienau, 1959, *Phil. Trans. Roy. Soc.* **251A**, 54.

I. I. Rabi, 1936, *Phys. Rev.* **49**, 324.

I. I. Rabi, 1937, *Phys. Rev.* **51**, 652.

III

STIMULATED EMISSION AND DIPOLE OSCILLATORS

3. Stimulated Emission and Dipole Oscillators

The phenomena which are principally responsible for the interaction of radiation with matter in everyday life are absorption and spontaneous emission. Light falling on matter is absorbed, leaving atoms in excited states which spontaneously emit radiation with a spread of frequency inversely proportional to the decay time. The situation is modified when the atomic decay times are sufficiently long and the radiation sufficiently intense, for then radiation may fall on *excited* atoms, resulting in stimulated emission rather than absorption. We saw in Chap. 2 that this third process was required to obtain the Planck blackbody radiation law in Einstein's derivation. It is stimulated emission which enables an atomic medium to amplify incident radiation, and therefore is an essential ingredient in light amplifiers and oscillators. In fact, the word LASER is an acronym for Light Amplification by Stimulated Emission of Radiation.

This chapter is concerned with the understanding of radiative processes and particularly of stimulated emission. The discussion centers about dipole oscillators, both quantum-mechanical and classical, which amplify or attenuate the electric fields they interact with. In Sec. 3-1 we show how the quantum theory of Chap. 2 predicts that atoms in superpositions of energy levels act like electric charges on springs, the model Lorentz successfully employed in 1880 to explain the anomalous index of refraction and absorption of media. With this physical picture in mind, we examine in Sec. 3-2 classical dipole oscillators which exhibit analogs of spontaneous emission, absorption, and stimulated emission. The stimulated processes of the classical oscillator depend on the oscillator's initial phase, which is, in general, independent of the incident radiation. An exception is an oscillator initially at rest, for an incident field induces in this dipole a phase giving absorption. This is part of the reason for the success of Lorentz's theory. The two-level atom of Chap. 2 shares this feature for *both* ground-state *and* excited-state atoms: unlike the situation for the classical dipole, an incident field induces in an excited atom a dipole with the right phase for emission. The failure of the classical model to similarly provide the emitting phase severely hurts its complete application to laser phenomena. As discussed at the end of the chapter, the difficulty arises because the oscillator is simple harmonic.

The quantum and classical dipoles are then contrasted as to the probabilistic interpretation of the former. Credence is given to the substitution of an *ensemble* of closely spaced quantum dipoles for the corresponding classical ensemble in Maxwell's equations. This kind of identification is fundamental in the semiclassical treatment (Chaps. 8–13) of the amplification and attenuation of radiation by matter. In that treatment, the dipoles appear as sources in Maxwell's equations for the field, thereby leading to an index of refraction and absorption or gain. Radiation by single dipoles and sheets of dipoles is considered, and properties of stimulated emission are summarized. The chapter closes with a discussion of classical anharmonic oscillators (weak spring) which, properly used, can provide a gain medium.

3-1. Quantum Electric Dipole

The classical Lorentz model of harmonically bound electrons gives a remarkably good account of the anomalous index of refraction and absorption. It is supposed that quantum electrons in atoms effectively behave like charges subject to the forced, damped oscillation of an applied electromagnetic field. It is not at all obvious that bound electrons behave in this way. Nevertheless, the average charge distribution does oscillate, as can be understood by the following simple argument. We know the probability density for the electron at any time, for this is given by the electron wave function as $\psi^*(\mathbf{r}, t)\psi(\mathbf{r}, t)$. Hence the effective charge density is

$$e\psi^*(\mathbf{r}, t)\psi(\mathbf{r}, t).$$

For example, consider a hydrogen atom initially in its ground state with the spherical distribution depicted in Fig. 3-1. Here the average electron charge is concentrated at the center of the sphere. Application of an electric field forces this distribution to shift with respect to the positively charged nucleus as shown in the figure. Subsequent removal of the field then causes the charged sphere to oscillate back and forth across the nucleus because of Coulomb attraction. This oscillating dipole acts something like a charge on a spring.

We can obtain a more quantitative feeling for this process by considering the two-level atom of Chap. 2. Suppose that at time $t = 0$ the atom is in its ground state with the wave function

(a) P = 0 (b) P ≠ 0

Figure 3-1. (a) Spherical electron charge distribution of hydrogen atom, indicated by plus sign with center of distribution at nucleus (1s energy state). (b) Application of electric field shifts distribution (center at −) relative to positively charged nucleus. Subsequent removal of field produces oscillating distribution, that is, an oscillating dipole like a charge on a spring.

$$\psi(\mathbf{r},\, 0) = u_b(\mathbf{r}).$$

Then, under the influence of the electric field, the wave function at time t later is given by the superposition of eigenfunctions:

$$\psi(\mathbf{r},\, t) = C_a(t)\exp(-i\omega_a t)u_a(\mathbf{r}) + C_b(t)\exp(-i\omega_b t)u_b(\mathbf{r}). \tag{1}$$

Inasmuch as the wave function is indeterminate up to a phase factor, we can without loss of generality write (1) as

$$\psi(\mathbf{r},\, t) = C_a(t)\exp[-i(\omega_a - \omega_b)t]\; u_a(\mathbf{r}) + C_b(t)\; u_b(\mathbf{r}). \tag{2}$$

For purposes of illustration, suppose that $u_a(\mathbf{r}) = u_{210}\,(\mathbf{r})$ and $u_b = u_{100}$, which have the unnormalized z dependences shown in Fig. 3-2a. At time $t = 2n\pi/\omega$, the exponential is unity and the wave function (2) becomes

$$\psi\left(\mathbf{r},\, \frac{2n\pi}{\omega}\right) = C_a u_a(\mathbf{r}) + C_b u_b(\mathbf{r}), \tag{3}$$

as depicted in Fig. 3-2*b*. At a time π/ω later, the exponential is -1 and

$$\psi\left[\mathbf{r},\, \frac{(2n+1)\pi}{\omega}\right] = -C_a u_a(r) + C_b u_b(r), \tag{4}$$

as shown in Fig. 3-2*c*. We see in Fig. 3-2*d* that the probability density $\psi^*\psi$ oscillates back and forth in time, generating an oscillating dipole. Quantitatively this dipole is given by the expectation value (1.5) as

$$\langle er\rangle = \wp C_a C_b^* \exp(-i\omega t) + \text{c.c.}, \tag{5}$$

where the matrix element $\wp$ is given by (2. 16) and the frequency difference $\omega = \omega_a - \omega_b$. Hence determination of the coefficients C_a and C_b or just their bilinear product $C_a C_b^*$ suffices to give the induced dipole moment.

The Rabi solution of Sec. 2-4 yields the values as follows. Choosing

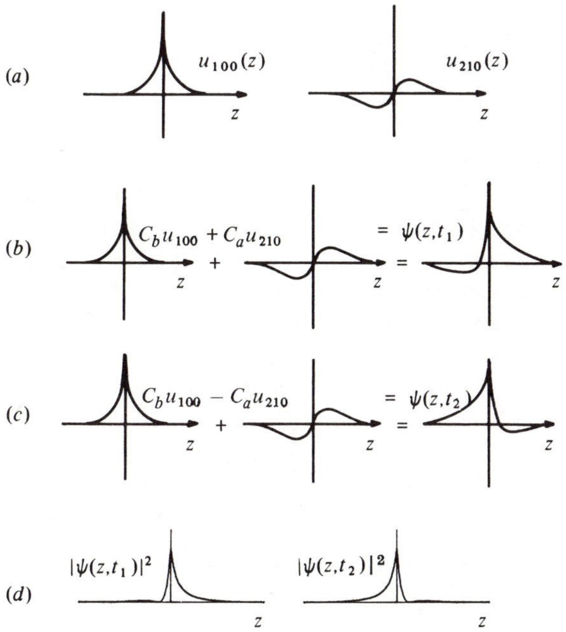

Figure 3-2. (a) z dependences for hydrogen eigenfunctions u_{210} and u_{100}. (b) Resulting z dependence of $\psi(r, t_1)$ for times $t_1 = 2n\pi/\omega$. (c) Dependence of $\psi(r, t_2)$, where $t_2 = (2n + 1)\pi/\omega = t_1 + \pi/\omega$. (d) Probability density $|\psi(r, t)|^2$ graphed for times $t = t_1, t_2$.

$C_a(0) = 1$, $C_b(0) = 0$ in Eqs. (2.59) and (2.60) for $C_b(t)$ and $C_a(t)$, we find

$$C_b(t) = \tfrac{1}{2}\,\wp E_0(\hbar\mu)^{-1}\,[\exp(i\mu_1 t) - \exp(i\mu_2 t)], \tag{6}$$

$$C_a(t) = \mu^{-1}\exp[i(\omega - \nu)t]\,[\mu_1\exp(i\mu_1 t) - \mu_2\exp(i\mu_2 t)], \tag{7}$$

which yield the product

$$C_a C_b^* = \tfrac{1}{2}\wp E_0(\hbar\mu^2)^{-1}\exp[i(\omega - \nu)t]\,\{\mu_1 + \mu_2 - \mu_1\exp(i\mu t) - \mu_2\exp(-i\mu t)\}. \tag{8}$$

Here the eigenvalues μ_1 and μ_2 are given by (2.58), and the difference $\mu = = \mu_1 - \mu_2$ by (2.61). The term in curly braces is just

$$-(\omega - \nu)\,(1 - \cos\mu t) - i\mu\sin\mu t.$$

Combining this with (8) and (5), we have the dipole moment:

$$\langle e\mathbf{r}\rangle = \hat{p}\wp^2 E_0(\hbar\mu^2)^{-1}\,[(\nu - \omega)(1 - \cos\mu t)\cos\nu t - \mu\sin\mu t\sin\nu t], \tag{9}$$

where the unit vector $\hat{p}$ gives the direction of $\langle e\mathbf{r}\rangle$. A similar calculation for an atom initially in the lower level gives (9) with an overall minus sign (see Prob. 3–1). We return to a discussion of this dipole after considering classical counterparts in Sec. 3-2.

3-2. Dipole Oscillators

In the preceding section, we saw that the atomic dipole behaves very much like a classical simple harmonic oscillator. With this in mind, we now consider the classical oscillator consisting of an electron on a spring, both oscillating freely and under the influence of an applied electromagnetic field. The amplification or absorption of this field is determined, and the differences between the classical and quantum dipoles are then enumerated.

Let the displacement from rest of an electron on a spring be given by x. The electron then has the equation of motion

$$\ddot{x}(t) + \omega^2 x(t) = 0 \tag{10}$$

with the solution

$$x(t) = x(0)\cos\omega t + \frac{\dot{x}(0)}{\omega}\sin\omega t. \tag{11}$$

Setting

$$x_0 = \left[x(0)^2 + \frac{\dot{x}(0)^2}{\omega^2}\right]^{1/2}, \qquad \tan\phi = -\frac{\dot{x}(0)}{\omega x(0)}, \tag{12}$$

we have the alternative form for (11):

$$x(t) = x_0 \cos(\omega t + \phi). \tag{13}$$

Classical electrodynamics tells us that an accelerating charge radiates an electromagnetic field with far-field electric and magnetic field values

$$\mathbf{E}(\mathbf{R}, t) = \frac{e}{4\pi\varepsilon_0 c^2}\left[\frac{\hat{n}\times(\hat{n}\times\dot{\mathbf{v}})}{R}\right]_{t-R/c}, \tag{14}$$

$$\mathbf{H}(\mathbf{R}, t) = \sqrt{\frac{\varepsilon_0}{\mu_0}}\,\hat{n}\times\mathbf{E}(\mathbf{R}, t), \tag{15}$$

(see Jackson, 1962, Sec. 14-2), where $\dot{\mathbf{v}}$ is evaluated at the retarded time $t - R/c$. This radiation has the characteristic dipole pattern depicted in Fig. 3-3 and causes the dipole to lose energy, that is, to be damped.

We can calculate the force $\mathbf{F}_{\text{rad}}$ corresponding to this damping by equating the work done by the force on the elecron in a time $\Delta t \gg 1/\omega$ to *minus* the energy radiated by the electron in that time, that is,

$$\int_t^{t+\Delta t} dt'\,\mathbf{F}_{\text{rad}}\cdot\mathbf{v} = -\int_t^{t+\Delta t} dt'\,(\text{power radiated}). \tag{16}$$

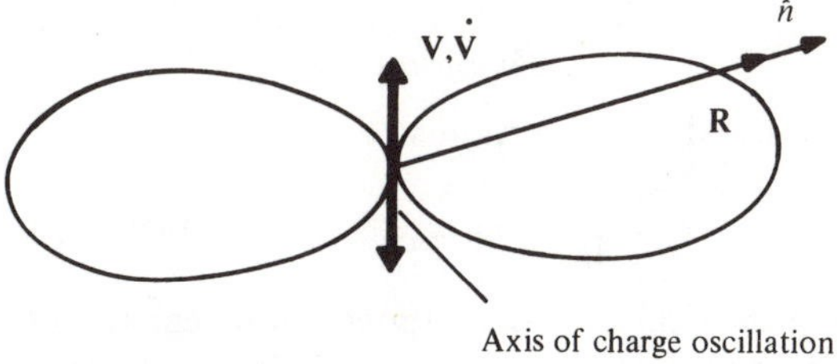

Figure 3-3. Oscillating dipole $\mathbf{p}$ *emits radiation in "butterfly" pattern given by Eq. (14).*

We calculate the power radiated as follows. The instantaneous electromagnetic energy flow is given by Poynting's vector:

$$\begin{aligned}\mathbf{S} = \mathbf{E} \times \mathbf{H} &= \sqrt{\frac{\varepsilon_0}{\mu_0}}\,\mathbf{E} \cdot \mathbf{E}\,\hat{n} \\ &= \frac{e^2}{16\pi^2\varepsilon_0{}^2c^4}\sqrt{\frac{\varepsilon_0}{\mu_0}}\frac{1}{R^2}(\hat{n} \times \dot{\mathbf{v}})^2\,\hat{n} \\ &= \frac{e^2\dot{\mathbf{v}}^2 \sin^2\theta}{16\pi^2\varepsilon_0 c^3 R^2}\,\hat{n}. \end{aligned} \tag{17}$$

The total power radiated is given by integration of $\mathbf{S}$ over a sphere surrounding the charge, namely,

$$\begin{aligned}\int \mathbf{S} \cdot d\mathbf{a} &= 4\pi \frac{2}{3}\frac{e^2\dot{\mathbf{v}}^2}{16\pi^2\varepsilon_0 c^3} \\ &= \frac{2}{3}\frac{e^2}{4\pi\varepsilon_0 c^3}\,\dot{\mathbf{v}}^2. \end{aligned} \tag{18}$$

This is Larmor's power formula for an accelerating charge. Substitution into (16) and integration by parts yields

$$\int_t^{t+\Delta t} dt'\,\mathbf{F}_{\text{rad}} \cdot \mathbf{v} = -\frac{2}{3}\frac{e^2}{4\pi\varepsilon_0 c^3}\left[\dot{\mathbf{v}} \cdot \mathbf{v}\Big|_t^{t+\Delta t} - \int_t^{t+\Delta t} dt'\,\ddot{\mathbf{v}} \cdot \mathbf{v}\right].$$

We choose Δt so that $\dot{\mathbf{v}} \cdot \mathbf{v}\Big|_t^{t+\Delta t} = 0$ and identify the remaining integrands as

$$\mathbf{F}_{\text{rad}} = \frac{2}{3}\frac{e^2}{4\pi\varepsilon_0 c^3}\,\ddot{\mathbf{v}} \simeq -\frac{2}{3}\frac{e^2\omega^2}{4\pi\varepsilon_0 c^3}\,\mathbf{v}, \tag{19}$$

where the approximation is valid for damping sufficiently small that the oscillation of the election remains essentially periodic with frequency ω. Including (19) in the equation of motion (10), we find the damped version:

$$\ddot{x}(t) + 2\Gamma\dot{x}(t) + \omega^2 x(t) = 0, \tag{20}$$

where the damping constant

$$\Gamma = \frac{1}{3}\frac{e^2\omega^2}{4\pi\varepsilon_0 c^3 m} = \frac{1}{3}\frac{\omega r_0}{c}\omega. \tag{21}$$

Here r_0 is the classical electron radius

$$r_0 = \frac{e^2}{4\pi\varepsilon_0 mc^2} \simeq 2.8 \times 10^{-15} \text{ meter.} \tag{22}$$

This decay is the classical analog of spontaneous emission (see Sec. 14-4 for the quantum-mechanical version, which differs considerably).

The solution of (20) is

$$x(t) = x_0 \exp(-\Gamma t) \cos(\omega t + \phi). \tag{23}$$

The spectrum of the emitted radiation can be determined by insertion of (23) into the radiated field expression (14) and Fourier transformation. We find the field amplitude

$$E(t) = E_r \exp(-\Gamma t) \cos(\omega t + \phi), \tag{24}$$

where

$$E_r = \frac{e\omega^2 x_0}{4\pi\varepsilon_0 c^2 R}. \tag{25}$$

This has the Fourier transform

$$\begin{aligned} E(\nu) &= \frac{E_r}{2\pi}\int_0^\infty dt \exp(-\Gamma t)\cos(\omega t + \phi)\exp(+i\nu t) \\ &= \frac{E_r}{4\pi}\left[\frac{1}{\Gamma + i(\omega - \nu)} + \frac{1}{\Gamma - i(\omega + \nu)}\right] \\ &\simeq \frac{E_r}{4\pi}\left[\frac{1}{\Gamma + i(\omega - \nu)}\right], \end{aligned} \tag{26}$$

where the last approximation, the rotating-wave approximation, is valid for $\omega \simeq \nu$. The spectrum is then given by

$$|E(\nu)|^2 = \frac{(E_r/4\pi)^2}{\Gamma^2 + (\omega - \nu)^2}, \tag{27}$$

which is a Lorentzian with full width at half maximum (halfwidth) of 2Γ (see Fig. 3-4).

Suppose now that a field $E_0 \cos \nu t$ acts on the charge-spring oscillator. The equation of motion (20) becomes

$$\ddot{x} + 2\Gamma\dot{x} + \omega^2 x = \frac{eE_0}{m}\cos \nu t = \frac{1}{2}\frac{eE_0}{m}\exp(-i\nu t) + \text{c.c.} \tag{28}$$

For simplicity we solve first for the steady-state solution:

$$x(t) = \tfrac{1}{2}X(t)\exp(-i\nu t) + \text{c.c.}, \tag{29}$$

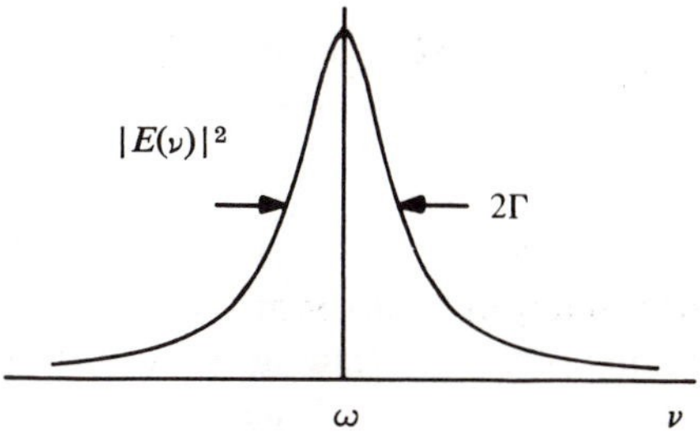

Figure 3-4. Radiated intensity versus frequency ν of Eq. (27).

and add in the transient solution (23) later. We can substitute (29) into (28) without the complex conjugates since an expression valid for $X \exp(-i\nu t)$ is also valid for the c.c. and for the sum of the two as well. We will have a number of occasions to utilize this simplification in the semiclassical theory of the laser. We find

$$(-\nu^2 - 2i\Gamma\nu + \omega^2)X = \frac{1}{2}\frac{eE_0}{m},$$

that is,

$$X = \frac{1}{2}\frac{eE_0}{m}\left(\frac{1}{\omega^2 - \nu^2 - 2i\Gamma\nu}\right) \simeq \frac{eE_0}{4m\nu}\left(\frac{1}{\omega - \nu - i\Gamma}\right), \tag{30}$$

where the approximation follows near resonance ($\omega \approx \nu$), since then $\omega + \nu \simeq 2\nu$.

The average rate of work done by the field on the atom is given by the power expression

$$\begin{aligned}\bar{P}(t) &= \frac{1}{2\pi/\nu}\int_0^{2\pi/\nu} d\tau\, \dot{x}(t+\tau)eE_0 \cos[\nu(t+\tau)] \\ &= \tfrac{1}{4}\, eE_0\, i\nu X^* + \text{c.c.} \\ &= \frac{1}{8}\frac{e^2E_0{}^2}{m\Gamma}\left[\frac{\Gamma^2}{(\omega-\nu)^2+\Gamma^2}\right] > 0.\end{aligned} \tag{31}$$

This shows that a steady-state forced oscillator absorbs radiation, dissipating it in the damping process. For the radiative damping this involves scattering the incident radiation in the doughnut pattern indicated in Fig. 3-3.

Consider now a transient solution under the influence of a field $E_0 \cos(\nu t + \phi_0)$, which acts starting at time $t = 0$. Furthermore, neglect damping and assume that the initial displacement and velocity are given by

$$x(0) = x_0 \cos\phi, \qquad \dot{x}(0) = -\omega x_0 \sin\phi. \tag{32}$$

The appropriate solution is a combination of homogeneous solutions like Eq. (13) with the undamped forced version (29) with (30), that is,

$$x(t) = x_0 \cos(\omega t + \phi) + \frac{eE_0}{m}\left[\frac{\cos(\nu t + \phi_0)}{\omega^2 - \nu^2} - \frac{\cos(\omega t + \phi_0)}{2\omega(\omega - \nu)} - \frac{\cos(\omega t - \phi_0)}{2\omega(\omega + \nu)}\right] \tag{33}$$

since the term in square brackets vanishes at $t = 0$. The power averaged over a cycle is then given along the lines leading to Eq. (31) as

$$\bar{P} = \tfrac{1}{4}\, eE_0 \exp[-i(\nu t + \phi_0)] \left\{ i\omega x_0 \exp[i(\omega t + \phi)] + i\frac{eE_0}{m}\left[\frac{\exp[i(\nu t + \phi_0)]}{(\omega^2 - \nu^2)/\nu} - \frac{1}{2}\frac{\exp[i(\omega t + \phi_0)]}{\omega - \nu} - \frac{1}{2}\frac{\exp[i(\omega t - \phi_0)]}{\omega + \nu}\right]\right\} + \text{c.c.}$$

$$= -\tfrac{1}{2}\, eE_0\, \omega x_0 \sin[(\omega - \nu)t + \phi - \phi_0] + \left[\frac{(eE_0)^2}{4m}\right] t\, \frac{\sin[(\omega - \nu)t]}{(\omega - \nu)t}. \tag{34}$$

For small t (or small E_0), this could have either sign, depending on the initial x_0 and relative phase angle $\phi - \phi_0$. If this angle is $\pi/2$, the average power is initially negative, implying stimulated emission. Later (34) the power goes positive and stays so for $t < \omega - \nu$. These conditions can be stated as

$$\bar{P} \begin{cases} > 0 & \text{(stimulated) absorption,} \\ < 0 & \text{stimulated emission.} \end{cases} \tag{35}$$

These alternatives comprise the classical counterparts to the quantum stimulated absorption and emission of Chap. 2. They differ in that the classical dipole has an initially definite phase for stimulated emission, whereas the quantum case acquires the correct phase under the influence of the applied field. Moreover, the quantum expectation value refers to an ensemble of dipoles rather than the single system in the classical case. Any given measurement of an atom yields a dipole, but only the average of many such measurements gives the expectation value. Alternatively, the response of many closely spaced dipoles provides the required average. It is the latter interpretation which is particularly suited to semiclassical theory, for no measurements as such are performed.

In Prob. 3–3 the response of a dipole to a spread of frequencies is calculated, for which only absorption occurs. This calculation also yields a classical expression for the Einstein B coefficient. Problem 3–4 calculates the power radiated by the forced dipole and gives an expression for the scattering cross-section.

We now calculate the energy flow along the direction of propagation due to both the incident field and the radiating dipole. We take the dipole to have the form

$$\mathbf{p} = \hat{x} p_0 \cos(\nu t + \phi). \tag{36}$$

This flow is given by the slowly varying part $\mathbf{S}_{sv}$ of Poynting's vector:

$$\mathbf{S}_{sv} = \mathbf{E}_t \times \mathbf{H}_t = (\mathbf{E} + \mathbf{E}_R) \times (\mathbf{H} + \mathbf{H}_R), \tag{37}$$

where the incident electric and magnetic fields are

$$\mathbf{E} = \hat{x} E_0 \cos(\nu t - Kz), \tag{38}$$

$$\mathbf{H} = \hat{y} \sqrt{\frac{\varepsilon_0}{\mu_0}}\, E_0 \cos(\nu t - Kz), \tag{39}$$

and the far-field retarded dipole fields are (Jackson, 1962, p. 271)

$$\mathbf{E}_R = \frac{K^2 p_0}{4\pi\varepsilon_0 R}[(\hat{n} \times \hat{x}) \times \hat{n}] \cos(\nu t - K\mathrm{R} + \phi), \tag{40}$$

$$\mathbf{H}_R = \sqrt{\frac{\varepsilon_0}{\mu_0}} \frac{K^2 p_0}{4\pi\varepsilon_0 R} (\hat{n} \times \hat{x}) \cos(\nu t - K\mathrm{R} + \phi). \tag{41}$$

We obtain the slowly varying part of (37) by taking only the difference terms in products like

$$\cos\theta_1 \cos\theta_2 = \tfrac{1}{2}\cos(\theta_1 - \theta_2) + \tfrac{1}{2}\cos(\theta_1 + \theta_2). \tag{42}$$

Further noting the identities

$$(\hat{n} \times \hat{x}) \times \hat{n} = -\,\hat{n}(\hat{n} \cdot \hat{x}) + \hat{x},$$
$$(\hat{n} \times \hat{x}) \times \hat{n} \times (\hat{n} \times \hat{x}) = \hat{n}[1 - (\hat{n} \cdot \hat{x})^2],$$
$$\hat{x} \times (\hat{n} \times \hat{x}) = \hat{n} - \hat{x}(\hat{x} \cdot \hat{n}),$$

we find

$$\mathbf{S}_{sv} = \frac{1}{2}\sqrt{\frac{\varepsilon_0}{\mu_0}}\, E_0{}^2 \hat{z} + \tfrac{1}{2}[\hat{n} + \hat{z} - (\hat{x} \cdot \hat{n})(\hat{x} + \hat{n} \times \hat{y})] \sqrt{\frac{\varepsilon_0}{\mu_0}} \left(\frac{K^2 p_0 E_0}{4\pi\varepsilon_0 R}\right)$$
$$\times \cos(KR - Kz - \phi) + \tfrac{1}{2}\hat{n}[1 - (\hat{n} \cdot \hat{x})^2] \sqrt{\frac{\varepsilon_0}{\mu_0}} \left(\frac{K^2 p_0}{4\pi\varepsilon_0 R}\right)^2. \tag{43}$$

The interference term between the incident and dipole fields becomes rapidly varying when

$$\pi \le (R - z)\,K = Kz\left[\left(\frac{x^2 + y^2}{z^2} + 1\right)^{1/2} - 1\right] \simeq Kz\frac{1}{2}\left(\frac{x^2 + y^2}{z^2}\right) = \pi\frac{z}{\lambda}\tan^2\theta,$$

that is, when the angle θ between $\mathbf{R}$ and $\hat{z}$ satisfies

$$\theta > \left(\frac{\lambda}{z}\right)^{1/2}. \tag{44}$$

For large z (many wavelengths λ), this restricts the slowly varying terms close to the z axis. A detector off axis measures no energy flow due to its finite area (Poynting vector contributions cancel out because of rapid oscillations over the detector face). Thus the principal energy flow occurs along the z axis. For values of $\hat{n} \cdot \hat{x} \ll 1$, (43) reduces to

$$\mathbf{S}_{sv} = \frac{1}{2}\sqrt{\frac{\varepsilon_0}{\mu_0}}\,\hat{z}\left[E_0{}^2 + 2\left(\frac{K^2 p_0 E_0}{4\pi\varepsilon_0 R}\right)\cos\left(KR - Kz + \phi\right)\right]. \tag{45}$$

On resonance, $p_0 = \frac{1}{4}\, e^2 E_0 / m\nu\Gamma$ from (30), and the phase $\phi = \pm\,\pi/2$ for emission and absorption [from (9) or (34)]. For these cases, (45) becomes

$$\mathbf{S}_{sv} = \tfrac{1}{2}\,\hat{z}\sqrt{\frac{\varepsilon_0}{\mu_0}}\left[E_0{}^2 \pm \left(\frac{K^2 e^2\, E_0{}^2}{8\pi\varepsilon_0 R m\nu\Gamma}\right)\sin\left(KR - Kz\right)\right], \tag{46}$$

which reveals that a detector on the z axis would register gain for stimulated emission (the plus sign) and loss for the absorption.

In laser problems, more than a single dipole is involved. A dipole sheet is a better model. For this (see Fig. 3-5), radiation from dipoles off the z axis travels farther to an observation point on that axis than radiation from a dipole on the axis, and consequently the two contribute in general with different phases. The total contribution from an entire sheet yields a phase shift of 90° even in the near field ($z < \lambda$), as shown in Appendix A. This converts the sine in Eq. (9) or (34) into a cosine, that is, the sheet contributes either in phase (stimulated emission) or 180° out of phase (absorption).

The in-phase characteristic is one important property of stimulated

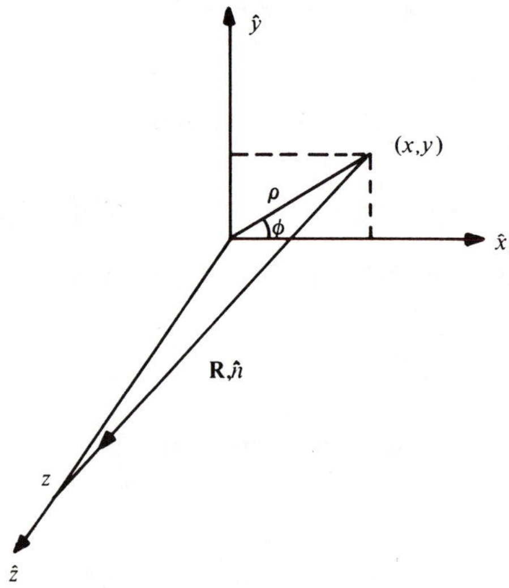

Figure 3-5. Axes for dipole sheet (lying in x-y plane) oscillating in x direction and radiating the field given by Eq. (A2). The polar coordinates (ρ, ϕ) are a useful alternative set for the Cartesian coordinates (x, y). R is a vector pointing from a representative dipole to the point of observation (0, 0, z) (on the z axis).

emission from a sheet of dipoles. Furthermore, the induced field has the same direction as the incident field, a second property of this stimulated emission. In the Rabi calculation of Eq. (9), we did not include damping mechanisms, which cause initial transients to decay. In later discussions of laser media, which include such damping and pumping to the upper level as well, we see a third important property, namely, that the stimulated emission has the same frequency as the incident radiation. These three properties—equality of phase, direction, and frequency of incident and stimulated fields—are completely obtained only for appropriately pumped media whose cross-sectional areas contain many similar dipoles.

We have noted that classical dipoles yield stimulated emission only when appropriately phased; the applied field does not induce the correct phase automatically, as it does in an excited atom. It is possible to use nonlinear (soft) springs with charges as in Fig. 3-6, which can yield net amplification of incident radiation. These springs get "softer" as the oscillation amplitudes increase and therefore acquire lower resonant frequencies. For an amplifying medium, the charges are set in motion with energies corresponding to frequencies slightly lower than the incident field frequency ν. Then charges with absorbing phases oscillate less rapidly, are driven further off resonance, change phase, and turn back toward ν in frequency. Charges with emitting phases begin to oscillate faster and are driven into resonance with the incident field. There exists an optimum interaction time for which gain is realized, as shown in Fig. 3-7 for one set of parameters. One can imagine a beam of such dipoles which pass through the incident field (as in a cavity). After the optimal time, the dipoles have passed completely through the field. This approach to stimulated emission leads to a generalization of the simple, forced harmonic oscillator known as the Duffing problem and has been investigated in the context of laser theory by Lamb (1965) and Borenstein and Lamb (1972).

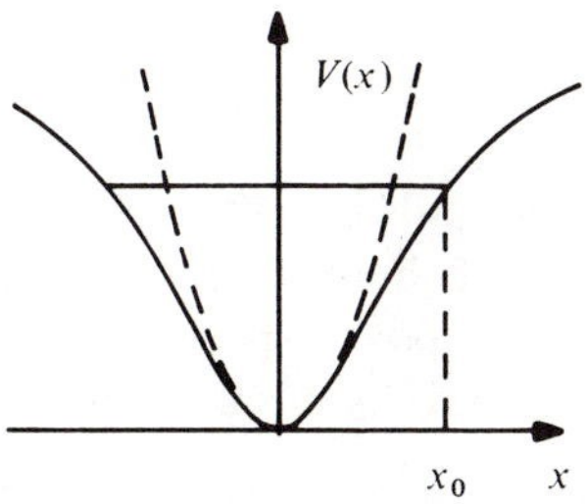

Figure 3-6. Harmonic (dashed line) and anharmonic (solid line) potentials. Inversion of the medium is difficult to obtain in harmonic cases, while for anharmonic excursions larger than x_0 gain can occur for appropriate "lifetimes" of the dipoles.

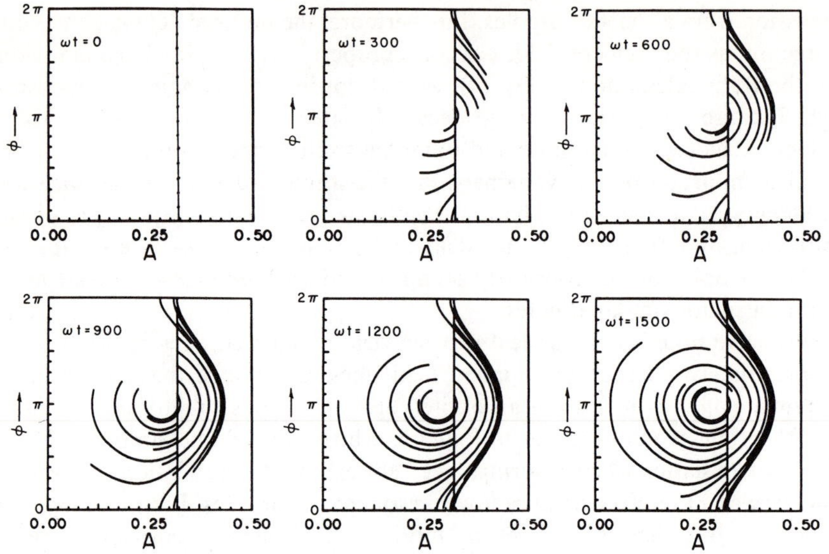

Figure 3-7. Time evolution of Duffing oscillators. Phase φ (modulo 2π) is plotted as a function of amplitude A stopped at times $\omega t = 0, 300, 600, 900, 1200, 1500$. Fifteen oscillators start at equally spaced initial phases between 0 and 2π with amplitude $A_0 = 0.32$. Maximum gain occurs for $\omega t \approx 900$ (a). From Borenstein and Lamb (1972).

Problems

3-1. Calculate the electric-dipole expectation value (5) for an atom initially in the lower level, using probability amplitudes (2.59) and (2.60).

3-2. Calculate the electric-dipole expectation value (5) for the damped probability amplitudes C_a and C_b of (2.48) with an initially excited atom $[C_a(0) = 1, C_b(0) = 0]$.

3-3. Given an oscillator equation of motion

$$\ddot{x} + \omega^2 x = \frac{e}{m} \sum_j E_j \cos(\nu_j t + \phi_j) \tag{47}$$

for forced operation at a spread of frequencies, show that

$$x(t) = \frac{e}{m}\sum_j E_j\left[\frac{\cos(\nu_j t + \phi_j)}{\omega^2 - \nu_j^2} - \frac{\cos(\omega t + \phi_j)}{2\omega(\omega - \nu_j)} - \frac{\cos(\omega t - \phi_j)}{2\omega(\omega + \nu_j)}\right] + x_0 \cos(\omega t + \phi). \tag{48}$$

Find the generalization of the monochromatic result (34) for the average power $\bar{P}$. Integrate this expression from 0 to t to obtain the energy gained by the oscillator in this time interval. Finally, convert the summation over j to an integral over ν and find a classical Einstein B coefficient.

3-4. Show that the power radiated by a forced simple harmonic oscillator (28) is given by

$$\int \mathbf{S}\cdot d\mathbf{a} = \frac{\pi}{3} r_0^2 \sqrt{\frac{\varepsilon_0}{\mu_0}}\, E_0^2 \frac{\nu^2}{(\omega - \nu)^2 + \Gamma^2}. \tag{49}$$

From this find the absorption cross section defined by

$$\text{Cross section for absorption} = \frac{\text{(energy/second) scattered}}{\text{(energy/second}\cdot\text{area) incident}}. \tag{50}$$

3-5. Calculate the full width at half maximum for the Lorentzian distribution:

$$I(\nu) = \frac{\gamma^2}{(\omega - \nu)^2 + \gamma^2} \tag{51}$$

and the Gaussian distribution:

$$I(\nu) = \exp[-(\omega - \nu)^2/(Ku)^2]. \tag{52}$$

What is the full width at the $1/e$ point for (52)? What are the normalization factors for (51) and (52), that is, the values of $\mathcal{N}$ such that

$$\mathcal{N}\int_{-\infty}^{\infty} d\nu\, I(\nu) = 1? \tag{53}$$

3-6. A very important method of checking equations consists of determining the dimensions of expressions. What are the dimensions of the following quantities:

$$\wp E_0, \quad \hbar\gamma, \quad \varepsilon_0 E_0^2, \quad \frac{\nu\wp^2}{(\hbar\varepsilon_0 Ku)}?$$

Here K is a wave number, and u is a speed.

3-7. Calculate the polarization of an inhomogeneously broadened medium, that is, one with a distribution of resonance frequencies given by

$$W(\omega) = (\sqrt{\pi}\,\Delta\omega)^{-1} \exp[-(\omega - \omega_0)^2/(\Delta\omega)^2].$$

Assume that the width parameter $\Delta\omega$ greatly exceeds the Rabi flopping frequency $\wp E_0/\hbar$ [so that $W(\omega) \simeq (\sqrt{\pi}\,\Delta\omega)^{-1}$ in the integrand], and suppose that the medium is initially excited

Ans:

$$P(t) = \int_{-\infty}^{\infty} d\omega\, W(\omega)\, \wp C_a C_b^* \exp(-i\omega t) + \text{c.c.}$$

$$= -\sqrt{\pi}\,\wp \left(\frac{\wp E_0}{\hbar\, \Delta\omega}\right) J_0\left(\frac{\wp E_0 t}{\hbar}\right) \sin \nu t. \tag{54}$$

References

Dipole oscillation and radiation are discussed in the following:

R. P. Feynman, R. B. Leighton, and M. Sands, 1965, *The Feynman Lectures on Physics,* Vol. I, Addison-Wesley Publishing Co., Reading, Mass., Chaps. 30–32.

J. D. Jackson, 1962, *Classical Electrodynamics,* John Wiley & Sons, New York, Chap. 9.

The classical maser problem is discussed by these authors:

M. Borenstein and W. E. Lamb, Jr., 1972, *Phys. Rev.* **A5**, 1298.

W. E. Lamb, Jr., 1965, in: *Lectures in Theoretical Physics,* ed. by C. Dewitt, A. Blandin, and C. Cohen-Tannoudji, Gordon and Breach, New York.

IV

CLASSICAL SUSTAINED OSCILLATORS

4. Classical Sustained Oscillators

In Chap. 3 we considered the free oscillation of an electron on a spring whose oscillations decay because of radiative damping. We also treated forced operation. In this chapter we consider *sustained* oscillators. These systems do not decay (although they do dissipate energy), and they choose their own oscillation frequencies in distinct contrast with the forced oscillators of Sec. 3-2. Examples of these devices are clocks, triode oscillator circuits, and, to some degree of approximation, masers and lasers.

In Sec. 4-1 we develop the equation of motion for an *RLC* oscillator with a negative, saturable resistance, utilizing the method of slowly varying amplitudes. This technique is of great value in laser theory. The results show how a sustained oscillator such as a vacuum-tube (or transistor) oscillator builds up to a constant amplitude (envelope) determined by the condition that the saturated gain equals the losses. We also discuss a variant of this equation which predicts relaxation oscillations, a phenomenon well known in the ruby laser as well as in circuit theory. The discussion is then extended to two-mode operation, in which interesting competition effects occur. The equations here describe many problems in nature, including the competition of biological species.

In Sec. 4-2 we consider frequency or mode locking in sustained oscillators, notably those in acoustics and circuit theory. After a brief history of the subject, we show how the triode oscillator equations of Sec. 4-1 are modified to include the effects of an injected signal. The unifying thread to the discussion is inherent in the concept of the forced oscillator: oscillation takes place at the forcing frequency (or frequencies for some multimode applications). The principal difference between locking phenomena and simple forced operation is that the locking oscillations are sustained independently from the *forcing* terms. Hence, for locking to occur, the forced frequencies must be quite close to the sustained frequencies—in distinct contrast with the usual forced oscillator behavior, which allows oscillation relatively far from the natural resonance frequencies. Here the locking mechanisms compete against the sustaining forces, which try to maintain different frequencies. Because of the sustained nature of the oscillations the locking invariably occurs in nonlinear systems, although the locking mechanisms themselves may not arise from nonlinearities. It is worth noting at the outset that not all coupling effects result in phase locking. In particular, the coupling between normal modes in the double-tank circuit of Fig. 4-2 does not result in locking. Rather the single frequency which is observed in the output of the circuit is due to *mode competition* in which one mode suppressed oscillation of the other.

4-1. The Sustained Oscillator

Suppose that a free oscillator has a net *negative*, saturable resistance as described by the equation of motion for its displacement $v(t)$ (we use v here for voltage inasmuch as our primary example is the triode oscillator circuit):

$$\ddot{v} - \frac{d}{dt}(av - \beta' v^3) + \omega^2 v = 0. \tag{1}$$

Here a is the linear net gain (i.e., the gain in excess of losses), β' is the saturation coefficient, and ω is the resonance frequency in the absence of dissipation or gain. Van der Pol obtained such an equation of motion in his treatment of the triode oscillator depicted in Fig. 4-1.

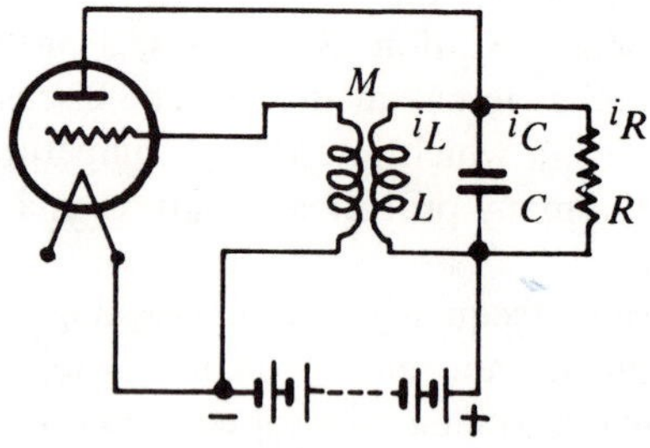

Figure 4-1. Triode oscillator circuit considered by Van der Pol in obtaining the equation of motion (1). Problem 4-1 deals with a derivation.

We can solve Eq. (1) approximately by an important technique called the method of slowly varying amplitude. In Sec. 4-2 we include a slowly varying phase as well, but in the present instance the generality is not needed. We suppose that the gain coefficient a is sufficiently small that the oscillator acts essentially like an undamped oscillator with frequency ω, whose amplitude $V(t)$ varies little in an optical period $(2\pi/\omega)$. Specifically we write

$$v(t) = \tfrac{1}{2} V(t) \exp(-i\omega t) + \text{c.c.} \tag{2}$$

In the approximation we keep only terms with exponential dependence $\exp(-i\omega t)$. This approximation is excellent at optical frequencies, but is often invalid at radio frequencies, for which various harmonics can exist relatively easily. Inasmuch as both $\dot{V}(t)$ and a are small quantities, we neglect their product along with $\ddot{V}(t)$. In these approximations, we have

$$\left(\frac{d}{dt}\right)^n V(t)\exp(-i\omega t) = [n(-i\omega)^{n-1}\,\dot{V} + (-i\omega)^n V] \exp(-i\omega t) \tag{3}$$

and, further, that the fundamental frequency (that at ω) term for v^3 is given by

$$v^3|_{\text{fund}} = \tfrac{3}{8} V^3 \exp(-i\omega t) + \text{c.c.} \tag{4}$$

Here the 3 results from the fact that there are three ways to obtain $V^3 \exp(-i\omega t)$ in the cube of (2). Substituting (2)–(4) into (1), we find

$$-2i\omega\dot{V} - \omega^2 V - (a - \beta V^2)(-i\omega V) + \omega^2 V = 0, \tag{5}$$

where $\beta = 3\beta'/4$. Solving for $\dot{V}(t)$, we find the slowly varying equation of motion:

$$\dot{V}(t) = \tfrac{1}{2}(a - \beta V^2)V(t). \tag{6}$$

For small $V(t)$, this equation is approximated by

$$\dot{V}(t) = \tfrac{1}{2}aV(t) \tag{7}$$

with the exponentially increasing solution

$$V(t) = V(0)\exp(\tfrac{1}{2}at). \tag{8}$$

For sufficiently large $V(t)$, saturation sets in, and the steady-state oscillation condition

$$\dot{V}(t) = 0 \tag{9}$$

occurs for the value

$$V^2 = \frac{a}{\beta}. \tag{10}$$

An analytic solution to (6) is discussed in the laser treatment of Sec. 8-3. Figure 8-6 depicts the predicted buildup from a small value to the steady-state value (10). Note that Eq. (1) and hence (6) does *not* build up from zero amplitude. In practice, some fluctuation gets things going. Such fluctuations are discussed in an unsustained, although in some ways analogous, problem of Brownian motion in Chap. 19.

In Eq. (1) we assumed that the resonant frequency $\omega \gg a$ in writing and using the decomposition (2) with $V(t)$ a slowly varying quantity. If, instead, $a \gg \omega$, buildup proceeds rapidly, overshooting the steady-state value, and the saturation gain goes negative, driving the amplitude back down. The process is a repetitive one and leads to a sequence of buildup-decay cycles called relaxation oscillations. Such oscillations are periodic, although not sinusoidal, and occur, for example, in multivibrators and car turn-signal blinkers. In the latter, the current for the light passes through a bimetallic strip switch, heating the strip and thus bending it and disconnecting the current path. Subsequent cooling restores the strip to its connected position, allowing current to flow again, and the cycle repeats. Our interest in relaxation oscillations is due to their occurrence in ruby laser operation. Qualitatively, this laser has an active medium whose gain increases until it exceeds the losses due to diffraction, mirror transmission, and other causes. Laser radiation then builds up and, for appropriate decay constants, drives the gain down to zero (equilibrates the atomic populations in the upper and lower levels). The losses then kill the electric field and the process starts all over again, something like the turn-signal blinker. A more quantitative discussion is given in the problems for Chap. 8. Relaxation oscillations are commonplace not only in physics, but also in physiology (heart beats) and economics (business cycles) as well. In fact, the expression "history repeats itself" is probably a description of a large-scale integrated relaxation oscillation phenomenon!

We now consider two-frequency operation. As for the single-frequency problem, the analysis here reveals in a relatively simple context some typical properties of corresponding operation in lasers. Hence a discussion is valuable for our later work (Chaps. 9-12). Van der Pol (1920) showed that the double-tank circuit in Fig. 4-2 is described by the equations

$$\ddot{v}_1 - \frac{d}{dt}(a_1 v_1 - \beta' v_1{}^3) + \omega_1{}^2 v_1 + k_1 \omega_1{}^2 v_2 = 0, \tag{11}$$

$$\ddot{v}_2 + a_2 \dot{v}_2 + \omega_2{}^2 v_2 + k_2 \omega_2{}^2 v_1 = 0. \tag{12}$$

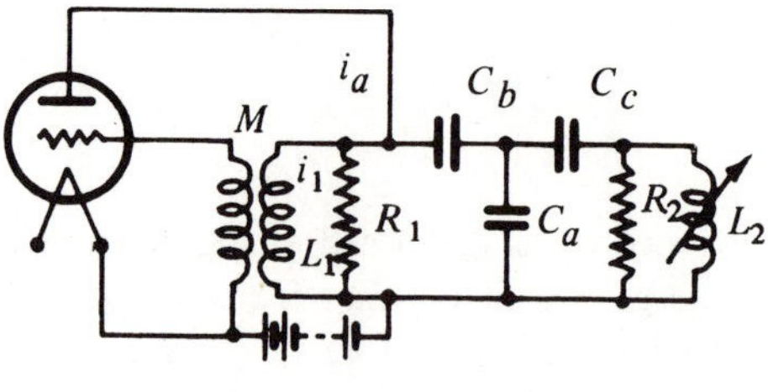

Figure 4-2. Double-tank circuit considered by Van der Pol in obtaining Eqs. (11) and (12).

These equations describe coupled oscillators in which there is a negative, saturable resistance analogous, for example, to a laser coupled to a passive Fabry-Perot cavity (e.g., scanning interferometer). Such coupled systems have, for sufficiently small a_1 and a_2, normal-mode frequencies given by the simultaneous solution of (11) and (12) with neglect of a_1, a_2, and β'. By setting

$$v_1 = v_{10}\exp(-i\Omega t), \qquad v_2 = v_{20}\exp(-i\Omega t), \tag{13}$$

we obtain the equation

$$\Omega^4 - (\omega_1^2 + \omega_2^2)\Omega^2 + \omega_1^2\omega_2^2(1 - k^2) = 0, \tag{14}$$

where $k^2 = k_1k_2$. Equation (14) gives the normal-mode frequencies

$$\Omega_{1,2}^2 = \tfrac{1}{2}(\omega_1^2 + \omega_2^2) \pm \tfrac{1}{2}[(\omega_1^2 - \omega_2^2)^2 + 4k^2\omega_1^2\omega_2^2]^{1/2}. \tag{15}$$

To solve (11) and (12) in the approximations of the single-mode treatment, we solve (11) for v_2 in terms of v_1 and its derivatives, and substitute the result into (12), thereby obtaining the fourth-order equation of motion:

$$\left[\left(\frac{d}{dt}\right)^2 + a_2\frac{d}{dt} + \omega_2^2\right]\left[\left(\frac{d}{dt}\right)^2 - \frac{d}{dt}(a_1 - \beta' v_1^2) + \omega_1^2\right]v_1$$
$$= k^2\omega_1^2\omega_2^2 v_1. \tag{16}$$

We further suppose that $v_1(t)$ is a superposition of the normal modes, namely,

$$v_1(t) = \tfrac{1}{2}V_1(t)\exp(-i\Omega_1 t) + \tfrac{1}{2}V_2(t)\exp(-i\Omega_2 t) + \text{c.c.}, \tag{17}$$

in which $V_1(t)$ and $V_2(t)$ vary little in oscillation periods $(2\pi/\Omega_1, 2\pi/\Omega_2)$. Hence derivatives of V_1 and V_2 higher than first order in (16) are neglected, as well as products of first-order derivatives with the small constants a_1, a_2, and β'. We further retain only the fundamental terms in v_1^3, as in the single-mode case. These are given by

$$v_1^3\,|_{\text{fund}} = \tfrac{3}{8}[V_1\exp(i\Omega_1 t) + V_2\exp(i\Omega_2 t)]$$
$$\times[V_1\exp(-i\Omega_1 t) + V_2\exp(-i\Omega_2 t)]^2 + \text{c.c.}, \tag{18}$$

in which the $\frac{1}{8}$ results from cubing the 1/2 in (17), and the 3 from the three possible ways of obtaining the bracketed expressions in (18) from the triple product v_1^3. Equation (18) simplifies to

$$v_1^3|_{\text{fund}} = \tfrac{3}{8}\{V_1^2 + V_2^2 + V_1V_2(\exp[i(\Omega_1 - \Omega_2)t] + \exp[i(\Omega_2 - \Omega_1)t])\}$$
$$\times [V_1\exp(-i\Omega_1 t) + V_2\exp(-i\Omega_2 t)] + \text{c.c.}$$
$$\cong \tfrac{3}{8}V_1(V_1^2 + 2V_2^2)\exp(-i\Omega_1 t)$$
$$+ \tfrac{3}{8}V_2(V_2^2 + 2V_1^2)\exp(-i\Omega_2 t) + \text{c.c.} \tag{19}$$

Note here that the "cross-coupling" terms (effect of one mode at the frequency of the other) result from two sources. One is the dc saturation, which is just like that a mode inflicts upon itself. The other pulsates at the intermode frequency, for example, $\exp[i(\Omega_1-\Omega_2)t]$, which interacts with a mode term

$V_1 \exp(-i\Omega_1 t)$ to yield a saturation of mode two. This pulsation effect is common to many nonlinear problems in physics, occurring in Raman and Brillouin scattering and for our particular purposes in the population pulsations of laser operation (Chaps. 9–12). Because of its generality, it is worth while to appreciate it here in this simple context.

We now substitute (17) and (19) into (16) without the complex conjugates (the equations are real and only the fundamental tones are kept). Inasmuch as we neglect time derivatives of the slowly varying amplitudes when multiplied by the small coefficients a_1, a_2, and β', we have (in the following, n, m = 1, 2, with $n \neq m$)

$$\left(\frac{d}{dt}\right)^k \{[a_1 V_n - \beta V_n(V_n^2 + 2V_m^2)] \exp(-i\Omega_n t)\}$$
$$= (-i\Omega_n)^k [a_1 V_n - \beta V_n(V_n^2 + 2V_m^2)] \exp(-i\Omega_n t), \tag{20}$$

where $\beta = 3\beta'/4$. We further note that in the slowly varying approximation of neglecting derivatives higher than first order,

$$\left(\frac{d}{dt}\right)^k [V_n \exp(-i\Omega_n t)] = [k(-i\Omega_n)^{k-1} \dot{V}_n + (-i\Omega_n)^k V_n] \exp(-i\Omega_n t). \tag{21}$$

With (20) and (21), the imaginary part of (16) for mode n reduces to [the real part vanishes because of (15)]

$$[4\Omega_n^3 - 2\Omega_n(\omega_1^2 + \omega_2^2)]\, \dot{V}_n + a_2\,(\Omega_n^3 - \Omega_r\omega_1^2)\, V_n$$
$$- (\Omega_n^3 - \Omega_n\omega_2^2)[a_1 - \beta(V_n^2 + 2V_m^2)]\, V_n = 0. \tag{22}$$

From (15), we note that

$$\Omega_1^2 + \Omega_2^2 = \omega_1^2 + \omega_2^2. \tag{23}$$

After further multiplying by $V_n/[\Omega_n(\Omega_n^2 - \Omega_m^2)]$ and transposing the gain and dissipation terms to the right-hand side, we find (n, m = 1, 2, with $m \neq n$)

$$\frac{d}{dt}(V_n^2) = V_n^2 \left(\frac{\Omega_n^2 - \omega_2^2}{\Omega_n^2 - \Omega_m^2}\right) [a_{0n} - \beta(V_n^2 + 2V_m^2)], \tag{24}$$

where the net gain coefficient

$$a_{0n} = a_1 - a_2 \left(\frac{\Omega_n^2 - \omega_1^2}{\Omega_n^2 - \omega_2^2}\right) = a_1 - a_2 \left[\frac{(\Omega_n^2 - \omega_1^2)^2}{k^2\omega_1^2\omega_2^2}\right]. \tag{25}$$

Of particular interest are the stationary solutions V_{ns} defined by values satisfying

$$\frac{d}{dt} V_n^2 = 0. \tag{26}$$

The solutions of slightly more general equations are discussed in some detail in Sec. 9-2 on two-mode laser operation. Here we consider some highlights in the present context. Four solutions are possible since (26) can be satisfied for combinations of each mode oscillating or not independently of the other. These solutions are as follows:

$$V_{1s}^2 = V_{2s}^2 = 0, \tag{27}$$

$$\left.\begin{aligned} V_{ns}^2 &= \frac{a_{0n}}{\beta}, \qquad V_{ms}^2 = 0 \\ V_{ns}^2 &= \frac{1}{3\beta}(2a_{0m} - a_{0n}) \end{aligned}\right\}, \quad m, n = 1, 2, \text{ with } m \neq n. \qquad \begin{matrix}(28)\\(29)\end{matrix}$$

Stable solutions are those for which small deviations from their values damp out in time, that is, a deviation ε_n, defined by

$$V_n^2(t) = V_{ns}^2 + \varepsilon_n(t), \tag{30}$$

must tend to zero in time. For example, with $a_{0n} > 0$, we have from (24)

$$\dot{\varepsilon}_n = \Theta_{nn}\varepsilon_n + \Theta_{nm}\varepsilon_m, \tag{31}$$

where the matrix

$$\Theta = -\beta \begin{pmatrix} V_{1s}^2\left[\dfrac{\Omega_1^2 - \omega_1^2}{\Omega_1^2 - \omega_2^2}\right] & 2V_{1s}^2\left[\dfrac{\Omega_1^2 - \omega_1^2}{\Omega_1^2 - \omega_2^2}\right] \\ 2V_{2s}^2\left[\dfrac{\Omega_2^2 - \omega_1^2}{\Omega_2^2 - \omega_2^2}\right] & V_{2s}^2\left[\dfrac{\Omega_2^2 - \omega_1^2}{\Omega_2^2 - \omega_2^2}\right] \end{pmatrix}. \tag{32}$$

This matrix must have negative eigenvalues if (29) is to be stable. As shown in Prob. 4-2, the eigenvalues are positive, leading to exponential buildup of the deviations ε_n, that is, (29) is unstable. As shown by a similar analysis, both solutions of (28) can be stable so that only one normal mode of the coupled-tank circuit can oscillate at a time. Which one succeeds depends on past history, that is, there is hysteresis. Similar bistable situations occur in the operation of some laser configurations, as depicted in the phase plane graph of Fig. 9-4. More commonly, however, both modes oscillate in laser operation, for the 2's in Eq. (24) are replaced by values less than unity. Note here that the factor of 2 rather than unity results from the pulsation coupling in Eq. (19). Simple saturation alone would have produced only a neutrally coupled system.

The coupling phenomena appearing here show up also in nonphysics contexts, such as the coexistence of two species, a host population and a parasite population. This problem was originally treated by Lotka (1925) and Volterra (1931). A detailed review and development of this application and other ones has been given by Goel, Maitra, and Montroll (1971). They point out (p. 231) that similar competition phenomena occur in "populations of biological species, political parties, businesses, countries, coupled reacting components in the atmosphere, bodies of water, organisms as a whole or part, components of nervous systems, elementary excitations in fluids (eddies in turbulent fluids)." We turn now to another widely occurring phenomenon: frequency or mode locking.

4-2. Frequency Locking

Probably the first reported observation of two sustained modes locking together occurred in 1665. Christiaan Huygens noticed that two pendulum clocks hanging close together on the same wall would, after a short while, tick precisely together for an arbitrarily long period of time. He described the observation in a letter to his father (Huygens, 1665):

> Being obliged to stay in my room for several days and also occupied in making observations on my two newly made clocks, I have noticed a remarkable effect which no one could have ever thought of. It is that these two clocks hanging next to one another separated by one or two feet keep an agreement so exact that the pendulums invariably oscillate together without variation. After admiring this for a while, I finally figured out that it occurs through a kind of sympathy: mixing up the swings of the pendulums, I have found that within a half hour they always return to consonance and remain so constantly afterwards as long as I let them go. I then separated them, hanging one at the end of the room and the other fifteen feet away, and noticed that in a day there was five seconds difference between them. Consequently, their earlier agreement must in my opinion have been caused by an imperceptible agitation of the air produced by the motion of the pendulums. The clocks are always shut in their boxes, each weighing a total of less than 100 pounds. When in consonance, the pendulums do not oscillate parallel to one another, but instead they approach and separate in opposite directions.

In a later letter, Huygens (1665) wrote that the coupling mechanism was a small vibration transmitted through the wall, and not movement of air.

Lord Rayleigh (1907) made similar observations about two driven tuning forks coupled by vibrations transmitted through the table on which both forks sat. He displayed the beat note using Lissajous figures, while varying the frequency of one fork toward that of the other. When the forks were uncoupled, he could make the beat note as small as desired. But when the forks were coupled by the table top (and in other ways), the beat note fell abruptly to zero after a critical point was reached and remained at zero for a region of tuning about the stationary fork frequency. Rayleigh further noted that, just before the critical point was attained, "at one part of the cycle, the changes are very slow and at the opposite part relatively quick." We will see shortly a simple mathematical representation of this locking and "slipping" behavior, a representation that applies to a variety of laser locking problems as well.

Locking in triode oscillator circuits was explained by Van der Pol (1927), who included in Eq. (1) an external electromotive force as given in

$$\ddot{v} - \frac{d}{dt}(av - \beta' v^3) + \omega^2 v = \nu^2 V_0 \sin \nu t. \tag{33}$$

He showed that, as he tuned the external frequency ν close to the oscillator frequency ω, the oscillator suddenly jumped to the external frequency. It is important to note that the beat note between the two frequencies vanishes not because the triode stops oscillating, but because it oscillates at the external frequency. This is different from the two-mode operation of Sec. 4-1, in which one normal mode actually suppresses oscillation of the other.

We can show the locking effect for (33) by utilizing the slowly varying amplitude approach again, this time including a slowly varying phase Ψ and oscillation at the external frequency ν (deviation from this value is included in Ψ). We write

$$v(t) = \tfrac{1}{2}V(t) \exp[-i(\nu t + \Psi)] + \text{c.c.} \tag{34}$$

Substituting (34) into (33) with the approximations enumerated in obtaining (5), we find the corresponding equation:

$$-2i\nu\dot{V} - (\nu + \dot{\Psi})^2 V - (a - \beta V^2)(-i\nu V) + \omega^2 V = i\nu^2 V_0 \exp(i\Psi). \tag{35}$$

Here the real part does not vanish as it did in (5), for we now allow for a dispersion effect, namely, that due to locking. Later we encounter in optical media the usual dispersion effects due to variation in index of refraction. Equating the real and imaginary parts of (35) separately to zero, we find the amplitude and phase determining equations:

$$\dot{V} = \tfrac{1}{2}(a - \beta V^2)V - \tfrac{1}{2}\nu V_0 \cos \Psi, \tag{36}$$

$$\dot{\Psi} = \omega - \nu + \tfrac{1}{2}\nu \frac{V_0}{V} \sin \Psi$$

$$= d + l \sin \Psi, \tag{37}$$

where we use d for the detuning term and l for the locking coefficient.

In general the coupled equations (36) and (37) must be solved numerically. We can obtain an approximate solution satisfactory for sufficiently small V_0 by decoupling the equations, that is, by simply using the value of Eq. (10) for V^2 in (36). The resulting time dependence of (37) depends critically on the relative magnitudes of the detuning factor d and the locking factor l. If $|d| > |l|$, Ψ changes monotonically in time; but as $|d|$ approaches $|l|$, part of the cycle varies rapidly (when d adds to $l \sin \Psi$) compared to the opposite part of the cycle, in which the terms subtract from one another. When $|d| = |l|$, the slow part of the cycle actually stops altogether and locking ensues. Locking continues throughout the region defined by $|d| < |l|$ for the value of Ψ which yields $\dot{\Psi} = 0$ in (37) in a stable fashion. A more complete discussion is given in Sec. 9-3, where mode locking of three-mode laser operation is discussed.

In addition to locking in multimode laser operation (yielding, e.g., trains of light pulses), locking occurs between the oppositely running waves in a ring laser (Chap. 11) and between oppositely directed circular polarizations in a Zeeman laser (Chap. 12). In Chap. 21 the discussion relates mode-locking phenomena to second-order phase transitions in, for example, ferromagnets and superconductivity.

Problems

4-1. Derive Eq. (1) from circuit equations for Van der Pol's triode oscillator depicted in Fig. 4-1 and operating at the inflection point of the triodes. Recall that the anode current i_a equals the sum of the currents $i_R + i_L + i_C$ through the resistor, capacitor, and inductor and that the voltage drops across these elements are equal. Let the deviation of the anode voltage v_a from its mean value v_b (the battery voltage) be denoted by v, and let i denote the corresponding current deviation. Finally choose a saturating triode characteristic

$$i = -a'v + r'v^2 + a'v^3. \tag{38}$$

Equation (1) follows with the substitutions $a = (a' - 1/R)/C$, $r' = 0$ (inflection point), $\beta = 3\beta'/C$, and $\omega^2 = 1/LC$. The complete derivation with an explanation of (38) is given in Van der Pol's (1934) review article.

4-2. Diagonalize matrix (32) and show thereby that solution (29) is unstable. Also, find the corresponding solution with the 2's in (24) replaced by $\frac{1}{2}$'s, and show that this solution is stable. The general theory is given in Sec. 9-2.

References

Historical References

C. Huygens, 1665, letters to his father, in: *Oeuvres Complètes de Christiaan Huygens,* Vol. 5 *Société Hollandaise des Sciences,* La Haye, Martinus Nijhoff, 1893, pp. 243–244.

A. J. Lotka, 1925, *Elements of Physical Biology,* Baltimore.

Rayleigh (Lord), 1907, *Phil. Mag.* **XIII**, 316–333.

B. van der Pol, 1920, *Radio Rev.* **1**, 704–754.

B. van der Pol, 1927, *Phil. Mag.* **3**, *65,*

V. Volterra, 1931, *Leçons sur la Théorie Mathématique de la Lutte pour la Vie,* Gauthier-Villars, Paris.

Review Articles

N. S. Goel, C. Maitra, and E. W. Montroll, 1971, *Rev. Mod. Phys.* **43**, 231.

Nicholas Minorsky, 1962, *Nonlinear Oscillations,* D. Van Nostrand, Princeton, N. J., Chap. 18.

M. Sargent III, 1973, *Appl. Phys.* **1**, 133.

B. van der Pol, 1934, *Proc. IRE* **22**, 1051.

V
AMMONIA BEAM MASER

5. Ammonia Beam Maser

To treat the laser in a realistic way, we need the density matrix concepts and formalism given in Chap. 7. This material makes it possible to incorporate dipole phase decays due, for example, to collisions and to treat multimode electric fields in a straightforward way. Nevertheless, it is gratifying to find that we can already determine the field intensity for a simplified model of the ammonia beam maser, the first device of the maser-laser genre. The maser was originally conceived in 1951 by Townes, who noted his thoughts on the back of an envelope during a period of contemplation on a bench in Franklin Park, Washington, DC.† The device was realized experimentally by Gordon, Zeiger, and Townes (1954). In this chapter, we treat the maser along the lines given by Lamb (1960). The analysis is not complicated by atomic (molecular) lifetimes, because they are long compared to the Rabi flopping time, which is the optimal transit time for molecules in the maser cavity. Furthermore, by considering single-mode, steady-state operation, we can assume by energy conservation (1) that the energy given up by the atoms due to stimulated emission is gained by the field, and (2) that this energy flow balances the cavity losses.

In Sec. 5-1 we discuss the nature of the ammonia molecule and point out the similarity between the two energy eigenfunctions of interest and the normal modes of a coupled oscillator problem. In Sec. 5-2 we discuss the maser itself, the method of obtaining more molecules in the higher energy level than in the lower (population inversion), the maser cavity characteristics, and finally the electric field intensity produced.

†For a colorful account of this and other laser-oriented tales, read Gunther's (1968) account.

5-1. The Ammonia Molecule

The active medium in the ammonia beam maser consists of ammonia molecules, which have the form depicted in Fig. 5-1. There are many more degrees of freedom in this molecule than in the hydrogen atom because of various vibrations, rotations, and molecular electron orbitals. These degrees of freedom result in a proliferation of energy levels and quantum numbers required to specify the levels. Maser action, however, involves only two levels, both characterized by the lowest electronic and vibrational quantum numbers and by the rotational numbers 3, 3. The two levels result from an additional degree of freedom, namely, that the nitrogen can be located on either side of the hydrogen plane. If this plane constituted a barrier of infinite height, there would be two degenerate (equal-energy) eigenfunctions corresponding to this degree of freedom. The barrier is not infinitely high, however, resulting in an energy splitting of 23,800 MHz (in frequency units), a microwave frequency with wavelength slightly greater than 1 cm.

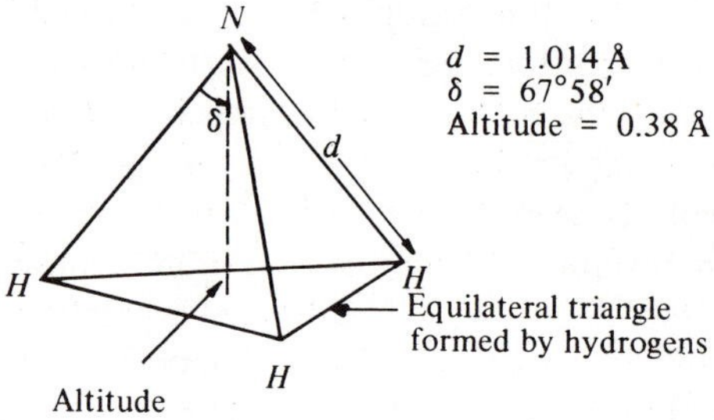

Figure 5-1. Diagram of ammonia (NH_3) molecule.

We can understand the energy difference intuitively by "eyeballing" an integration of an approximate Schrödinger's eigenvalue equation for this problem. We describe the tunneling of the nitrogen through the hydrogen plane by the one-dimensional equation

$$\frac{d^2}{dx^2} u(x) = \frac{2m}{\hbar^2} [V(x) - \hbar\omega] u(x), \tag{1}$$

where the barrier energy $V(x)$ is depicted in Fig. 5-2*a*, m is an effective mass, $\hbar\omega$ is an energy eigenvalue, and $u(x)$ is the corresponding eigenfunction. When the eigenvalue energy exceeds the barrier energy $V(x)$ and $u(x)$ is positive, the second derivative of u given by Eq. (1) is *negative*, leading to the humps in Fig. 5-2*b*. These regions of the x axis are the classical allowed ones

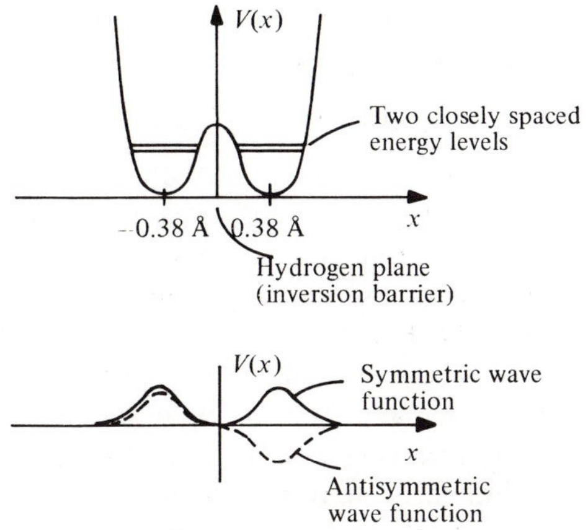

Figure 5-2. (a) Qualitative potential well for nitrogen in ammonia molecule. The coordinate x is the distance of the nitrogen from the "hydrogen plane." The two minimums in V(x) correspond to the locations of the nitrogen on either side of this plane for x = ±0.38 A. (b) Qualitative forms of wave functions for nitrogen subject to potential well in NH_3. Solid line depicts eigenfunction, which is symmetric with respect to inversion [$u_b(x) = u_b(-x)$, same for nitrogen on either side of hydrogen plane]. The antisymmetric eigenfunction [$u_a(x) = -u_a(-x)$] is given by the dashed line.

for a particle. When the barrier energy $V(x)$ is larger than $\hbar\omega$, and $u(x)$ is positive, however, the second derivative of $u(x)$ is *positive.* If $\hbar\omega$ is just right, $u(x)$ decays to zero as x approaches $\pm\infty$, and between the wells it dips down but always remains above the x axis. A slightly different energy $\hbar\omega$ would be associated with a divergent eigenfunction. This correct value leads to the symmetric wave function [i.e., $u(-x) = u(x)$] sketched in Fig. 5-2*b*. For a somewhat larger energy $\hbar\omega$, $u(x)$ has a *steeper slope* when it crosses the central barrier, so that $u(x)$ crosses, in turn, the x axis at $x = 0$. At this point, $u(x)$ becomes negative, reversing the sign of the second derivative to a *negative* value, and for the correct energy the antisymmetric [$u(-x) = -u(x)$] eigenfunction results. Thus we see that the antisymmetric wave function has the higher energy. Still higher energies lead to oscillations in the well. We recommend that the reader try his hand at understanding the Hermite-Gaussian functions of Sec. 1-3 by means of the "eyeball" method.

Note that the function $\psi_L(x)$, for which the nitrogen is most probably located in the left well, is not an energy eigenfunction, for it is weakly coupled to a similar function $\psi_R(x)$ for the right well [$\psi_L(x)$ is not orthogonal to $\psi_R(x)$]. The wave function $\psi(x, t)$, which is initially given by $\psi_L(x)$, oscillates back and forth between ψ_L and ψ_R. If, however, the wave function is initially given by the symmetric combination:

$$u_b(x) = 2^{-1/2}[\psi_L(x) + \psi_R(x)] \tag{2}$$

or the antisymmetric combination:

$$u_a(x) = 2^{-1/2}[\psi_L(x) - \psi_R(x)], \tag{3}$$

it remains so indefinitely, for these are the energy eigenfunctions.

The situation is analogous to two weakly coupled pendulums. If one pendulum is oscillating alone at some time, its energy gradually transfers to the other until the situation is reversed, as indicated in Fig. 5-3*a*. The energy continues to transfer back and forth between the pendulums at a rate determined by the coupling between them. However, if the two pendulums oscillate with the same amplitude and in phase with one another, they continue this way unchanged in time (Fig. 5-3*b*). This corresponds to the symmetric combination (2). Similarly, if they oscillate with the same amplitude and precisely out of phase, they also continue to oscillate unchanged in time. This latter mode corresponds to the antisymmetric combination (3), which has higher energy since the spring increases the restoring force.

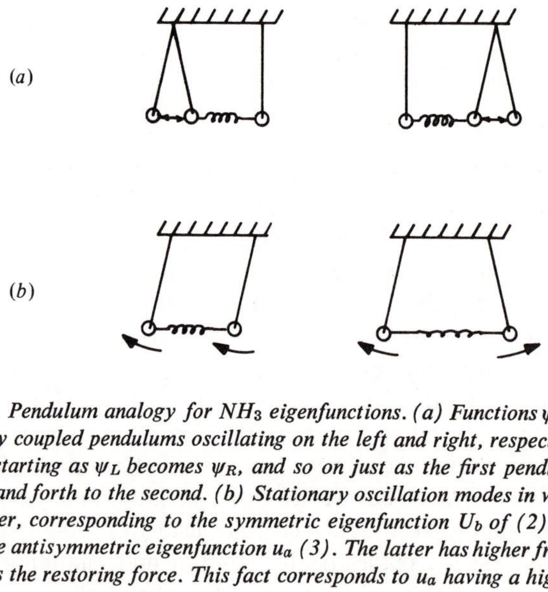

Figure 5-3. Pendulum analogy for NH_3 eigenfunctions. (a) Functions ψ_L and ψ_R correspond to weakly coupled pendulums oscillating on the left and right, respectively. In time a wave function starting as ψ_L becomes ψ_R, and so on just as the first pendulum oscillation transfers back and forth to the second. (b) Stationary oscillation modes in which pendulums oscillate together, corresponding to the symmetric eigenfunction U_b of (2), or against one another, like the antisymmetric eigenfunction u_a (3). The latter has higher frequency, for the spring increases the restoring force. This fact corresponds to u_a having a higher energy than u_b.

Thus we have what reduces for our purpose to a two-level problem. For maser action, the active medium must sustain an electric field and this requires an electric dipole. That the ammonia molecule has an electric dipole when in a superposition of eigenfunctions u_a and u_b can be seen as follows. Because filling the three remaining $2p$ electron states in nitrogen is favored energetically (has lower energy), the valence electrons in the molecule tend to spend more of their time in the vicinity of the nitrogen than in the hydrogen plane. This effect gives the latter a net positive charge, and the former a negative charge, provided the molecule is described approximately by a function ψ_L or ψ_R for which the nitrogen is probably found on one side of the hydrogen plane. As in Sec. 3-1, we write the wave function as a superposition of energy eigenstates:

$$\psi(x, t) = C_a(t)u_a(x) \exp(-i\omega_a t) + C_b(t)u_b(x) \exp(-i\omega_b t). \qquad (4)$$

Discarding the overall phase factor $\exp(-i\omega_b t)$ as before and choosing the probability amplitudes

$$C_a = C_b = 2^{-1/2} \qquad (5)$$

for the sake of discussion, we have the particular wave function

$$\psi(x, t) = 2^{-1/2}[u_a \exp(-i\omega t) + u_b]. \qquad (6)$$

At time $t = 0$, this reduces to

$$\psi(x, 0) = \psi_L(x) = 2^{-1/2}(u_a + u_b) \qquad (7)$$

from Eq. (2). Similarly at time $t = \pi/\omega$ the wave function is

$$\psi\left(x, \frac{\pi}{\omega}\right) = -\psi_R(x) = 2^{-1/2}(-u_a + u_b). \qquad (8)$$

Hence we see that in a superposition of energy eigenstates the probability amplitude for the negatively charged nitrogen oscillates back and forth through the positively charged hydrogen plane, just as the electron wave function (3.2) in hydrogen oscillated back and forth across the nucleus as depicted in Fig. 3-2. Here it is very difficult to calculate the explicit position dependence of the eigenfunctions.† Consequently we will not plot accurate curves as in Fig. 3-2. However, the result is qualitatively the same: the ammonia molecule can have an oscillating electric-dipole moment‡ with an associated radiated electric field (3.14) and a nonzero electric-dipole interaction energy. In the next section we apply the theory of Chaps. 2 and 3 to the ammonia molecule as it interacts with the electromagnetic field in a maser.

5-2. Maser Operation

To make a maser, we inject ammonia molecules in their upper level u_a into a maser cavity resonant at the frequency Ω, and remove them after the electric

† A good discussion of the theoretical techniques for NH_3 and other hydrides is given by Slater (1968).

‡ Note that the energy eigenfunctions u_a and u_b have quadrupole moments, but no dipole moments. This too is similar to the hydrogen $1s$ and $2p$ levels, neither of which has a dipole moment.

field has induced transitions to the lower level u_b. By energy conservation, the energy given up by the molecules is gained by the inducing electric field. Optimally, the cavity frequency Ω should equal the molecular frequency ω and the molecules should be removed after time $t = \pi/\mu$, which, according to the flopping formula (2.68), is the time at which the probability for a transition to the lower level is unity.

The apparatus for accomplishing these procedures is diagrammed in Fig. 5-4. The collimator increases the flux of molecules up to three times that for the same-size hole.† A typical collimator is the Zacharias nozzle, consisting of many small tubes with length small compared to the mean free path of the molecules and large compared to the tube diameters. The probability P_a that a molecule arrives at the focuser in the upper energy level relative to the probability P_b for the lower level is given by the Boltzmann distribution

$$\frac{P_a}{P_b} = \exp[-\hbar(\omega_a - \omega_b)/k_B T], \tag{8a}$$

in which k_B is Boltzmann's constant, and T is the absolute temperature. For a molecular temperature of 300°K, the thermal energy is $\frac{1}{40}$eV while 30,000 MHz corresponds to (1.24 eV corresponds to 1 μm or to 3×10^8 MHz) 10^{-4} eV $\ll \frac{1}{40}$ eV. Hence molecules arrive at the focuser with essentially equal probability for being in the upper and lower energy levels. The focuser then has the effect of a Stern-Gerlach apparatus with an obstruction in the path of the lower energy state, that is, it passes only molecules in the upper energy level.

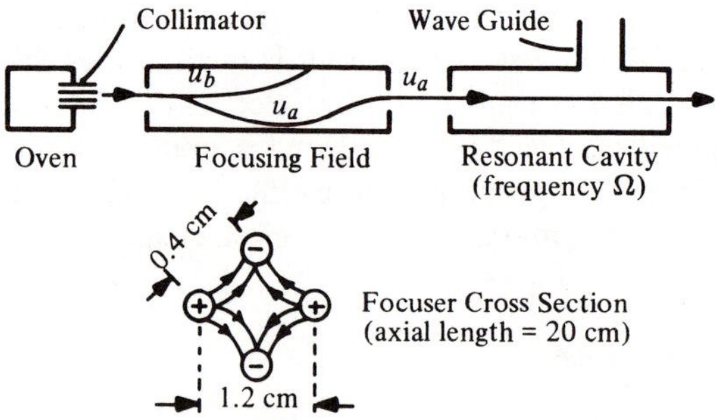

Figure 5-4. Diagram of NH_3 maser. Molecules produced by oven are collimated, screened into state u_a (higher energy state), and passed through maser cavity, where they give up their energy to the electric field in accordance with the Rabi flopping formula. Typical focuser form and dimensions are indicated.

† See Sears (1953), p. 240.

To understand this, note that a NH_3, two-level Hamiltonian *including* a dc electric field $E(r)$ has the slightly shifted eigenvalues (Prob. 5-2)

$$\hbar\omega_a' \simeq \hbar\omega_a + \frac{\wp^2 E^2(r)}{\hbar\omega} \tag{9}$$

$$\hbar\omega_b' \simeq \hbar\omega_b - \frac{\wp^2 E^2(r)}{\hbar\omega} \tag{10}$$

and the approximate eigenfunctions

$$u_a'(x) = Z^{-1/2}\left[u_a(x) + \frac{\wp E}{\hbar\omega} u_b(x)\right] \tag{11}$$

$$u_b'(x) = Z^{-1/2}\left[u_b(x) - \frac{\wp E}{\hbar\omega} u_a(x)\right], \tag{12}$$

in which the normalization factor Z is given by

$$Z = 1 + \left(\frac{\wp E}{\hbar\omega}\right)^2. \tag{13}$$

The focuser in Fig. 5-4 creates the electric field

$$E(r) = 2\zeta r. \tag{14}$$

As for the magnetic force expression (1.35), the consequent energy variations in Eqs. (9) and (10) result in forces here radially oriented with values

$$F_r = -\frac{\partial}{\partial r}(\hbar\omega_\alpha') = \mp\left(\frac{8\wp^2\zeta^2}{\hbar\omega}\right)r, \qquad \alpha = a, b, \tag{15}$$

for which the negative term corresponds to the higher energy level. Equation (15) gives the equation of motion for a molecule:

$$m\ddot{r} = F_r = -kr, \tag{16}$$

where the "spring" constant $k = \pm 8\wp^2\zeta^2/\hbar\omega$. For positive k (upper level) this gives simple harmonic motion, whereas for negative k (lower level) it produces an exponentially increasing deviation. Specifically, for the initial conditions $r = r_0$, $\dot{r} = 0$, we find (Prob. 5-3)

$$\begin{aligned} r &= r_0 \cos\sqrt{k/m}\, t, && k > 0, \text{ upper state,} \\ r &= r_0 \cosh\sqrt{-k/m}\, t, && k < 0, \text{ lower state,} \end{aligned} \tag{17}$$

as sketched in Fig. 5-4. Inasmuch as the eigenfunction $u_a'(x) \approx u_a(x)$ since $\wp E \ll \hbar\omega$, the focuser prepares molecules in the upper level for maser operation.

The condition for oscillation is that the energy lost in the cavity per unit time (loss) is equal to the energy gained per unit time (saturated gain) from the NH_3 molecules, that is,

$$\text{Loss} = \text{saturated gain}. \tag{18}$$

We write the loss in terms of an important ratio Q called the cavity quality factor and defined by

$$Q = \Omega \frac{\text{energy stored in field}}{\text{energy lost/second}} . \tag{19}$$

The energy stored is given by (see Prob. 5-4 for cavity shape, etc.)

$$\text{Energy stored in field} = \tfrac{1}{2}\int(\varepsilon_0 E^2 + \mu_0 H^2)\, dV = \tfrac{1}{8}\,\varepsilon_0 E_0{}^2 V, \tag{20}$$

where V is the volume of the maser cavity. With (19) and (20), we have the energy lost per second:

$$\text{Loss} = \frac{1}{8}\frac{\Omega}{Q}\,\varepsilon_0 E_0{}^2 V. \tag{21}$$

The energy gained per second is

$$\text{Saturated gain} = \frac{Nv}{l_z}\,|C_b|^2{}_{\text{exit}}\,\hbar\Omega. \tag{22}$$

Here N is the number of molecules in the cavity, v is the axial (z-direction) velocity of the molecules, and l_z is the cavity length; hence Nv/l_z is the number of molecules leaving the cavity per second. Furthermore $|C_b|^2{}_{\text{exit}}$ is the probability that a molecule exits in the lower level, thereby giving up energy $\hbar\nu$. The frequency ν of the electric field is close to the cavity frequency (for central tuning, $\nu = \Omega = \omega$), although a quantitative relationship cannot be determined from our present, simple analysis. Hence we take $\nu = \Omega$ and substitute (21) and (22) into (18) to obtain

$$\frac{Nv}{l_z}\,|C_b|_{\text{exit}}{}^2\,\hbar\Omega = \frac{1}{8}\frac{\Omega}{Q}\,\varepsilon_0 E_0{}^2 V.$$

Further substituting the Rabi flopping probability (2.68) with $t = l_z/v$, the transit time, and rearranging, we find

$$\frac{1}{Q} = \frac{2N\wp^2}{l_x l_y \varepsilon_0 \hbar v}\cdot\frac{\sin^2(\mu l_z/2v)}{(\mu l_z/2v)^2}, \tag{23}$$

in which the $l_x l_y l_z = V$. This is a transcendental equation for the field intensity $E_0{}^2$, written in terms of the atomic and cavity parameters. By definition of the flopping frequency μ of (2.61),

$$\mu^2 = \left(\frac{\wp E_0}{\hbar}\right)^2 + (\Omega - \omega)^2, \tag{24}$$

which implies that the curves of $\wp E_0/\hbar$, that is, amplitude, versus detuning $(\Omega - \omega)$ are semicircles (see Fig. 5-5). Note that, for sufficient detuning, oscillation ceases, and note also that the greatest intensity occurs for central tuning ($\Omega = \omega$). A numerical method for solving (23) is given in Prob. 5-5.

We see that the beam of ammonia molecules plays here the same role as the negative resistance played in the classical sustained oscillators of Chap. 4,

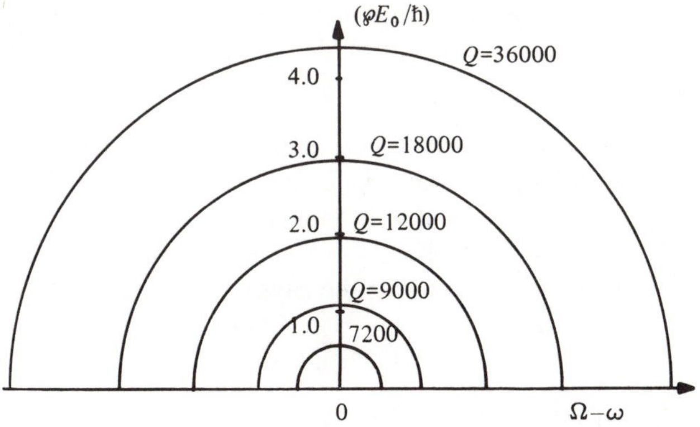

Figure 5-5. Curves of maser amplitude $\wp E_0/\hbar$ versus cavity detuning for various cavity Q's.

namely, that of overcoming oscillator losses. Furthermore, the molecular response saturates, thereby limiting the buildup of oscillations, as did the negative resistance. These properties are common to all the sustained oscillators that we will study. Finally we note that the molecular beam could in principle be replaced by a beam of classical, anharmonic oscillators, mentioned at the end of Sec. 3-2 and discussed by Borenstein and Lamb (1972). In practice quantum systems (molecules, atoms, etc.) are used for maser gain media, but in principle there is nothing inevitably quantum mechanical about the basic maser concept.

Problems

5-1. Explain the curvature of the first two Hermite-Gaussian functions in Fig. 1-3, using the "eyeball" integration method of Sec. 5-1.

5-2. Show that the two-level NH_3 Hamiltonian with eigenvalues $\hbar\omega_a$ and $\hbar\omega_b$ acquires the new eigenvalues (9) and (10) when supplemented adiabatically by a dc electric field $E(r)$ limited in magnitude by the condition $\wp E \ll \hbar\omega$. Further show that the corresponding eigenfunctions are (11) and (12). Hint: Represent the supplemented Hamiltonian $\mathscr{H}$ by the matrix

$$\mathscr{H} = \begin{pmatrix} \hbar\omega_a & \wp E \\ \wp E & \hbar\omega_b \end{pmatrix} \tag{25}$$

and use the eigenvalue equation

$$\mathscr{H} u'(x) = \lambda u'(x), \tag{26}$$

where $u'(x)$ is an eigenfunction (to be determined) of $\mathscr{H}$ written as a column vector. Diagonalize (25) by setting the determinant

$$\text{Det}\,(\mathscr{H} - \lambda \mathscr{I}) = 0, \tag{27}$$

where $\mathscr{I}$ is the identity matrix. Then substitute the resulting eigenvalues separately back into (26) to obtain the eigenfunctions (11) and (12).

5-3. Derive (17) for the molecular deviations in the focuser of Fig. 5-4.

5-4. Suppose that a resonator has a cross-sectional area A, length L, and end transmission (loss) $\mathscr{T}$. Show that the cavity Q (19) is given by

$$Q = \frac{\nu L}{\mathscr{T} c}. \tag{28}$$

What is the value for the NH_3 maser ($L = c/\nu$, $\mathscr{T} = 10^{-4}$)? What is it for an optical cavity with $L = 1\text{m}$ and $\mathscr{T} = 0.01$?

5-5. Calculate the intensity vs detuning curves in Fig. 5-5 as follows. The radius of the circles is given by $[\wp E_0(\omega)/\hbar]^2$, the intensity value at central tuning ($\Omega = \omega$). From (24), this is given, in turn, by μ^2, where $\mu = (2v/l_z)x$ is determined by the transcendental equation (23). We write (23) for definiteness as

$$\frac{6000}{Q} = \frac{\sin^2 x}{x^2} = f(x). \tag{29}$$

The problem is to find the value $x = a$ which satisfies (29) for a given Q. To do this use the Newton-Raphson numerical method on any available digital computer. The method inverts the first-order approximation

$$f(x) \approx f(a) + (x - a)\,\frac{df(x)}{dx} \tag{30}$$

to find the desired value a as

$$a \approx x - \frac{f(x) - f(a)}{df/dx}. \tag{31}$$

One then guesses a value for x to start with and iterates this formula (uses the computed value as the guess value a number of times). If the initial guess value was good enough, the process converges yielding the results in Fig. 5-5.

References

M. Borenstein and W. E. Lamb, Jr., 1972, *Phys. Rev.* **A5**, 1298.

R. P. Feynman, R. B. Leighton, and M. Sands, *The Feynman Lectures on Physics,* Addison-Wesley Publishing Co., Reading, Mass., 1965, Chaps. 8–9 of Vol. III.

M. Gunther, 1968, "Lasers: the Light Fantastic," *Playboy,* February.

J. P. Gordon, H. J. Zeiger, and C. H. Townes, 1955, *Phys. Rev.* **99**, 1264; **95,** 282L (1954).

W. E. Lamb, Jr., 1960, "Quantum Mechanical Amplifiers," in: *Lectures in Theoretical Physics,* Vol. 2, ed. by W. E. Brittin and B. W. Downs, Interscience Publishers, New York.
A preliminary version of this work appears as an Appendix in J. C. Helmer's Ph. D. thesis, Stanford University, 1957.

F. W. Sears, 1953, *An Introduction to Thermodynamics, the Kinetic Theory of Gases, and Statistical Mechanics,* 2nd ed., Addison-Wesley Publishing Co., Reading, Mass., p. 240.

VI
THE STATE VECTOR

6. The State Vector

At this point one might have noticed that both the wave function $\psi(\mathbf{r}, t)$ and the expansion coefficients $C_k(t)$ have been referred to as probability amplitudes. In this chapter, we introduce the Dirac notation, a way of writing vectors which reveals that $\psi(\mathbf{r}, t)$ and $C_k(t)$ are both expansion coefficients of the state vector $|\psi(t)\rangle$, a vector in an abstract space. The absolute squares of the expansion coefficients are the probabilities of finding the state vector in corresponding eigenstates. The Dirac notation also provides a compact, versatile means for considering any (and all) representatives (energy, position, etc.) of a state vector.

We also discuss the principal ways of handling time dependence in quantum mechanics, namely, the Schrödinger, interaction, and Heisenberg pictures. The difference between the Schrödinger and interaction pictures amounts to the use in the former of an integrating factor, a practice which may be inconvenient. Both pictures deal with a wave function, state vector, or density matrix in the description of the time dependence of a problem. The Heisenberg picture has a different spirit, however, for problems are treated with operator equations of motion. All three pictures must yield, of course, the same expectation value for a given observable, since this value is the connection between theory and experiment. To distinguish the probability amplitudes in the Schrödinger and interaction pictures, we have used $C_k(t)$ for the slowly varying amplitudes of the latter, and $c_k(t)$ for the rapidly varying amplitudes of the Schrödinger picture. Similarly, with operator equations of motion, we have used capital letters for slowly varying operators and lower-case letters for their rapidly varying counterparts.

In Sec. 6-1 the Dirac notation is defined and compared to more standard vector representations, such as $(\hat{x}, \hat{y})$ for two dimensions. In Sec. 6-2 the Dirac and matrix representations of operators are given. In Sec. 6-3 the state vector is expanded in various complete sets of eigenvectors and its relationship to the wave function is given. In Sec. 6-4 the Schrödinger, interaction, and Heisenberg pictures are discussed. The formalism in this chapter is used in Chap. 7 on the density matrix, to a minor degree in Chaps. 8–13 on the semiclassical theories of lasers and pulse propagation, and a great deal in Chaps. 14–20 on fully quantum-mechanical theories. Our approach here is informal in nature; the reader can find more complete treatments in the references given at the end of the chapter.

6-1. Dirac Notation

A standard way of representing a vector $\mathbf{v}$ in a two-dimensional space is by means of the unit vectors $\hat{x}$ and $\hat{y}$ as follows:

$$\mathbf{v} = v_x\hat{x} + v_y\hat{y}. \tag{1}$$

This is depicted in Fig. 6-1*a*. The vector $\mathbf{v}$ can be written equivalently in Dirac notation as

$$|v\rangle = v_x|x\rangle + v_y|y\rangle \tag{2}$$

as depicted in Fig. 6-1*b*. In representation (1), the x component of $\mathbf{v}$, v_x, is given by the dot product

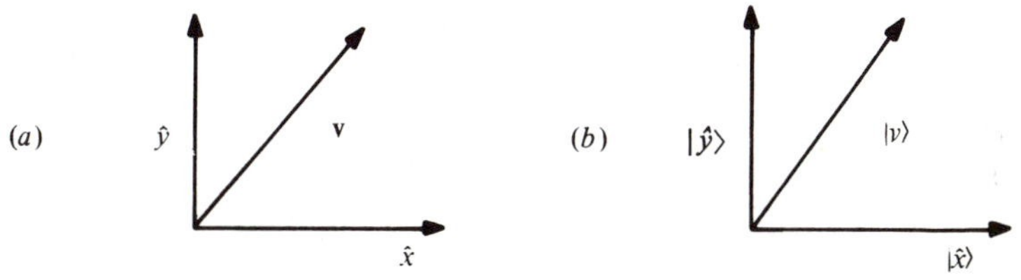

Figure 6-1. (a) Standard representation of the vector $\mathbf{v}$. *(b) Equivalent representation in Dirac notation.*

$$v_x = \hat{x} \cdot \mathbf{v}. \tag{3}$$

In Dirac notation, this dot product is written as the bracket

$$v_x = \langle x|v\rangle. \tag{4}$$

Hence (1) could have been written as

$$\mathbf{v} = \hat{x}(\hat{x} \cdot \mathbf{v}) + \hat{y}(\hat{y} \cdot \mathbf{v}). \tag{5}$$

In Dirac notation, (2) becomes, with (4) and $v_y = \langle y|v\rangle$,

$$|v\rangle = |x\rangle\langle x|v\rangle + |y\rangle\langle y|v\rangle. \tag{6}$$

It is convenient to introduce the outer product of two vectors **a** and **b** written as the dyad **ab**. One of several ways to apply this to a vector **v** is by means of the dot product:

$$\mathbf{ab} \cdot \mathbf{v} = \mathbf{a}\,(\mathbf{b} \cdot \mathbf{v}),$$

which yields a new vector. In particular we can define the identity diadic

$$\mathscr{I} = \hat{x}\hat{x} + \hat{y}\hat{y}, \tag{7}$$

which, when dotted into **v**, just gives **v**:

$$\mathscr{I} \cdot \mathbf{v} = \hat{x}\,(\hat{x} \cdot \mathbf{v}) + \hat{y}(\hat{y} \cdot \mathbf{v}) = \mathbf{v}. \tag{8}$$

Similarly, in Dirac notation

$$\mathscr{I} = |x\rangle\langle x| + |y\rangle\langle y|, \tag{9}$$

so that with $|v\rangle = \mathscr{I}\,|v\rangle$, we regain (6) for $|v\rangle$.
Note that (9) is an expression of completeness of the unit vectors $|x\rangle$ and $|y\rangle$ in two dimensions.

More generally in an N- (or infinite-) dimensional space, a vector can be expanded in the complete set of unit vectors $\{\hat{e}_k\} \equiv \{\hat{e}_1, \hat{e}_2, \ldots, \hat{e}_k, \ldots\}$ as

$$\mathbf{v} = \sum_k \hat{e}_k(\hat{e}_k \cdot \mathbf{v}). \tag{10}$$

These unit vectors are orthonormal (orthogonal and normalized), that is,

$$\hat{e}_k \cdot \hat{e}_j = \delta_{kj} = \begin{cases} 1, & k = j, \\ 0, & k \neq j. \end{cases}$$

In Dirac notation this can be written as

$$|v\rangle = \sum_k |k\rangle\langle k|v\rangle, \tag{11}$$

for which the Dirac unit vectors have the inner product

$$\langle k|j\rangle = \delta_{kj}. \tag{12}$$

Similarly to (9), the identity operator is written as

$$\mathscr{I} = \sum_k |k\rangle\langle k|. \tag{13}$$

The reader will identify $(\hat{e}_k \cdot \mathbf{v})$ and $\langle k|v\rangle$ as expansion coefficients or amplitudes of the vector **v** in the bases $\{\hat{e}_k\}$ and $\{|k\rangle\}$, respectively. The bracket notation for the dot product yields a convenient nomenclature for the vectors involved:

$$\langle a|b\rangle \longleftrightarrow \overset{\text{bra c ket}}{\langle a| \cdot |b\rangle} \tag{14}$$

that is, vectors like $\langle a|$ are called "bras" and those like $|b\rangle$ are called "kets."

One refers verbally to $|n\rangle$ as "n ket" and to $\langle\psi|$ as "ψ bra." The two kinds of vectors are *adjoints* of one another, that is,

$$\langle n| = |n\rangle^\dagger.$$

In general the expansion coefficients in quantum mechanics are complex. In this connection, the complex conjugate of a bracket (inner product) is given by

$$\langle k|v\rangle^* = \langle v|k\rangle. \tag{15}$$

This feature differs from the real vector space of $\hat{x}$ and $\hat{y}$.

6-2. Dirac and Matrix Representations of Operators

In general an operator $\mathcal{O}$ maps a vector $|\xi\rangle$ into a new vector $|\zeta\rangle$ in the vector space, that is,

$$|\zeta\rangle = \mathcal{O}|\xi\rangle. \tag{16}$$

The inverse $\mathcal{O}^{-1}$ (if it exists) of the operator $\mathcal{O}$ has the property

$$\mathcal{O}^{-1}|\zeta\rangle = \mathcal{O}^{-1}\mathcal{O}|\xi\rangle = |\xi\rangle. \tag{17}$$

This is true for all non-null vectors in the space, and hence

$$\mathcal{O}^{-1}\mathcal{O} = \mathcal{O}\mathcal{O}^{-1} = \mathcal{I}. \tag{18}$$

In Eq. (13) the identity operator is written as the sum of outer products of vectors. Similarly, the general operator $\mathcal{O}$ can be written as

$$\mathcal{O} = \sum_n\sum_m \mathcal{O}_{nm}|n\rangle\langle m|. \tag{19}$$

where the matrix elements $\mathcal{O}_{nm}$ are given by

$$\mathcal{O}_{nm} = \langle n|\mathcal{O}|m\rangle. \tag{20}$$

This is easy to see, for by insertion of identity (13) before and after $\mathcal{O}$ in (16), we have

$$\begin{aligned}|\zeta\rangle &= \mathcal{I}\;\mathcal{O}\;\mathcal{I}\;|\xi\rangle \\ &= \sum_n|n\rangle\langle n|\;\mathcal{O}\;\sum_m|m\rangle\langle m|\xi\rangle \\ &= \sum_n\sum_m \mathcal{O}_{nm}|n\rangle\langle m|\xi\rangle.\end{aligned}$$

Since this is true for all $|\xi\rangle$, $\mathcal{O}$ is given by (19).

It is sometimes convenient to specify the expansion coefficients $\langle m|\xi\rangle$ and matrix elements simply as column vector and matrix arrays. Thus $|\xi\rangle$ is represented by

$$|\xi\rangle \leftrightarrow \begin{pmatrix} \langle 1|\xi\rangle \\ \langle 2|\xi\rangle \\ \vdots \\ \langle n|\xi\rangle \\ \vdots \end{pmatrix} \tag{21}$$

and the operator $\mathcal{O}$ by

$$\mathcal{O} \leftrightarrow \begin{pmatrix} \mathcal{O}_{11} & \mathcal{O}_{12} & \cdots & \mathcal{O}_{1m} & \cdots \\ \mathcal{O}_{21} & & & & \\ \vdots & & & & \\ \mathcal{O}_{n1} & \mathcal{O}_{n2} & \cdots & \mathcal{O}_{nm} & \cdots \\ \vdots & \vdots & & \vdots & \end{pmatrix}, \tag{22}$$

where the $\mathcal{O}_{nm}$ are given by (20). Here, of course, the complex numbers comprising these arrays manifestly depend on the set of basis vectors $|n\rangle$. The basis vectors themselves have the simple representatives

$$|1\rangle \leftrightarrow \begin{pmatrix} 1 \\ 0 \\ 0 \\ \vdots \end{pmatrix}, \quad |2\rangle \leftrightarrow \begin{pmatrix} 0 \\ 1 \\ 0 \\ \vdots \end{pmatrix}, \quad \ldots, \quad |n\rangle \leftrightarrow \begin{pmatrix} 0 \\ \vdots \\ 0 \\ 1 \\ \vdots \end{pmatrix}, \quad \ldots . \tag{23}$$

The inner product of two vectors $|\xi\rangle$, $|\zeta\rangle$ has the Dirac form $\langle\zeta|\xi\rangle$ and corresponds to the matrix product

$$\begin{aligned} \langle\zeta|\xi\rangle &= (\langle\zeta|1\rangle\langle\zeta|2\rangle \ldots \langle\zeta|n\rangle \ldots) \begin{pmatrix} \langle 1|\xi\rangle \\ \langle 2|\xi\rangle \\ \vdots \\ \langle n|\xi\rangle \\ \vdots \end{pmatrix} \\ &= \sum_n \langle\zeta|n\rangle\langle n|\xi\rangle, \end{aligned} \tag{24}$$

a relation which follows also by insertion of the identity between $\langle\xi|$ and $|\zeta\rangle$.

Operators which represent observables in quantum mechanics have the important property of Hermiticity. Specifically, if the ket $|\zeta\rangle = \mathcal{O}|\xi\rangle$ corresponds to the bra $\langle\zeta| = \langle\xi|\mathcal{O}$ for all $|\xi\rangle$, the operator $\mathcal{O}$ is Hermitean. In general. $\langle\zeta| = \langle\xi|\mathcal{O}^\dagger$, where $\mathcal{O}^\dagger$ is the adjoint of $\mathcal{O}$. Consequently a Hermitean operator equals its adjoint and is sometimes called self-adjoint. Alternatively, the matrix representation of a Hermitean operator equals its conjugate transpose, that is, $\mathcal{O}^*_{ji} = \mathcal{O}_{ij}$. Note that the adjoint of the operator product $\mathcal{O}\mathcal{Q}$ is given by $(\mathcal{O}\mathcal{Q})^\dagger = \mathcal{Q}^\dagger\mathcal{O}^\dagger$. Not all operators used in quantum mechanics are Hermitean. In Chap. 14, we use the annihilation and creation operators a and $a^\dagger$ in the quantization of the radiation field. However, these operators represent physical observables only in Hermitean combinations, such as the sum $\mathscr{E}\,(a + a^\dagger)$, the electric field operator.

A very important problem in quantum mechanics (and elsewhere) is the determination of the vectors $|\xi\rangle = |\phi\rangle$ that an operator $\mathcal{O}$ maps into constant multiples of themselves, that is,

$$\mathcal{O}\,|\phi\rangle = \lambda|\phi\rangle. \tag{25}$$

This is the Dirac form for the eigenvalue equation [e.g., (1.10)].

6-3. Expansions of the State Vector

The state vector $|\psi\rangle$ for a quantum-mechanial system contains all (possible) information about that system. It obeys the Schrödinger equation

$$i\hbar|\dot{\psi}\rangle = \mathscr{H}\,|\psi\rangle, \tag{26}$$

which is closely related to Eq. (1.6) for the wave function. The expectation value of an operator $\mathcal{O}$ is given by

$$\langle\mathcal{O}\rangle = \langle\psi|\mathcal{O}|\psi\rangle, \tag{27}$$

which corresponds to Eq. (1.4) in terms of the wave function. The state vector belongs to an abstract vector space having complex scalar expansion coefficients. A set of vectors which spans this space is generated by the eigenvectors of all operators which commute with one another. For example, the space for the simple harmonic oscillator (Sec. 1-3) is spanned by the eigenvectors $|n\rangle$ of the Hamiltonian $\mathscr{H}$, that is, the $|n\rangle$ for which

$$\mathscr{H}\,|n\rangle = (n + \tfrac{1}{2})\,\hbar\Omega|n\rangle. \tag{28}$$

In terms of energy eigenstates, the state vector $|\psi(t)\rangle$ has the expansion

$$|\psi(t)\rangle = \sum_n C_n \exp(-i\omega_n t)|n\rangle, \tag{29}$$

which corresponds to (1.13)

The expectation value (27) with (29) is given by

$$\langle\mathcal{O}\rangle = \sum_n\sum_m C_n C_m{}^* \exp[-i(\omega_n - \omega_m)t]\mathcal{O}_{mn}, \tag{30}$$

which is, of course, just the value obtained from (1.4) with the wave function (1.13). Under ideal experimental conditions, the expectation value is the average value of a set of measurements for an operator made on identically prepared systems (same state vector). The spread of measured values about this value is characterized by the mean-square deviation:

$$\sigma = \sqrt{\langle\mathcal{O}^2\rangle - \langle\mathcal{O}\rangle^2}. \tag{31}$$

Because measurements give real numbers, the expectation values of operators corresponding to physical observables must be real. This is true provided the operators are Hermitean, a fact which follows immediately from the definition:

$$\langle\psi|\mathcal{O}|\psi\rangle = [(\langle\psi|\mathcal{O}^{+})|\psi\rangle]^{*} = \langle\psi|\mathcal{O}|\psi\rangle^{*}.$$

In this sequence, the first equality is true for any linear operator; the second requires the Hermiticity.

To determine the relationship between the state vector and the wave function, let us consider a one-dimensional problem (like the simple harmonic oscillator of Sec. 1-3) and expand $|\psi\rangle$ in the eigenstates $|x\rangle$ of the position operator, states which also form a complete set. Specifically, the eigenstate $|x\rangle$ is the state of being at position x and corresponds to the eigenvalue x of the position operator. Because these states are continuous, expansions consist of continuous summations, that is, integrals. Thus the state vector $|\psi\rangle$ is written

$$|\psi\rangle = \int dx\,|x\rangle\langle x|\psi\rangle. \tag{32}$$

The ordinary Schrödinger wave function $\psi(x)$ is just the expansion coefficient in this basis:

$$\psi(x) \equiv \langle x|\psi\rangle. \tag{33}$$

This is the reason for calling $\psi(x)$ a probability amplitude. Because a continuous basis is involved and $\psi^*\psi$ is a probablity density, $\psi(x)$ might better be called a "probability density" amplitude. The basis vectors $|x\rangle$ are orthonormal in the special sense

$$\langle x'|x\rangle = \delta(x' - x), \tag{34}$$

where $\delta(x' - x)$ is the Dirac delta function. This "function" has the same effect on a continuous summation over x' that the Kronecker delta has on a discrete summation: only the term with $x' = x$ survives. The completeness relation for the position states can be written as

$$\int dx\,|x\rangle\langle x| = \mathcal{I}. \tag{35}$$

The expansion coefficients $\langle k|\psi\rangle$ themselves can be expanded in terms of any complete orthonormal basis, that is,

$$\langle k|\psi\rangle = \sum_{a}\langle k|a\rangle\langle a|\psi\rangle. \tag{36}$$

In fact, the identity given by the completeness relations (12) and (35) can be inserted into the middle of any inner product $\langle n|m\rangle$ or between any operator and vector. In particular the wave function has the energy eigenstate expansion

$$\begin{aligned}\psi(x, t) &= \langle x|\psi(t)\rangle = \sum_{n}\langle x|n\rangle\langle n|\psi(t)\rangle \\ &= \sum_{n} C_n \exp(-i\omega_n t)u_n(x),\end{aligned} \tag{37}$$

where the eigenfunctions $u_n(x)$ are defined by

$$u_n(x) \equiv \langle x | n \rangle \tag{38}$$

and the expansion coefficients

$$\langle n | \psi(t) \rangle = C_n \exp(-i\omega_n t).$$

Similarly, three-dimensional problems can be described in terms of the three-dimensional position states $|\mathbf{r}\rangle$, often labeled by the spherical coordinates r, θ, and ϕ, that is, $|r\,\theta\,\phi\rangle$. For these states the identity is written as

$$\mathscr{I} = \int d^3r |\mathbf{r}\rangle\langle\mathbf{r}|. \tag{39}$$

In terms of the $|\mathbf{r}\rangle$, the state vector is written as

$$|\psi\rangle = \int dr \int r\,d\theta \int r \sin\theta\,d\phi |r\,\theta\,\phi\rangle\langle r\,\theta\,\phi\,|\psi\rangle, \tag{40}$$

where the wave function $\psi(\mathbf{r})$ is defined by

$$\psi(\mathbf{r}) = \langle \mathbf{r} | \psi \rangle. \tag{41}$$

In particular, for the hydrogen atom, the quantum vector space is spanned by the $|\mathbf{r}\rangle$ and by the eigenvectors of the Hamiltonian $\mathscr{H}$, the square of the orbital angular momentum $|\mathbf{L}|^2$, and the z component of the angular momentum, L_z. These eigenvectors are labeled by the radial, polar, and azimuthal quantum numbers n, l, and m and are written in Dirac notation as $|n\;l\;m\rangle$. One of the advantages of Dirac notation over the $\hat{e}_k$ variety is the simplicity with which several quantum numbers can be used to label eigenvectors. In terms of the $|n\;l\;m\rangle$, the state vector is written as

$$\begin{aligned} |\psi\rangle &= \sum_n\sum_l\sum_m |n\;l\;m\rangle\langle n\;l\;m|\psi\rangle \\ &= \sum_n\sum_l\sum_m C_{nlm} \exp(-i\omega_{nlm}t)|nlm\rangle. \end{aligned} \tag{42}$$

The wave function of Sec. 1-4 can be expanded as

$$\begin{aligned} \psi(\mathbf{r}, t) = \langle\mathbf{r}|\psi(t)\rangle &= \sum_n\sum_l\sum_m \langle\mathbf{r}|n\;l\;m\rangle\langle n\;l\;m|\psi(t)\rangle \\ &= \sum_n\sum_l\sum_m C_{nlm} \exp(-i\omega_{nlm}t)\, u_{nlm}(\mathbf{r}), \end{aligned}$$

which is just (1.33) with the eigenfunction defined by

$$u_{nlm}(\mathbf{r}) \equiv \langle\mathbf{r}|n\;l\;m\rangle. \tag{43}$$

We can project the Schrödinger equation (26) for the state vector $|\psi(t)\rangle$ onto the $|\mathbf{r}\rangle$ basis and obtain a Schrödinger equation for the wave function by using (39) as follows:

$$\begin{aligned} i\hbar\frac{\partial}{\partial t}\langle\mathbf{r}|\psi(t)\rangle &= \langle\mathbf{r}|\int d^3r'\,\mathscr{H}\,|\mathbf{r}'\rangle\langle\mathbf{r}'|\psi(t)\rangle \\ &= \int d^3r'\langle\mathbf{r}|\mathscr{H}\,|\mathbf{r}'\rangle\langle\mathbf{r}'|\psi(t)\rangle. \end{aligned} \tag{44}$$

The term $\langle \mathbf{r}|\mathscr{H}|\mathbf{r}'\rangle$ is given on purely physical grounds, that is, an assumption based on experience, by

$$\langle \mathbf{r}|\mathscr{H}|\mathbf{r}'\rangle = \mathscr{H}(\mathbf{r}, \nabla^2)\delta(\mathbf{r} - \mathbf{r}'), \tag{45}$$

which reduces (44) to the Schrödinger wave equation (1.6):

$$i\hbar \frac{d}{dt}\psi(\mathbf{r}, t) = \mathscr{H}\psi(\mathbf{r}, t).$$

Similarly, we can insert identity (39) before and after the operator in the expectation value (27) and obtain the earlier definition (1.4) in terms of the wave function.

In general, the bases for the expansion of state vectors have infinite dimensionality. However, we have found that in several cases the expansion coefficients for all but two basis vectors vanished, leaving a two-level system. We now use this simple case to illustrate the formalism of Secs. 6-2 and 6-3. The state vector is written as

$$|\psi(t)\rangle = C_a \exp(-i\omega_a t)|a\rangle + C_b \exp(-i\omega_b t)|b\rangle, \tag{46}$$

which corresponds to the wave function

$$\psi(\mathbf{r}, t) = \langle \mathbf{r}|\psi(t)\rangle = C_a \exp(-i\omega_a t)u_a(\mathbf{r}) + C_b \exp(-i\omega_b t)u_b(\mathbf{r}). \tag{47}$$

The matrix forms for the unit vectors $|a\rangle$ and $|b\rangle$ are

$$|a\rangle \longleftrightarrow \begin{pmatrix} 1 \\ 0 \end{pmatrix}, \qquad |b\rangle \longleftrightarrow \begin{pmatrix} 0 \\ 1 \end{pmatrix}, \tag{48}$$

which give

$$\begin{aligned} |\psi(t)\rangle \longleftrightarrow C_a(t)\exp(-i\omega_a t)\begin{pmatrix} 1 \\ 0 \end{pmatrix} + C_b(t)\exp(-i\omega_b t)\begin{pmatrix} 0 \\ 1 \end{pmatrix} \\ = \begin{pmatrix} C_a(t)\exp(-i\omega_a t) \\ C_b(t)\exp(-i\omega_b t) \end{pmatrix}. \end{aligned} \tag{49}$$

The two-level Hamiltonian of Chap. 2 can be written as

$$\mathscr{H} = \hbar\omega_a|a\rangle\langle a| + \mathscr{V}_{ab}|a\rangle\langle b| + \mathscr{V}_{ba}|b\rangle\langle a| + \hbar\omega_b|b\rangle\langle b|, \tag{50}$$

which is represented by the matrix

$$\mathscr{H} = \begin{pmatrix} \mathscr{H}_{aa} & \mathscr{H}_{ab} \\ \mathscr{H}_{ba} & \mathscr{H}_{bb} \end{pmatrix} = \begin{pmatrix} \hbar\omega_a & \mathscr{V}_{ab} \\ \mathscr{V}_{ba} & \hbar\omega_b \end{pmatrix}. \tag{51}$$

Thus the matrix form of the Schrödinger equation is

$$i\hbar \frac{d}{dt}\begin{pmatrix} C_a \exp(-i\omega_a t) \\ C_b \exp(-i\omega_b t) \end{pmatrix} = \begin{pmatrix} \hbar\omega_a & \mathscr{V}_{ab} \\ \mathscr{V}_{ba} & \hbar\omega_b \end{pmatrix}\begin{pmatrix} C_a \exp(-i\omega_a t) \\ C_b \exp(-i\omega_b t) \end{pmatrix}. \tag{52}$$

The expectation value (27) is given by

$$\langle \mathcal{O} \rangle = C_a C_a^* \mathcal{O}_{aa} + C_b C_b^* \mathcal{O}_{bb} + \{C_a C_b^* \exp[-i(\omega_a - \omega_b)t] \mathcal{O}_{ba} + \text{c.c.}\} \tag{53}$$

The expectation value for which we will have the most use is that for the dipole moment operator $e\mathbf{r}$. This is most easily calculated in terms of the position states. Diagonal matrix elements of $e\mathbf{r}$ between eigenstates of the Hamiltonian generally vanish, whereas off-diagonal elements may not (see Prob. 1-2). In particular, for the two-level system we have been considering (46), the matrix elements are as follows:

$$\begin{aligned} e\mathbf{r}_{aa} &= \langle a | e\mathbf{r} | a \rangle = e \int d^3r\, u_a^*(\mathbf{r})\, \mathbf{r}\, u_a(\mathbf{r}) = 0, \\ e\mathbf{r}_{ab} &= e\mathbf{r}_{ba}^* = \langle a | e\mathbf{r} | b \rangle = e \int d^3r\, u_a^*(\mathbf{r})\, \mathbf{r}\, u_b(\mathbf{r}), \\ e\mathbf{r}_{bb} &= \langle b | e\mathbf{r} | b \rangle = e \int d^3r\, u_b^*(\mathbf{r})\, \mathbf{r}\, u_b(\mathbf{r}) = 0, \end{aligned} \tag{54}$$

which can be written in matrix form as

$$e\mathbf{r} = e \begin{pmatrix} 0 & \mathbf{r}_{ab} \\ \mathbf{r}_{ba} & 0 \end{pmatrix}. \tag{55}$$

The expectation value for $e\mathbf{r}$ in state (46) is

$$\begin{aligned} \langle e\mathbf{r} \rangle &= \langle \psi | e\mathbf{r} | \psi \rangle \\ &= e C_a C_b^* \exp[-i(\omega_a - \omega_b)\, t]\, \mathbf{r}_{ba} + \text{c.c.}, \end{aligned} \tag{56}$$

which is the same as (3.5) with $\mathbf{r}_{ba} = \mathbf{r}_{ab} \to \wp$.

6-4. The Schrödinger, Interaction, and Heisenberg Pictures

In laser physics the interaction of radiation with matter typically involves a Hamiltonian $\mathcal{H}$, given by the sum of an "unperturbed" term $\mathcal{H}_0$ and an interaction energy $\mathcal{V}$, that is,

$$\mathcal{H} = \mathcal{H}_0 + \mathcal{V}. \tag{57}$$

The corresponding Schrödinger equation (26) is

$$|\dot{\psi}(t)\rangle = -\frac{i}{\hbar} \mathcal{H} |\psi(t)\rangle = -\frac{i}{\hbar} (\mathcal{H}_0 + \mathcal{V}) |\psi(t)\rangle. \tag{58}$$

This equation can be integrated formally to give

$$|\psi(t)\rangle = \exp(-i\mathcal{H}\, t/\hbar) |\psi(0)\rangle. \tag{59}$$

In any problem we are interested ultimately in the expectation values of certain operators which represent observables. Such a value for the general operator $\mathcal{O}$ is ordinarily a function of time:

$$\langle\mathcal{O}\rangle = \langle\mathcal{O}\rangle(t) = \langle\psi(t)|\mathcal{O}(0)|\psi(t)\rangle \tag{60}$$

since $|\psi(t)\rangle$ is a function of time. Here we have written $\mathcal{O}(0)$ to indicate that the operator $\mathcal{O}$ itself has not changed in the time t. Equation (60) is the Schrödinger picture way of writing the expectation value.

We can decompose (60) in two other particularly useful ways, the interaction and Heisenberg picture forms. The first of these assigns only the time dependence created by the interaction energy to the state vector. For this, Eq. (60) reads

$$\begin{aligned}\langle\mathcal{O}\rangle &= \langle\psi(0)\exp(i\mathscr{V}t/\hbar)|\exp(i\mathscr{H}_0 t/\hbar)\mathcal{O}(0)\exp(-i\mathscr{H}_0 t/\hbar)|\exp(-i\mathscr{V}t/\hbar)\psi(0)\rangle \\ &= \langle\psi^I(t)|\mathcal{O}^I(t)|\psi^I(t)\rangle, \end{aligned}\tag{61}$$

in which the interaction picture state vector $|\psi^I(t)\rangle$ is given by

$$|\psi^I(t)\rangle = \exp(-i\mathscr{V}t/\hbar)|\psi(0)\rangle. \tag{62}$$

By differentiating (62) with respect to time, we see that $|\psi^I(t)\rangle$ obeys the equation of motion

$$|\dot{\psi}^I(t)\rangle = -\frac{i}{\hbar}\mathscr{V}|\psi^I(t)\rangle, \tag{63}$$

which is somewhat simpler than (58) but requires the additional (although usually straightforward) calculation of $\mathcal{O}^I(t)$.

The Heisenberg picture way of writing $\langle\mathcal{O}\rangle$ of (60) is to use $|\psi(0)\rangle$ and place the complete time dependence in (59) in the operator, that is,

$$\begin{aligned}\langle\mathcal{O}\rangle &= \langle\psi(0)|\exp(i\mathscr{H}t/\hbar)\ \mathcal{O}(0)\exp(-i\mathscr{H}t/\hbar)|\psi(0)\rangle \\ &= \langle\psi(0)|\mathcal{O}(t)|\psi(0)\rangle.\end{aligned}\tag{64}$$

The Heisenberg operator $\mathcal{O}(t)$, defined by

$$\mathcal{O}(t) = \exp(i\mathscr{H}t/\hbar)\ \mathcal{O}(0)\exp(-i\mathscr{H}t/\hbar), \tag{65}$$

obeys the equation of motion [by direct differentiation of (65)]

$$\frac{d}{dt}\mathcal{O}(t) = \frac{i}{\hbar}[\mathscr{H}\mathcal{O} - \mathcal{O}\mathscr{H}] = \frac{i}{\hbar}[\mathscr{H}, \mathcal{O}]. \tag{66}$$

The second bracketed expression in (66) is a short-hand notation for the first and is called the "commutator" of $\mathscr{H}$ and $\mathcal{O}$. Considerable use of the Heisenberg picture is made in Chaps. 19 and 20 on the Langevin approach to system-reservoir interaction and laser operation. The interaction picture is employed extensively in Chaps. 14–18. The Schrödinger picture is used in Chaps. 8–13 on the semiclassical approach, in which the presence of decay terms [as in Eqs. (2.46) and (2.47)] nullifies much of the simplification otherwise gained in the interaction picture.

We conclude this section with a discussion of the Schrödinger and interaction picture probability amplitudes. As explained in the introduction, the $C_n(t)$ are the interaction picture probability amplitudes with Schrödinger picture counterparts $c_n(t)$, defined by

$$c_n(t) = \exp(-i\omega_n t)C_n(t). \tag{67}$$

Here the complete time dependence of the system is described by the $c_n(t)$, while only that due to the interaction energy is given by $C_n(t)$. The equation of motion for the $C_n(t)$ is (2.3). It is easy to show (Prob. 6-2) that the corresponding equation for the $c_n(t)$ is

$$\dot{c}_n = -i\omega_n c_n - \frac{i}{\hbar}\sum_k \langle n|\mathscr{V}|k\rangle c_k. \tag{68}$$

The interaction picture state vector (62) has the decomposition

$$|\psi^I(t)\rangle = \sum_n C_n(t)|n\rangle. \tag{69}$$

The Schrödinger picture state vector is written as

$$|\psi(t)\rangle = \sum_n c_n(t)|n\rangle = \sum_n C_n(t)\exp(-i\omega_n t)|n\rangle. \tag{70}$$

In particular, we have the two-level Schrödinger picture probability amplitudes

$$c_a(t) = C_a(t)\ \exp(-i\omega_a t), \tag{71}$$

$$c_b(t) = C_b(t)\ \exp(-i\omega_b t). \tag{72}$$

Problems

6-1. What is the expectation value of the projection operator $|k\rangle\langle k|$ for the state vector

$$|a\rangle = \sum_n \exp(-\tfrac{1}{2}|a|^2)\frac{a^n}{\sqrt{n!}}|n\rangle? \tag{73}$$

This is the coherent state of Chap. 15. It provides the link between the quantum theory of radiation and the classical electromagnetic field.

6-2. Show that the equation of motion for $c_n(t)$ is Eq. (68).

6-3. Show, for the unperturbed Hamiltonian $\mathscr{H}_0$ given by

$$\mathscr{H}_0 = \begin{pmatrix} \hbar\omega_a & 0 \\ 0 & \hbar\omega_b \end{pmatrix}, \tag{74}$$

that

$$\exp(i\mathscr{H}_0 t/\hbar) = \begin{pmatrix} \exp(i\omega_a t) & 0 \\ 0 & \exp(i\omega_b t) \end{pmatrix}. \tag{75}$$

6-4. Using (75), show that the interaction energy

$$\mathscr{V} = \begin{pmatrix} 0 & \mathscr{V}_{ab} \\ \mathscr{V}_{ba} & 0 \end{pmatrix} \tag{76}$$

has the interaction picture version (61)

$$\begin{aligned} \mathscr{V}^I &= \exp(i\mathscr{H}_0 t/\hbar)\,\mathscr{V}\exp(-i\mathscr{H}_0 t/\hbar) \\ &= \begin{pmatrix} 0 & \mathscr{V}_{ab}\exp(i\omega t) \\ \mathscr{V}_{ba}\exp(-i\omega t) & 0 \end{pmatrix}. \end{aligned} \tag{77}$$

6-5. Write the Heisenberg equation of motion (66) for the spin-flip operator σ defined by (1.41), given the two-level Hamiltonian

$$\mathscr{H} = \begin{pmatrix} \hbar\omega/2 & ga \\ ga^\dagger & -\hbar\omega/2 \end{pmatrix}. \tag{78}$$

6-6. A useful way of writing the time dependence of the state vector $|\psi(t)\rangle$ is by means of the "U matrix," which satisfies the Schrödinger equation

$$i\hbar\frac{d}{dt}U(t, t_0) = \mathscr{H}U(t, t_0) \tag{79}$$

with the initial condition

$$U(t_0, t_0) = 1. \tag{80}$$

Show that

$$U(t, t_0) = \exp[-i\mathscr{H}(t - t_0)/\hbar] \tag{81}$$

and hence that

$$|\psi(t)\rangle = U(t, t_0)|\psi(t_0)\rangle, \tag{82}$$

$$U(t, t_0) = U(t, t_1)\,U(t_1, t_0). \tag{83}$$

6-7. Compute the root-mean-square deviation (31) for the electric-dipole moment operator (55) with the state vector (46).

References

In addition to the references of Chap. 1, see the following:

P. A. M. Dirac, 1958, *The Principles of Quantum Mechanics,* 4th ed., Clarendon Press, Oxford.

W. H. Louisell, 1964, *Radiation and Noise in Quantum Electronics,* McGraw-Hill Book Co., New York.

A. Messiah, 1961, *Quantum Mechanics,* North-Holland Publishing Co., Amsterdam.

VII

THE DENSITY MATRIX

7. The Density Matrix

In Chap. 2 we solved for the probability amplitudes C_a and C_b explicitly. The results of our discussion, such as the probability of a transition or the value of the induced dipole moment, were invariably expressed in terms of bilinear combinations of the amplitudes, such as $C_aC_a^*$ and $C_aC_b^*$. In fact, the expectation value of any observable involves bilinear combinations. Hence an alternative formulation of quantum mechanics consists of dealing with the bilinear quantities directly. The method involves organizing the quantities in a matrix form called the density matrix. There are two principal advantages in doing so. First, the resulting mathematics is often simpler. Second, when the wave function for an ensemble of systems is not known but the probabilities for having various different wave functions are known, the ensemble can be described by a weighted sum of individual density matrices. In particular, the amplitudes cannot themselves describe simply certain common statistical phenomena, such as the effects of elastic collisions on the induced dipole moments of atoms.

In Sec. 7-1 we introduce the density matrix with its equation of motion for the two-level systems that we have been considering. The discussion is important for much of our later work and lends itself naturally to the generalization in Sec. 7-2. This second section defines the density matrix for a mixture, that is, an ensemble of systems whose complete wave function is not known. The general formalism is illustrated with application to simple cases, such as a mixture of atoms having various excitation times and filtered thermal radiation. In Sec. 7-3 the density matrix equations of motion are modified phenomenologically to account for decay due to spontaneous emission and collisions. We show an additional decay of the off-diagonal term ρ_{ab} resulting from collision phenomena which do not affect c_a or c_b individually. In Sec. 7-4 perturbational solutions of the equations of motion are derived. In Sec. 7-5 a pictorial representation of the density matrix along the lines of the Bloch equations is given. This chapter concludes our presentation of quantum-mechanical preliminaries. Additional formalism is developed when required.

7-1. The Pure-Case Density Matrix

A quantum-mechanical system for which the wave function is known is called a pure case. A system for which less information is available is a mixed case and is discussed in the next section. In most of this chapter and in the semiclassical laser theory (Chaps. 8–13) we use the Schrödinger picture (see Sec. 6-4). Hence for the two-level system we write the wave function with Schrödinger picture amplitudes c_a and c_b, defined by (6.71) and (6.72). We have

$$\psi(\mathbf{r}, t) = c_a(t)u_a(\mathbf{r}) + c_b(t)u_b(\mathbf{r}). \tag{1}$$

The equations of motion for the c_a and c_b are given by (6.68) as

$$\dot{c}_a = -i\omega_a c_a - \frac{i}{\hbar} \mathscr{V}_{ab} c_b, \tag{2}$$

$$\dot{c}_b = -i\omega_b c_b - \frac{i}{\hbar} \mathscr{V}_{ba} c_a. \tag{3}$$

The density matrix corresponding to this wave function is defined by the billinear products

$$\begin{aligned} \rho_{aa} &\equiv c_a c_a^*, \text{ probability of being in upper level,} \\ \rho_{ab} &\equiv c_a c_b^*, \text{ proportional to the complex dipole moment,†} \\ \rho_{ba} &\equiv c_b c_a^* = \rho_{ab}^*, \\ \rho_{bb} &\equiv c_b c_b^*, \text{ probability of being in lower level;} \end{aligned} \tag{4}$$

or in matrix notation

$$\rho = \begin{pmatrix} c_a c_a^* & c_a c_b^* \\ c_b c_a^* & c_b c_b^* \end{pmatrix} = \begin{pmatrix} \rho_{aa} & \rho_{ab} \\ \rho_{ba} & \rho_{bb} \end{pmatrix}.$$

This matrix can also be written as the outer product

$$\rho = \psi\psi^* = \begin{pmatrix} c_a \\ c_b \end{pmatrix} (c_a^* \quad c_b^*) = \begin{pmatrix} c_a c_a^* & c_a c_b^* \\ c_b c_a^* & c_b c_b^* \end{pmatrix}. \tag{5}$$

In terms of the density matrix, the expectation value (6.60) of an operator $\mathscr{O}$ is given by

$$\langle \mathscr{O} \rangle = (\rho_{aa}\mathscr{O}_{aa} + \rho_{ab}\mathscr{O}_{ba}) + (\rho_{ba}\mathscr{O}_{ab} + \rho_{bb}\mathscr{O}_{bb}). \tag{6}$$

† Provided an electric-dipole transition is allowed between u_a and u_b.

In particular the dipole moment is given in the u_a, u_b basis by

$$\langle er \rangle = \wp(\rho_{ab} + \rho_{ba}). \tag{7}$$

Equation (6) is just the trace of the matrix product $\rho\mathcal{O}$:

$$\langle \mathcal{O} \rangle = \sum_i \Big[\sum_j \rho_{ij}\,\mathcal{O}_{ji}\Big] = \sum_i (\rho\mathcal{O})_{ii} = \mathrm{Tr}(\rho\mathcal{O}). \tag{8}$$

We show in the following section that the expectation value of an operator is given by Eq. (8) even when the system is described by the most general density matrix.

We can derive the equations of motion for the elements of the density matrix from those [Eqs. (2) and (3)] for the probability amplitudes. Proceeding one element at a time, we have

$$\begin{aligned}\dot\rho_{aa} &= \dot c_a c_a{}^* + c_a \dot c_a{}^* \\ &= (-i\omega_a c_a - \frac{i}{\hbar}\mathscr{V}_{ab} c_b) c_a{}^* + c_a(i\omega_a c_a{}^* + \frac{i}{\hbar}\mathscr{V}_{ba} c_b{}^*) \\ &= -\frac{i}{\hbar}\mathscr{V}_{ab}\rho_{ba} + \text{c.c.}\end{aligned} \tag{9}$$

It is not surprising to find the complex conjugate in this equation, for probabilities are real. Similarly, we find

$$\dot\rho_{bb} = \frac{i}{\hbar}\mathscr{V}_{ab}\rho_{ba} + \text{c.c.} \tag{10}$$

This value is equal in magnitude and opposite in sign to that in (9) for ρ_{aa}, which expresses the fact that, when the system loses probability for being in the upper level, it gains probability for being in the lower. Mathematically, the result follows from the normalization condition, $\rho_{aa} + \rho_{bb} = 1$.

The off-diagonal element ρ_{ab} obeys the equation of motion

$$\begin{aligned}\dot\rho_{ab} &= \dot c_a c_b{}^* + c_a \dot c_b{}^* \\ &= (-i\omega_a c_a - \frac{i}{\hbar}\mathscr{V}_{ab} c_b)\, c_b{}^* + c_a\,(i\omega_b c_b{}^* + \frac{i}{\hbar}\mathscr{V}_{ba}{}^* c_a{}^*) \\ &= -i\omega\rho_{ab} + \frac{i}{\hbar}\mathscr{V}_{ab}(\rho_{aa} - \rho_{bb}),\end{aligned} \tag{11}$$

where the frequency difference $\omega = \omega_a - \omega_b$. The term $-i\omega\rho_{ab}$ results from similar frequency terms in (2) and (3) and leads to a rapid time variation at frequency ω. The term is missing in the interaction picture (see Prob. 7-4). The equation for ρ_{ba} follows from the relation

$$\rho_{ba} = \rho_{ab}{}^*. \tag{12}$$

In particular, if the interaction Hamiltonian is the electric-dipole energy considered earlier (2.16), the equations of motion (9), (10), and (11) become in the rotating-wave approximation

$$\dot{\rho}_{aa} = \frac{i}{2\hbar} \wp E_0 \exp(-i\nu t)\, \rho_{ba} + \text{c.c.}, \tag{13}$$

$$\dot{\rho}_{ab} = -i\omega\rho_{ab} - \frac{i}{2\hbar} \wp E_0 \exp(-i\nu t)\, (\rho_{aa} - \rho_{bb}), \tag{14}$$

$$\dot{\rho}_{bb} = -\frac{i}{2\hbar} \wp E_0 \exp(-i\nu t)\, \rho_{ba} + \text{c.c.} \tag{15}$$

We close this section by noting that the density matrix (5) is a particular representation of the density operator

$$\rho = |\psi\rangle\langle\psi|. \tag{16}$$

For example, $\rho_{ab} = \langle a|\rho|b\rangle = \langle a|\psi\rangle\langle\psi|b\rangle = c_a c_b{}^*$. This equation is the general definition of the density operator for a pure case with state vector $|\psi\rangle$.

7-2. The Mixed-Case Density Matrix

We have seen how the density matrix can be used in place of the state vector in formulating the quantum mechanics of a single system. The matrix approach has an advantage in lending itself to a simple pictorial representation, as discussed in Sec. 7-5. However, the primary reason for using the density matrix is that it facilitates the treatment of interacting quantum systems. In fact, most often we do not even know the state vector for a many-particle problem. Instead we know only certain statistical properties. These properties are conveniently incorporated into the density matrix formalism by definition of the general density operator:

$$\rho = \sum_{\psi} P_\psi |\psi\rangle\, \langle\psi|. \tag{17}$$

Here the summation can be discrete or continuous, and P_ψ is that fraction of the systems which has the state vector $|\psi\rangle$. The summation can take the form of several summations and integrals.

For state vectors of the form

$$|\psi\rangle = \sum_n c_n |n\rangle, \tag{18}$$

Eq. (17) reduces to

$$\rho = \sum_{\psi} P_\psi \sum_n\sum_m c_n c_m{}^* |n\rangle\, \langle m| = \sum_n\sum_m \rho_{nm} |n\rangle\langle m|. \tag{19}$$

To illustrate (17), let us consider two-level state vectors defined by

$$|\psi_j(t)\rangle = c_{aj}(t)|a\rangle + c_{bj}(t)|b\rangle. \tag{20}$$

For these state vectors, ρ of (17) reads as

$$\rho = \sum_j P_j |\psi_j\rangle\langle\psi_j| \tag{21}$$

with the matrix representation

$$\rho = \sum_j P_j \begin{pmatrix} |c_{aj}|^2 & c_{aj}c_{bj}^* \\ c_{bj}c_{aj}^* & |c_{bj}|^2 \end{pmatrix}. \tag{22}$$

In particular, suppose that all the state vectors are identical except for a uniform distribution of phases ϕ_j between the c_a and c_b amplitudes, that is, suppose that

$$|\psi_j\rangle = c_a|a\rangle + \exp(i\phi_j)\, c_b|b\rangle \tag{23}$$

with the probability factors $P_j = 1/N$ for N systems. Then the off-diagonal element ρ_{ab} becomes

$$\begin{aligned}\rho_{ab} &= \sum_j P_j c_{aj} c_{bj}^* \\ &= c_a c_b^* \frac{1}{N} \sum_{j=1}^{N} \exp(-i\phi_j) \\ &= c_a c_b^* \frac{1}{2\pi} \int_0^{2\pi} d\phi \exp(-i\phi) \\ &= 0,\end{aligned} \tag{24}$$

that is, with mixtures, the coherence between upper and lower levels can cancel out. In the next section, we see this effect in the more rapid decay of ρ_{ab} relative to the diagonal elements.

Thermal radiation in a single mode (monochromatic) can also be simply described by Eq. (17), in which the individual states are photon number states $|n\rangle$ [formally just harmonic oscillator states corresponding to (1.23) and discussed in Chap. 14], and the probabilities P_ψ are given by the Boltzmann distribution:

$$P_\psi = P_n = \exp(-n\hbar\omega/k_B T)\,[1 - \exp(-\hbar\omega/k_B T)]. \tag{25}$$

The density operator for this is

$$\rho = [1 - \exp(-\hbar\omega/k_B T)] \sum_n \exp(-n\hbar\omega/k_B T)|n\rangle\langle n|. \tag{26}$$

Here too the off-diagonal elements vanish.

The expectation value of an operator $\mathcal{O}$ is still given by

$$\langle \mathcal{O} \rangle = \mathrm{Tr}\,(\rho\mathcal{O}), \tag{27}$$

for, in terms of the P_ψ,

$$\begin{aligned}\langle \mathcal{O} \rangle &= \sum_\psi P_\psi \langle\psi|\mathcal{O}|\psi\rangle \\ &= \sum_\psi P_\psi \sum_k \langle\psi|\mathcal{O}|k\rangle\langle k|\psi\rangle \\ &= \sum_k \sum_\psi P_\psi \langle k|\psi\rangle\langle\psi|\mathcal{O}|k\rangle\end{aligned}$$

$$= \sum_k (\rho\mathcal{O})_{kk}$$

$$= \mathrm{Tr}(\rho\mathcal{O}). \tag{28}$$

Note that the basis $|k\rangle$ need only be complete; we could equally well expand the trace in position states

$$\langle\mathcal{O}\rangle = \mathrm{Tr}\,(\rho\mathcal{O}) = \int d^3r\,\langle\mathbf{r}|\rho\mathcal{O}|\mathbf{r}\rangle.$$

The equation of motion of the density operator (and hence that of its matrix elements) is easily determined from Schrödinger's equation

$$\dot\rho = \sum_\psi P_\psi\{|\dot\psi\rangle\langle\psi| + |\psi\rangle\langle\dot\psi|\} = -\frac{i}{\hbar}\sum_\psi P_\psi\{\mathcal{H}|\psi\rangle\langle\psi| - |\psi\rangle\langle\psi|\mathcal{H}\}$$

$$= -\frac{i}{\hbar}[\mathcal{H}, \rho], \tag{29}$$

where the commutator

$$[\mathcal{H}, \rho] = \mathcal{H}\rho - \rho\mathcal{H}. \tag{30}$$

The final step in Eq. (29) follows only if the Hamiltonian $\mathcal{H}$ is the same for all $|\psi\rangle$. In our laser discussions (e.g., Chap. 8), we add contributions from atoms excited at different times, places, and states. Inasmuch as all these systems have the same equations of motion, we can use (29) for a density matrix (17) which includes summations over the times, places, and states of excitation. By dealing with this sum of density matrices directly, we save considerable effort over both the single-system density matrix approach and the state vector method. Summations of statistical averages for systems with different equations of motion must be performed after the equations have been integrated. Note that the ijth matrix element of (29) is

$$\dot\rho_{ij} = -\frac{i}{\hbar}\langle i|\mathcal{H}\rho - \rho\mathcal{H}|j\rangle$$

$$= -\frac{i}{\hbar}\sum_k \{\langle i|\mathcal{H}|k\rangle\langle k|\rho|j\rangle - \langle i|\rho|k\rangle\langle k|\mathcal{H}|j\rangle\}$$

$$= -\frac{i}{\hbar}\sum_k \{\mathcal{H}_{ik}\rho_{kj} - \rho_{ik}\mathcal{H}_{kj}\}. \tag{31}$$

This formula is useful in the treatment of many-level problems.

7-3. Decay Phenomena

In Sec. 2-3 we included phenomenological decay constants γ_a and γ_b to account for the effects of finite atomic lifetimes. These effects are negligible for the ammonia maser because the lifetimes of the molecules are long compared to the reciprocal of the Rabi flopping frequency and the cavity transit time. In that maser, actually, the transit time itself plays the role of a lifetime. At optical frequencies, however, spontaneous emission, inelastic collisions (gaseous media), and other mechanisms force excited atoms to decay much more rapidly and often to modify the atomic response to radiation in important ways. In fact, most lasers depend on decay from the lower laser level to prevent atoms that have given up energy from subsequently absorbing radiation, much as the ammonia beam maser requires removal of lower-level molecules from the cavity. In this section, we modify the density matrix equations of motion to include lifetime effects for both the levels themselves and for the off-diagonal element ρ_{ab}, which represents coherence between the levels and yields the electric dipole.

We include decay in Eqs. (2) and (3) as in (2.46) and (2.47), obtaining

$$\dot{c}_a = -(i\omega_a + \tfrac{1}{2}\gamma_a)c_a - \frac{i}{\hbar}\,\mathscr{V}_{ab}c_b, \tag{32}$$

$$\dot{c}_b = -(i\omega_b + \tfrac{1}{2}\gamma_b)c_b - \frac{i}{\hbar}\,\mathscr{V}_{ba}c_a, \tag{33}$$

where γ_a and γ_b are the decay constants of the probabilities $|c_a|^2$ and $|c_b|^2$ (see Fig. 2-6). The component equations of motion for the density matrix can then be determined from (32) and (33) as derived without decay rates in Sec. 7-1. We find

$$\dot{\rho}_{aa} = -\gamma_a\rho_{aa} - \frac{i}{\hbar}[\mathscr{V}_{ab}\rho_{ba} - \text{c.c.}], \tag{34}$$

$$\dot{\rho}_{bb} = -\gamma_b\rho_{bb} + i\hbar^{-1}[\mathscr{V}_{ab}\rho_{ba} - \text{c.c.}], \tag{35}$$

$$\dot{\rho}_{ab} = -(i\omega + \gamma_{ab})\rho_{ab} + i\hbar^{-1}\mathscr{V}_{ab}(\rho_{aa} - \rho_{bb}), \tag{36}$$

where

$$\gamma_{ab} = \tfrac{1}{2}(\gamma_a + \gamma_b). \tag{37}$$

These equations of motion can be written in the single equation

$$\dot{\rho} = -\tfrac{1}{2}(\Gamma\rho + \rho\Gamma) - i\hbar^{-1}[\mathscr{H}, \rho], \tag{38}$$

where in general the decay operator Γ has the matrix representation

$$\Gamma_{ij} = \gamma_i\delta_{ij}. \tag{39}$$

Elastic collisions between atoms in a gas or between phonons and atoms in a solid can cause ρ_{ab} to decay separately from the diagonal elements. Specifically, if during an interaction the energy levels are merely shifted slightly without a change of state (e.g., distant van der Waals interaction), the decay rate for ρ_{ab} is increased without much change in γ_a and γ_b. This is due to the fact that the phase of the radiating atomic dipole is shifted in a somewhat random fashion, and the contributions of a collection of such dipoles tend to average to zero. We can gain semiquantitative understanding of this process by considering the following discussion, couched in terms of phonon interactions in ruby.†

The active atom in ruby is the Cr^{3+} ion, which is surrounded with O^{2-} atoms. At room temperatures, all atoms are vibrating, with the result that the energy levels in the Cr^{3+} ions experience random Stark shifts (see Prob. 5-2). For simplicity we assume that this phenomenon can be expressed mathematically by adding a random shift $\delta\omega(t)$ to the energy difference ω. Ignoring other perturbations for simplicity, we can write the equation of motion for the off-diagonal element ρ_{ab} as

$$\dot{\rho}_{ab} = -[i\omega + i\,\delta\omega(t) + \gamma_{ab}]\,\rho_{ab}. \tag{40}$$

Integrating (40) formally, we have

$$\rho_{ab}(t) = \rho_{ab}(0)\exp\left[-(i\omega + \gamma_{ab})\,t - i\int_0^t dt'\,\delta\omega(t')\right]. \tag{41}$$

We now perform an ensemble average of (41) over the random variations in $\delta\omega(t)$. This average affects only the $\delta\omega(t)$ factor. Expanding the second part of the exponential term by term, we have

$$\left\langle \exp\left[-i\int_0^t dt'\,\delta\omega(t')\right]\right\rangle$$

$$= \langle 1 - i\int_0^t dt'\,\delta\omega(t') - \frac{1}{2}\int_0^t dt'\int_0^t dt''\,\delta\omega(t')\,\delta\omega(t'') + \cdots$$

$$+ \frac{(-i)^{2n}}{(2n)!}\int_0^t dt_1 \ldots \int_0^t dt_{2n}\,\delta\omega(t_1)\ldots\delta\omega(t_{2n}) + \cdots. \tag{42}$$

The function $\delta\omega(t)$ is as often positive as negative, as suggested in Fig. 7-1. Hence the ensemble average $\langle\delta\omega(t)\rangle$ is zero.† Furthermore, averages of products $\langle\delta\omega(t)\,\delta\omega(t')\rangle$ are zero as well, unless $t \simeq t'$, in which case the product is mostly positive $[(-1)^2 = 1]$. Assuming that variations in $\delta\omega(t)$ are rapid compared to other changes (which occur in times like $1/\gamma_{ab}$), we take

$$\langle\delta\omega(t)\,\delta\omega(t')\rangle = 2\gamma_{ph}\,\delta(t - t'). \tag{43}$$

This is called the Markoff approximation and is discussed further in Chap. 19. Similarly (for Gaussian statistics) the $2n$th correlation $\langle\delta\omega(t_1)\ldots\delta\omega(t_{2n})\rangle$ is given by the sum of all distinguishable products of pairs like (43), for only when the random functions coincide in pairs is the entire product positive with nonvanishing ensemble average. The number of combinations of $2n$ terms in pairs is given by $\binom{2n}{2}$, that for the remaining $2n - 2$ terms is $\binom{2n-2}{2}$, and

† For more exact treatments of collision phenomena, see Berman (1972) and references cited therein.

† Actually in some problems (collisions in gases) a frequency shift as well as damping can occur. This shift modifies ω just as γ_{ph} changes γ.

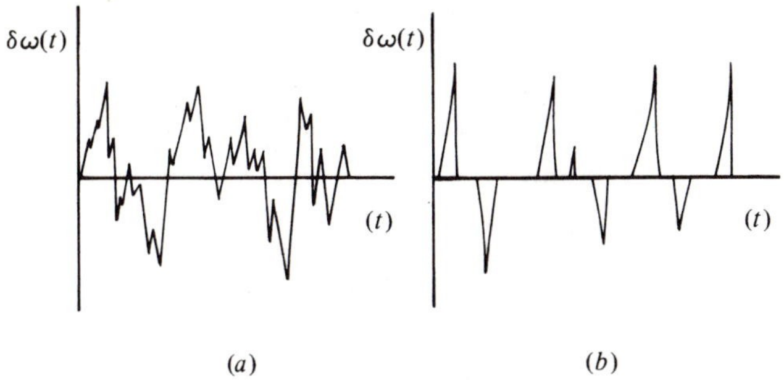

Figure 7-1. (a) Possible time dependence of random frequency shift $\delta\omega(t')$ imposed upon frequency difference ω for Cr^{3+} in Al_2O_3 lattice. This shift could also occur for atoms in gaseous state due to "soft" (i.e., elastic) collisions. Because $\delta\omega(t)$ is as often positive as negative, the integral $\int_0^t dt' \langle\delta\omega(t')\rangle$ vanishes. (b) Possible dependence for collision fluctuations. Same random characteristics hold in our simple model.

so forth. Hence the total number of ways of breaking $2n$ terms into products of n pairs is

$$\binom{2n}{2}\binom{2n-2}{2}\cdots\binom{2}{2} = \frac{(2n)!}{2^n}. \tag{44}$$

Since permutation of the n pairs in a given product does not lead to distinguishable combinations, we divide (44) by $n!$ This gives for the $2n$th term in (42)

$$\frac{(-1)^n}{(2n)!}(2\gamma_{ph})^n\frac{(2n)!}{2^n n!}\int_0^t dt_1 \ldots \int_0^t dt_{2n}\, \delta(t_1 - t_2) \ldots \delta(t_{2n-1} - t_{2n})$$

$$= \frac{(-\gamma_{ph}t)^n}{n!}. \tag{45}$$

The $(2n + 1)$th term vanishes since it is given by the sum of products of n pairs multiplied by a lone random function with zero average. Therefore

$$\left\langle \exp\left[-i\int_0^t dt'\, \delta\omega(t')\right]\right\rangle = \sum_{n=0}^{\infty}\frac{(-\gamma_{ph}t)^n}{n!} = \exp(-\gamma_{ph}t), \tag{46}$$

which gives for the average of (41)

$$\rho_{ab}(t) = \exp[-(i\omega + \gamma_{ab} + \gamma_{ph})t]\, \rho_{ab}(0). \tag{47}$$

Defining a new decay rate

$$\gamma = \gamma_{ab} + \gamma_{ph}, \tag{48}$$

differentiating (47), and including $\mathscr{V}_{ab}$, we have the modified equation of motion:

$$\dot{\rho}_{ab} = -(i\omega + \gamma)\rho_{ab} + i\hbar^{-1}\mathscr{V}_{ab}(\rho_{aa} - \rho_{bb}). \tag{49}$$

We use this in place of (36) in our laser calculations. Equation (49) is an average equation with respect to collisions, whereas (40) included fluctuations due to collisions. Our later treatment (Appendix I) of spontaneous emission reveals that the semiclassical equations (34)–(36), too, are average equations, in which *vacuum* fluctuations have been averaged over.

It is interesting to note that at low temperatures the Cr^{3+} ions "freeze" irregularly into place in the crystal lattice and the decay rate γ becomes still smaller. Different ions are subject to different Stark shifts and hence have different resonant frequencies. Hence the atomic medium as a whole responds to a range of frequencies considerably larger than that for a single atom, as suggested in Fig. 7-2. This kind of response is called "inhomogeneously broadened," as contrasted to the kind at higher temperatures (termed "homogeneously broadened," inasmuch as every active atom has essentially the same resonance frequency and linewidth). The two kinds of broadening overlap to some degree in a real medium, but one kind is often responsible for most of the linewidth. An intermediate situation is met when the temperature of ruby is somewhere between that of liquid helium and room temperature, and the linewidth is due to approximately equal homogeneous and inhomogeneous interactions. We emphasize right away that the two sources of broadening are very different physically, the inhomogeneous one being a dynamical influence which can be effectively reversed (e.g., photon echo of Chap. 13), whereas the

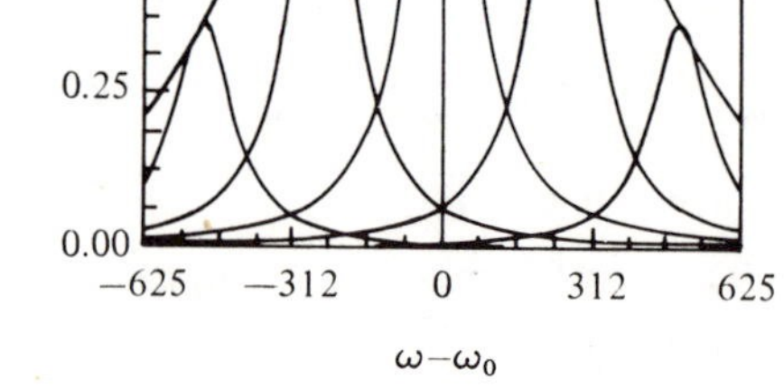

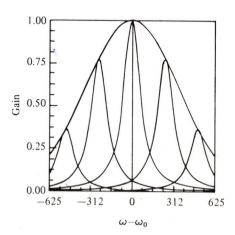

Figure 7-2. Graph showing individual atomic response curves superimposed on inhomogeneously broadened line for possible laser medium. The homogeneous contribution to the medium line width is 150 MHz; the inhomogeneous contribution, 1000 MHz. Hence the medium is primarily inhomogeneously broadened.

homogeneous is an irreversible influence. In Chap. 10 we see in greater detail just how differently the two sources are represented mathematically.

A transition similar to that occurring in ruby for temperature changes takes place in gaseous media for pressure changes. Although gas atoms such as neon generally have the same resonance frequencies in their rest frames, they see Doppler-shifted electric fields (as discussed in Chap. 10) and hence respond as a group to a range of frequencies, that is, are inhomogeneously broadened. At low pressures, the linewidth of a single atom is almost completely due to spontaneous emission and is usually small compared to average Doppler shifts. As the pressure is increased, however, collisions broaden the atomic response homogeneously and ultimately mask out the Doppler effect altogether. We see in Chap. 10 on gas lasers and in Chap. 13 on pulse propagation the degree to which atomic response is inhomogeneously broadened can be determined experimentally by measuring, for example, intensity versus detuning profiles in gases and photon echoes in general. Note that although the considerations discussed here apply qualitatively to many laser media, interesting counterexamples do exist, e.g., Dicke narrowing in gases (Dicke, 1953) and Doppler effects in solids (Lamb, 1939).

7-4. Integration of the Equations of Motion

Most of our use of the density matrix depends upon integration of its equations of motion. We will see various methods for doing this in subsequent chapters. One method, however, is of sufficiently general importance that we include it here. This is the perturbational method used for the probability amplitudes in Chap. 2. For simplicity, we consider the undamped equations of motion (9)–(11) and defer the damped case to Prob. 7-7. Our solution takes the form of an expansion in powers of the electric field interaction:

$$\rho(t) = \rho^{(0)}(t) + \rho^{(1)}(t) + \rho^{(2)}(t) + \cdots, \tag{50}$$

where $\rho^{(n)}(t)$ is proportional to $(\wp E_0/\hbar)^n$. We choose the initial condition $c_a(0) = 0$, $c_b(0) = 1$, which in density matrix notation is

$$\rho(t=0) = \begin{pmatrix} 0 & 0 \\ 0 & 1 \end{pmatrix}, \tag{51}$$

that is, $\rho_{aa}(0) = 0 = \rho_{ab}(0) = \rho_{ba}(0)$, $\rho_{bb}(0) = 1$. From the equations of motion, we see that the zeroth-order solution (no powers in the electric field interaction $\wp E_0$) is just the initial condition,

$$\rho^{(0)}(t) = \begin{pmatrix} 0 & 0 \\ 0 & 1 \end{pmatrix}. \tag{52}$$

Choosing the sinusoidally varying electric field in (2.16), we find no first order contribution for the diagonal elements ρ_{aa} and ρ_{bb} since $\rho_{ab}^{(0)}(t) = \rho_{ba}^{(0)}(t) = 0$. The off-diagonal elements have first-order contributions inasumch as $\rho_{bb}^{(0)}(0) = 1$. Multiplying both sides of (11) by the integrating factor $\exp(i\omega t)$, we find

$$\frac{d}{dt}[\rho_{ab}\exp(i\omega t)] = \frac{i}{\hbar}\mathscr{V}_{ab}\exp(i\omega t)(\rho_{aa} - \rho_{bb}). \tag{53}$$

Integrating this equation from time $t = 0$ to time t, using the probability difference

$$\rho_{aa} - \rho_{bb} \approx \rho_{aa}^{(0)} - \rho_{bb}^{(0)} = -1, \tag{54}$$

and the electric-dipole perturbation energy (2.16), we find

$$\rho_{ab}(t) \approx \rho_{ab}(0)\exp(-i\omega t) + \frac{i}{2\hbar}\wp E_0 \times \left\{\exp(-i\nu t)\int_0^t dt' \exp[-i(\omega - \nu)(t - t')] + \exp(i\nu t)\int_0^t dt' \exp[-i(\omega + \nu)(t - t')\right\}.$$

The first term is zero, and the third can be neglected in the rotating-wave approximation. The remaining integration is easily performed and yields

$$\rho_{ab}^{(1)}(t) = \frac{1}{2}\left(\frac{\wp E_0}{\hbar}\right)\exp(-i\nu t)(\omega - \nu)^{-1}\{1 - \exp[-i(\omega - \nu)t]\}. \tag{55}$$

This is just what Eqs. (2.22) and (2.27) give for $C_a C_b^* \exp(-i\omega t)$. Note that ρ_{ab} has a positive frequency dependence $[\exp(-i\nu t)]$, and its complex conjugate ρ_{ba} has a negative frequency dependence $[\exp(i\nu t)]$.

Substitution of (55) and its complex conjugate into the equations of motion for the diagonal elements gives the second-order transition probability:

$$\rho_{aa}(t) \approx \rho_{aa}^{(2)}(t) = \frac{1}{4}\left(\frac{\wp E_0}{\hbar}\right)^2 \frac{\sin^2[(\frac{1}{2}(\omega - \nu)t]}{[\frac{1}{2}(\omega - \nu)]^2}, \tag{56}$$

and

$$\rho_{bb}(t) = 1 - \rho_{aa}(t). \tag{57}$$

Note that (56) is just what Eq. (2.30) gives for $|C_a^{(1)}(t)|^2$.

When the density matrix is initially diagonal, as for (51), contributions to the diagonal elements occur in even orders, for example,

$$\rho_{aa}(t) = \rho_{aa}^{(0)} + \rho_{aa}^{(2)}(t) + \rho_{aa}^{(4)}(t) + \cdots, \tag{58}$$

and those to off-diagonal elements in odd orders:

$$\rho_{ab}(t) = \rho_{ab}^{(1)}(t) + \rho_{ab}^{(3)}(t) + \cdots. \tag{59}$$

This observation enables us to write the formal integrals

$$\rho_{ab}{}^{(2n+1)}(t) = -\frac{i}{2\hbar}\wp E_0 \exp(-i\nu t)\int_0^t dt' \exp[-i(\omega - \nu)(t - t')] \times [\rho_{aa}{}^{(2n)}(t') - \rho_{bb}{}^{(2n)}(t')], \tag{60}$$

$$\rho_{aa}{}^{(2n)}(t) = \frac{i}{2\hbar}\wp E_0 \int_0^t dt' [\exp(-i\nu t')\,\rho_{ba}{}^{(2n-1)} + \text{c.c.}]. \tag{61}$$

In writing (60) and (61), we used the rotating-wave approximation and capitalized on the time dependences of off-diagonal elements:

$$\rho_{ab}(t) = \rho_{ab}'(t)\exp(-i\nu t), \tag{62}$$

$$\rho_{ba}(t) = \rho_{ab}(t)^* = \rho_{ba}'(t)\exp(i\nu t), \tag{63}$$

where ρ_{ab}' and ρ_{ba}' are slowly varying compared to optical frequency variations. These relations follow from (53).

7-5. Vector Model of Density Matrix

It is possible to make the density matrix equations of motion (34), (35), and (49) resemble those for a magnetic dipole undergoing precession in a magnetic field. The model has value not only in solving the equations, but also in providing a physical picture of the density matrix in motion. The equations we derive here are equivalent to the Bloch equations appearing in nuclear magnetic resonance, and our approach constitutes a fundamental derivation of those equations. We see how one two-level system (e.g., an atom) is similar to another, namely, the spin-$\frac{1}{2}$ magnetic dipole. A difference exists, however, in that the atom really has many levels leading to three decay constants, whereas the Bloch equations have only two. For some laser media, this is a poor approximation. On the other hand, when γ_a and γ_b are small compared to γ and reciprocals of other times, the Bloch model is accurate and may be easier to use (see Prob. 7-8).

We suppose that the perturbing energy $\mathscr{V}_{ab}$ is given in the rotating-wave approximation by

$$\mathscr{V}_{ab} = -\tfrac{1}{2}\wp E_0 \exp(-i\nu t). \tag{64}$$

We further go into an (not *the*) interaction picture by multiplying both sides of (49) by exp $(i\nu t)$, thereby obtaining

$$\frac{d}{dt}[\rho_{ab}\exp(i\nu t)] = -[i(\omega - \nu) + \gamma]\rho_{ab}\exp(i\nu t) - \frac{1}{2}\frac{i}{\hbar}\wp E_0(\rho_{aa} - \rho_{bb}). \tag{65}$$

We introduce the real quantities

$$R_1 = \rho_{ab}\exp(i\nu t) + \text{c.c.}, \tag{66}$$

$$R_2 = i\rho_{ab}\exp(i\nu t) + \text{c.c.}, \tag{67}$$

that is,

$$R_1 - iR_2 = 2\rho_{ab} \exp(i\nu t) \tag{68}$$

and

$$R_3 = \rho_{aa} - \rho_{bb}. \tag{69}$$

These quantities vary little in an optical frequency period and are the components of the vector $\mathbf{R}$, given by

$$\begin{aligned} \mathbf{R} &= R_1\hat{e}_1 + R_2\hat{e}_2 + R_3\hat{e}_3 \\ &= \mathrm{Tr}(\rho'\sigma), \end{aligned} \tag{70}$$

where ρ' is an interaction picture density matrix, defined by

$$\rho' = \begin{pmatrix} \rho_{aa} & \rho_{ab} \exp(i\nu t) \\ \rho_{ba} \exp(-i\nu t) & \rho_{bb} \end{pmatrix}. \tag{71}$$

We then write the real and imaginary parts of (65) as

$$\dot{R}_1 = -(\omega - \nu)R_2 - \gamma R_1, \tag{72}$$

$$\dot{R}_2 = (\omega - \nu)R_1 - \gamma R_2 + \frac{\wp E_0}{\hbar} R_3. \tag{73}$$

We further assume equal-level decay constants:

$$\gamma_a = \gamma_b = \frac{1}{T_1}. \tag{74}$$

The equation for the probability difference R_3 of (69) is then given by (34) and (35) with (64) and (74) as

$$\begin{aligned} \dot{R}_3 &= -\frac{R_3}{T_1} + \left[\frac{i}{\hbar} \wp E_0 \rho_{ba} \exp(-i\nu t) + \text{c.c.}\right] \\ &= -\frac{R_3}{T_1} - \frac{\wp E_0}{\hbar} R_2. \end{aligned} \tag{75}$$

The decay constant γ in (72) and (73) is usually called $1/T_2$ in this context; T_2 is called the *transverse* relaxation time since the transverse directions are involved. Correspondingly, T_1 is called a *longitudinal* relaxation time. If, in addition to (74), $\gamma = 1/T_2 = 1/T_1$, the three equations of motion (72), (73), and (75) have the simple, combined form

$$\dot{\mathbf{R}} = -\gamma\mathbf{R} + \mathbf{R} \times \mathcal{B}, \tag{76}$$

where the effective force $\mathcal{B}$ is given by

$$\mathcal{B} = \frac{\wp E_0}{\hbar} \hat{e}_1 - (\omega - \nu)\hat{e}_3. \tag{77}$$

The time dependence of $\mathbf{R}$ as given by (76) is well known from classical mechanics. The $\mathbf{R}$ vector precesses clockwise about the effective field $\mathcal{B}$ with diminishing magnitude. The precessions for resonance and slightly off resonance are depicted in Fig. 7-3. On resonance, **R** precesses about the $\hat{e}_1$ axis in a major circle.

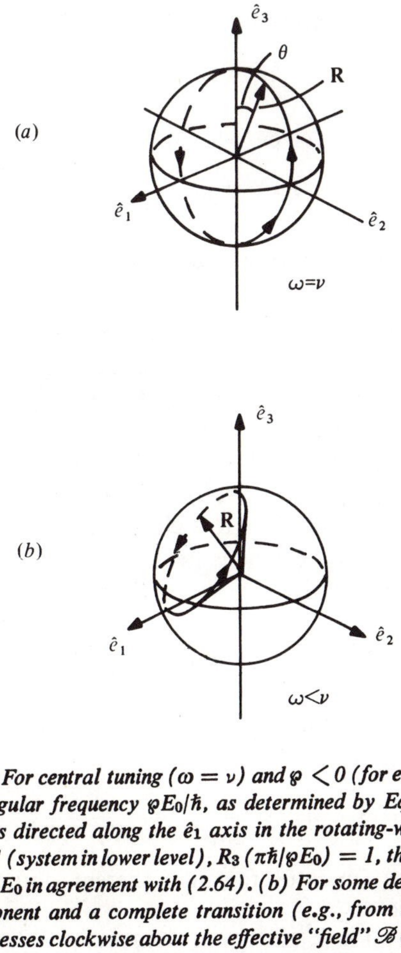

*Figure 7-3. (a) For central tuning ($\omega = \nu$) and $\wp < 0$ (for electron), **R** precesses clockwise about $\hat{e}_1$ at angular frequency $\wp E_0/\hbar$, as determined by Eq. (76). The electric-dipole interaction energy is directed along the $\hat{e}_1$ axis in the rotating-wave approximation. For an initial $R_3(0) = -1$ (system in lower level), $R_3(\pi\hbar/\wp E_0) = 1$, that is, the atom makes a transition in a time $\pi\hbar/\wp E_0$ in agreement with (2.64). (b) For some detuning ($\omega \neq \nu$), **R** acquires a nonzero $\hat{e}_1$ component and a complete transition (e.g., from upper to lower level) never occurs. Then **R** precesses clockwise about the effective "field" $\mathscr{B}$ of (77) with frequency $|\mathscr{B}| = [(\omega - \nu)^2 + (\wp E_0/\hbar)]^2$, thus tracing out a cone in the rotating frame.*

Physically **R** pointing along $\hat{e}_3$ ($R_3 = 1$, $R_2 = R_1 = 0$) represents a system in its upper level ($\rho_{aa} = 1$, $\rho_{bb} = 0$). Similarly **R** points along $-\hat{e}_3$ for a system in its lower level. The on-resonance precession about $\hat{e}_1$ is just a vector embodiment of the Rabi flopping of Eq. (2.64).

Our discussion has been presented in an interaction picture. For the vector model, this amounts to being on a merry-go-round rotating at frequency ν (!). We can return to the Schrödinger picture (get off the merry-go-round) by means of the transformation

$$\dot{\mathbf{r}} = \left(\frac{d\mathbf{R}}{dt}\right)_{\text{space}} = \left(\frac{d\mathbf{R}}{dt}\right)_{\text{body}} + \nu\hat{e}_3 \times \mathbf{R}. \tag{78}$$

We then find the rapidly rotating vector $\mathbf{r}$, which obeys the equation

$$\dot{\mathbf{r}} = -\frac{\mathbf{r}}{T_1} + (\mathbf{r} \times \mathbf{b}), \tag{79}$$

where the effective field $\mathbf{b}$ is given by

$$\mathbf{b} = \frac{\wp E_0}{\hbar}\hat{e}_1 - \omega\hat{e}_3. \tag{80}$$

The vector $\mathbf{r}$ rotates counterclockwise about $\hat{e}_3$ at approximately the frequency ω.

In the case of the spin-$\frac{1}{2}$ particle, $\wp E_0$ is replaced by μH in the interaction energy $\mathscr{V}_{ab}$. Otherwise, the precession equations are the same. What is particularly relevant about the spin case is that the abstract axes $\hat{e}_1$, $\hat{e}_2$, and $\hat{e}_3$ actually coincide with the real-life axes $\hat{x}$, $\hat{y}$, and $\hat{z}$, for an electron with spin up (along $\hat{z}$) is in the upper state (along $\hat{e}_3$). The reader should bear in mind that in the first instance the equations refer to the probabilistic density matrix, and only for an ensemble of spins can $\mathbf{R}$ itself be identified with a macroscopic classical dipole precessing in a magnetic field.

Problems

7-1. Show that the average photon number corresponding to the thermal distribution (26) is

$$\langle n \rangle = \sum_n n\,\rho_{nn} = \frac{1}{\exp(\hbar\omega/k_B T) - 1} \tag{81}$$

7-2. What is the density operator for the coherent state (15.11)? What is its phase-diffused value?

7-3. Show for the general density operator (17) that

$$\text{Tr}\,(\rho^2) \leqq 1 \tag{82}$$

and that equality holds only for a pure case.

7-4. Show that the equations of motion (13)–(15) for a density matrix ρ' in the interaction picture are

$$\dot{\rho}_{aa}' = \frac{1}{2}\frac{i}{\hbar}\wp E_0 \rho_{ba}' \exp[i(\omega - \nu)t] + \text{c.c.} = -\dot{\rho}_{bb}' \tag{83}$$

$$\dot{\rho}_{ab}' = -\frac{1}{2}\frac{i}{\hbar}\wp E_0 \exp[i(\omega - \nu)t](\rho_{aa}' - \rho_{bb}'). \tag{84}$$

7-5. Compute the root-mean-square deviation (6.31) for the dipole moment operator:

$$er = \begin{pmatrix} 0 & \wp \\ \wp & 0 \end{pmatrix} \tag{85}$$

using the density matrix (5).

7-6. Show that for $\mathbf{R}(0) = \hat{e}_3$ the equation of motion

$$\dot{\mathbf{R}} = \frac{\wp E_0}{\hbar}\mathbf{R} \times \hat{e}_1$$

has the solutions

$$R_2(t) = -\sin[(|\wp| E_0/\hbar)t],$$
$$R_3(t) = \cos[(\wp E_0/\hbar)t],$$

for $\wp < 0$. Give a physical interpretation of what is happening.

7-7. Solve the damped density matrix equations of motion (34), (35), and (49) to second order in the electric-dipole perturbation energy (2.16). Use the rotating-wave approximation.

7-8. If $\gamma_a \neq \gamma_b$, one must consider equations of motion for the sum $R_0 = \rho_{aa} + \rho_{bb}$ as well as the difference $R_3 = \rho_{aa} - \rho_{bb}$. Show that if $\dot{R}_0 \simeq 0$, the time constant T_1 of (74) is generalized to

$$T_1 = (\gamma_a^{-1} + \gamma_b^{-1})/2. \tag{86}$$

This approximation can lead to a factor of two error for systems with $\gamma_b \gg \gamma_a$, such as found in many laser media (see Sargent, Toschek and Danielmeyer (1976)).

7-9. Show that $T_1 = 1/\gamma_a$ for upper level decay to ground lower level. Note that for $\gamma_{ph} \simeq 0$, $T_2 = 2/\gamma_a > T_1$.

References

P. R. Berman, 1972, *Phys. Rev.* **A5**, 927.

F. Bloch, 1946, *Phys. Rev.* **70**, 460.

R. H. Dicke, 1954, *Phys. Rev.* **93**, 99.

U. Fano, 1957, *Rev. Mod. Phys.* **29**, 74.

R. P. Feynman, F. L. Vernon, and R. W. Hellwarth, 1957, *J. Appl. Phys.* **28**, 49.

W. E. Lamb, Jr., 1939, *Phys. Rev.* **55**, 234 (A).

W. E. Lamb, Jr., 1964 "Theory of Optical Maser Oscillators," in: *Proceedings of the International School of Physics*, Course XXXI, 78, esp. Sec. 3.

M. Sargent III, P. E. Toschek, H. G. Danielmeyer, 1976, *Appl. Phys.* **11**, 55.

D. ter Haar, 1961, *Rept. Progr. Phys.* **24**, 304.

VIII

SEMICLASSICAL LASER THEORY

8. Introduction to Semiclassical Laser Theory

In Chap. 5, the steady-state solution for the intensity in the ammonia beam maser was determined by energy conservation, that is, the energy given up by the molecules (gain) was equated to the losses in the maser cavity. Although attractive in its simplicity, this method provides no clue to the time dependence of the intensity or to mode pulling (index of refraction) and cannot be used for treating multimode operation. These and other limitations can be avoided by describing maser action by a classical electromagnetic field governed by Maxwell's equations and by demanding that the field be self-consistent, that is, that the field $\mathbf{E}$ inducing polarization of the active medium be equal to the resulting field $\mathbf{E}'$ as depicted in Fig. 8-1.

$$\mathbf{E}(\mathbf{r},t)\xrightarrow[\text{mechanics}]{\text{quantum}}\langle \mathbf{p}_i\rangle\xrightarrow[\text{summation}]{\text{statistical}}\mathbf{P}(\mathbf{r},t)\xrightarrow[\text{equations}]{\text{Maxwell's}}\mathbf{E}'(\mathbf{r},t)$$

self-consistency

Figure 8-1. Electric field $\mathbf{E}$ *assumed in cavity induces microscopic dipole moments* $(\mathbf{p}_i)$ *in the active medium according to the laws of quantum mechanics. These moments are then summed to yield the macroscopic polarization of the medium* $\mathbf{P}(\mathbf{r}, t)$*, which acts as a source in Maxwell's equations. The condition of self-consistency then requires that the assumed field* $\mathbf{E}$ *equal the reaction field* $\mathbf{E}'$.

Here the field induces electric-dipole moments in the medium according to the laws of quantum mechanics. The density matrix is used to facilitate the statistical summations involved in obtaining the macroscopic polarization of the medium from the individual dipole moments. It is used, furthermore, to describe decay phenomena which are not easily expressable in terms of probability amplitudes directly. This semiclassical approach to laser theory is remarkably good for many problems of interest. However, questions of laser linewidth, buildup from vacuum (no field), and photon statistics require a fully quantum-mechanical treatment, that is, both the field and the atoms are quantized (Chaps. 14–20).

In this chapter we carry out the semiclassical approach along lines given by Lamb (1964) for a single longitudinal mode of a scalar, plane wave electric field and a homogeneously broadened medium consisting of two-level atoms. The scalar field theory applies, for example, to lasers with Brewster windows (see Fig. 8-2), which ensure oscillation of a linearly polarized field. A more correct model, which includes a vector field and a many-level atomic medium, is given in Chap. 12, where our current model is evaluated. The discussion here applies to a single transverse (Fox and Li) mode. A brief development of these modes is given in Appendix B and the problems.

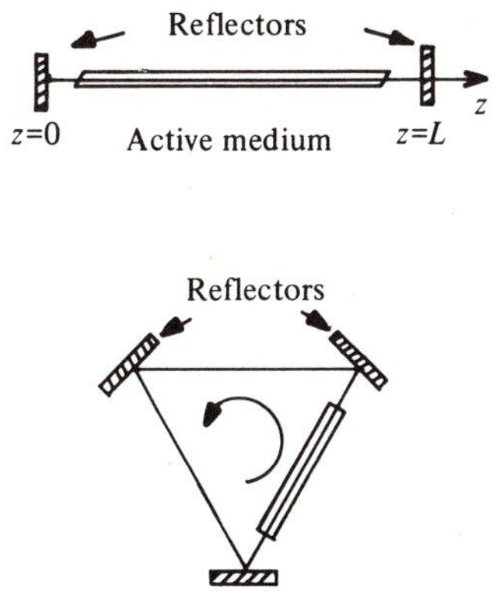

Figure 8-2. (a) Diagram of laser, showing reflectors in plane perpendicular to laser (z) axis and active medium between reflectors. Brewster windows are sketched on the ends of the active medium to help enforce conditions that only one polarization component of the electric field exists, as is assumed in this chapter. (b) Corresponding unidirectional ring configuration. Usually both running waves oscillate in a ring laser; the unidirectional situation can be achieved by insertion of a device with high loss for one running wave in the cavity.

In Sec. 8-1 we use the Maxwell equations to relate derivatives of the electric field to the polarization of the medium. In this part of the analysis we consider a multimode field, for the calculation is as easy to follow as that for a single mode and the generality is useful in later chapters. In Sec. 8-2 the polarization of the medium induced by a single-mode field is calculated in the rate equation approximation. In Sec. 8-3 this polarization is inserted into the field "self-consistency" equations found in Sec. 8-1 to yield amplitude- and frequency-determining equations. The laser threshold condition, the tuning dependence of the steady-state intensity, and mode pulling effects are included in the discussion.

8-1. Electromagnetic Field Equations

We describe the electromagnetic radiation in the laser cavity (Fig. 8-2) by Maxwell's equations in mks units:

$$\left.\begin{aligned} \operatorname{div} \mathbf{D} &= 0 \qquad \operatorname{curl} \mathbf{E} = -\frac{\partial \mathbf{B}}{\partial t} \\ \operatorname{div} \mathbf{B} &= 0 \qquad \operatorname{curl} \mathbf{H} = \mathbf{J} + \frac{\partial \mathbf{D}}{\partial t} \end{aligned}\right\}, \tag{1}$$

where

$$\mathbf{D} = \varepsilon_0 \mathbf{E} + \mathbf{P}, \qquad \mathbf{B} = \mu_0 \mathbf{H}, \qquad \mathbf{J} = \sigma \mathbf{E}. \tag{2}$$

Here, to avoid a complicated boundary value problem, we assume the presence of a lossy medium with conductivity σ, which we adjust to give damping due to diffraction and reflector transmission. Substituting (2) with the time derivative of the curl **H** equation into the curl of the curl **E** equation, we have the wave equation

$$\operatorname{curl}\operatorname{curl} \mathbf{E} + \mu_0 \sigma \frac{\partial \mathbf{E}}{\partial t} + \mu_0 \varepsilon_0 \frac{\partial^2 \mathbf{E}}{\partial t^2} = -\mu_0 \frac{\partial^2 \mathbf{P}}{\partial t^2}. \tag{3}$$

The curl curl **E** can be simplified by the expansion

$$\begin{aligned} \operatorname{curl}\operatorname{curl} \mathbf{E} &= -\nabla^2 \mathbf{E} + \nabla(\operatorname{div} \mathbf{E}) = -\nabla^2 \mathbf{E} + \varepsilon_0^{-1} \nabla[\operatorname{div} \mathbf{D} - \operatorname{div} \mathbf{P}] \\ &= -\nabla^2 \mathbf{E}, \end{aligned}$$

where the last equality results from the first Maxwell equation and the assumption that $\nabla \cdot \mathbf{P} \approx 0$ (see Prob. 8–13). When we consider solid-state media for which the host dispersion cannot be neglected, this assumption will be reappraised. Fox and Li (1961) have shown that variations in the field intensity transverse to the laser axis are slowly varying compared to optical wavelengths, and hence we neglect x and y derivatives, that is,

$$\operatorname{curl}\operatorname{curl} \mathbf{E} = -\frac{\partial^2 \mathbf{E}}{\partial z^2}. \tag{4}$$

The time dependence of the wave equation can be separated from the spatial dependence by expansion of the electric field in normal modes of the cavity. Inasmuch as the passive cavity is a Fabry-Perot interferometer without side walls, there is a continuum of these modes. Nevertheless, only certain discrete modes achieve appreciable magnitude within the cavity,† namely, those with circular frequencies

$$\Omega_n = \frac{n\pi c}{L} = K_n c, \tag{5}$$

where L is the length of the cavity, c is the velocity of light, n is a large integer typically on the order of 10^6, and K_n is the corresponding wave number. In our discussion we usually take the (unnormalized) normal modes to have sinusoidal z dependence:

$$U_n(z) = \sin K_n z. \tag{6}$$

We also have occasion to use the simpler, unidirectional (running-wave) mode functions:

$$U_n(z) = \exp(iK_n z). \tag{7}$$

The electric field is then written as

$$E(z, t) = \frac{1}{2}\sum_n \mathscr{A}_n(t) U_n(z) + \text{c.c.},$$

where $\mathscr{A}_n(t)$ is the complex electric field amplitude of mode n. We further set

$$\mathscr{A}_n(t) = E_n(t) \exp\{-i[\nu_n t + \phi_n(t)]\},$$

where the amplitude coefficient $E_n(t)$ and phase $\phi_n(t)$ vary little in an optical frequency period, and $\nu_n + \dot{\phi}_n$ is the oscillation frequency of the mode. We will see that this frequency is not necessarily equal to the passive cavity frequency Ω_n due to the dispersion of the active medium. In this chapter we consider two-level atoms. These contribute to a single polarization component of the electric field, and we thus ignore the vector character of the field. Our expansion for the electric field then becomes

$$E(z, t) = \frac{1}{2}\sum_n E_n(t) \exp[-i(\nu_n t + \phi_n)]\, U_n(z) + \text{c.c.} \tag{8}$$

With this form for the field, the induced polarization of the medium can be written as

$$P(z, t) = \frac{1}{2}\sum_n \mathscr{P}_n(t) \exp[-i(\nu_n t + \phi_n)]\, U_n(z) + \text{c.c.}, \tag{9}$$

where $\mathscr{P}_n(t)$ is the complex, slowly varying component of the polarization for mode n. The real part of $\mathscr{P}_n$ is in phase with the electric field and, as we shall see shortly, results in dispersion due to the medium. The imaginary part is in quadrature with the electric field and results in gain or loss.

Substituting (8) and (9) without the complex conjugates into the wave equa-

† See Sec. 21-2 for a philosophical discussion of this point.

tion (3), using (4), projecting onto $U_n(z)$, and multiplying through by $\exp[i(\nu_n t + \phi_n)]$, we have

$$K_n^2 E_n + \mu_0\sigma[\dot{E}_n - i(\nu_n + \dot{\phi}_n)E_n] + \mu_0\varepsilon_0\{\ddot{E}_n - i\ddot{\phi}E_n - i(\nu_n + \dot{\phi}_n)\dot{E}_n - i(\nu_n + \dot{\phi}_n)[\dot{E}_n - i(\nu_n + \dot{\phi}_n)E_n]\} = \mu_0(\nu_n + \dot{\phi}_n)^2\mathscr{P}_n + \text{terms with } \dot{\mathscr{P}}_n \text{ and } \ddot{\mathscr{P}}_n.$$

Because E_n, ϕ_n, and $\mathscr{P}_n$ vary little in an optical frequency period and the losses are small and the medium dilute, we neglect terms containing $\ddot{E}_n$, $\ddot{\phi}_n$, $\ddot{\mathscr{P}}_n$, $\dot{E}_n\dot{\phi}_n$, $\sigma\dot{E}_n$, $\sigma\dot{\phi}_n$, $\dot{\phi}_n\mathscr{P}_n$ and $\dot{\mathscr{P}}_n$, and obtain

$$\Omega_n^2 E_n - i\left(\frac{\sigma}{\varepsilon_0}\right)\nu_n E_n - 2i\nu_n\dot{E}_n - (\nu_n + \dot{\phi}_n)^2 E_n = \nu_n^2\varepsilon_0^{-1}\mathscr{P}_n,$$

where we have used $\Omega_n = K_n c$ and multiplied through by c^2. Adjusting the fictional conductivity σ to yield the desired Q's (5.19) of the modes, we set

$$\sigma = \varepsilon_0 \frac{\nu}{Q_n}, \tag{10}$$

where $\nu \simeq \nu_n$.† We note also that, since $\mathscr{P}_n$ is very small, the error made in replacing $\nu_n^2\mathscr{P}_n$ by $\nu_n\nu\mathscr{P}_n$ is negligible and this substitution is a convenient simplification for later multimode treatments. Further noting that $\Omega_n^2 - (\nu_n + \dot{\phi}_n)^2 \simeq 2\nu_n(\Omega_n - \nu_n - \dot{\phi}_n)$, dividing by $2\nu_n$, and equating real and imaginary parts, we have the self-consistency equations

$$\dot{E}_n + \frac{1}{2}\frac{\nu}{Q_n}E_n = -\frac{1}{2}\frac{\nu}{\varepsilon_0}\operatorname{Im}(\mathscr{P}_n), \tag{11}$$

$$\nu_n + \dot{\phi}_n = \Omega_n - \frac{1}{2}\frac{\nu}{\varepsilon_0}E_n^{-1}\operatorname{Re}(\mathscr{P}_n). \tag{12}$$

These are two of our basic working equations.

Let us consider their physical meaning. In the absence of an active medium, $\mathscr{P}_n = 0$, and the equations imply that the mode intensities E_n^2 decrease exponentially in time with constant ν/Q_n, and that the mode oscillation frequencies are just those of the passive cavity. More generally, there exists a polarizaton defined by Fourier components of the susceptibility χ,

$$\mathscr{P}_n = \varepsilon_0\chi_n E_n = \varepsilon_0(\chi_n' + i\chi_n'')E_n. \tag{13}$$

Here the χ_n depend on both the mode amplitudes and the frequencies—a dependence that results in saturation and coupling effects for the former, and dispersive (actually pulling) phenomena for the latter. Substituting (13) into (11) and (12), we obtain the mode amplitude- and frequency-determining equations:

† The frequency ν is sometimes used in place of ν_n when the difference between various ν_n produces negligible effect. This is a useful simplification in multimode theory.

$$\dot{E}_n = -\frac{1}{2}\frac{\nu}{Q_n}E_n - \tfrac{1}{2}\nu\chi_n''E_n, \tag{14}$$

$$\nu_n + \dot{\phi}_n = \Omega_n - \tfrac{1}{2}\nu\chi_n'. \tag{15}$$

The energy per unit volume h_n for a given mode is proportional to the field amplitude squared, that is,

$$h_n \propto E_n^2, \tag{16}$$

which has the equation of motion [from (14)]

$$\dot{h}_n = \underbrace{-\left(\frac{\nu}{Q_n}\right)h_n}_{\text{losses}} \underbrace{- \nu\chi_n'' h_n}_{\text{gain}}. \tag{17}$$

This states that the time rate of change of the energy equals the difference between the cavity losses and the gain ($\chi_n'' < 0$) from the medium. In steady state, $\dot{h}_n = 0$, and we recover the oscillation condition that the *saturated gain equals* the *loss*, as postulated (5.18) in connection with the ammonia beam maser on the basis of energy conservation.

The frequency relation (15) implies that the oscillation frequency of the nth mode, ν_n, is shifted from the passive cavity frequency by the amount $-\frac{1}{2}\nu\chi_n'$. We can relate these variables to an index of refraction $\eta(\nu_n)$ by combining (15) with the definition

$$\eta(\nu_n) = \frac{K_n c}{\nu_n} = \frac{\Omega_n}{\nu_n} = \frac{\Omega_n}{\Omega_n - \frac{1}{2}\nu\chi_n'}\ . \tag{18}$$

Since we are treating a dilute active medium ($\chi_n' \ll 1$), we can set $\nu/\Omega_n \simeq 1$ and keep only the first two terms of the geometric series for (18). We find

$$\eta(\nu_n) \cong 1 + \tfrac{1}{2}\chi_n', \tag{19}$$

a result familiar in elementary electromagnetic theory. We give specific solutions for (17) and (19) in Sec. 8-3 for both amplifying (laser) and attenuating media.

8-2. Polarization of the Medium

We assume that the laser medium consists of two-level homogeneously broadened atoms having a resonance transition frequency ω as in Chaps. 2 and 7. An atom excited to the state α ($= a$ or b), at time t_0 and place z, is described by the density matrix $\rho(\alpha, z, t_0, t)$. Atoms are excited to the state α at the rate $\lambda_\alpha(z, t_0)$ atoms per second per unit volume. The level probabilities decay with constants γ_a and γ_b, as discussed in Sec. 7-4 and shown in Fig. 8-3. The polarization matrix element ρ_{ab} decays with the constant γ of Eq.

(7.48). This model illustrates simply some basic laser principles.† It does not apply well, however, to media which are inhomogeneously broadened (see Chap. 10) or those whose lower state is the ground state like ruby (see Prob. 8-6 for a more accurate approach).

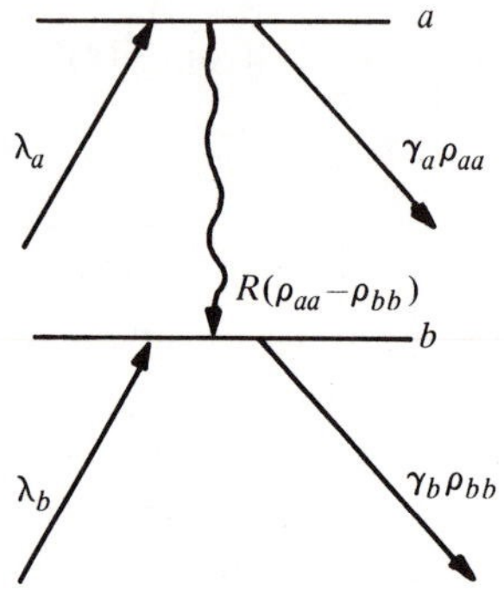

Figure 8-3. Energy-level diagram for two-level atoms comprising the active medium of the laser. The resonance frequency of the transition from the lower (b) to the upper (a) level is ω; the numbers of atoms per unit volume per unit time excited to the a and b levels are λ_a and λ_b, respectively; the numbers decaying are $\gamma_a \rho_{aa}$ and $\gamma_b \rho_{bb}$, respectively; and the number making transitions from a to b is $R(\rho_{aa} - \rho_{bb})$, according to the rate equations (33) and (34).

The macroscopic polarization $P(z, t)$ of (9) for this medium is given by the contributions of all atoms at z at time t regardless of their initial states and times of excitation, that is,

$$P(z, t) = \sum_a \int_{-\infty}^{t} dt_0\, \lambda_a(z, t_0)\, \langle er \rangle$$

$$= \wp \sum_a \int_{-\infty}^{t} dt_0\, \lambda_a(z, t_0) \rho_{ab}(a, z, t_0, t) + \text{c.c.} \tag{20}$$

As noted in Sec. 3-1, a single atom does not yield the electric-dipole expectation value $\langle er \rangle$, a quantity which refers to an ensemble. The contribution of many closely spaced atoms implicit in (20) does, however, yield appropriate ensembles for the use of $\langle er \rangle$. The slowly varying, complex polarization $\mathscr{P}_n(t)$ in the field self-consistency equations (11) and (12) is given by identification of the positive frequency parts in (9) and (20) [those that go like

† It describes what are often called "four-level" homogeneously broadened lasers, where the extra two levels are the ground state and a diffuse pump level, usually many levels, lying above the upper laser level a.

$\exp(-i\nu t)$], projection onto the nth Fourier component, and multiplication by 2 exp [$i(\nu_n t + \phi_n)$], that is,

$$\mathscr{P}_n(t) = 2\wp \exp[i(\nu_n t + \phi_n)]\frac{1}{\mathscr{N}_n}\int_0^L dz\, U_n{}^*(z)$$

$$\times \sum_a \int_{-\infty}^t dt_0\, \lambda_a(z, t_0)\rho_{ab}\,(a, z, t_0, t), \tag{21}$$

where the normalization factor

$$\mathscr{N}_n = \int_0^L dz\,|\,U_n(z)\,|^2. \tag{22}$$

The equations of motion for the single-atom matrix $\rho(a, z, t_0, t)$ are given by (7.34), (7.35), and 7.49). One could solve for the induced dipoles by the iteration method given in Sec. 7-4 and use (21) to obtain the complex polarization $\mathscr{P}_n(t)$.It is easier, however, to integrate equations of motion for a "population" matrix which includes the summations over initial times and states of excitation. Specifically, we define the population matrix†

$$\rho(z, t) = \sum_a \int_{-\infty}^t dt_0\, \lambda_a(z, t_0)\, \rho(a, z, t_0, t) \tag{23}$$

and calculate its equation of motion by differentiating with respect to time t. Two t dependences exist: the upper limit of integration over t_0 and that of the single-atom density matrix. We have

$$\frac{d\rho(z, t)}{dt} = \sum_a \lambda_a(z, t)\, \rho(a, z, t, t) + \sum_a \int_{-\infty}^t dt_0\, \lambda_a(z, t_0)\dot{\rho}(a, z, t_0, t). \tag{24}$$

By definition,

$$\rho_{ij}(a, z, t, t) = \delta_{ia}\delta_{ja},$$

and the first term can be replaced by the operator with matrix representation

$$\begin{pmatrix} \lambda_a & 0 \\ 0 & \lambda_b \end{pmatrix}. \tag{25}$$

The second term has components identical to the right-hand sides of the equations of motion for the pure-case density matrix [see Eqs. (7.34), (7.35), and (7.49)]. This result follows because the Hamiltonian does not depend on a or b and the excitations $\lambda_a(z, t_0)$ vary slowly enough to be evaluated at time t. We have the component equations of motion for the population matrix $\rho(z, t)$:

$$\dot{\rho}_{ab} = -(i\omega + \gamma)\rho_{ab} + i\hbar^{-1}\mathscr{V}_{ab}(z, t)(\rho_{aa} - \rho_{bb}), \tag{26}$$

$$\dot{\rho}_{aa} = \lambda_a - \gamma_a\rho_{aa} - (i\hbar^{-1}\mathscr{V}_{ab}\rho_{ba} + \text{c.c.}), \tag{27}$$

$$\dot{\rho}_{bb} = \lambda_b - \gamma_b\rho_{bb} + (i\hbar^{-1}\mathscr{V}_{ab}\rho_{ba} + \text{c.c.}), \tag{28}$$

†Note that the diagonal element $\rho_{aa}\,(z, t)$ is the population of the ath level.

where, of course, $\rho_{ba} = \rho_{ab}^*$. In terms of $\rho(z, t)$, the complex polarization (21) is given by

$$\mathscr{P}_n(t) = 2\exp[i(\nu_n t + \phi_n)]\frac{1}{\mathscr{N}_n}\int_0^L dz\, U_n^*(z)\, \wp\rho_{ab}(z, t). \tag{29}$$

Hence, by determining $\rho_{ab}(z, t)$, it is possible to obtain amplitude- and frequency-determining equations by combining (29) with the self-consistency equations (11) and (12). We now proceed to do this for a single-mode field in the rate equation approximation.

The off-diagonal element ρ_{ab} is given by the formal integral of (26) [see Eq. (7.53) for method]:

$$\rho_{ab}(z, t) = i\hbar^{-1}\int_{-\infty}^{t} dt' \exp[-(i\omega + \gamma)(t - t')] \times \mathscr{V}_{ab}(z, t')[\rho_{aa}(z, t') - \rho_{bb}(z, t')]. \tag{30}$$

The perturbation energy $\mathscr{V}_{ab} = -\wp E(z, t)$ for a single-mode field of (8) is given in the rotating-wave approximation by

$$\mathscr{V}_{ab} = -\tfrac{1}{2}\wp E_n(t)\exp[-i(\nu_n t + \phi_n)]U_n(z). \tag{31}$$

The integration (30) can be simply performed provided the amplitude E_n, phase ϕ_n, and population difference $\rho_{aa} - \rho_{bb}$ do not change appreciably in the time $1/\gamma$, for then these terms can be factored outside the integral. These conditions comprise the rate equation approximation, for, as we will see shortly, they lead to rate equations for the atomic populations. The value of ρ_{ab} so obtained can then be substituted into the equations of motion (27) and (28), which, in steady state, determine the population difference. This difference can be substituted, in turn, back into the equation for ρ_{ab}, thus determining the complex polarization.

Performing the integration (30) in these approximations, we have

$$\rho_{ab}(z, t) = -\tfrac{1}{2}i\wp\hbar^{-1}E_n\exp[-i(\nu_n t + \phi_n)]U_n(z)\left[\frac{\rho_{aa} - \rho_{bb}}{i(\omega - \nu_n) + \gamma}\right]. \tag{32}$$

Substituting this into the equations of motion for ρ_{aa} and for ρ_{bb}, we find the rate equations

$$\dot\rho_{aa} = \lambda_a - \gamma_a\rho_{aa} - R(\rho_{aa} - \rho_{bb}), \tag{33}$$

$$\dot\rho_{bb} = \lambda_b - \gamma_b\rho_{bb} + R(\rho_{aa} - \rho_{bb}), \tag{34}$$

where the rate constant

$$R = \frac{1}{2}\left(\frac{\wp}{\hbar}\right)^2 E_n^2|U_n|^2\gamma[(\omega - \nu_n)^2 + \gamma^2]^{-1} = \frac{1}{2}\left(\frac{\wp E_n}{\hbar}\right)^2|U_n|^2\gamma^{-1}\mathscr{L}(\omega - \nu_n), \tag{35}$$

and where the dimensionless Lorentzian $\mathscr{L}(\omega - \nu_n)$ is defined by

$$\mathscr{L}(\omega - \nu_n) = \frac{\gamma^2}{\gamma^2 + (\omega - \nu_n)^2}, \tag{36}$$

The flow of populations given by (33) and (34) is depicted in Fig. 8-3. In steady state, $\dot{\rho}_{aa} = \dot{\rho}_{bb} = 0$, and (33) and (34) yield [divide (33) by γ_a and (34) by γ_b and subtract the resulting equations from one another]

$$\rho_{aa} - \rho_{bb} = \frac{N(z)}{1 + R/R_s}, \tag{37}$$

where the saturation parameter

$$R_s = \frac{\gamma_a \gamma_b}{2\gamma_{ab}} = \tfrac{1}{2} T_1, \tag{38}$$

and the unsaturated population difference

$$N(z) = \lambda_a \gamma_a^{-1} - \lambda_b \gamma_b^{-1}. \tag{39}$$

In (37), we see that the population difference is given by that in the absence of the field, $N(z)$, divided by the factor $1 + R/R_s$, which increases as the intensity of the electric field increases. In particular, if $R = R_s$, the population difference is one half of its zero-field value. For a rate constant R with $U_n^2(z) = \sin^2 K_n z$ dependence, the population difference $\rho_{aa} - \rho_{bb}$ has this dependence as well. Specifically, the sinusoidally varying electric field intensity burns holes into the population difference spaced one-half wavelength apart, as shown in Fig. 8-4. At the nodes of the field, the population difference has the zero-field value.

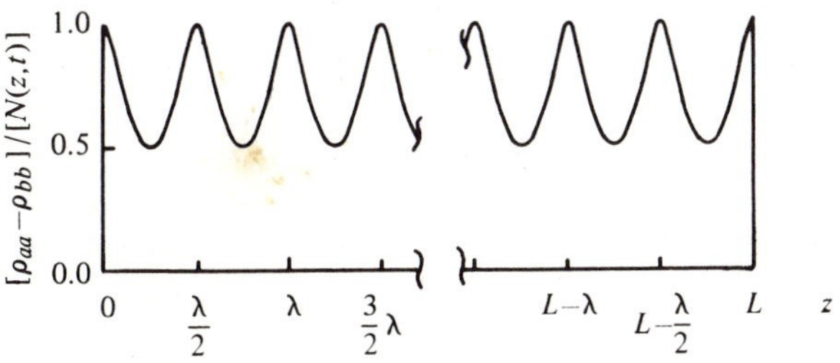

Figure 8-4. Normalized population difference versus axial coordinate. Spatial holes burned by laser field in this djfference are clearly depicted. Equation (37) was used with $R/R_s = \sin^2 K_n z$. In terms of the dimensionless intensity I_n of (45), $R/R_s = I_n(2\gamma_{ab}/\gamma)\, \mathscr{L}(\omega - \nu_n) \sin^2 K_n z$. Hence, for $2\gamma_{ab} = \gamma$ and central tuning ($\omega = \nu_n$), this figure corresponds to $I_n = 1$, which is not an uncommon value in lasers. For further interpretation and consequences, see Sargent (1976a, b).

Combining (32) and (37) with (29), we have the complex polarization

$$\mathscr{P}_n(t) = -\wp^2\hbar^{-1}E_n\frac{(\omega - \nu_n) + i\gamma}{(\omega - \nu_n)^2 + \gamma^2}\frac{1}{\mathscr{N}_n}\int_0^L dz\,\frac{|U_n(z)|^2N(z)}{1 + R(z)/R_s}. \tag{40}$$

The integral over z can be evaluated exactly [Eq. (47)] provided the unsaturated population difference $N(z)$ varies little in an optical wavelength.

Greater insight can be gained by supposing that E_n and hence R are small enough to allow the saturation denominator $1 + R/R_s$ to be expanded in a geometric series and integrated term by term. To first order in E_n (linear in E_n), the complex polarization is given by (40) with R set equal to zero, that is, by

$$\mathscr{P}_n(t) \simeq \mathscr{P}_n^{(1)}(t) = -\wp^2\hbar^{-1}\,\bar{N}E_n(t)\frac{(\omega - \nu_n) + i\gamma}{(\omega - \nu_n)^2 + \gamma^2}. \tag{41}$$

Here we have assumed that $N(z, t)$ is slowly varying in a wavelength so that [in the standing wave case, note that $|U_n(z)|^2 = \frac{1}{2}(1 - \cos 2K_n z)$]

$$\bar{N} = \frac{1}{\mathscr{N}}\int_0^L dz\,|U_n(z)|^2N(z, t) = \frac{1}{L}\int_0^L dz\,N(z, t). \tag{42}$$

This linear approximation contains no saturation terms and consequently is valid only for threshold, that is, the minimum excitation for which oscillation can occur. Expanding the denominator of the integrand to one higher order (linear in R) and performing the integration over $|U_n|^4$ using the identity

$$|U_n|^4 = \tfrac{1}{4}(1 - \cos 2K_nz)^2 = \tfrac{3}{8} - \tfrac{1}{2}\cos 2K_nz + \tfrac{1}{8}\cos 4K_nz, \tag{43}$$

we have the result accurate to third order:

$$\begin{aligned}\mathscr{P}_n(t) &\simeq \mathscr{P}_n^{(1)}(t) + \mathscr{P}_n^{(3)}(t)\\ &= -\wp^2\hbar^{-1}E_n\bar{N}\left[\frac{(\omega - \nu_n) + i\gamma}{(\omega - \nu_n)^2 + \gamma^2}\right]\left[1 - \frac{3}{2}\frac{\gamma_{ab}\gamma I_n}{(\omega - \nu_n)^2 + \gamma^2}\right],\end{aligned} \tag{44}$$

also derivable from perturbation theory. Here I_n is the dimensionless intensity

$$I_n = \frac{1}{2}\frac{\wp^2}{\hbar^2\gamma_a\gamma_b}E_n^2. \tag{45}$$

Considering the second bracket in (44) as part of a geometric series, we can "resum" to obtain

$$\mathscr{P}_n(t) \simeq -\wp^2\hbar^{-1}E_n\bar{N}\left\{\frac{(\omega - \nu_n) + i\gamma}{(\omega - \nu_n)^2 + \gamma^2\,[1 + (3/2)(\gamma_{ab}/\gamma)I_n]}\right\}. \tag{46}$$

Here we see that the gain $\mathrm{Im}(\mathscr{P}_n)$ at resonance is decreased from the first-order result by the factor $[1 + (3/2)\,(\gamma_{ab}/\gamma)I_n]$, a saturation effect, and that the response is power broadened from the width γ to $\gamma[1 + (3/2)(\gamma_{ab}/\gamma)I_n]^{1/2}$. The integral in (40) can be evaluated exactly with the result

$$\mathscr{P}_n(t) = -\wp^2\hbar^{-1}\,\bar{N}\left[\frac{(\omega - \nu_n) + i\gamma}{(\omega - \nu_n)^2 + \gamma^2}\right]f(w)E_n, \tag{47}$$

where

$$w = 2\frac{\gamma_{ab}}{\gamma} I_n \mathscr{L}(\omega - \nu_n) \tag{48}$$

and

$$f(w) = \frac{2}{w}[1 - (1 + w)^{-1/2}], \tag{49}$$

which gives (44) when expanded to first order in w.

8-3. Single-Mode Operation

Combining the complex polarization (44) with the self-consistency equations (11), we have the single-mode amplitude-determining equation

$$\dot{E}_n = E_n(\alpha_n - \beta_n I_n), \tag{50}$$

where the "net-gain" coefficient α_n and the self-saturation coefficient β_n are given in Table 8-1, and where I_n is the dimensionless intensity of (45). Multiplying (50) by $E_n\wp^2/\hbar^2\gamma_a\gamma_b$, we obtain the equation of motion for I_n:

$$\dot{I}_n = 2I_n(\alpha_n - \beta_n I_n). \tag{51}$$

Furthermore, from (41) and (12), we have the frequency-determining equation

$$\nu_n + \dot{\phi}_n = \Omega_n + \sigma_n - \rho_n I_n, \tag{52}$$

where the mode pulling and pushing† coefficients σ_n and ρ_n are given in Table 8-1. The third-order theory given here is reasonably accurate for excitations as high as 20% above threshold, although observable deviation from the exact calculation [from (47)] appears.

Setting the linear net-gain coefficient $\alpha_n = 0$, we determine the threshold condition (at resonance):

$$\frac{\wp^2 \bar{N}_T}{\varepsilon_0 \hbar \gamma} = \frac{1}{Q_n}, \tag{53}$$

where N_T is the value of the population inversion N at threshold. This can be used to express the coefficients in terms of the relative excitation

$$\mathfrak{N} = \frac{\bar{N}}{\bar{N}_T}, \tag{54}$$

† In previous work (Lamb, 1963, 1964) ρ_n has the opposite sign from that used here. The change is made to emphasize the fact that the third-order term subtracts from the first-order terms, that is, ρ_n results in pushing, rather than pulling, of the mode frequency. This is due to the fact that the first-order term gives an overestimate of the pulling by using the unsaturated population difference.

Table 8-1 Coefficients in Intensity-and Frequency-Determining Equations (51) and (52). The threshold condition allows one to write the first-order factor as $F_1 = \frac{1}{2}(\nu/Q_n)\,\mathfrak{N}$. The constant F_3 is chosen so that saturation terms are expressed in terms of the dimensionless intensity (45). Other symbols are defined by equations as follows: the electric dipole matrix element $\wp$, by (2.16); the atomic dipole decay constant γ, by (7.48); the average population inversion density $\bar{N}$, by (42); the mode Q's, Q_n, by (10) with (5.19): and ν is an optical frequency with can be set equal to the atomic resonance frequency ω.

Coefficient	Physical Context
$\alpha_n = \mathscr{L}(\omega - \nu_n)F_1 - \frac{1}{2}\nu/Q_n$	Linear net gain
$\beta_n = \mathscr{L}^2(\omega - \nu_n)F_3$	Saturation
$\sigma_n = [(\omega - \nu_n)/\gamma]\mathscr{L}(\omega - \nu_n)F_1$	Linear mode pulling
$\rho_n = [(\omega - \nu_n)/\gamma]\mathscr{L}^2(\omega - \nu_n)F_3$	Mode pushing
$\mathscr{L}(\omega - \nu_n) = \gamma^2[\gamma^2 + (\omega - \nu_n)^2]^{-1}$	Dimensionless Lorentzian
$F_1 = \frac{1}{2}\nu\wp^2(\varepsilon_0\hbar\gamma)^{-1}\bar{N}$	First-order factor
$F_3 = (3/2)(\gamma_{ab}/\gamma)F_1$	Third-order factor

a parameter more easily measured than the excitation $\bar{N}$.

Setting $\dot{I}_n = 0$ in the intensity-determining equation (51), we have the steady-state solution

$$\begin{aligned} I_n &\equiv \frac{1}{2}\frac{\wp^2}{\hbar^2\gamma_a\gamma_b}E_n{}^2 \\ &= \frac{\alpha_n}{\beta_n} \\ &= \frac{2}{3}\frac{\mathscr{L}(\omega - \nu_n) - 1/\mathfrak{N}}{(\gamma_{ab}/\gamma)\,\mathscr{L}^2(\omega - \nu_n)}, \end{aligned} \tag{55}$$

where the (dimensionless) Lorentzian $\mathscr{L}(\omega - \nu_n)$ is defined by (36). This intensity versus cavity detuning $\Delta\Omega$ is plotted in Fig. 8-5.

One might ask how it is possible to know what oscillation frequency ν is to be used for calculating the intensity, for ν_n itself depends on the intensity according to Eq. (52). A good approximation results from taking $\nu_n = \Omega_n$ in the calculation of the coefficients. Improved values for the coefficients can be obtained by calculating ν_n from (52), using approximate values for the coefficients, and then recalculating the coefficients with this ν_n. This procedure can be iterated, and typically the first calculation gives an answer within 1% of the exact answer and the second step a result within 0.1%. Hence the iteration is seldom carried more than one step.

The field equations for single-mode operation, Eqs. (51) and (52), do not depend on the location of the active medium along the laser (z) axis, for only $\bar{N}$, the average population difference, appears. In two-and higher-mode operation this is no longer the case.

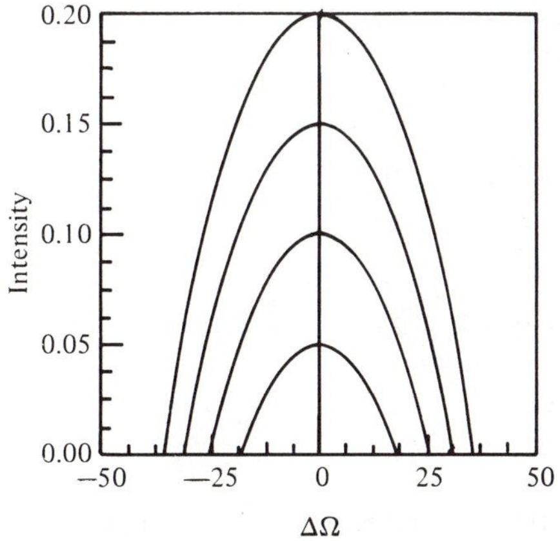

Figure 8-5. Graph of a single-mode (dimensionless) intensity (55) versus detuning ($\omega = \nu$) for homogeneously broadened ($\gamma = 2\pi \times 100$ MHz, $\gamma_{ab} = 2\pi \times 55.55$ MHz), stationary atoms. The relative excitation $\mathfrak{N}$ of Eq. (54) are (in order of increasing maxima) 1.05, 1.10, 1.15, and 1.2.

The intensity equation of motion (51) can be integrated as discussed in Prob. 8-12 with the solution

$$I_n(t) = \frac{a_n[I_0/(a_n - \beta_n I_0)] \exp(2a_n t)}{1 + \beta_n[I_0/(a_n - \beta_n I_0)] \exp(2a_n t)}, \tag{56}$$

where $I_0 \equiv I_n(0)$. As would be expected from the equation of motion, initially this yields exponential gain $I_n(t) = I_0 \exp(2a_n t)$ for small I_0, and it yields the steady-state result (55) as $t \to \infty$. The time development is illustrated in Fig. 8-6 for several values of the ratio a_n/β_n.

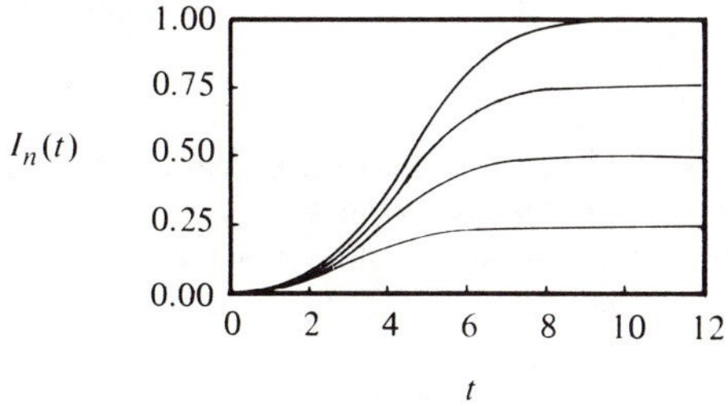

Figure 8-6. Buildup of dimensionless intensity (45) from the small value $I_n(0) = 0.01$. Equation (56) was used. Time t is given in units of $2a_n$. The curves correspond (in order of increasing final value) to $a_n/\beta_n = 0.25$, 0.50, 0.75, and 1.0.

The frequency-determining equation (52) predicts a pulling of the oscillation frequency ν_n from the passive cavity frequency Ω_n toward line center. Specifically,

$$\nu_n = \frac{\Omega_n + \mathscr{S}\omega}{1+\mathscr{S}}, \tag{57}$$

where the "stablization factor"

$$\mathscr{S} = \frac{F_1}{\gamma}\mathscr{L}(\omega - \nu_n)\left[1 - \frac{3}{2}\frac{\gamma_{ab}}{\gamma} I_n \mathscr{L}(\omega - \nu_n)\right].$$

In steady-state, the condition $\alpha_n = \beta_n I_n$ gives $\mathscr{S} = \nu/2Q_n\,\gamma$, so that Eq. (57) can be written as

$$\nu_n = \frac{\gamma\Omega_n + [\frac{1}{2}(\nu/Q_n)]\,\omega}{\gamma + [\frac{1}{2}(\nu/Q_n)]}. \tag{58}$$

Equation (58) can be interpreted as a center of mass equation in which the oscillation frequency ν_n assumes the average value of Ω_n and ω with weights γ and $\nu/2Q_n$, respectively. For $\nu/2Q_n \ll \gamma$ (usually the case), $\nu_n \simeq \Omega_n$, but pulled slightly toward the atomic frequency ω (mode pulling). The index of refraction determined by (58) is

$$\eta(\nu_n) = \frac{\Omega_n}{\nu_n} \simeq 1 + \frac{\nu}{2Q_n}\frac{1}{\gamma}\left(1 - \frac{\omega}{\Omega_n}\right), \tag{59}$$

which is drawn in Fig. 8-7. Note that the curve for an absorbing medium has the opposite sign from that in Fig. 8-7 and results for $\mathfrak{N} < 1$.

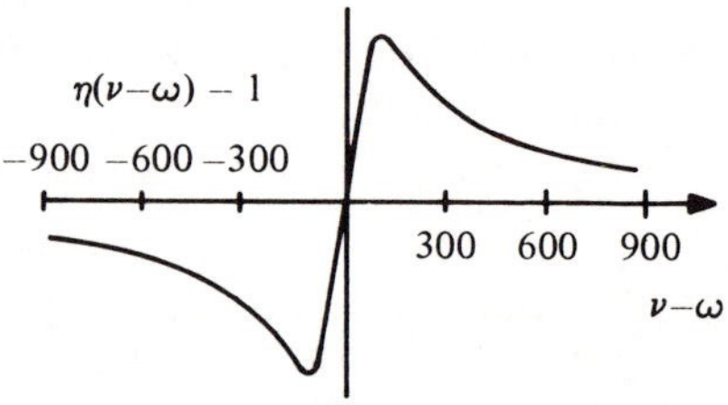

Figure 8-7. Index of refraction $\eta(\nu)$ of (59) versus frequency ν for a gain medium ($\bar{N} > 0$). An inverted curve results for an absorptive medium ($\bar{N} < 0$). The curve is evaluated in linear approximation (valid near threshold) for $\gamma = 2\pi \times 100$ MHz.

Problems

8-1. Stating each approximation when made, derive the self-consistency equations (11) and (12), writing the electric field and polarization in the forms

$$E(z, t) = \sum_n E_n(t) \cos(\nu_n t + \phi_n) U_n(z),$$
$$P(z, t) = \sum_n [C_n(t) \cos(\nu_n t + \phi_n) + S_n(t) \sin(\nu_n t + \phi_n)] U_n(z), \qquad (60)$$

where the slowly varying coefficients C_n and S_n are related to the complex polarization by

$$\mathscr{P}_n(t) = C_n(t) + iS_n(t). \qquad (61)$$

Note that $C_n = \text{Re}(\mathscr{P}_n)$ is the in-phase coefficient, and $S_n = \text{Im}(\mathscr{P}_n)$ is the in-quadrature coefficient.

8-2. Using the steady-state population inversion (37), show that the linear gain increases as γ_a decreases, but that the saturation (R/R_s) increases too. This behavior is observed, for example, in the He-Ne laser, where the 3.39 μm line has large linear gain accompanied by large saturation, while the 6328 Å line has lower gain and saturation.

8-3. Fill in the steps in arriving at the equations of motion (26), (27), and (28), using (24) and (7.34), (7.35), and (7.49).

8-4. Calculate $\mathscr{P}_n$, using the running-wave perturbation mode (7) in (40).

Ans.:

$$\mathscr{P}_n(t) = -i\frac{\wp^2 \bar{N}}{\hbar\gamma} E_n\left(1 - i\frac{\omega - \nu_n}{\gamma}\right)\frac{\gamma^2}{(\omega - \nu_n)^2 + \gamma^2(1 + 2I_n\gamma_{ab}/\gamma)}. \qquad (62)$$

8-5. Using the running-wave result (62), calculate the steady-state intensity I_n. Note that on resonance $I_n = (\mathfrak{N} - 1)\gamma/2\gamma_{ab}$, which ultimately increases linearly in the relative excitation. Expand your result to third-order, and show that this tends to the constant value $I_n \to \gamma/(2\gamma_{ab}\mathscr{L}^2)$. In general third-order theories overestimate the saturation.

8-6. Write rate equations for a two-level medium with the ground state as the lower level. Include pump, decay from the upper level to the lower, and stimulated emission. *Note*: Pump *depletes* lower level and can saturate. This type of system describes ruby approximately. Comment on how it differs from the model of Sec. 8-2. Solve for the steady state, noting that the total number of atoms is constant.

Ans.: Like Eq. (37) with $R_s^{-1} = (\Lambda_a + \gamma_a)/2$, where $\Lambda_a \rho_{bb}$ is the rate of pumping (like λ_a) to the upper level. This kind of medium is often called a "three-level" medium, for pumping actually occurs to additional levels lying above level *a*. These levels decay so rapidly to *a*, however, that the pumping appears to come directly from the ground state. Hence the simplification to two levels. Note that $T_1 = 1/(\gamma_a + \Lambda_a)$, whereas $T_2 = 1/\gamma = 2/\gamma_a$ (NMR terminology) if $\gamma_{pn} \ll \gamma_a$.

8-7. Write and solve (in steady state) the rate equations for upper-level decay to the lower level only, and decay of the lower level to a ground state below. This scheme describes the continuous operation, for example, of a He-Cd laser.

Ans.:

$$\rho_{aa} - \rho_{bb} = \frac{(\lambda_a/\gamma_a) - (\lambda_a + \lambda_b)/\gamma_b}{1 + R/\gamma_a}. \tag{63}$$

8-8. Show that the cavity Q_n (quality factor) of Eq. (5.19) is given for mode n of the Fabry-Perot in Fig. 8-2a as

$$Q_n = \frac{2\pi L}{f_n \lambda_n}, \tag{64}$$

where f_n is the fractional loss per pass through the cavity. Assuming that the energy decays quasi-exponentially in time and that $\Delta\nu$ is the frequency separation between points of half-power, show further that the Q is approximately equal to the resolving power, defined as

$$\text{Resolving power} \equiv \frac{\lambda}{\Delta\lambda} = \frac{\nu}{\Delta\nu}. \tag{65}$$

8-9. Show that a large field which varies little in a time $1/\gamma$ equilibrates the populations, that is, forces $\rho_{aa} \to \rho_{bb}$. This process is part of the reason for spiking (not the same as mode locking of Sec. 9-3) in ruby, as mentioned in Sec. 4-1. Explain qualitatively why a delayed high Q can therefore lead to a shorter, sharper pulse output from a ruby laser (Q switching). The delayed high Q can be obtained by high-speed rotation of one mirror or, more commonly, by a saturable absorber. The latter stops absorbing (saturates) for sufficiently large fields because of this population equilibration process. It can simultaneously generate side bands for mode locking (see Secs. 9-3 through 9-5).

8-10. Show that the spherical wave

$$\Phi(r) = \frac{\exp(iKr)}{r}$$

satisfies the scalar Helmholtz equation

$$\nabla^2\Phi(r) + K^2\Phi(\mathrm{r}) = 0,$$

where the radial distance $r^2 = (x - x_0)^2 + (y - y_0)^2 + (z - z_0)^2$. In particular, consider the choice

$$r = [x^2 + y^2 + (z - iz_0)^2]^{1/2}.$$

Show, for $z^2 + z_0{}^2 \gg x^2 + y^2$, that $\Phi(r)$ is approximated by the Gaussian beam

$$\Phi(r) = \left(\frac{1}{z - iz_0}\right) \exp\left\{iK\left[z + \frac{z}{2(z^2 + z_0{}^2)}(x^2 + y^2)\right]\right\}\exp(Kz_0)$$
$$\times \exp\left[-\frac{1}{2}\frac{Kz_0}{z^2 + z_0{}^2}(x^2 + y^2)\right]. \tag{66}$$

Figure B-5 gives a diagram of this wave form. It is fairly typical of the light distribution in a laser cavity. [Derivation due to H. Kogelnik; see Gordon (1968)].

8-11. Show that the beam waist ($1/e$ point), $w_s(z)$, of the Gaussian beam (66) is given by

$$w_s^2(z) = w_0^2\left(1 + \frac{z^2}{z_0^2}\right), \tag{67}$$

where the waist w_0 at the focal point ($z = 0$) satisfies

$$w_0^2 = w_s^2(0) = \frac{2z_0}{K} = \frac{z_0\lambda}{\pi}. \tag{68}$$

Show further that the surfaces of constant phase are spherical with the radius of curvature given by

$$R_c = z + \frac{z_0^2}{z}. \tag{69}$$

8-12. Show that Eq. (56) for $I_n(t)$ is the general solution of the equation of motion (51). *Hint*: Divide both sides of (51) by its right-hand side, multiply through by dt, and integrate from time $t = 0$ to t. The left-hand side can be integrated by partial fractions.

8-13. Show that $\nabla \cdot \mathbf{P} \simeq 0$ in a laser medium. Hint: note that the inducing field varies slowly in the transverse directions.

8-14. Show that the steady-state intensity for the exact polarization (47) is given by

$$I_n = \frac{\mathfrak{N}\mathscr{L}(\omega - \nu_n) - \frac{1}{4} - \frac{1}{4}\sqrt{8\mathfrak{N}\mathscr{L}(\omega - \nu_n) + 1}}{(\gamma_{ab}/\gamma)\mathscr{L}(\omega - \nu_n)}$$

References

The following list of books and review articles on semiclassical laser theory is far from exhaustive. See also the references for Chap. 10.

Laser Handbook, ed. by F. T. Arecchi and E. O. Schulz-DuBois, North-Holland Publishing Co., Amsterdam, 1972. See specific chapters cited below.

Lasers and Light, with introduction by A. L. Schawlow, W. H. Freeman and Co., San Francisco, 1969.

J. P. Gordon, 1968, "Elements of Laser Theory," in: *Laser Technology and Applications*, ed. by S. L. Marshall, McGraw-Hill Book Co., New York.

H. Haken, 1970, "Laser Theory," in: *Encyclopedia of Physics*, Vol. XXV/2c, ed. by S. Flügge, Springer-Verlag, Berlin. Also Chap. A3 in *Laser Handbook*, *op. cit.*

H. Kogelnik, 1966, "Modes in Optical Resonators," in: *Lasers*, ed. by M. Levine, Marcel Dekker, New York, Vol. I, p. 296.

W. E. Lamb, Jr., 1963, "Theory of Optical Maser Oscillators," in: *Proceedings of the International School of Physics*, Course XXXI, 78. See also chapter in *Quantum Optics and Electronics*, ed. by C. DeWitt, A. Blandin, and C. Cohen-Tannoudji, Gordon and Breach, New York, 1965, pp. 331–381.

M. Lax, 1968, in: *Brandeis University Summer Institute in Theoretical Physics*, 1966: *Statistical Physics, Phase Transitions and Superfluidity*, New York.

M. Sargent III, and M. O. Scully, 1972, "Theory of Laser Operation," Chap. A2 in *Laser Handbook, op. cit.*

A. E. Siegman, 1971, *An Introduction to Lasers and Masers,* McGraw-Hill Book Co., New York.

A. Yariv, 1971, *Introduction to Optical Electronics,* Holt, Rinehart and Winston, New York.

There is a list of laser books in the following: A. E. Siegman, 1971, *Appl. Opt.* **10**, A38. Some basic papers are listed blow. See Chaps. 17 and 20 for basic papers treating fully quantized theories of the laser.

G. D. Boyd and J. P. Gordon, 1961, *Bell Syst. Tech. J.* **40**, 489.

A. G. Fox and T. Li, 1961, *Bell Syst. Tech. J.* **40**, 453.

H. Haken and H. Sauermann, 1963, *Z. Phys.* **173**, 261; **176**, 47.

W. E. Lamb, Jr., 1964, *Phys. Rev.* **134**, A1429.

A. M. Prokhorov, 1958, *Sov. Phys. JETP* **34**, 1658.

A. L. Schawlow and C. H. Townes, 1958, *Phys. Rev.* **112**, 1940.

H. Statz and G. A. deMars, 1960, in: *Quantum Electronics,* ed. by C. H. Townes, Columbia University Press, New York, p. 530.

IX

MULTIMODE OPERATION

9. Multimode Operation

When two or more modes oscillate in a laser, beat frequencies develop because of nonlinearities in the medium. Specifically, the population difference $\rho_{aa} - \rho_{bb}$ contains pulsations at multiples of the intermode frequency. We encountered similar pulsations in Van der Pol's double-tank circuit (Sec. 4-1). The pulsations act very much like a modulator generating side bands on impressed signals. The process is analogous to the Raman effect; in fact, the physics is sufficiently similar that much of the algebra encountered here carries over with little modification to the Raman effect.

The rate equation approximation of Chap. 8 often neglects the pulsations ($\rho_{aa} - \rho_{bb}$ is assumed to vary little in a time $1/\gamma$) and hence is inadequate for a general treatment of multimode operation. Instead, we can represent the multimode field by the superposition† given by Eq. (8.8):

$$E(z, t) = \tfrac{1}{2} \sum_n E_n(t) \exp\{-i[\nu_n t + \phi_n(t)]\} U_n(z) + \text{c.c.} \tag{1}$$

Equations of motion for the amplitudes E_n and phases ϕ_n contain gain and saturation terms encountered in the single-mode case and, in addition, intermode coupling terms resulting from the nonlinearities in the polarization. The derivation of these equations can be carried out in a straightforward fashion provided the field is not too intense and the mode amplitudes and phases vary little in atomic lifetimes. The latter condition is akin to the rate equation approximation but is considerably less restrictive since the total field envelope and the population difference can have rapid pulsations even for constant E_n's and ϕ_n's.

In Sec. 9-1 we derive the polarization of the atomic medium, using third-order perturbation theory. The resulting complex polarization components

†It is also possible to use a time-domain representation along lines described in Chap. 13.

$\mathscr{P}_n$ are substituted into the self-consistency equations (8.11) and (8.12) to yield multimode amplitude- and frequency-determining equations. The form of the equations is quite general and can be used to describe a wide variety of laser configurations, as illustrated in later chapters. The single-mode operation predicted by this theory is the same as that given by the third-order expansion in the rate equation approach of Chap. 8. In Sec. 9-2 we consider general two-mode operation in some detail. Specific numerical results are given in Chap. 10 on gas lasers, Chap. 11 on the ring laser, and Chap. 12 on the Zeeman laser. In Sec. 9-3 three-mode operation is discussed with particular emphasis on mode locking. In Sec. 9-4 general multimode operation is considered in both free-running and mode-locked conditions. The chapter closes with Sec. 9-5 on forced locking.

9-1. Polarization of the Medium

In this section, we integrate the population matrix equations of motion (8.26), (8.27), and (8.28), using the perturbation theory discussed in Sec. 7–4, here with the multimode perturbation energy $\mathscr{V}_{ab} = -\wp E(z, t)$. For field (1) this energy is given in the rotating-wave approximation by

$$\mathscr{V}_{ab}(z, t) = -\tfrac{1}{2}\wp \sum_n E_n(t) \exp[-i(\nu_n t + \phi_n(t))]U_n(z). \tag{2}$$

No nonzero contributions exist for the off-diagonal elements in zeroth-order ($\mathscr{V}_{ab} = 0$), but from (8.27) and (8.28) the diagonal elements do contribute. For example, muliplying (8.27) by the integrating factor $\exp(\gamma_a t)$, we have

$$\frac{d[\rho_{aa}^{(0)} \exp(\gamma_a t)]}{dt} = \lambda_a \exp(\gamma_a t).$$

Integrating this from $-\infty$ to t, we obtain

$$\rho_{aa}^{(0)}(z, t) = \lambda_a \gamma_a^{-1}. \tag{3}$$

With a similar result for $\rho_{bb}^{(0)}$, we have the zeroth-order population difference:

$$N(z, t) \equiv \rho_{aa}^{(0)} - \rho_{bb}^{(0)} = \lambda_a \gamma_a^{-1} - \lambda_b \gamma_b^{-1}. \tag{4}$$

Using this difference to evaluate the formal integral (8.30) for the off-diagonal element ρ_{ab} in first order, we have

$$\begin{aligned}\rho_{ab}^{(1)} &= i\hbar^{-1} \int_{-\infty}^{t} dt' \exp[-(i\omega + \gamma)(t - t')]\,\mathscr{V}_{ab}(t')[\rho_{aa}^{(0)} - \rho_{bb}^{(0)}] \\ &= -\tfrac{1}{2} i\wp\hbar^{-1} N(z, t) \sum_\sigma E_\sigma(t) \exp[-i(\nu_\sigma t + \phi_\sigma)]U_\sigma(z) \\ &\quad \times \int_{-\infty}^{t} dt' \exp\{-[i(\omega - \nu_\sigma) + \gamma](t - t')\}\end{aligned}$$

$$= -\tfrac{1}{2}\wp i\hbar^{-1} N(z, t) \sum_\sigma E_\sigma \exp[-i(\nu_\sigma t + \phi_\sigma)]\, U_\sigma(z)\, \mathscr{D}(\omega - \nu_\sigma), \tag{5}$$

where

$$\mathscr{D}_x(\Delta\omega) = \frac{1}{\gamma_x + i\Delta\omega}, \qquad x = a, b, \text{ or missing}, \tag{6}$$

is a convenient abbreviation for a frequently occurring complex denominator. In the second line of (5), we assumed that the E_σ, ϕ_σ, and $N(z, t)$ vary little in a time $1/\gamma$ and can be factored outside the time integration. Substituting (5) into (8.27), we find the upper-level population equation:

$$\dot\rho_{aa} \simeq \dot\rho_{aa}^{(0)} + \dot\rho_{aa}^{(2)} = \lambda_a - \gamma_a\rho_{aa} - \Bigg[\frac{1}{4}\left(\frac{\wp}{\hbar}\right)^2 N \sum_\rho\sum_\sigma E_\rho E_\sigma$$

$$\times \exp\{i[(\nu_\rho - \nu_\sigma)t + \phi_\rho - \phi_\sigma]\}\, \mathscr{D}(\omega - \nu_\sigma)\, U_\rho^* U_\sigma + \text{c.c.}\Bigg]. \tag{7}$$

Here we see that the population ρ_{aa} contains pulsations at the intermode beat frequencies. In the rate equation approximation of Sec. 8-2, we neglect these pulsations and obtain an expression like (7) with $N(z, t)$ replaced by the population difference $\rho_{aa} - \rho_{bb}$ and the double summation replaced by a single summation (see Prob. 9-1). Inasmuch as the pulsations can lead to important mode coupling terms, this approximation is often not particularly good. Integrating (7), we obtain the second-order contribution

$$\rho_{aa}^{(2)} = -\frac{1}{4}\left(\frac{\wp}{\hbar}\right)^2 N \sum_\rho\sum_\sigma E_\rho E_\sigma U_\rho^* U_\sigma \exp\{i[(\nu_\rho - \nu_\sigma)t + \phi_\rho - \phi_\sigma]\}$$

$$\times \mathscr{D}_a(\nu_\rho - \nu_\sigma)\, \mathscr{D}(\omega - \nu_\sigma) + \text{c.c.} \tag{8}$$

Here we have assumed that the mode amplitude and phases vary little in times $1/\gamma_a$ and $1/\gamma_b$. This is generally a good approximation for the He-Ne laser but not, for example, for the ruby laser, where $\gamma_a \simeq 2\pi$ kHz ($= 10^3$ sec^{-1}). Similarly the contribution

$$\rho_{bb}^{(2)} = -\rho_{aa}^{(2)} \quad \text{with } \gamma_a \to \gamma_b. \tag{9}$$

Equation (8) can be written more simply in full by noting that an interchange of the indices ρ and σ in the complex conjugate term results in an expression identical to that written explicitly, except that the factor $\mathscr{D}(\omega - \nu_\sigma)$ is replaced by $\mathscr{D}(\nu_\rho - \omega)$. Combining this observation with (9), we have the second-order contribution to the population difference:

$$\rho_{aa}^{(2)} - \rho_{bb}^{(2)} = -\frac{1}{4}\left(\frac{\wp}{\hbar}\right)^2 N \sum_\rho\sum_\sigma E_\rho E_\sigma U_\rho^* U_\sigma \exp\{i[(\nu_\rho - \nu_\sigma)\, t + \phi_\rho - \phi_\sigma]\}$$

$$\times [\mathscr{D}_a(\nu_\rho - \nu_\sigma) + \mathscr{D}_b(\nu_\rho - \nu_\sigma)]\,[\mathscr{D}(\omega - \nu_\sigma) + \mathscr{D}(\nu_\rho - \omega)]. \tag{10}$$

Substituting (10) into the formal integral (8.30) for ρ_{ab}, we integrate to obtain the third-order contribution:

$$\rho_{ab}^{(3)} = \frac{1}{8} i \left(\frac{\wp}{\hbar}\right)^3 N \sum_\mu\sum_\rho\sum_\sigma E_\mu E_\rho E_\sigma U_\mu U_\rho^* U_\sigma$$

$$\times \exp\{-i[(\nu_\mu - \nu_\rho + \nu_\sigma)t + \phi_\mu - \phi_\rho + \phi_\sigma]\} \mathscr{D}(\omega - \nu_\mu + \nu_\rho - \nu_\sigma)$$

$$\times [\mathscr{D}_a(\nu_\rho - \nu_\sigma) + \mathscr{D}_b(\nu_\rho - \nu_\sigma)][\mathscr{D}(\omega - \nu_\sigma) + \mathscr{D}(\nu_\rho - \omega)]. \tag{11}$$

In combining (11) with (8.29) for $P_n(t)$, it is necessary to calculate the integral

$$\frac{\int_0^L dz\, N(z) U_n^*(z) U_\mu(z) U_\rho^*(z) U_\sigma(z)}{\int_0^L dz\, |U_n(z)|^2}. \tag{12}$$

For the standing-wave laser† the normal modes are $U_n(z) = \sin K_n z$, and the product of sines can be simplified by using trigonometric identities as follows:

$$(\sin K_n z \sin K_\mu z)(\sin K_\rho z \sin K_\sigma z)$$
$$= \tfrac{1}{4}\{\cos[(K_n - K_\mu)z] - \cos[(K_n + K_\mu)z]\}$$
$$\times \{\cos[(K_\rho - K_\sigma)z] - \cos[(K_\rho + K_\sigma)z]\}.$$

The products of cosines can by similarly expanded, and we keep only terms which are slowly varying in z, that is, we neglect terms like $\cos(K_n + K_\mu + K_\rho - K_\sigma)z$, for these vanish in the integral (12) over z. The product of sines becomes

$$\tfrac{1}{8}\{\cos[(K_n - K_\mu + K_\rho - K_\sigma)z] + \cos(K_n - K_\mu - K_\rho + K_\sigma)z]$$
$$+ \cos[(K_n + K_\mu - K_\rho - K_\sigma)z]\}.$$

Further note that the complex polarization $\mathscr{P}_n(t)$ of (8.29) with $\rho_{ab}^{(3)}$ of (11) includes the phase factor $\exp[-i(\nu_n - \nu_\mu + \nu_\rho - \nu_\sigma)t]$. Since E_n in (11) cannot respond to frequency variations much larger than the cavity band width ν/Q, we retain only terms for which $\nu_n - \nu_\mu + \nu_\rho - \nu_\sigma \simeq 0$, that is, terms for which

$$n = \mu - \rho + \sigma. \tag{13}$$

The product of sines in (12) becomes in these approximations

$$\tfrac{1}{8}\{1 + \cos[2(K_\sigma - K_\rho)z] + \cos[2(K_\mu - K_\rho)z]\}$$

and by (11) the third-order contribution to the complex polarization

$$\mathscr{P}_n^{(3)} = \tfrac{1}{16} i \wp^4 \hbar^{-3} \bar{N} \sum_\mu \sum_\rho \sum_\sigma E_\mu E_\rho E_\sigma \exp(i\Psi_{n\mu\rho\sigma})$$

$$\times [1 + \bar{N}^{-1}(N_{2(\sigma-\rho)} + N_{2(\mu-\rho)})] \mathscr{D}(\omega - \nu_\mu + \nu_\rho - \nu_\sigma)$$

$$\times [\mathscr{D}_a(\nu_\rho - \nu_\sigma) + \mathscr{D}_b(\nu_\rho - \nu_\sigma)][\mathscr{D}(\omega - \nu_\sigma) + \mathscr{D}(\nu_\rho - \omega)], \tag{14}$$

where the relative phase angles

$$\Psi_{n\mu\rho\sigma} = (\nu_n - \nu_\mu + \nu_\rho - \nu_\sigma)t + \phi_n - \phi_\mu + \phi_\rho - \phi_\sigma \tag{15}$$

† The unidirectional ring laser is discussed in Prob. 9-2.

and the spatial Fourier components

$$N_{2l} = \frac{1}{L}\int_0^L dz\, N(z)\cos[(2l\pi/L)z]. \tag{16}$$

Here $N_0 = \bar{N}$ by Eq. (8.42).

Combining the complex polarization

$$\mathscr{P}_n(t) = \mathscr{P}_n^{(1)}(t) + \mathscr{P}_n^{(3)}(t) \tag{17}$$

determined by Eqs. (5) and (14) with the self-consistency equations (8.11) and (8.12) we have the multimode amplitude- and frequency-determining equations:

$$\dot{E}_n = \alpha_n E_n - \sum_\mu\sum_\rho\sum_\sigma E_\mu E_\rho E_\sigma \operatorname{Im}\{\vartheta_{n\mu\rho\sigma}\exp(i\Psi_{n\mu\rho\sigma})\}, \tag{18}$$

$$\nu_n + \dot{\phi}_n = \Omega_n + \sigma_n - \sum_\mu\sum_\rho\sum_\sigma E_\mu E_\rho E_\sigma E_n^{-1} \operatorname{Re}\{\vartheta_{n\mu\rho\sigma}\exp(i\Psi_{n\mu\rho\sigma})\}, \tag{19}$$

where the various coefficients are defined with their physical roles in Table 9-1. In using (18) and (19), we neglect first-order components with phase angles like $(\nu_2 - \nu_1)t + \phi_2 - \phi_1$, for these vary rapidly compared to E_n and ϕ_n for typical laser parameters and average to zero in time, as discussed for Eq. (13).

TABLE 9-1. *Definitions of Multimode Saturation Coefficients Appearing in Amplitude and Frequency Equations (18)–(25)*

Other variables are given in Table 8-1. The factor $F_3 = (3/2)\ (\gamma_{ab}/\gamma)F_1$ is chosen for use with dimensionless intensities defined by (8.45). For β_n, see Table 8-1.

Coefficient	Physical Content
$\vartheta_{n\mu\rho\sigma} = \frac{1}{4}i[\frac{1}{2}\wp/\hbar]^2 F_1\gamma[1 + (N_{2(\rho-\sigma)} + N_{2(\rho-\mu)})/\bar{N}]$ $\times \mathscr{D}(\omega - \nu_\mu + \nu_\rho - \nu_\sigma)[\mathscr{D}_a(\nu_\rho - \nu_\sigma) + \mathscr{D}_b(\nu_\rho - \nu_\sigma)]$ $\times [\mathscr{D}(\omega - \nu_\sigma) + \mathscr{D}(\nu_\rho - \omega)]$	General saturation term
$\theta_{nm} = \operatorname{Im}(\vartheta_{nnmm} + \vartheta_{nmmn})[2\hbar^2\gamma_a\gamma_b/\wp^2]$ $= F_3\frac{1}{3}[2 + N_{2(n-m)}/\bar{N}][\mathscr{L}(\omega - \nu_n)\mathscr{L}(\omega - \nu_m)$	Cross-saturation ($n \neq m$) (Hole burning part)
$+ \operatorname{Re}\{\gamma\mathscr{D}(\omega - \nu_n)[\gamma_a\gamma_b/(\gamma_a + \gamma_b)][\mathscr{D}_a(\nu_m - \nu_n)$ $+ \mathscr{D}_b(\nu_m - \nu_n)]\frac{1}{2}\gamma[\mathscr{D}(\omega - \nu_n) + \mathscr{D}(\nu_m - \omega)]\}]$	(Population pulsation part)
$\tau_{nm} = \operatorname{Re}(\vartheta_{nnmm} + \vartheta_{nmmn})[2\hbar^2\gamma_a\gamma_b/\wp^2]$	Cross pushing
$= F_3\frac{1}{3}[2 + N_{2(n-m)}/\bar{N}][(\omega - \nu_n)/\gamma]\mathscr{L}(\omega - \nu_n)$	(Hole burning part)
$\mathscr{L}(\omega - \nu_m) - \operatorname{Im}\{\ldots\}$	(Population pulsation part, { } as for θ_{nm})
$\mathscr{D}_x(\Delta\omega) = (\gamma_x + i\,\Delta\omega)^{-1}$	Complex frequency denominator ($x = a, b$, or blank)
$\Psi_{n\mu\rho\sigma} = (\nu_n - \nu_\mu + \nu_\rho - \nu_\sigma)t + \phi_n - \phi_\mu + \phi_\rho - \phi_\sigma$	Relative phase angle

9-2 Two-Mode Operation

In this section we discuss the simplest case of multimode operation, that for two modes. Although the discussion is made with specific reference to the homogeneously broadened, standing-wave (sin Kz) laser, many of the results apply also to other two-mode problems, such as those encountered in the gas, ring, and Zeeman lasers (Chaps. 10, 11, and 12) and in van der Pol's double-tank circuit of Sec. 4-1. The mathematical similarity of these problems reflects the fact that the physical phenomena involved are analogous. In particular, each problem has both saturated gain and mode coupling. In turn, the mode coupling arises from two sources, a direct saturation source providing competition for the gain and a pulsation source for which the nonlinear gain medium acquires pulsations at the intermode frequency. As mentioned in Chap. 4, these pulsations interact with a mode to produce frequency *shifted* outputs, oscillating at both the frequency of the *other* mode and at a new frequency. The first of these shifted "tones" couples the modes and, for example, in van der Pol's problem results in *twice* as much cross-saturation (effect of one mode on the other) as self-saturation. The second shifted tone has no importance in two-mode operation but is responsible for multimode self-locking, as discussed in Sec. 9-3. In the laser media, the pulsations occur in the population difference $\rho_{aa} - \rho_{bb}$ and hence are called "population pulsations." The two sources of coupling are identified in the θ's of Tables 9-1, 10-2, 11-1, and 12-1.

When two modes are above threshold, the amplitude equations reduce to

$$\dot{E}_1 = E_1(\alpha_1 - \beta_1 I_1 - \theta_{12} I_2), \tag{20}$$

$$\dot{E}_2 = E_2(\alpha_2 - \beta_2 I_2 - \theta_{21} I_1), \tag{21}$$

where the self-saturation coefficients β_1 and β_2 and the cross-saturation coefficients θ_{12} and θ_{21} are given in Table 9-1, and I_n are the dimensionless intensities $\frac{1}{2}(\wp^2/\hbar^2\gamma_a\gamma_b)E_n{}^2$. We have further neglected terms containing the phase angles $\Psi_{1121} = (\nu_2 - \nu_1)t + \phi_2 - \phi_1$, $\Psi_{1212} = 2\Psi_{1121}$, $\Psi_{1112} = -\Psi_{1211}$, for the terms average to zero in time for typical laser parameters, as discussed for Eq. (13). This assumption must be reappraised, however, if $\nu_2 \cong \nu_1$. The hole burning part of the cross-saturation coefficient θ_{12} resembles a self-saturation coefficient β_1 with the Lorentzian squared replaced by the product of two Lorentzians and the factor 3 replaced by $2 + N_2/\bar{N}$. To the extent that the Lorentzians are equal, that $N_2 = \bar{N}$, and that population pulsations are negligible, the third-order coefficients have the same value.

In this connection, Fig. 9-1 helps in visualizing the integral (16), which defines N_2. There it is shown for $N(z) > 0$ that, when the active medium is located in the middle of the cavity, N_2 is negative, and when in the ends, N_2 is positive. Hence the coupling between modes is greater for the latter location.

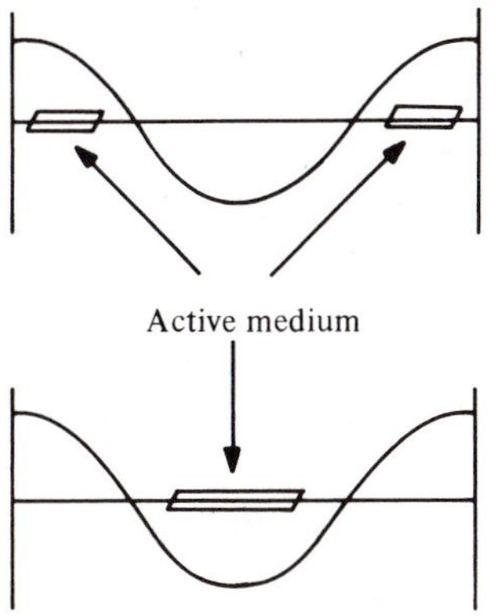

Figure 9-1. (a) Active medium located in ends of cavity yields a positive value for $N_2 = L^{-1}\int_0^L dz\, N(z)\cos(2\pi z/L)$, for the cosine is positive over nonzero values of $N(z)$. (b) Active medium located in middle of cavity yields $N_2 < 0$, for the cosine is negative over nonzero values of $N(z)$. Here we have assumed that $N(z)$ is itself positive (inverted medium), which is not, however, an assumption of the theory. As the medium is located closer to the mirrors, N_2 approaches $\bar{N} = L^{-1}\int_0^L dz\, N(z)$ in value. For grating interpretation, see Sargent (1976).

If the medium is uniformly distributed throughout the cavity, N_2 vanishes, for the cosine averages to zero over one cycle. Here the coupling is intermediate in strength between the values for the other two cases. The terms with nonzero phase angles neglected above are all proportional to N_2 and hence vanish identically for a uniformly distributed medium.

Corresponding to the amplitude equations (20) and (21), we have [from (19)] the two-mode frequency-determining equations

$$\nu_1 + \dot{\phi}_1 = \Omega_1 + \sigma_1 - \rho_1 I_1 - \tau_{12} I_2, \tag{22}$$

$$\nu_2 + \dot{\phi}_2 = \Omega_2 + \sigma_2 - \rho_2 I_2 - \tau_{21} I_1, \tag{23}$$

where the coefficients are defined in Table 9-1.

Multiplying (20) and (21) by $E_1\wp^2/(\hbar^2\gamma_a\gamma_b)$ and $E_2\wp^2/(\hbar^2\gamma_a\gamma_b)$ respectively, we find the equations of motion for the dimensionless intensities [or, equivalently, for the mode energies (8.16)]:

$$\dot{I}_1 = 2I_1(\alpha_1 - \beta_1 I_1 - \theta_{12} I_2), \tag{24}$$

$$\dot{I}_2 = 2I_2(\alpha_2 - \beta_2 I_2 - \theta_{21} I_1). \tag{25}$$

Stationary states are the solutions to these equations, for which

$$\dot{I}_1 = \dot{I}_2 = 0. \tag{26}$$

There are four possibilities: I_1 can be zero or nonzero, and, independently, I_2 can be zero or nonzero. In determining laser operation, solutions which are both physical, that is, nonnegative, and stable are required. If both net-gain coefficients a are negative, it is clear that no oscillation can occur. If one a is negative and the other positive, the mode with positive a oscillates with the single-mode solution $I_n = a_n/\beta_n$.† If both a's are positive, it is not so obvious what the stable solutions are and a more systematic approach must be used.

A method which deals correctly with any possible oscillation is the small vibrations analysis. For this, we consider small deviations of the intensities about a stationary solution

$$I_1 = I_1^{(s)} + \varepsilon_1, \tag{27}$$

$$I_2 = I_2^{(s)} + \varepsilon_2, \tag{28}$$

where the superscript (s) denotes the stationary-state value. The condition for a stable solution is, then, that the small deviations

$$\varepsilon_1 \to 0, \qquad \varepsilon_2 \to 0, \qquad \text{as } t \to \infty. \tag{29}$$

Armed with this approach, let us now consider stationary states for $a_1, a_2 > 0$. We must consider all stationary solutions, for more than one might be both physical and stable.

Suppose that $I_1^{(s)} = 0$ and $I_2^{(s)} = a_2/\beta_2$. Using the intensities (27) and (28) in the equations of motion (24) and (25), we have

$$\dot{\varepsilon}_1 = 2\varepsilon_1\left(a_1 - \theta_{12}\frac{a_2}{\beta_2}\right) + O(\varepsilon^2), \tag{30}$$

$$\dot{\varepsilon}_2 = -2\left(\frac{a_2}{\beta_2}\right)(\beta_2\varepsilon_2 + \theta_{21}\varepsilon_1) + O(\varepsilon^2), \tag{31}$$

where the expression $O(\varepsilon^2)$ represents terms of order ε^2, which can be neglected for sufficiently small deviations. Equation (30) requires that the deviation $\varepsilon_1 \to 0$ in time provided the effective a,

$$a_1' = a_1 - \theta_{12}\frac{a_2}{\beta_2}, \tag{32}$$

is negative. Here a_1' is the gain that mode 1 experiences at the start of oscillation in the presence of mode 2 oscillating with intensity a_2/β_2. This gain is less than that with no modes oscillating, for some of the excited atoms available for I_1 in the absence of oscillation contribute to I_2, that is, I_2 partly saturates the laser medium, thereby reducing the gain for I_1. In addition, population pulsations couple the modes. If $a_1' < 0$ but $a_1 > 0$, we say that I_2 *inhibits* oscillation of I_1 (mode inhibition). If a_1 is sufficiently large to over-

† Provided that the θ's are positive (virtually always the case, but see Prob. 9-3 for a counterexample).

come the competition by I_2, I_1 builds up. In this case the deviation ε_1 does not decay in time, and the solution $I_1^{(s)} = 0$, $I_2^{(s)} = a_2/\beta_2$ is unstable. Similar discussion applies to the solution $I_1^{(s)} = a_1/\beta_1$, $I_2^{(s)} = 0$.

If neither intensity is zero, condition (26) for the stationary state implies that the parenthesized expressions in (24) and (25) must be zero, that is,

$$\beta_1 I_1^{(s)} + \theta_{12} I_2^{(s)} = a_1, \tag{33}$$

$$\beta_2 I_2^{(s)} + \theta_{21} I_1^{(s)} = a_2. \tag{34}$$

These equations are linear and have the solutions

$$I_1^{(s)} = \frac{(a_1/\beta_1) - (\theta_{21}/\beta_1)(a_2/\beta_2)}{1 - C} = \frac{a_1'/\beta_1}{1 - C}, \tag{35}$$

$$I_2^{(s)} = \frac{(a_2/\beta_2) - (\theta_{12}/\beta_2)(a_1/\beta_1)}{1 - C} = \frac{a_2'/\beta_2}{1 - C}, \tag{36}$$

where the "coupling constant"

$$C \equiv \frac{\theta_{12}\theta_{21}}{\beta_1\beta_2}. \tag{37}$$

In the limit of small coupling, $C \to 0$, $a_1' \to a_1$, and $a_2' \to a_2$, and we see from (35) and (36) that the modes oscillate independently with single-mode steady-state intensities a_1/β_1 and a_2/β_2. If one effective a is negative, the solution is unphysical and must be discarded.

Substituting these solutions with (27) and (28) into (24) and (25), we find

$$\dot{\varepsilon}_1 = -2I_1^{(s)}(\beta_1\varepsilon_1 + \theta_{12}\varepsilon_2), \tag{38}$$

$$\dot{\varepsilon}_2 = -2I_2^{(s)}(\beta_2\varepsilon_2 + \theta_{21}\varepsilon_1). \tag{39}$$

In matrix form this is

$$\frac{d}{dt}\begin{pmatrix}\varepsilon_1\\ \varepsilon_2\end{pmatrix} = \Theta\begin{pmatrix}\varepsilon_1\\ \varepsilon_2\end{pmatrix}, \tag{40}$$

where the stability matrix

$$\Theta = \frac{-2}{1 - C}\begin{pmatrix} a_1' & a_1'(\theta_{12}/\beta_1) \\ a_2'(\theta_{21}/\beta_2) & a_2' \end{pmatrix} \tag{41}$$

Appropriate linear combinations of the ε's diagonalize Θ, Specifically

$$\dot{\varepsilon}_1' = \lambda_1\varepsilon_1', \qquad \dot{\varepsilon}_2' = \lambda_2\varepsilon_2',$$

where λ_1 and λ_2 are the eigenvalues of the Θ matrix. Since the original ε's are, in turn, linear combinations of ε_1' and ε_2', all relax to zero in time provided the eigenvalues λ_1 and λ_2 are negative. To find these values, we set

$$\Theta\begin{pmatrix}\varepsilon_1\\ \varepsilon_2\end{pmatrix} = \lambda\begin{pmatrix}\varepsilon_1\\ \varepsilon_2\end{pmatrix},$$

which can be true if and only if the determinant

$$\mathrm{Det}\,(\Theta - \lambda\mathscr{I}) = 0,$$

where $\mathscr{I}$ is the identity matrix. This yields the eigenvalues

$$\lambda_{1,2} = -\frac{a_1' + a_2'}{1 - C} \pm \sqrt{\left(\frac{a_1' + a_2'}{1 - C}\right)^2 - 4\frac{a_1'a_2'}{1 - C}}. \tag{42}$$

For positive (physical) intensities (35) and (36), there are two possibilities: (*a*) $C < 1$ and $a_1' > 0$, $a_2' > 0$; and (*b*) $C > 1$ and $a_1' < 0$, $a_2' < 0$. The former has negative eigenvalues (42) and is stable, for the magnitude of the square root in (42) is smaller than the leading term. In fact the difference between squares of the latter and the former is

$$\left(\frac{a_1' + a_2'}{1 - C}\right)^2 - \left[\left(\frac{a_1' + a_2'}{1 - C}\right)^2 - \frac{4a_1'\,a_2'}{1 - C}\right] = \frac{4a_1'\,a_2'}{1 - C}.$$

Conversely, the second possibility has one positive eigenvalue and is unstable, for the magnitude of the square root in (42) exceeds that of the leading term. Because both effective a's are negative, both single-mode solutions are stable. The solution which oscillates depends on the intitial conditions, as is shown in Fig. 9-4. If $C = 1$, any solution satisfying (33) or equivalently (34) exhibits neutral equilibrium for these yield eigenvalues with zero values. We summarize our results in Table 9-2.

TABLE 9-2. *Conditions for Which Stationary Solutions for the Two-Mode Equations (24) and (25) are Stable.*

We have assumed that the cross-saturation coefficients θ_{12}, θ_{21} are positive and that the intensities I_1, $I_2 \geq 0$ (negative intensity solutions are nonphysical). Here, for example, the effective a, $a_1' = a_1 - \theta_{12}(a_2/\beta_2)$, is the "linear" net-gain coefficient for I_1 when I_2 is oscillating alone. The coupling constant $C = (\theta_{12}\,\theta_{21})/(\beta_1\beta_2)$.

I_1	I_2	Conditions for Stability	
0	0	$a_1 < 0$, $a_2 < 0$	
$\frac{a_1}{\beta_1}$	0	$a_1 > 0$, $a_2' < 0$	a_1' and $a_2' < 0$, or $C > 1$
0	$\frac{a_2}{\beta_2}$	$a_1' < 0$, $a_2 > 0$	
$\frac{a_1'/\beta_1}{1 - C}$	$\frac{a_2'/\beta_2}{1 - C}$	$C < 1$	
I_1	$\frac{a_1 - \beta_1 I_1}{\theta_{12}}$	$C = 1$	

As we have seen, the degree of coupling between the modes is conveniently expressed in terms of the coupling constant C. We define three regions for coupling strengths as follows:

$$
\begin{aligned}
C &< 1 \quad \text{weak coupling,} \\
C &= 1 \quad \text{neutral coupling,} \\
C &> 1 \quad \text{strong coupling.}
\end{aligned} \tag{43}
$$

Note that weak coupling does not necessarily imply that both modes oscillate; one mode may inhibit oscillation of the other, as expressed by a negative effective a. Examples of this mode inhibition are given in Figs. 9-2, 10-6, 11-2, and 12-4*a*. Neutral coupling implies that the two lines (33) and (34) are parallel. As shown in Prob. 9-5, any combination of intensities on those lines is stable.

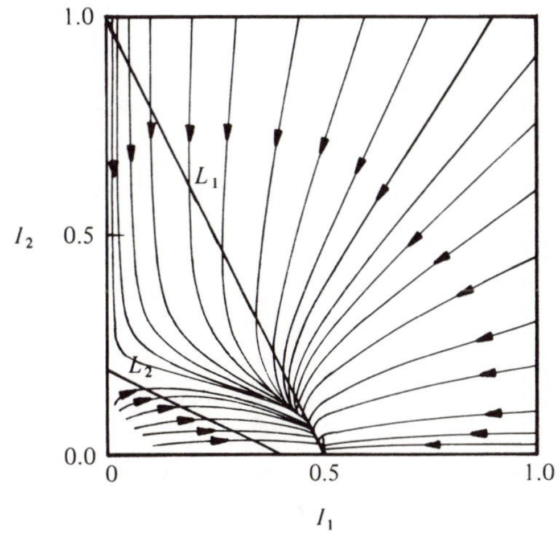

Figure 9-2. Phase curves showing the transient behavior of two-mode oscillation. In time, intensity pairs follow the phase curves in the direction of the arrows until they reach the steady-state value $I_1 = 0.5$, $I_2 = 0$. The straight lines L_1 and L_2 of Eqs. (33) and (34) are taken to have coefficients $a_1 = 1$, $a_2 = 0.4$, $\beta_1 = \beta_2 = 2$, $\theta_{12} = \theta_{21} = 1$. The slope of a phase curve is zero when it crosses line L_2 and infinite when it crosses L_1. Although both modes are above threshold, the favored I_1 oscillation is able to quench the I_2 oscillation. From Lamb (1964).

The time dependences of the intensities predicted by (24) and (25) are illustrated by Figs. 9-2, 9-3, and 9-4, which show stable steady-state solutions commensurate with the conclusions listed in Table 9-2.

It is interesting to note that, for the homogeneously broadened stationary atoms with small population pulsations ($\nu_2 - \nu_1 \gg \gamma_a, \gamma_b$), $C \cong [\frac{2}{3}(1 +$

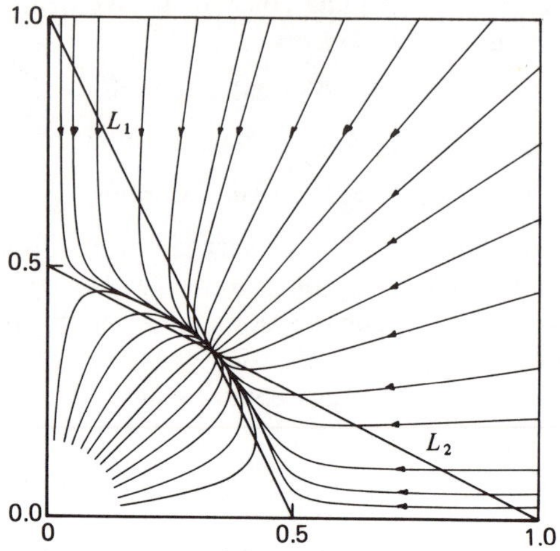

Figure 9-3. Diagram similar to Fig. 9-2, except that the gain parameter for the second mode has been raised to $a_2 = 1$. Simultaneous oscillations at both frequencies occur at the single stable steady state. Both Figs. 9-2 and 9-3 correspond to "weak" coupling ($C < 1$). From Lamb (1964).

$\frac{1}{2} N_2/\bar{N})]^2$, which is independent of tuning and is ordinarily less than unity for all gain media of nonzero length, that is, weak coupling (but see Prob. 9-4). This might be surprising, for with a homogeneously broadened line both modes might be expected to obtain gain from the same atoms and thus, with additional coupling from pulsation terms, to compete strongly ($C > 1$). In fact, however, the modes do not share all atoms because they burn different standing-wave holes in the population inversion [see remarks following (8.39)] as a result of their different oscillation frequencies. As we will see in the next chapter (Fig. 10-5), two modes within a homogeneous linewidth in a predominately Doppler broadened medium do compete strongly, for there the atoms move through several wavelengths of the field and see both modes.

We recall from Sec. 4-1 that van der Pol considered a two-tank circuit whose normal modes satisfied Eqs. (4.24), which have the same form as the intensity equations (24) and (25) with $\theta_{12} = \theta_{21} = 2\beta_1 = 2\beta_2$, that is, strong coupling occurs with the time development indicated in Fig. 9-4. As mentioned also in Sec. 4-1, similar coupled equations appear in the descriptions of many phenomena of both natural and sociological origin.

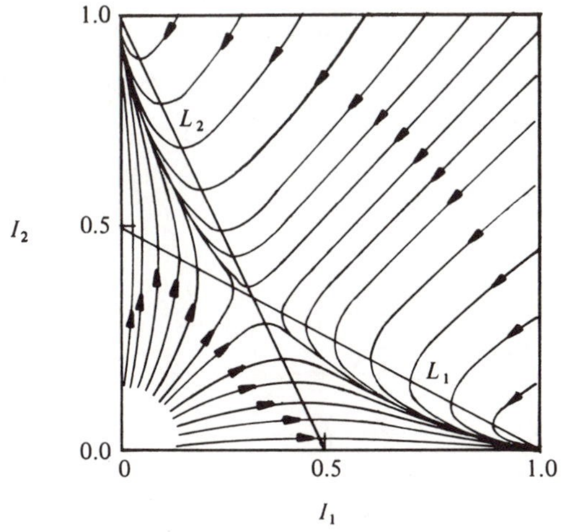

Figure 9-4. Phase curves showing the transient behavior of two-mode oscillation when the straight lines L_1 and L_2 of Eqs. (33) and (34) are taken to have the coefficients $\alpha_1 = \alpha_2 = 1$, $\beta_1 = \beta_2 = 1$, $\theta_{12} = \theta_{21} = 2$ (strong coupling). There are two possible stable steady states, each corresponding to single-frequency operation. The particular state reached depends on the initial conditions. Hysteresis phenomena would occur if the parameters characterizing the oscillator were slowly changed. From Lamb (1964).

9-3. Locking of Beat Frequencies Between Three Modes

When three modes are oscillating, a detector ordinarily registers two distinct low-frequency beat notes given by $\nu_2 - \nu_1$ and $\nu_3 - \nu_2$ which are approximately equal to $c/2L$. In addition there is a very-low-frequency (even audio) tone resulting from the beat between these beats, namely, $(\nu_3 - \nu_2) - (\nu_2 - \nu_1)$, which is generated by the nonlinearities in the medium. The last of these is represented in the theory by very slowly-varying relative phase angles such as

$$\Psi \equiv \Psi_{2123} = (2\nu_2 - \nu_1 - \nu_3)t + 2\phi_2 - \phi_1 - \phi_3. \tag{44}$$

In fact, the ν's are close to the Ω's for which $2\Omega_2 - \Omega_1 - \Omega_3 = 0$. These terms result from nonlinear beating between frequencies in the medium to produce "combination" tones very nearly equal in frequency to the normal mode frequencies Ω_n. For example, frequency ν_2 beats with ν_3, forming the difference frequency $\nu_2 - \nu_3$ in the population difference (population pulsation), which, in turn, interacts with mode 2 to give the tone

$$2\nu_2 - \nu_3 \equiv \nu_1', \tag{45}$$

which is very nearly equal to ν_1 and contributes in third-order to the complex polarization for mode 1. This is illustrated by Fig. 9-5 and follows from the trigonometric identities applying to the nonlinear active medium:

$$\cos \nu_2 t \cos \nu_2 t \cos \nu_3 t = \tfrac{1}{2}[\cos (2\nu_2 t) + 1] \cos \nu_3 t$$
$$= \tfrac{1}{4} \cos [(2\nu_2 + \nu_3)t] + \tfrac{1}{4} \cos [(2\nu_2 - \nu_3)t] + \tfrac{1}{2} \cos \nu_3 t.$$

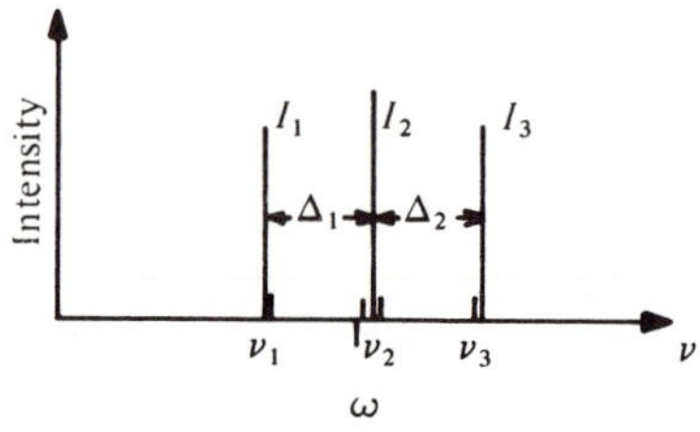

Figure 9-5. Diagram of three-mode operation, showing the intensities I_1, I_2, and I_3 and the corresponding oscillation frequencies ν_1, ν_2, and ν_3. As the cavity is tuned so that Ω_2 approaches line center, mode 2 sees less and less dispersion ($\nu_2 \to \Omega_2 \to \omega$) and the side modes see oppositely directed dispersion. Hence the beat notes Δ_1 between mode 2 and mode 1 and Δ_2 between mode 3 and mode 2 approach one another. In mode-locked operation $\Delta_1 = \Delta_2$. The small lines drawn next to the mode intensities represent combination tones. Note that the placement of combination tones here is suggestive only.

This difference tone is associated with the relative phase angle $\Psi_{1232} = -\Psi$. It acts very much like the injected signal in van der Pol's triode oscillator circuit of Sec. 4-2. There the oscillator "locked" onto the injected signal when the frequency difference between the two was sufficiently small. Similarly ν_1' tempts mode 1 to oscillate at frequency ν_1', that is, to frequency-lock so that

$$\nu_1 \equiv \nu_1' = 2\nu_2 - \nu_3. \tag{46}$$

This condition can be written as

$$\nu_2 - \nu_1 = \nu_3 - \nu_2, \tag{47}$$

that is, the beat notes between adjacent modes are equal. This situation is usually called mode locking, and lasers oscillating in this fashion are said to be mode locked. This type differs from other kinds of mode locking, such as that for oppositely directed running-wave modes in the ring laser (Chap. 11), in that it is the beat frequencies between adjacent modes which lock to the same frequency, rather than the frequencies of the modes themselves. In this section we investigate the three-mode locking problem analytically and indicate the conditions under which locking takes place. As we will see in later chapters, much of the analysis presented here can be applied also to other simple mode-locking problems; in fact, van der Pol's injected signal was a special case. More involved problems may be solved by means of techniques developed in the next section or by an approach in the time (Chap. 13) rather than the frequency domain.

Specializing the amplitude- and frequency-determining equations to the three-mode case, we find

$$\dot{E}_1 = E_1\left(\alpha_1 - \sum_{m=1}^{3} \theta_{1m} I_m\right) - \mathrm{Im}\{\vartheta_{1232} \exp(-i\Psi)\}\, E_2^2 E_3, \tag{48}$$

$$\dot{E}_2 = E_2\left(\alpha_2 - \sum_{m=1}^{3} \theta_{2m} I_m\right) - \mathrm{Im}\{(\vartheta_{2123} + \vartheta_{2321}) \exp(i\Psi)\}\, E_1 E_2 E_3, \tag{49}$$

$$\dot{E}_3 = E_3\left(\alpha_3 - \sum_{m=1}^{3} \theta_{3m} I_m\right) - \mathrm{Im}\{\vartheta_{3212} \exp(-i\Psi)\}\, E_2^2 E_1, \tag{50}$$

$$\nu_1 + \dot{\phi}_1 = \Omega_1 + \sigma_1 - \sum_{m=1}^{3} \tau_{1m} I_m - \mathrm{Re}\{\vartheta_{1232} \exp(-i\Psi)\}\, E_2^2 E_3/E_1, \tag{51}$$

$$\nu_2 + \dot{\phi}_2 = \Omega_2 + \sigma_2 - \sum_{m=1}^{3} \tau_{2m} I_m - \mathrm{Re}\{(\vartheta_{2123} + \vartheta_{2321}) \exp(i\Psi)\}\, E_1 E_3, \tag{52}$$

$$\nu_3 + \dot{\phi}_3 = \Omega_3 + \sigma_3 - \sum_{m=1}^{3} \tau_{3m} I_m - \mathrm{Re}\{\vartheta_{3212} \exp(-i\Psi)\}\, E_2^2 E_1/E_3, \tag{53}$$

where the coefficients are defined in Tables 8-1 and 9-1. Here the right-hand sides of Eqs. (48)–(53) do not depend on the phases individually, but only on the combination Ψ. Using definition (44), we find the equations of motion for Ψ:

$$\dot{\Psi} = d + l_s \sin \Psi + l_c \cos \Psi, \tag{54}$$

where the unlocked beat frequency

$$d = 2\sigma_2 - \sigma_1 - \sigma_3 - \sum_{m=1}^{3} (2\tau_{2m} - \tau_{1m} - \tau_{3m}) E_m^2, \tag{55}$$

and the locking coefficients

$$l_s = \mathrm{Im}\left\{2E_1E_3(\vartheta_{2123} + \vartheta_{2321}) + \left(\frac{\vartheta_{1232}E_3}{E_1} + \frac{\vartheta_{3212}E_1}{E_3}\right)E_2^2\right\}, \tag{56}$$

$$l_c = \mathrm{Re}\left\{-2E_1E_3(\vartheta_{2123} + \vartheta_{2321}) + \left(\frac{\vartheta_{1232}E_3}{E_1} + \frac{\vartheta_{3212}E_1}{E_3}\right)E_2^2\right\}. \tag{57}$$

We have not written the general expressions out more explicitly, for they are sufficiently complicated to obscure analytic insight. In the form given here, they are easily programmable for the computer, which is required in the final analysis in any case.

Equation (54) can be written in the simpler form

$$\dot{\Psi} = d + l \sin(\Psi - \Psi_0), \tag{58}$$

where the locking coefficient

$$l = l_s\left(1 + \frac{l_c^2}{l_s^2}\right)^{1/2} \tag{59}$$

and the phase angle

$$\Psi_0 = -\tan^{-1}(l_c/l_s). \tag{60}$$

This can be shown by expanding $l \sin(\Psi - \Psi_0)$ and identifying the coefficients of $\sin \Psi$ and $\cos \Psi$ with l_s and l_c.

Equations (48), (49), (50), and (58) determine the electric field in three-mode operation. If variations in the field amplitudes E_n can be neglected, we can take the quantities d and l to be constant and treat (58) independently of (48), (49), and (50). This is called the "decoupled approximation." The solution of (58) depends markedly on the relative magnitudes of d and l. For $|d| > |l|$, $\dot{\Psi}$ cannot vanish for any choice of Ψ. Hence Ψ changes monotonically in time with a time-varying low-frequency beat note $\dot{\Psi}(t)$. As shown in Prob. 9-6, the time average of this note $\overline{\Delta\nu}$ is given by

$$\overline{\Delta\nu} = d\left(1 - \frac{l^2}{d^2}\right)^{1/2}. \tag{61}$$

For values of $|d|$ only slightly larger than $|l|$, part of the cycle is slow [when $l \sin(\Psi - \Psi_0)$ subtracts from d] and part is fast. This is called a "slipping" phenomenon and was observed by Lord Rayleigh as he tuned two sustained tuning forks close together (see Sec. 4-2).

When $|d|$ actually equals $|l|$, the slow part of the cycle stops altogether, for the time derivative in (58) vanishes. The values for which this occurs are given by the solution of (58) with $\dot{\Psi} = 0$, namely,

$$\Psi^{(s)} = \begin{cases} \Psi_1 = \Psi_0 - \sin^{-1}(d/l), & (62) \\ \Psi_2 = \Psi_0 + \pi + \sin^{-1}(d/l). & (63) \end{cases}$$

By definition of Ψ (44), the condition $\dot{\Psi} = 0$ implies that the beat frequency between modes 1 and 2 equals that between modes 2 and 3, that is, locking exists. Furthermore, over the range for which $|d| \leq |l|$, the modes remain locked together. This is the simplest case of (beat frequency) mode locking in the multimode laser.

The stability of the solutions can be determined using a small vibrations analysis as discussed for two-mode operation (Prob. 9-7). We find that $\Psi^{(s)}$ is stable provided that the coefficient

$$l \cos(\Psi^{(s)} - \Psi_0) < 0. \tag{64}$$

As the second mode is tuned toward line center, it sees less and less dispersion (see Fig. 9-5). Specifically, the linear pulling σ_2 and the nonlinear pushing ρ_2 approach zero (see Table 8-1). The cross-pushing coefficient $\tau_{21} \to -\tau_{23}$. Hence, if $E_1 = E_3$, as is likely because of the symmetry of the problem, the medium has little effect on ν_2. Furthermore, the dispersions for modes 1 and 3 become equal and opposite ($\sigma_1 = -\sigma_3$, $\rho_1 = -\rho_3$, etc.). Consequently for nearly central tuning of mode 2, the d term in the equation of motion for $\dot{\Psi}$(58) approaches zero. Taking $E_1 = E_3$ in (56) and (57), we find with some algebra that

$$l_s \to \tfrac{1}{8}\nu\wp^4(\gamma_a{}^2\gamma^2\varepsilon_0\hbar^3)^{-1}\bar{N}\left(1 + 2\frac{N_2}{\bar{N}}\right)\mathscr{L}(\Delta)\mathscr{L}_a(\Delta)$$
$$\times\left\{\left[2\gamma_a - \frac{\Delta^2(1 - \gamma_a/\gamma)}{\gamma}\right]E_1{}^2\right.$$
$$\left.+\tfrac{1}{2}\mathscr{L}(\Delta)\left(2\gamma_a - \frac{3\Delta^2}{\gamma} - \frac{\Delta^4}{\gamma^3}\right)E_2{}^2\right\} + (\text{same with } \gamma_a \to \gamma_b), \tag{65}$$

$$l_c \to 0, \tag{66}$$

where $\Delta = \Delta_1 = \Delta_2$ is the intermode frequency difference. We see from (59) that $l \cong l_s$, $\Psi_0 \cong 0$, and that the possible stationary states are

$$\Psi_1 = -\sin^{-1}(d/l), \tag{67}$$

$$\Psi_2 = \pi + \sin^{-1}(d/l). \tag{68}$$

For central tuning of mode 2, $d = 0$ and $\Psi = 0$ or π for $l < 0$ or $l > 0$, respectively. Note that proper placement of the medium in the cavity can change the sign of the factor $(1 + 2N_2/\bar{N})$ in (65) and therefore that of l.

The value $\Psi = 0$ requires that

$$\phi_2 - \phi_1 = \phi_3 - \phi_2.$$

Choosing the time origin so that $\phi_2 = \phi_1$, we have that

$$\phi_1 = \phi_2 = \phi_3, \tag{69}$$

that is, at points periodic in time spaced by the interval $2\pi/\Delta$, the field phasors $E_n \exp[-i(\nu_n t + \phi_n)]$ add constructively.† At other times they tend to cancel one another, as shown in Fig. 9-6*a*. This is three-mode pulsing. The time width of the pulses are inversely proportional to the mode spacing Δ (as is the interval between pulses), but the degree to which there is destructive interference between pulses is limited by the small number of modes. The phase relation (69) is termed AM in analogy with signals encountered in AM radio.

The second value $\Psi = \pi$ leads to the phase relation

$$\phi_1 = \phi_2 = \pi + \phi_3, \tag{70}$$

†Here we take $E_n \geq 0$ for all modes. This constructive interference is shown without appeal to the time origin in Sec. 9-4.

$$I_{AM} = \tfrac{1}{4}(E_1 e^{-i\Delta t} + E_2 + E_3 e^{i\Delta t})^2$$

(a)

$$I_{FM} = \tfrac{1}{4}(E_1 e^{-i(\Delta t + \pi)} + E_2 + E_3 e^{i\Delta t})^2$$

(b)

(c)

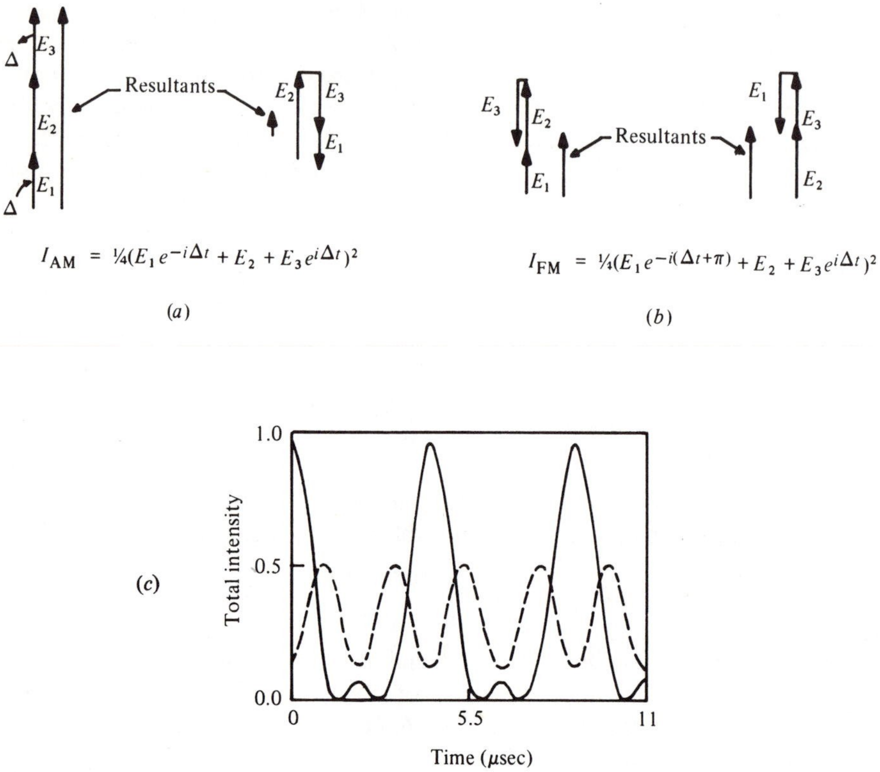

Figure 9-6. (a) AM phasor diagram. (b) FM phasor diagram. (c) Possible total intensities in time (AM—solid line, FM—dashed line).

which is termed FM in analogy with signals in FM radio (see Sec. 9-5). In general, there are always some variations in the total intensity (squared sum of phasors), even for FM phases (see Fig. 9-6b), unless the amplitudes E_n are given by appropriate Bessel functions; but the "best," that is, the highest, pulses occur for AM phases since only for these do all phasors add at points in time. The time development for the two cases is illustrated in Fig. 9-6c. A comparison for three-mode operation in the gas laser case is given in Fig. 10-7.

A rigorous steady-state solution to the amplitude and phase equations (48), (49), (50), and (58) requires finding the zeros to these equations. This transcendental problem can be accomplished with the use of a generalized Newton-Raphson numerical method.

9-4. Analysis of General Multimode Operation

In this section, we consider first solutions to the amplitude- and frequency-determining equations (18) and (19) in the free-running approximation, that is, we neglect all terms with relative phase angles which are not identically zero. Physically this corresponds to a situation in which not only terms containing $\exp[i(\nu_2 - \nu_1)t]$, but also those with $\exp(i\Psi)$, are assumed to average to zero in time. This approximation precludes discussion of mode locking in which the second kind of phase angle is constant in time. Nevertheless, it is useful to consider the approximation, for it yields average-value solutions in free-running operation (modes "dance" around the solutions in time), and provides first approximations in attacking mode-locked problems. Furthermore, the method arises naturally in other contexts such as in the two-mode Zeeman laser calculation (see Chap. 12). We then consider mode-locked operation for which the nonzero phase angles play an important role. This kind of mode locking is a generalization of the three-mode locking of Sec. 9-3. and consists of the equality of beat frequencies between adjacent modes. Note that this does *not* imply pulsing, for the phase relationships between modes are still arbitrary. An important example of mode-locked operation in which pulsing is totally absent is forced locked FM operation (see Sec. 9-5).

Free-Running Operation

Linear intensity equations like (24) and (25) for two-mode operation can be obtained for the multimode case in the free-running approximation. The relative phase angles $\Psi_{n\mu\rho\sigma}$ of (15) vanish identically only if the subscripts obey the relations $n = \mu$, $\rho = \sigma$, or the relations $n = \sigma$, $\rho = \mu$ (or both). The amplitude product $E_\mu E_\rho E_\sigma$ in (18) reduces to $E_n E_\rho^2$ and $E_\rho^2 E_n$, respectively. Multiplying the general amplitude equation (18) by $E_n \wp^2/(\hbar^2 \gamma_a \gamma_b)$ and keeping only terms with these amplitude products, we obtain the intensity equations of motion

$$\dot{I}_n = 2I_n(a_n - \sum_m \theta_{nm} I_m), \tag{71}$$

where the saturation coefficients θ_{nm} are given, for example, in Table 9-1.

For N-mode operation, the stationary solutions ($\dot{I}_n = 0$ for all n) are 2^N in number, for any one of the N modes can be zero or nonzero independently of the others. The physical ($I_n > 0$ for all n), stable stationary solutions can be determined by a generalization of the small vibrations analysis presented for two-mode operation. In our discussion it is convenient to introduce two sets of subscript values for n and m: the set indexing nonzero intensities $\mathfrak{N}$ and that indexing zero intensities $\mathfrak{Z}$. In this terminology, the 2^N stationary solutions of (71) have the form

$$I_n^{(s)} = 0, \qquad n \in \mathfrak{Z}, \tag{72}$$
$$I_n^{(s)} = \sum_m (\theta^{-1})_{nm} a_m, \quad n, m \in \mathfrak{N},$$

where θ^{-1} is the inverse of the truncated θ matrix whose indices belong to $\mathfrak{N}$.

The stable solutions are those for which deviations ε_n from the $I_n^{(s)}$ relax to zero in time. Inserting

$$I_n = I_n^{(s)} + \varepsilon_n \tag{73}$$

into (71), we find

$$\dot{\varepsilon}_n = -2I_n^{(s)} \sum_m \theta_{nm}\varepsilon_m + O(\varepsilon^2) \tag{74}$$

for n, $m \in \mathfrak{N}$ and

$$\dot{\varepsilon}_n = 2\varepsilon_n \left[a_n - \sum_m \theta_{nm} I_m^{(s)}\right] + O(\varepsilon^2) \tag{75}$$

for $n \in \mathfrak{Z}$, $m \in \mathfrak{N}$. The small deviations $\varepsilon_n \to 0$ as $t \to \infty$ provided that the eigenvalues of the matrix in (74) with elements $-2I_n^{(s)}\theta_{nm}$ are negative and the effective a's

$$a_n' = a_n - \sum_m \theta_{nm} I_m^{(s)}, \quad n \in \mathfrak{Z}, m \in \mathfrak{N} \tag{76}$$

are negative.

The meaning of these restrictions is discussed earlier in some detail for the two-mode problem and is essentially the same here. We note now only that the solutions and the tests for being stable and physical can be programmed for the computer to yield solutions for any desired set of laser parameters. Some results are given in Chap. 10 for gas lasers. The reader is referred also to Sargent, Lamb, and Fork (1967) for Zeeman lasers and to O'Bryan and Sargent (1973) for multimode gas lasers.

Locking of Beat Frequencies Between *N* Modes

In the three-mode locking problem, we saw that locking occurs if

$$\dot{E}_n = \dot{\Psi}_{n\mu\rho\sigma} = 0,$$

for which the four relative phase angles $\Psi_{n\mu\rho\sigma}$ could be written in terms of Ψ_{2123}. In the N-mode problem, there are many more than four $\Psi_{n\mu\rho\sigma}$ and they are not all equal to $\pm\, \Psi_{2123}$. In fact, there are $N - 2$ linearly independent relative phase angles, which we take to have the form

$$\Psi_j = (2\nu_j - \nu_{j-1} - \nu_{j+1})t + 2\phi_j - \phi_{j-1} - \phi_{j+1}. \tag{77}$$

This might be surprising, for there are N ϕ's. However, in general, as in the three-mode problem, one phase angle is arbitrary and another is determined by the choice of time origin. Mode locking determines only $N - 2$ ϕ's (e.g., one ϕ for $N = 3$).

The condition for locking can therefore be written in terms of the $N - 2$ Ψ_j's as

$$\dot{E}_n = \dot{\Psi}_j = 0, \quad \text{for } n = 1, \ldots, N; j = 2, \ldots, N - 1. \tag{78}$$

The equations of motion for the Ψ_j can be determined from appropriate combinations of the frequency-determining equations (18). The locking condition $\dot{\Psi} = 0$ implies the frequency spacing

$$\nu_n = \nu_q + (n - q)\, \Delta, \quad \text{for } n, q = 1, \ldots, N, \tag{79}$$

where $\Delta \simeq c/2L$ is the beat frequency between adjacent modes. This follows by induction since $\dot{\Psi}_2 = 0$ gives $\nu_3 - \nu_2 = \nu_2 - \nu_1 = \Delta$, $\dot{\Psi}_3 = 0$ gives $\nu_4 - \nu_3 = \nu_3 - \nu_2 = \Delta$, and so forth. Hence, for $\dot{E}_n = 0$, the electric field (1) repeats the value $E(z, t)$ at times $t + 2\pi n/\Delta$, yielding periodic output.

In particular, the special case $\Psi_j = 0$ for all j implies that adjacent phases ϕ_j differ by a constant amount

$$\phi_{j+1} - \phi_j = \phi_j - \phi_{j-1} \equiv \delta$$

or equivalently

$$\phi_n = \phi_q + (n - q)\delta, \quad \text{for all } n, q = 1, \ldots, N. \tag{80}$$

The electric field $E(z, t)$ of (1) which satisfies (79) and (80) is given by

$$E(z, t) = \tfrac{1}{2} \exp[-i(\nu_q t + \phi_q)] \sum_n E_n \exp[-i\,(n - q)\,(\Delta t + \delta)]\; U_n(z) + \text{c.c.} \tag{81}$$

We see here that by proper choice of the time origin the δ can be dropped, that is, the phases ϕ_j can be set equal to one another. At points in time spaced $2\pi/\Delta$ apart the phasors $E_n \exp[-i(n - q)\,(\Delta t + \delta)]$ add constructively, and at other times they add destructively, thus yielding a train of pulses. As discussed in Sec. 9-3, this is sometimes called AM locking. More quantitatively, we recognize the phasor sum in (81) as the product of an infinite-length "comb" of Dirac delta functions $\delta(\nu - \nu_0 - n\Delta)$ with a smooth envelope that coincides with the values E_n at the corresponding frequencies ν_n. The inverse Fourier transform of this product is a convolution in time of the transform of the envelope with a comb of delta functions $\delta(t - 2L/c)$. This convolution consists of an infinite train of pulses spaced $2L/c$ apart, each pulse having the envelope transform for shape. In particular, a square envelope gives $(\sin x)/x$ pulses, a Gaussian envelope gives Gaussian pulses, and so forth. Note in particular that for the Gaussian pulses the pulse width is inversely proportional to the frequency width. This is a useful fact for estimating the minimum pulse width that a given medium can generate. For the two-mirror laser, we set $U_n(z) = \sin K_n z$ and break the sine into exponentials, thereby finding for (81)

$$E(z, t) = -\tfrac{1}{4} i \exp[-i(\nu_q t + \phi_q - K_q z)] \sum_n E_n \exp[-i(n - q)(\Delta t - \frac{\pi}{L} z)]$$

$$+ \tfrac{1}{4} i \exp[-i(\nu_q t + \phi_q + K_q z)] \sum_n E_n \exp[-i(n - q)(\Delta t + \frac{\pi}{L} z)]$$

$$+ \text{c.c.} \tag{82}$$

Thus we have two infinite trains of pulses moving in opposite directions with adjacent pulses in a given train spaced a distance $2L$ apart. The effect is that of a pulse bouncing back and forth in the cavity, for at the mirrors ($z = 0$, L) an outgoing pulse of one running train coincides with an incoming pulse of the oppositely running train.

The problem of finding locked solutions reduces to that of determining the stable, stationary solutions satisfying (78). This can be done for any desired set of laser parameters by use of a many-function Newton-Raphson numerical method.

9-5. Forced Locking of Multimode Beat Frequencies

The locking of beat frequencies discussed so far has resulted from "self-injected signals" generated by field interactions with population pulsations in the medium. The pulsations play the role of a modulator, creating side bands which either contribute to mode coupling or act as injected signals.

It is also possible to inject signals through the use of a polarization external to the medium, such as a device which varies the optical path length or loss of the cavity at roughly the intermode frequency. These modulators cause each laser mode E_n to generate side bands. The effects can be made substantially larger than those resulting from population pulsations and can control laser oscillation quite remarkably, even forcing FM oscillation with amplitudes E_n very different from free-running values.

We describe† the phase and loss modulations by a time-varying, complex susceptibility $\Delta\chi(z, t)$, which produces a polarization

$$\Delta P(z, t) = \varepsilon_0 \Delta\chi(z, t) \tfrac{1}{2} \sum_n E_n(t) \exp[-i(\nu_n t + \phi_n)] U_n(z) + \text{c.c.} \tag{83}$$

having the explicit time-domain form

$$\Delta P(z, t) = \varepsilon_0 \cos(\nu_M t)\, \Delta\chi'(z)\, E(z, t)$$

$$+ \varepsilon_0 \nu^{-1}(1 + \cos \nu_M t)\, \Delta\chi''(z) \left(\frac{\partial}{\partial t}\right) E(z, t). \tag{84}$$

Here ν_M is the modulation frequency, $\Delta\chi'(z)$ yields a frequency modulation,

† With notational changes, much of the analysis presented here is synthesized from the work of Harris and McDuff (1965, 1967).

and $\Delta\chi''(z)$ gives a loss modulation. The loss term is proportional to $1 + \cos \nu_M t$ rather than $\cos\nu_M t$ alone, so that it remains positive (a negative term gives gain). Using the field definition (8.8), we have

$$\Delta P(z, t) = \tfrac{1}{2}\varepsilon_0\{[\Delta\chi'(z) + i\,\Delta\chi''(z)]\cos \nu_M t + i\,\Delta\chi''(z)\} \times \sum_q E_q(t)\exp[-i(\nu_q t + \phi_q)]\,U_q(z) + \text{c.c.} \tag{85}$$

Writing this in the Fourier-analyzed form

$$\Delta P(z, t) = \tfrac{1}{2}\sum_n \Delta\mathscr{P}_n(t)\exp[-i(\nu_n t + \phi_n)]\,U_n(z) + \text{c.c.}, \tag{86}$$

we identify the complex polarization component

$$\Delta\mathscr{P}_n(t) = \varepsilon_0 \exp[i(\nu_n t + \phi_n)]\frac{1}{\mathscr{N}}\int_0^L dz\,U_n^*(z) \times \{[\Delta\chi'(z) + i\,\Delta\chi''(z)]\cos\nu_M t + i\,\Delta\chi''(z)\} \times \sum_q E_q(t)\exp[-i(\nu_q t + \phi_q)]\,U_q(z). \tag{87}$$

For sufficiently high cavity Q, the amplitudes E_n do not vary much in time $2\pi/\Delta$ ($\Delta \cong \frac{1}{2}c/L$). Hence we neglect terms in (87) which vary at intermode frequencies. Further assuming that the modulation susceptibility varies little in an optical wavelength, we can take for the two-mirror laser $\sin K_n z \sin K_q z \cong \frac{1}{2}\cos[(K_n - K_q)z] = \frac{1}{2}\cos[(n - q)\,\pi z/L]$ in the integrand, for the sum term $\frac{1}{2}\cos(K_n + K_q)z$ nearly averages out in the z integration. With these considerations, the complex polarization (87) reduces to

$$\Delta\mathscr{P}_n(t) = i\varepsilon_0\,\overline{\Delta\chi''}\,E_n + \tfrac{1}{2}\varepsilon_0\,\overline{\Delta\chi_1}\,\{E_{n+1}\exp[-i(\nu_{n+1} - \nu_n - \nu_M)t - i(\phi_{n+1} - \phi_n)] + E_{n-1}\exp[i(\nu_n - \nu_M - \nu_{n-1})t + i(\phi_n - \phi_{n-1})]\}, \tag{88}$$

where the averaged complex susceptibilities

$$\overline{\Delta\chi} = \frac{1}{L}\int_0^L dz\,\Delta\chi(z),$$
$$\overline{\Delta\chi_1} = \frac{1}{L}\int_0^L dz\,\Delta\chi(z)\cos(\pi z/L). \tag{89}$$

It is furthermore convenient to take

$$\nu_n = \Omega_q + (n - q)\nu_M, \tag{90}$$

recognizing that this can result in nonzero values for the phase derivatives $\dot\phi_n$. The complex polarization becomes

$$\Delta\mathscr{P}_n(t) = i\varepsilon_0\,\overline{\Delta\chi''}\,E_n + \tfrac{1}{2}\varepsilon_0\,\overline{\Delta\chi_1}\,\{E_{n+1}\exp[-i(\phi_{n+1} - \phi_n)] + E_{n-1}\exp[i(\phi_n - \phi_{n-1})]\}. \tag{91}$$

Calling the contribution from the active medium $\varepsilon_0\chi_n E_n$ (8. 13), we have the total complex polarization:

$$\mathscr{P}_n(t) = \varepsilon_0 \chi_n E_n + \Delta \mathscr{P}_n(t), \tag{92}$$

which with the self-consistency equations (8. 11) and (8. 12) yields

$$\dot{E}_n + \tfrac{1}{2}\nu(Q_n^{-1} + \chi''_n) E_n = -\tfrac{1}{2}\nu\overline{\Delta\chi''} E_n - \tfrac{1}{4}\nu \operatorname{Im}\{E_{n+1} \overline{\Delta\chi_1} \exp[-i(\phi_{n+1} - \phi_n)] + E_{n-1} \overline{\Delta\chi_1} \exp[i(\phi_n - \phi_{n-1})]\}, \tag{93}$$

$$[\dot{\phi}_n - (n - q)\, \Delta\nu + \tfrac{1}{2}\nu\chi_n'] E_n = -\tfrac{1}{2}\nu \operatorname{Re}\{E_{n+1} \overline{\Delta\chi_1} \exp[-i(\phi_{n+1} - \phi_n)] + E_{n-1} \overline{\Delta\chi_1} \exp[i(\phi_n - \phi_{n-1})]\}. \tag{94}$$

Here the modulation detuning

$$\Delta\nu = \Delta\Omega - \nu_M = \frac{1}{2}\frac{c}{L} - \nu_M. \tag{95}$$

These equations are generalizations of the amplitude- and frequency-determining equations (18) and (19) with $-\tfrac{1}{2}\nu\chi_n' E_n$ and $-\tfrac{1}{2}\nu\chi_n'$ given by the right-hand sides of (18) and (19), respectively. The linear coupling terms resulting from the modulation appear on right-hand sides of (93) and (94). Hence stationary solutions can be determined for desired parameters by using a Newton-Raphson numerical method.

Some analytical progress can be made by assuming that the medium is sufficiently simple. Harris and McDuff (1965, 1967) have considered a number of cases. To illustrate the formalism and to motivate the FM solution, we specialize to an ideal medium capable of supporting an infinity of modes with no dispersion. For example, we can take $\gamma \to \infty$ in a_n of Table 8-1 and $\bar{N}/\gamma$ constant. For this,

$$-\chi_n'' = Q^{-1}, \qquad \chi_n' = 0. \tag{96}$$

Consider further that only the length is modulated, for which the loss susceptibility $\Delta\chi''$ vanishes. Looking for stationary solutions $\dot{E}_n = 0$ and $\dot{\phi}_n$ constant, we find that (93) and (94) reduce to

$$\dot{E}_n = 0 = E_{n+1} \sin(\phi_{n+1} - \phi_n) - E_{n-1} \sin(\phi_n - \phi_{n-1}), \tag{97}$$

$$[\dot{\phi}_n - (n - q)\, \Delta\nu] E_n = -\tfrac{1}{2}\nu\, \overline{\Delta\chi_1'}\, [E_{n+1} \cos(\phi_{n+1} - \phi_n) + E_{n-1} \cos(\phi_n - \phi_{n-1})]. \tag{98}$$

Equation (97) can be easily satisfied by equal phases:

$$\phi_{n+1} = \phi_n. \tag{99}$$

The derivatives $\dot{\phi}_n$ are then independent of the index n. With these considerations, (98) reduces to

$$[\dot{\phi} - (n - q)\, \Delta\nu] E_n = -\tfrac{1}{2}\, \nu\, \overline{\Delta\chi_1}' \quad (E_{n+1} + E_{n-1}). \tag{100}$$

This is a three-term recursion relation which satisfies the Bessel function relation

$$2k\Gamma^{-1} J_k(\Gamma) = J_{k+1}(\Gamma) + J_{k-1}(\Gamma), \tag{101}$$

provided we take

$$\dot{\phi} = l\,\Delta\nu, \tag{102}$$

where l is an integer, $k = n - q - l$; the field amplitudes

$$E_n = J_{n-q-l}(\Gamma); \tag{103}$$

and we define the "modulation depth" [see Eq. (107) for meaning]

$$\Gamma = \nu\,\overline{\Delta\chi_1{}'}\,(\Delta\nu)^{-1}. \tag{104}$$

For $l = 0$, the electric field (8.8) becomes

$$E(z, t) = \tfrac{1}{2}\exp[-i(\nu_q t + \phi_q)] \sum_n J_{n-q}(\Gamma)\exp[-i(n-q)\nu_M t] \times \sin K_n z + \text{c.c.} \tag{105}$$

At first glance, one might conclude from this formula that pulsing occurs, for at points in time spaced by $2\pi/\nu_M$ it appears that all the phasors $J_{n-q}(\Gamma)$ $\exp[i(n-q)\nu_M t]$ add at periodic intervals in time. However, Bessel functions with negative integer subscripts

$$J_{-k}(\Gamma) = (-1)^k J_k(\Gamma), \tag{106}$$

and hence positive odd-numbered modes are precisely out of phase (Fig. 9-7) with their negative counterparts, in agreement with our earlier definition of FM phases (Sec. 9-4). Here, for the sake of agreement with standard terminology, we have let some E_n be negative, although ordinarily π is added to ϕ_n to obtain the same thing.

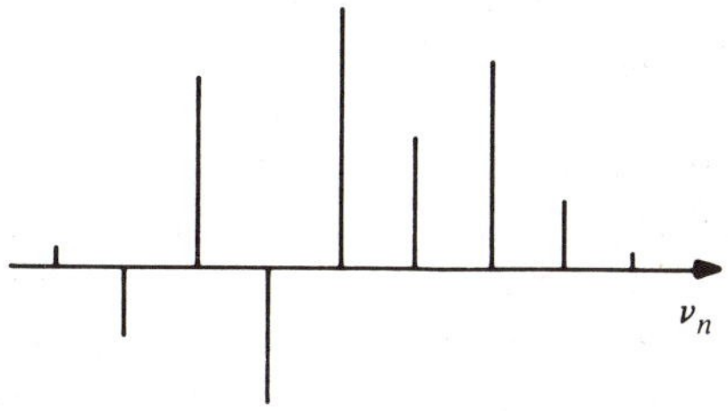

Figure 9-7. Diagram depicting FM phasors at a particular time. Phasors opposite one another with respect to a center phasor are alternately in and out of phase with one another.

With the use of a trigonometic addition formula, the electric field (1) can be written as the sum of two waves, one traveling forward along the z axis and one backward. Further making use of the Bessel identity†

† See, for example, Eq. (6), Sec. 5.12, of *Advanced Mathematics in Physics and Engineering*, by A. Bronwell, McGraw-Hill, New York, 1953.

$$\exp(i\Gamma \sin \theta) = \sum_{k=-\infty}^{\infty} \exp(ik\theta) J_k(\Gamma),$$

we can sum the series in (105):

$$E(z, t) = \tfrac{1}{2} \sin [\Omega_q t + q\pi z/L + \Gamma \sin (\nu_M t + \pi z/L)] - \tfrac{1}{2} \sin [\Omega_q t - q\pi z/L + \Gamma \sin (\nu_M t - \pi z/L)]. \quad (107)$$

We see that solution (105) consists of two oppositely directed running waves with the frequency-modulated frequency (we differentiate arguments with respect to time)

$$\Omega_q + \Gamma \nu_M cos(\nu_M t \pm \pi z/L). \quad (108)$$

The peak-to-peak frequency variation is thus $2\Gamma\nu_M$.

Inasmuch as some fraction of one running wave emerges from a given reflector, the output field for this single FM "mode" is one of these FM waves.

In addition to the $l = 0$ solution, there are other FM oscillations with carrier frequencies Ω_{q+l}. The corresponding side bands of the various oscillations are close to one another in frequency (within ν/Q) and compete strongly for typical values of the modulation depth Γ, expecially when compared to ordinary mode competition due to nonlinearities in the medium. When a real laser medium is involved (see Ammann, McMurtry, and Oshman, 1965; Kuizenga and Siegman, 1970), deviation from both the FM phases and the Bessel function amplitudes can occur, but the laser output is very close to ideal FM oscillation.

The first mode-locked pulse train was observed by Hargrove, Fork, and Pollack (1964), who introduced a loss modulator into the laser cavity containing a He-Ne active medium. The loss was modulated at the intermode frequency $c/2L$, allowing radiation at periodic points in time to pass unattenuated, that is, allowing a pulse to bounce back and forth. Trains of picosecond pulses were observed later in ruby by Mocker and Collins (1965) and in Nd: Glass by DeMaria, Stetser, and Heyman (1966), both groups with saturable absorbers in their laser cavities. It is tempting to think of this nonlinear absorber as a side-band generator like the active modulators, for it produces combination tones. In the model commonly used, patterned after that of Cutler (1955), the saturable absorber acts as an "expander," successively creating side bands farther and farther apart, resulting in shorter and shorter pulses. However, a model like that of Letokhov (1969) appears to be more accurate. For this the noise in the initial laser radiation consists of random fluctuations or "minipulses." A group of these concentrating sufficient energy in the short relaxation time of the saturable absorber effectively "opens" the absorber, allowing radiation to pass unattenuated for a short time. Groups with smaller energy are absorbed. The most energetic packet builds up at the expense of the others and bounces back and forth in the cavity to produce the output pulse train. In contrast to the Cutler model, the circulating pulse starts with approximately the same short duration that it ultimately attains. The modal picture with side band injected signals and so forth offers some insight here, but the temporal model gives a more satisfying explanation. A frequency-domain treatment along the lines of Sec. 9-4 is further complicated by the large number of modes (~1000) and changes in the mode amplitudes ($E_n \neq 0$), although the frequency and phase conditions (79) and (80) still hold.

Another interesting kind of multimode locking occurs between the transverse Fox and Li modes discussed in Appendix B. As shown there, these modes are given approximately by Hermite-Gaussian functions. An appropriate superposition of these functions produces a Schrödinger packet, that is, a Gaussian packet of constant width that scans back and forth across the mirror face. Such superpositions also occur elsewhere in laser physics, for the wave function corresponding most closely to the classical electric field is an analogous (but physically unrelated) packet consisting of photon number states, as discussed in Chap. 15. The scanning "transverse mode-locked" operation has been observed by Auston (1968). For further discussion of this and other mode-locked configurations the reader is referred to the book by Smith, Duguay and Ippen (1973).

Problems

9-1. Use the rate equation approximation (REA) of Sec. 8-2 in the evaluation of the integral (8.30) for ρ_{ab} with the multimode perturbation energy (2). Use the result in (8.27) and (8.28) for the population equations of motion. Here drop the pulsation terms to be consistent with the REA. Finally, solve for the "steady-state" population difference.

Ans.:

$$\rho_{aa} - \rho_{bb} = \frac{N(z)}{1 + 2(\gamma_{ab}/\gamma)\sum_n I_n \,|U_n(z)|^2 \mathscr{L}(\omega - \nu_n)} \tag{109}$$

9-2. (*a*) Evaluate the complex polarization $\mathscr{P}_n(t)$ given by (17), using the running-wave (unidirectional) mode functions $U_n(z) = \exp(iK_n z)$. (*b*). Calculate the linear and saturation coefficients corresponding to those in Table 9-1.

9-3. (*a*) Write rate equations for the three-level problems depicted in the adjoining figure. (*b*) Solve for the two mode intensities, assuming both are centrally tuned. Note that the cascade case has *negative* θ's. What does this mean physically?

9-4. Show that a judicious choice of gain and absorption for $N(z)$ can yield

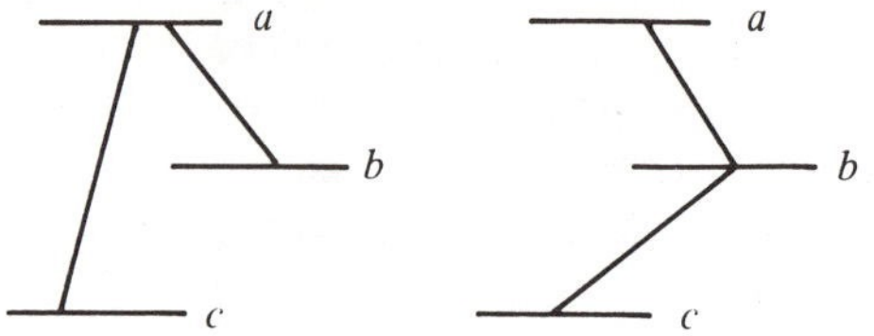

Competitive Cascade

strong coupling in the homogeneously broadened two-level laser (Chap. 8 laser with two modes).

9-5. Show that for neutral coupling lines (33) and (34) are parallel. Show that any solution satisfying (33) is neutrally stable, that is, deviations on the line are permitted, but other deviations return to the line.

9-6. Show that the average beat frequency $\overline{\Delta\nu}$ for a system near locking is given by (61). *Hint*: Equation (58) has the implicit solution

$$t = \int_0^t dt' = \int_{\Psi_0}^{\Psi} \frac{d\Psi'}{d + l \sin(\Psi' - \Psi_0)}, \tag{110}$$

an integral given in several integral tables. Equation (110) gives the beat frequency period when $\Psi = \Psi_0 + 2\pi$. The average beat frequency is then $2\pi/t(\Psi_0 + 2\pi)$, which yields (61).

9-7. Show along the lines of the two-mode stability analysis that the stable solution $\Psi^{(s)}$ of (62) or (63) is that for which (64) holds.

9-8. Prove that the Fourier transform of an infinite train of Dirac delta functions $\sum_n \delta(\nu - n\Delta)$ is another infinite train $\sum_k \delta(t - 2\pi k/\Delta)$. *Hint:* Use the geometric series.

References

The references of Chap. 8 are relevant here also. Additional references are as follows:

E. O. Ammann, B. J. McMurtry, and M. K. Oshman, 1965, *IEEE J. Quantum Electron.* **QE-4**, 263.

D. H. Auston, 1968, *IEEE J. Quantum Electron.* **QE-4**, 420, 417.

C. C. Cutler, 1955, *Proc. IRE* **43**, 140.

A. J. DeMaria, D. A. Stetser, and H. Heynau, 1966, *Appl. Phys. Letters* **8**, 174.

J. Hambenne and M. Sargent III, 1976, *Phys. Rev.* **A13**, 784, 797.

L. E. Hargrove, R. L. Fork, and M. A. Pollack, 1964, *Appl. Phys. Letters* **5**, 4.

S. E. Harris and O. P. McDuff, 1965, *IEEE J. Quantum Electron.* **QE-4**, 245.

S. E. Harris and O. P. McDuff, 1967, *IEEE J. Quantum Electron.* **QE-3**, 101.

D. J. Kuizenga and A. E. Siegman, 1970, *IEEE J. Quantum Electron.* **QE-6**, 673, 694, 709, 803.

V. S. Letokhov, 1969, *Sov. Phys. JETP* **28**, 562.

H. W. Mocker and R. J. Collins, 1965, *Appl. Phys. Letters* **7**, 270.

C. L. O'Bryan III and M. Sargent III, 1973, *Phys. Rev.* **A8**, 3071.

M. Sargent III, W. E. Lamb, Jr., and R. L. Fork, 1967, *Phys. Rev.* **164**, 450.

P. W. Smith, M. A. Duguay and E. P. Ippen, 1974, *Mode Locking of Lasers*, Pergamon Press, Oxford.

E. B. Treacy, 1968, *Phys. Letters* **28A**, 34.

For further interpretation of mode competition see Les Houches lectures of M. Sargent III, 1976, *op. cit.*, p. 114.

X

GAS LASER THEORY

10. GAS LASER THEORY

In Chaps. 8 and 9, we considered a laser medium consisting of homogeneously broadened, stationary atoms. When the lasing atoms move as in a gas, they see an electric field with shifted frequency due to the Doppler effect, as shown in Fig. 10-1. One might argue that it is merely necessary to average the complex polarization $\mathscr{P}_n$ or the susceptibility χ_n over the frequency range corresponding to the velocity distribution, that is, calculate a new $\mathscr{P}_n$, for which

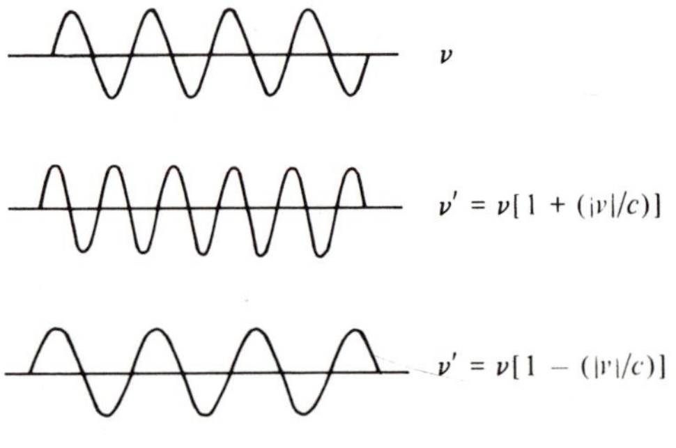

Figure 10-1. Drawing showing how a traveling wave with oscillation frequency ν appears Doppler downshifted to an atom moving the same direction as the running wave and upshifted to one moving in the opposite direction.

$$\mathscr{P}_n(t) = \int_{-\infty}^{\infty} d\omega\, W(\omega)\, \mathscr{P}_n(\omega, t). \tag{1}$$

Here the frequency $\omega = \omega_0 + Kv$, where ω_0 is the frequency at atomic line center, v is the component of velocity along the laser (z) axis, and the frequency distribution is determined by a Maxwell-Boltzmann velocity distribution

$$W(\omega) = (\sqrt{\pi}\, Ku)^{-1} \exp[-(\omega - \omega_0)^2/(Ku)^2]$$

or

$$W(v) = (\sqrt{\pi}\, u)^{-1} \exp[-(v/u)^2]. \tag{2}$$

Here u is the most probable speed of the atom, and K is the wave number. The simple recipe (1) is, in fact, valid for an inhomogeneously broadened medium† consisting of stationary atoms such as ruby at low temperatures. However, in the standing-wave laser, the atoms not only see Doppler-shifted frequencies, but also move through the standing-wave electric field, effectively seeing an amplitude-modulated field. Equivalently, each atom sees two frequencies, for the standing-wave field is the sum of oppositely directed running waves, one of which appears to be Doppler upshifted, one downshifted. Hence it is necessary to consider both the frequency shift and the time dependence of the atomic z coordinate when calculating the polarization of a gaseous medium.

In Sec. 10-1, we calculate the polarization of a gaseous laser medium subject to a single-mode electric field, using the rate equation approximation along the lines followed for the stationary case (Chap. 8). The polarization is then inserted into the field self-consistency equations (8.11) and (8.12) to yield single-mode amplitude- and frequency-determining equations. It is shown that for suitable laser parameters the steady-state solution for the intensity exhibits a dip as the cavity is tuned through the atomic line center (Lamb dip). The notion of hole burning developed by Bennett (1962) is used to provide a physical interpretation of this dip. In Sec. 10-2 multimode phenomena predicted by the perturbation theory of Appendix D are discussed, in which the strong interaction between modes symmetrically placed with respect to line center plays an important role. In Sec. 10-3 we give results from the single-mode strong-signal theory of Appendix E. This theory is based on the work of Stenholm and Lamb (1969) and includes some observations by Feld and Feldman (1970). It is used to test the validity of the third-order and rate equation approximations.

10-1. Polarization of a Doppler Broadened Medium

The macroscopic polarization $P(z, t)$ induced in a medium consisting of moving atoms is contributed to by all atoms regardless of their velocity or their places of excitation. Hence, in forming $P(z, t)$ from the density matrix

†Different atoms have different line centers (see Sec. 7-3).

for a single atom, it is necessary to include integrals over the velocity distribution and excitation positions in addition to those over states and times of excitation. On the assumption that collisions do not take place within the lifetime of an atom, it moves from its excitation position z_0 to a place z', where it interacts at time t' with the electric field. The two positions are related by

$$z' = z_0 + v\,(t' - t_0). \tag{3}$$

In adding up contributions from atoms excited at places z_0, we require that at time t the atoms have moved to position z. This can be accomplished by integrating over a Dirac delta function $\delta(z - z_0 - vt + vt_0)$. Specifically the macroscopic polarization of (8. 9) is given by

$$P(z, t) = \wp \int_{-\infty}^{\infty} dv \sum_a \int_{-\infty}^{t} dt_0 \int_0^L dz_0\, \lambda_a(z_0, t_0, v)$$
$$\times\, \rho_{ab}(a, z_0, t_0, v, t)\, \delta\,(z - z_0 - vt + vt_0) + \text{c.c.}, \tag{4}$$

where $\lambda_a(z_0, t_0, v)$ is the number of atoms excited to the eigenstate $|a\rangle$ at time t_0 and position z_0 and with z component of velocity v.

This can also be written more conveniently as

$$P(z, t) = \wp \int_{-\infty}^{\infty} dv\, \rho_{ab}\,(z, v, t) + \text{c.c.}, \tag{5}$$

where the population matrix

$$\rho\,(z, v, t) = \sum_a \int_{-\infty}^{t} dt_0 \int_0^L dz_0\, \lambda_a(z_0, t_0, v)\, \rho\,(a, z_0, t_0, v, t)$$
$$\times\, \delta\,(z - z_0 - vt + vt_0), \tag{6}$$

for the matrix (6) has component equations of motion identical in form to (8.26), (8.27), and (8.28) for the stationary atom population matrix $\rho(z, t)$ of (8.23). In fact, taking the time derivative of (6) as was done for $\rho(z, t)$ we have

$$\frac{\partial}{\partial t}\rho(z, v, t) = \begin{pmatrix} \lambda_a & 0 \\ 0 & \lambda_b \end{pmatrix}$$
$$+ \sum_a \int_{-\infty}^{t} dt_0 \int_0^L dz_0\, \lambda_a(z_0, t_0, v)\, \dot\rho\,(a, z_0, t_0, v, t)\, \delta\,(z - z_0 - vt + vt_0)$$
$$- v \sum_a \int_{-\infty}^{t} dt_0 \int_0^L dz_0\, \lambda_a(z_0, t_0, v)\, \rho\,(a, z_0, t_0, v, t)\, \delta'\,(z - z_0 - vt + vt_0),$$

where the excitation matrix occurs as in (8.25). The expression $v(\partial/\partial z)\,\rho(z, v, t)$ is the negative of the last term. Hence the convectional derivative $[\partial/\partial t + v(\partial/\partial z)]\,\rho(z, v, t)$ has the simpler value

$$\left(\frac{\partial}{\partial t} + v\frac{\partial}{\partial z}\right)\rho\,(z, v, t) = \begin{pmatrix} \lambda_a & 0 \\ 0 & \lambda_b \end{pmatrix}$$

$$+ \sum_a \int_{-\infty}^{t} dt_0 \int_0^L dz_0\, \lambda_a(z_0, t_0, v)\, \dot{\rho}\,(a, z_0, t_0, v, t)$$

$$\times\, \delta(z - z_0 - vt + vt_0). \tag{7}$$

The equations of motion (7.34), (7.35), and (7.49) for the single-atom density matrix can be substituted into (7). The decay constants γ_a, γ_b, and γ are independent of a, t_0, and z_0 and can be factored outside the integrals over these variables. Furthermore, the interaction Hamiltonian of (8.31):

$$\mathscr{V}_{ab}(t') = -\tfrac{1}{2}\,\wp \sum_n E_n(t') \exp[-i(\nu_n t' + \phi_n)]\, U_n(z') \tag{8}$$

does not depend on these variables either, for the intermediate position z' given by (3) is also given by

$$z' = z - v(t - t'), \tag{9}$$

which does not depend on t_0. Hence $\rho(z, v, t)$ has the component equations of motion

$$\left(\frac{\partial}{\partial t} + v\frac{\partial}{\partial z}\right) \rho_{ab} = -(i\omega + \gamma)\rho_{ab} + \frac{i}{\hbar}\, \mathscr{V}_{ab}(z, t)\,(\rho_{aa} - \rho_{bb}), \tag{10}$$

$$\left(\frac{\partial}{\partial t} + v\frac{\partial}{\partial z}\right) \rho_{aa} = \lambda_a - \gamma_a\rho_{aa} - \left\{\frac{i}{\hbar}\mathscr{V}_{ab}\rho_{ba} + \text{c.c.}\right\}, \tag{11}$$

$$\left(\frac{\partial}{\partial t} + v\frac{\partial}{\partial z}\right) \rho_{bb} = \lambda_b - \gamma_b\rho_{bb} + \left\{\frac{i}{\hbar}\mathscr{V}_{ab}\rho_{ba} + \text{c.c.}\right\}. \tag{12}$$

Comparing Eq. (5) for $P(z, t)$ with (8.9) involving the complex polarization component $\mathscr{P}_n(t)$, we find [as for (8.21)]

$$\mathscr{P}_n(t) = 2\,\wp \exp(i\nu_n t + i\phi_n) \int_{-\infty}^{\infty} dv \frac{1}{\mathscr{N}} \int_0^L dz\, U_n{}^*(z)\, \rho_{ab}(z, v, t), \tag{13}$$

where $\mathscr{N}$ is the normalization factor (8.22).

We now solve the equations of motion (10), (11), and (12) for $\rho(z, v, t)$ in the rate equation approximation for a single-mode field. This comprises a first approximation to the strong-signal theory of Appendix E and reveals much of the qualitative aspects of single-mode operation of a gas laser.

Rate Equation Solution and Hole Burning

The equations of motion (10)–(12) have the form

$$\left(\frac{\partial}{\partial t} + v\frac{\partial}{\partial z}\right) f(z, v, t) = g(z, v, t). \tag{14}$$

This equation has the formal solution

$$f(z, v, t) = \int_{-\infty}^{t} dt'\, g(z', v, t'), \tag{15}$$

with z' given by (9), a fact immediately verifiable by substitution of (15) into (14). In particular, Eq. (10) has the formal solution

$$\rho_{ab}(z, v, t) = \frac{i}{\hbar}\int_{-\infty}^{t} dt' \exp[-(i\omega + \gamma)(t - t')] \\ \times \mathscr{V}_{ab}(z', t')\,[\rho_{aa}(z', v, t') - \rho_{bb}(z', v, t')]. \tag{16}$$

Specializing the perturbation energy (8) to single-mode operation and making the rate equation approximation of Sec. 8-2 (i.e. assuming that the population difference $\rho_{aa} - \rho_{bb}$, the mode amplitude E_n, and the phase ϕ_n vary little in the time $1/\gamma$), we find for (16)

$$\rho_{ab}(z, v, t) = -\tfrac{1}{2} i \frac{\wp}{\hbar} E_n \exp[-i(\nu_n t + \phi_n)]\,[\rho_{aa}(z, v, t) - \rho_{bb}(z, v, t)] \\ \times \int_{-\infty}^{t} dt'\, U_n(z') \exp\{-[i(\omega - \nu_n) + \gamma](t - t')\}. \tag{17}$$

Unlike the stationary atom case, this approach is at best approximate here, for the atoms actually see two frequencies $\nu_n \pm Kv$, which beat together, causing population pulsations much like those for two-mode operation. These pulsations do not appear in the macroscopic polarization because of the velocity integral, but their presence modifies the final result. We will see precisely how the pulsations can be treated in the strong-signal theory (Appendix E) and that the rate equation approximation is, in fact, reasonably accurate for some gas laser parameters.

Writing the two-mirror normal-mode function $U_n(z') = \sin K_n z'$ as the difference of exponentials and performing the simple time integrals in (17), we find

$$\rho_{ab}(z, v, t) = -\tfrac{1}{4} i \frac{\wp}{\hbar} E_n \exp[-i(\nu_n t + \phi_n)]\,[\rho_{aa} - \rho_{bb}] \\ \times \frac{1}{i}[\exp(iK_n z)\,\mathscr{D}(\omega - \nu_n + Kv) - \exp(-iK_n z)\,\mathscr{D}(\omega - \nu_n - Kv)], \tag{18}$$

where $\mathscr{D}(\Delta\omega) = 1/(\gamma + i\,\Delta\omega)$ as in Eq. (9.6). Here, for convenience, particularly in later multimode calculations, we have replaced $K_n v$ by Kv, where $K = \omega/c$. The approximation is excellent since for appreciable values of $W(v)$, the difference $|K_n - K|\,v$ is negligible compared to γ. Further expanding $\exp(\pm iK_n z) = \cos K_n z \pm i \sin K_n z$, we note that, for a symmetric velocity distribution and population difference $[W(-v) = W(v)]$, the $\cos K_n z$ terms will cancel in the integration over v in (13). Assuming this symmetry, we therefore drop the $\cos K_n z$ term and are left with

$$\rho_{ab}(z, v, t) = -\tfrac{1}{4} i \frac{\wp}{\hbar} E_n \exp[-i(\nu_n t + \phi_n)]\, U_n(z)\, (\rho_{aa} - \rho_{bb})$$
$$\times [\mathscr{D}(\omega - \nu_n + Kv) + \mathscr{D}(\omega - \nu_n - Kv)]. \tag{19}$$

Substituting ρ_{ab} and its complex conjugate into the equation of motion given by (11) for the population ρ_{aa}, we obtain the rate equation

$$\dot{\rho}_{aa} = -\gamma_a \rho_{aa} + \lambda_a - R(\rho_{aa} - \rho_{bb}), \tag{20}$$

where the rate constant

$$R(v) = \frac{1}{8}\left(\frac{\wp E_n}{\hbar}\right)^2 \gamma^{-1} [\mathscr{L}(\omega - \nu_n - Kv) + \mathscr{L}(\omega - \nu_n + Kv)], \tag{21}$$

and the Lorentzian $\mathscr{L}(\Delta\omega) = \gamma^2/[\gamma^2 + (\Delta\omega)^2]$. Here we have replaced $|U_n(z)|^2$ by the average value $\frac{1}{2}$ for a standing-wave field, which corresponds more closely to the field that rapidly moving atoms experience. The term ρ_{bb} satisfies a similar equation. Combining these equations as in Sec. 8-2, we solve for the steady state:

$$\rho_{aa} - \rho_{bb} = \frac{N(z, v, t)}{1 + R(v)/R_s}, \tag{22}$$

where the population difference in the absence of field oscillation

$$N(z, v, t) = \lambda_a(z, v, t)\, \gamma_a^{-1} - \lambda_b(z, v, t)\, \gamma_b^{-1} \tag{23}$$

and the saturation parameter R_s is given by (8.38). We will assume an excitation mechanism which factors as

$$N(z, v, t) = W(v)\, N(z, t), \tag{24}$$

for which the velocity distribution $W(v)$ is Maxwellian (2) with most probable speed u. Treatment of systems with different velocity distributions for upper and lower levels (Rhodes and Szöke, 1972) requires some straightforward changes in the following. For appreciable values of $R(v)$, the population difference (22) is reduced below the zero-field value $N(z, v, t)$. As noted in Sec. 8-2, static atoms at the nodes of the field do not interact with the field whereas atoms at the antinodes interact strongly. This leads to spatial holes (see Fig. 8-4). In the present case (moving atoms) these spatial holes should be washed out to some extent, as we have assumed in (21), for atoms may move through several wavelengths of the field in their lifetimes (see Fig. 10-12 for an exact illustration of this "washout").

In addition to spatial holes, the Lorentzians in (21) show that holes are burned in the plot of $\rho_{aa} - \rho_{bb}$ versus v. Off resonance ($\nu_n \neq \omega$), one of the Lorentzians is peaked at the detuning value $\omega - \nu_n = +Kv$ and one at $\omega - \nu_n = -Kv$, thereby burning *two* holes, as shown in Fig. 10-2. On resonance ($\omega = \nu_n$), the peaks coincide and a *single* hole is burned. To understand this physically, note than an atom interacts strongly with an electric field which

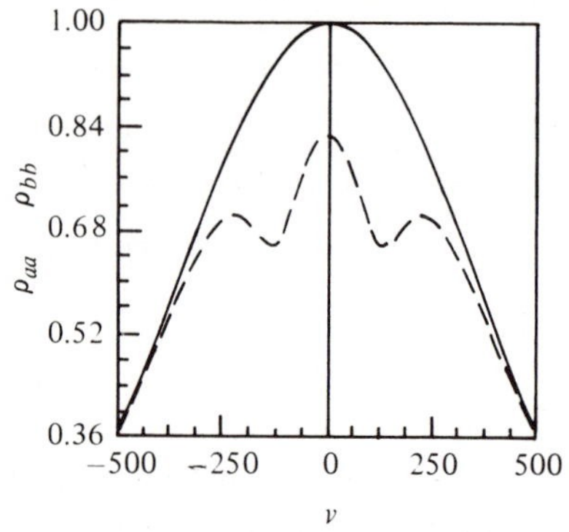

Figure 10-2. Graph of unsaturated (solid) and saturated (dashed) population differences $\rho_{aa} - \rho_{bb}$ versus axial component of velocity for a Maxwellian velocity distribution. The formula used is Eq. (22). Drawing depicts holes burned into population difference by the electric field intensity for $v = \pm c(1 - \omega/\nu)$.

appears resonant in the rest frame of the atom. Specifically, the standing-wave electric field can be expanded as

$$\begin{aligned}\tfrac{1}{2}E_n \exp[-i(\nu_n t + \phi_n)] \sin K_n z + \text{c.c.} \\ = -i\tfrac{1}{4}E_n \{\exp[-i(\nu_n t + \phi_n - K_n z)] \\ - \exp[-i(\nu_n t + \phi_n + K_n z)]\} + \text{c.c.},\end{aligned} \tag{25}$$

that is, as the sum of two oppositely directed running waves. For $\nu_n > \omega$, an atom moving along the z axis sees the first of the running waves in (25) "stretched out" or Doppler downshifted as shown in Fig. 10-1. If the resonant frequency in its rest frame is given by

$$\omega = \nu_n \left(1 - \frac{|v|}{c}\right), \tag{26}$$

the atom interacts strongly. Comparison of (25) with (18) reveals that $\mathscr{D}(\omega - \nu_n + Kv)$ and hence the Lorentzian $\mathscr{L}(\omega - \nu_n + Kv)$ in (21) result from this running wave, as would be expected since this Lorentzian is peaked for the frequency relationship (26). Similarly, an atom moving with velocity $-v$ sees the second traveling wave downshifted, interacts strongly if the atom with velocity v did, and produces the $\mathscr{L}(\omega - \nu_n - Kv)$ in (21). Thus, for the standing wave (25), *two* holes are burned about atoms whose speeds $|v|$ satisfy (26). In contrast, both running waves of a resonant field ($\nu_n = \omega$) interact with the same ensemble, that for $v = 0$, thereby burning a *single* hole. We return to these points shortly, for the relative amounts of saturation resulting from hole burning on and off line center can have a profound effect on the detuning dependence of the intensity.

Substituting the steady-state solution for the population difference (22) into Eq. (19) for ρ_{ab} and using (13) for the complex polarization $\mathscr{P}_n(t)$, we find

$$\mathscr{P}_n(t) = -\tfrac{1}{2}i\wp^2\hbar^{-1}\bar{N}E_n \int_{-\infty}^{\infty} dv\ W(v)\ [\mathscr{D}(\omega - \nu_n - Kv) + \mathscr{D}(\omega - \nu_n + Kv)] \left[1 + \frac{R(v)}{R_s}\right]^{-1}, \tag{27}$$

which can be solved in closed form (see Prob. 10-12). Here we only expand the bracketed denominator in the truncated geometric series $1 + R$, that is, to third order in the electric field amplitude.

To first order in E_n, $R = 0$ and (27) becomes

$$\mathscr{P}_n^{(1)} = -\wp^2(Ku\hbar)^{-1}E_n\bar{N}Z[\gamma + i(\omega - \nu_n)], \tag{28}$$

where the plasma dispersion function (of complex argument, "upsilon")

$$Z(\upsilon) = iK\pi^{-1/2} \int_{-\infty}^{\infty} dv \exp[-(v/u)^2]\ (\upsilon + iKv)^{-1}. \tag{29}$$

Here we note that the same value is given for $(\upsilon + iKv)^{-1}$ as for $(\upsilon - iKv)^{-1}$. Another common definition of this important function is

$$Z(\upsilon) = iKu \int_0^{\infty} d\tau' \exp[-\upsilon\tau' - \tfrac{1}{4}(Ku)^2\tau'^2], \tag{30}$$

which results from performing the velocity integration in (27) before the time integration in (16). Appendix C and Probs. 10-2 and 10-3 give useful properties and values of the function. Although in general it must be evaluated numerically, it simplifies for small average velocities ($u \to 0$):

$$Z(\upsilon)_{u\to 0} = iKu\upsilon^{-1} \tag{31}$$

and the Doppler limit ($Ku \gg \gamma$)

$$Z\,(i\,\Delta\omega)_{(Ku\gg\gamma)} = \exp(-\xi^2)\ [i\sqrt{\pi} - 2\int_0^{\xi} dx \exp(x^2)], \tag{32}$$

where $\xi = \Delta\omega/Ku$. Both limiting cases are most easily derived by the use of (30). The second case is the Laplace transform of a Gaussian and requires completion of the square in τ'.

The third-order contribution to the polarization (term linear in R)

$$\mathscr{P}_n^{(3)} = \tfrac{1}{8}\wp^4(\hbar^3 Ku\gamma)^{-1}\ E_n^3\ \bar{N}\{\ iKu \int_{-\infty}^{\infty} dv\ W(v)\ \mathscr{D}(\omega - \nu_n + Kv) \times [\mathscr{L}(\omega - \nu_n + Kv) + \mathscr{L}(\omega - \nu_n - Kv)]\}/R_s, \tag{33}$$

in which we have used the evenness of v of the integrand to replace $\mathscr{D}(\omega - \nu_n - Kv)$ by $\mathscr{D}(\omega - \nu_n + Kv)$, yielding a multiplicative factor of 2. The velocity integral can be reduced to a sum of plasma dispersion functions by separating the denominators into partial fractions and using the definition (29). The term in curly braces in (33) becomes

$$\tfrac{1}{4}\gamma\left\{-2\frac{\partial Z(\upsilon)}{\partial \upsilon} + \upsilon^{-1}[Z(\upsilon) + Z(\upsilon)] + \gamma^{-1}[Z(\upsilon^*) + Z(\upsilon)] - i(\omega - \nu_n)^{-1}[Z(\upsilon^*) - Z(\upsilon)]\right\} \tag{34}$$

$$= \tfrac{1}{2}\gamma\left\{\frac{Z(\upsilon)}{\upsilon} - \frac{\partial Z(\upsilon)}{\partial \upsilon} + \frac{iZ_r(\upsilon)}{\omega - \nu_n} + \frac{iZ_i(\upsilon)}{\gamma}\right\},$$

for which the derivative

$$\frac{\partial Z(\upsilon)}{\partial \upsilon} = -iK\pi^{-1/2}\int_{-\infty}^{\infty} dv \exp[-(v/u)^2](\upsilon + iKv)^{-2}, \tag{35}$$

and the complex frequency

$$\upsilon = \gamma + i(\omega - \nu_n). \tag{36}$$

In the Doppler limit (32) this derivative is zero, and (34) reduces to

$$\tfrac{1}{2}i\sqrt{\pi}\exp[-(\omega - \nu_n)^2/(Ku)^2][\gamma\mathscr{D}(\omega - \nu_n) + 1]. \tag{37}$$

Hence the third-order contribution to the polarization

$$\mathscr{P}_n^{(3)} = \frac{1}{8}\left(\frac{\sqrt{\pi}\ \bar{N}\wp^4\gamma_{ab}}{\hbar^3 K u\gamma\gamma_a\gamma_b}\right) E_n^3 \exp[-(\omega - \nu_n)^2/(Ku)^2] \times\left\{\frac{\omega - \nu_n}{\gamma}\mathscr{L}(\omega - \nu_n) + i[1 + \mathscr{L}(\omega - \nu_n)]\right\}. \tag{38}$$

Combining the self-consistency relation (8.11) with the first- (28) and third-order (38) contributions as for (8.44), we find the dimensionless steady-state ($\dot{E}_n = 0$) intensity:

$$I_n = \frac{1}{2}\frac{\wp^2}{\hbar^2\gamma_a\gamma_b}E_n^2 = \frac{\alpha_n}{\beta_n} = 4\frac{1 - \exp[(\omega - \nu_n)^2/(Ku)^2]\mathfrak{N}^{-1}}{(\gamma_{ab}/\gamma)[1 + \mathscr{L}(\omega - \nu_n)]} \tag{39}$$

where various coefficients are defined in Table 10-1, and the threshold condition determined by $\alpha_n = 0$:

$$\bar{N}_T = \varepsilon_0\hbar K u(\wp^2 Q\sqrt{\pi})^{-1} \tag{40}$$

has been used to express the intensity (39) in terms of the relative excitation $\mathfrak{N}$ (8.54).

In (39) we see that, if the ratio of the Doppler width Ku to the decay rate γ and the relative excitation $\mathfrak{N}$ are large enough (often $\mathfrak{N} = 1.1$ is sufficient), the tuning dependence of the numerator is less marked than that of the Lorentzian in the denominator and a dip in intensity occurs as the cavity is tuned through line center. A number of intensity tuning profiles are drawn in Fig. 10-3 in the Doppler limit approximations used in obtaining (39). More quantitatively, a dip is observed provided

TABLE 10-1 *Coefficients Appearing in Intensity- and Frequency-Determining Equations (8.51) and (8.52) for Medium with Rapidly Moving Atoms (Doppler width $Ku \gg$ decay rate γ)*

Non-Doppler limit is given in Appendix D. F_3 is chosen for use with the dimensionless intensity I_n of (39). If the relative excitation $\mathfrak{N}$ of (8.54) is to be used, $F_1 = \frac{1}{2}(\nu/Q_n)\,\mathfrak{N}$.
†A slightly improved value of β_n [$\beta_n \rightarrow \beta_n \exp - (\omega - \nu_n)^2/(Ku)^2$] was used in Eq. (39).

Coefficient	Physical Context
$\alpha_n = \exp[-(\omega-\nu_n)^2/(Ku)^2]F_1 - \frac{1}{2}\nu/Q_n$	Linear net gain
$\beta_n = [1 + \mathscr{L}(\omega-\nu_n)]F_3$	Self-saturation†
$\sigma_n = -2\exp(-\xi^2)\int_0^\xi dx\, \exp(x^2)\, F_1,\ \xi = (\nu_n - \omega)/Ku$	Linear mode pushing
$\rho_n = [(\omega-\nu_n)/\gamma]\,\mathscr{L}(\omega-\nu_n)F_3$	Self-pushing
$\mathscr{L}_x(\Delta\omega) = \gamma_x^2/[\gamma_x^2 + (\Delta\omega)^2],\quad x = a, b,$ or missing	Dimensionless Lorentzian
$F_1 = \frac{1}{2}\nu\sqrt{\pi}\,[\wp^2/(\hbar\varepsilon_0 Ku)]\,\bar{N}$	First-order factor
$F_3 = \frac{1}{4}(\gamma_{ab}/\gamma)F_1$	Third-order factor

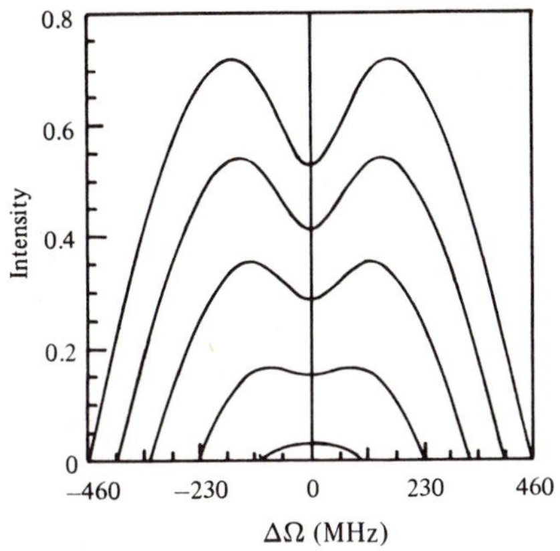

Figure 10-3. Single-mode dimensionless intensity of Eq. (39). Doppler width (Ku) at 1/e point is $2\pi \times 1010$ MHz, the decay constants $\gamma = 2\pi = 80$ MHz, $\gamma_{ab} = 2\pi \times 50$ MHz, and the relative excitation $\mathfrak{N}$ = 1.01, 1.05, 1.10, 1.15, and 1.20.

$$\mathfrak{N} > 1 + 2\left(\frac{\gamma}{Ku}\right)^2, \tag{41}$$

a relation that can be verified setting $[\partial^2 I/\partial\nu^2]_{\nu=\omega} > 0$ (Prob. 10-5).

We can understand the dip by writing the gain $-\frac{1}{2}(\nu/\varepsilon_0)\,\mathscr{P}_n$ in the $\dot{E}_n$ equation (8.11) as

$$a_g E_n = -\frac{\nu}{2\varepsilon_0}\,\mathrm{Im}\mathscr{P}_n. \tag{42}$$

Equation (27) gives for (41)

$$a_g = \frac{\nu\wp^2\bar{N}}{4\hbar\gamma u\varepsilon_0\sqrt{\pi}}\int_{-\infty}^{\infty} dv$$
$$\times\ \frac{\exp(-v^2/u^2)\,[\mathscr{L}(\omega-\nu_n+Kv)+\mathscr{L}(\omega-\nu_n-Kv)]}{1+\frac{1}{2}\,(\gamma_{ab}/\gamma)I_n\,[(\mathscr{L}(\omega-\nu_n+Kv)+\mathscr{L}(\omega-\nu_n-Kv)]} \tag{43}$$

We recall from the discussion associated with Eq. (25) that each Lorentzian in (27) and therefore in the denominator of (43) results from saturation by one running wave of the standing-wave field. Furthermore, we see from (18) that the Lorentzians in the numerators are similarly associated with their own running waves. For the sake of simplicity, suppose that there is sufficiently sharp homogeneous response, $\gamma \ll Ku$, that the numerator Lorentzians in (43) act like Dirac delta functions. Then, off resonance ($\nu_n \neq \omega$), this gives the saturated gain

$$a_g(\nu_n \neq \omega) = \left(\frac{\nu\wp^2\bar{N}}{4\hbar\gamma u\varepsilon_0\sqrt{\pi}}\right)\left[\frac{2}{1+\frac{1}{2}\,(\gamma_{ab}/\gamma)I_n}\right]\exp[-(\nu_n-\omega)^2/(Ku)^2], \tag{44}$$

for which only *one* (denominator) Lorentzian contributes to saturation at a time. On resonance, *both* Lorentzians contribute, giving

$$a_g(\nu_n = \omega) = \left(\frac{\nu\wp^2\bar{N}}{4\hbar\gamma u\varepsilon_0\sqrt{\pi}}\right)\left[\frac{2}{1+1(\gamma_{ab}/\gamma)I_n}\right]. \tag{45}$$

We see that, for a given intensity I_n, on resonance both running waves interact with the same ensemble (that centered at $v = 0$) and saturate the gain twice as much as is the case off resonance, where the waves interact with different ensembles (those centered about $\pm\, v$). Hence the saturated gain a_g satisfying the steady-state oscillation condition (5.18) given by $\dot{E}_n = 0$,

$$\text{Loss} = \frac{1}{2}\frac{\nu}{Q_n} = a_g\,, \tag{46}$$

occurs for a smaller intensity on resonance than off, that is, there is a dip in intensity versus detuning at line center. Of course, the transition between on and off resonance does not happen instantaneously, since the ensembles for $\pm v$ merge together for detunings less than γ. This merging is represented in third order by the Lorentzian in (39). As would be expected from this discussion, this Lorentzian is missing in the intensity formula corresponding to (39) for the unidirectional ring laser (single running wave), as shown in Prob. 10-1.

The dip has been utilized quite successfully in both stabilization schemes (see Hall, 1973) and spectroscopy (Brewer, 1972). In particular long-term stabilities of 1 part in 10^{12} have been reported through the use of the very sharp and stable *inverse* dip encountered when a methane *absorption* cell is placed within a He-Ne laser cavity operating at 3.39 μm. For this the absorption saturates, producing a peak instead of a dip, and stability is achieved by tuning the cavity for the maximum power output. The spectroscopy schemes center about the fact that the width of absorption and gain dips is determined primarily by the homogeneous decay constant γ rather than the Doppler width Ku. Hence it is possible to measure lifetimes $(1/\gamma)$ which were hitherto masked by the much larger Doppler broadening.

The intensity can be calculated without the Doppler limit by using (34) or more exactly by perturbation theory. However, only for $\gamma/(Ku) \sim 1$ is appreciable deviation from the Doppler limit noticed. This result is particularly felicitous, for it justifies using the relatively simple Doppler limit in most numerical calculations involving, for example, the He-Ne laser. It must be remembered, however, that the third-order expansion we have used departs badly from the exact results for relative excitations much larger than 1.1. A remarkably good alternative to this theory is numerical evaluation of the velocity integral for the complex polarization (or its imaginary part if only the amplitude is of interest) in (27). We come back to this in Sec. 10-3 and turn now to a discussion of multimode operation predicted by perturbation theory.

10-2. Multimode Phenomena

In steady state, the rate equation approximation [defined following (8.31)] can be exact for a single-mode field interacting with homogeneously broadened atoms (Sec. 8-2). In the corresponding gas laser case (Sec. 10-1), however, the approximation is not as good, because atoms moving with appreciable z components of velocity v see two Doppler-shifted frequencies whose difference can be greater than the atomic decay constant γ. The population difference $\rho_{aa}(z, v, t) - \rho_{bb}(z, v, t)$ can then acquire pulsations with period $2\pi/2Ku$, which can be short compared to the time $1/\gamma$ and compromise the rate equation approximation. In Sec. 10-3 it is shown that the approximation is quite accurate in spite of the fact that the pulsations are neglected. In multimode operation, however, the population pulsations resulting from nonlinear beating between modes can greatly modify the mode coupling and even cause mode locking. In order to allow for these complications we do not make the rate equation approximation, but instead use perturbation theory. Although the procedure is similar to that in Sec. 9-1, it requires careful consideration of the time dependence introduced into the normal-mode function $\sin K_n z$ by atomic motion. In this section we show how this time dependence enters the basic perturbation equations. The equations derived are useful for the ring laser treatment (Chap. 11) as well as for the standing-wave laser. The details of the calculation with use of the multimode energy (8) are given in

Appendix D. Multimode phenomena peculiar to Dopper-broadened media are then discussed with numerical results.

In the multimode treatment of the homogeneously broadened laser, we performed time integrals for low-order contributions to the population matrix and then substituted the results back into the equations of motion (8.26)–(8.28) to obtain higher-order terms. Because of the time dependence of the normal-mode functions, the algebra is simpler here if the time integrations are performed after the substitutions. Furthermore, it is convenient to write the multiple time integrals in terms of differences like

$$\tau' = t - t'. \tag{47}$$

Specifically, the formal integral for the off-diagonal element ρ_{ab} given by (16) becomes

$$\rho_{ab}(z, v, t) = \frac{i}{\hbar}\int_0^\infty d\tau' \exp[-(i\omega + \gamma)\tau']\, \mathscr{V}_{ab}(z', t') \times [\rho_{aa}(z', v, t') - \rho_{bb}(z', v, t')], \tag{48}$$

where $z' = z - v\tau'$ and $t' = t - \tau'$. The zeroth-order approximation to the equations of motion (10)–(12) yields the population difference

$$\rho_{aa}{}^{(0)} - \rho_{bb}{}^{(0)} = \gamma_a{}^{-1}\lambda_a - \gamma_b{}^{-1}\lambda_b \equiv N(z, v, t) \tag{49}$$

as before (9.3), except that the excitation rates λ_a are now functions of velocity (24). The difference $N(z, v, t)$ is constant in time unless the excitation level is varied and can always be factored outside the time integral (48). Hence the first-order contribution to ρ_{ab} is given by

$$\rho_{ab}{}^{(1)}(z, v, t) = \frac{i}{\hbar} N(z, v, t) \int_0^\infty d\tau' \exp[-(i\omega + \gamma)\tau']\, \mathscr{V}_{ab}(z', t'). \tag{50}$$

Similarly, the second-order contribution to the population of the upper level is given by

$$\begin{aligned}\rho_{aa}{}^{(2)}(z, v, t) &= -\frac{i}{\hbar}\int_0^\infty d\tau' \exp(-\gamma_a\tau')\, \mathscr{V}_{ab}(z', t')\, \rho_{ba}{}^{(1)}(z', v, t') + \text{c.c.} \\ &= -\hbar^{-2}N\int_0^\infty d\tau' \int_0^\infty d\tau'' \exp(-\gamma_a\tau')\, \mathscr{V}_{ab}(z', t') \\ &\quad\times \mathscr{V}_{ba}(z'', t'') \exp[(i\omega - \gamma)\tau''] + \text{c.c.},\end{aligned} \tag{51}$$

where the time difference $\tau'' = t' - t''$, $z'' = z - v(\tau' + \tau'')$ and $t'' = t - \tau' - \tau''$. From (11) and (12), $\rho_{bb}{}^{(2)} = -\rho_{aa}{}^{(2)}$ with $\gamma_a \to \gamma_b$. Hence the second-order population difference is

$$\begin{aligned}\rho_{aa}{}^{(2)} - \rho_{bb}{}^{(2)} &= -\hbar^{-2}N\int_0^\infty d\tau' \int_0^\infty d\tau'' [\exp(-\gamma_a\tau') + \exp(-\gamma_b\tau')] \\ &\quad\times \{\mathscr{V}_{ab}(z', t')\, \mathscr{V}_{ba}(z'', t'') \exp[(i\omega - \gamma)\tau''] + \text{c.c.}\}.\end{aligned} \tag{52}$$

From (48) the third-order contribution to ρ_{ab} is

$$\rho_{ab}^{(3)}(z, v, t) = +i\hbar^{-1}\int_0^\infty d\tau' \exp[-(i\omega+\gamma)\tau'] \mathscr{V}_{ab}(z', t')[\rho_{aa}^{(2)}(z', v, t') - \rho_{bb}^{(2)}(z', v, t')]$$

$$= -i\hbar^{-3}N\int_0^\infty d\tau' \int_0^\infty d\tau'' \int_0^\infty d\tau''' \exp[-(i\omega+\gamma)\tau'] \mathscr{V}_{ab}(z', t') \times [\exp(-\gamma_a\tau'') + \exp(-\gamma_b\tau'')] \{\mathscr{V}_{ab}(z'', t'') \times \mathscr{V}_{ba}(z''', t''') \exp[(i\omega-\gamma)\tau'''] + \text{c.c.}\}. \quad (53)$$

This contribution is conveniently represented in Fig. 10-4 by a perturbation "tree." There terms connected in series (lines vertically) are multiplied together and then added to other such products. We call the products "limbs." The value of the tree is that it allows us to visualize a fairly complicated expression and even to give its value in tables without further algebra. This is particularly important in more complicated problems such as the Zeeman ring laser. In evaluating the perturbation energies $\mathscr{V}_{ab}$ of (8) for Fig. 10-4, it is very helpful to remember, for example, that $t''' = t - \tau' - \tau'' - \tau'''$. Hence the frequency associated with $\mathscr{V}_{ab}(z''', t''')$ shows up in all three τ in-

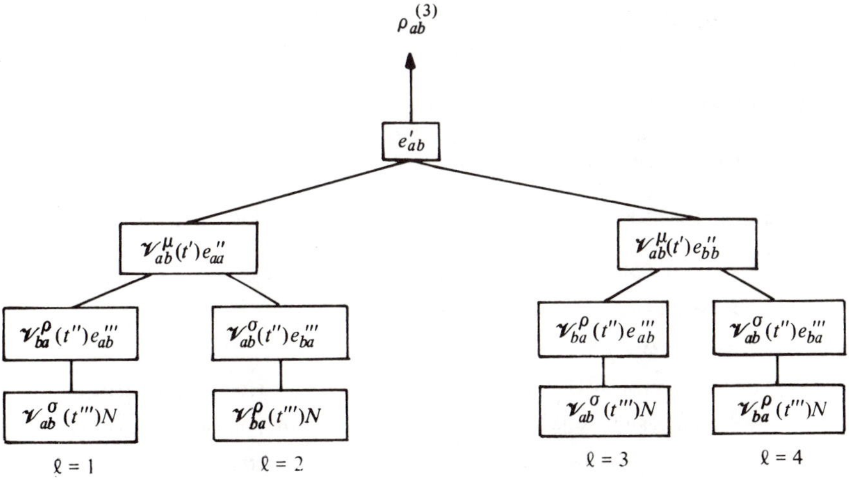

Figure 10-4. Perturbation tree for calculation of the third-order matrix element $\rho_{ab}^{(3)}(z, v, t)$. The interaction energy $\mathscr{V}_{ab}$ is evaluated at earlier times and positions. The Greek superscript is the multimode summation index which we use for that level interaction. The Symbols $e'_{aa} = (i/\hbar)\int_0^\infty d\tau' \exp(-\gamma_a\tau')$, $e'_{\alpha\beta} = -e'_{\beta\alpha}{}^* = (i/\hbar)\int_0^\infty d\tau' \exp[-(i\omega+\gamma)\tau']$, and $N = \rho_{aa}^{(0)} - \rho_{bb}^{(0)} = N(z, v, t)$. The tree limbs (products of four terms connected by descending lines) are labeled by l. The third-order contribution to $\rho_{ab}^{(3)}$ is given by the sum of the tree limbs.

tegrals, whereas that associated with $\mathscr{V}_{ab}(z', t')$ appears only in the integral over τ'.

In Appendix D the first-order (50) and third-order (53) integrals are evaluated for the multimode perturbation energy (8) and combined with (13) to give the complex polarization $\mathscr{P}_n(t)$. The resulting amplitude- and frequency-determining equations have the same forms as those for the homogeneously broadened case (9.18) and (9.19). The new coefficients are given in Appendix D for the general case and in Table 10-2 for the Doppler limit. The general case has been programmed for the digital computer. Let us now consider some results for two-, three-, and ten-mode operation.

Two-Mode Operation

Two-mode operation is described by the intensity equations (9.24) and (9.25), which are discussed in some detail in Sec. 9-2. We recall that the coupling between the modes comes both from competition for the same active medium by saturation and from population pulsations. The latter are generated by the beating of the modes in the nonlinear population difference and yield shifted frequencies on interaction with the modes. A shift from one to the other mode contributes to coupling; a shift in the other direction yields a combination tone which plays a role in self-mode locking (see Sec. 9-3).

TABLE 10-2 *Gas Laser Multimode Saturation Coefficients for the Amplitude- and Frequency-Determining Equations (9.18)–(9.25). Other Coefficients are given in Table 10-1.*

Coefficient	Physical Context
$\vartheta_{n\mu\rho\sigma} = \frac{1}{4}i(\frac{1}{2}\wp/\hbar)^2\,[\mathscr{D}_a(\nu_\rho - \nu_\sigma) + \mathscr{D}_b(\nu_\rho - \nu_\sigma)]$	General saturation term
$\times\,[\mathscr{D}(\omega - \frac{1}{2}\nu_\mu + \frac{1}{2}\nu_\rho - \nu_\sigma)\,N_{2(\rho-\sigma)}/\bar{N}$	
$+\,\mathscr{D}(-\frac{1}{2}\nu_\mu - \frac{1}{2}\nu_\sigma + \nu_\rho)]F_1$	
$\theta_{nm} = [\mathscr{L}(\omega - \frac{1}{2}\nu_n - \frac{1}{2}\nu_m) + \mathscr{L}(\frac{1}{2}\nu_m - \frac{1}{2}\nu_n)]F_3$	Hole burning part and
$+\,\frac{1}{2}(\gamma_a\gamma_b\gamma/\gamma_{ab})\,F_3\,\mathrm{Re}\{[\mathscr{D}_a(\nu_m - \nu_n) + \mathscr{D}_b(\nu_m - \nu_n)]$	population pulsation part of
$\times\,[\mathscr{D}(\omega - \nu_n)\,N_{2(m-n)}/\bar{N} + \mathscr{D}(\frac{1}{2}\nu_m - \frac{1}{2}\nu_n)]\}$	cross saturation
$\theta_{12} = \theta_{21} = [1 + \mathscr{L}(\frac{1}{2}\Delta)]F_3$	Symmetrically tuned
$+\,\frac{1}{2}(\gamma_a\gamma_b/\gamma_{ab}\gamma)\,F_3\mathscr{L}(\frac{1}{2}\Delta)\,(N_2/\bar{N} + 1)$	
$\times\,\{\mathscr{L}_a(\Delta)\,[(\gamma/\gamma_a) - \frac{1}{2}(\Delta/\gamma_a)^2] + \gamma_a \rightarrow \gamma_b\}$	$(\nu_2 - \omega = \omega - \nu_1)\,\theta$
$\tau_{nm} = F_3/2\gamma\,[(2\omega - \nu_n - \nu_m)\,\mathscr{L}(\omega - \frac{1}{2}\nu_n - \frac{1}{2}\nu_m)$	Cross-pushing term
$+\,(\nu_m - \nu_n)\,\mathscr{L}(\frac{1}{2}\nu_m - \frac{1}{2}\nu_n)]$	
$+\,\frac{1}{2}(\gamma_a\gamma_b\gamma/\gamma_{ab})F_3\,\mathrm{Im}\{\;\}$	(for $\mathrm{Im}\{\;\}$, see θ_{nm})

In the present instance, we see from Tables 10-1 and 10-2 that the saturation contribution called hole burning is the same for the self- (β) *and* cross- (θ) saturation coefficients when the modes are symmetrically turned about line center ($\omega - \nu_1 = \nu_2 - \omega$). With neglect of the population pulsation terms, this gives *neutral* coupling, in the terminology of Sec. 9-2. The reason for this rather sizable interaction stems from the fact that each mode is amplified by a pair of velocity groups $v = \pm(\omega - \nu_n)/K$ (see discussion in Sec. (10-1), and for symmetric tuning the pair for one mode coincides with that for the other. The population pulsation contribution to θ_{12} in Table 10-2 is usually negative, for ordinarily the squared beat frequency $\Delta^2 > \gamma_a\gamma$ and the excitation $\bar{N} > -N_2$. Provided these conditions are satisfied, the coupling constant $C = \theta_{12}\theta_{21}(\beta_1\beta_2)^{-1} < 1$ and the coupling is weak (see Sec. 9-2). However, for long cavities the intermode spacing can become small enough that Δ^2 is less than $2\gamma_a\gamma$. Alternatively, N_2 can be made less than $-\bar{N}$ by choosing $N(z) > 0$ in the middle of the cavity and $N(z) < 0$ in the ends (see Fig. 9-1). In either case, the coupling constant C becomes greater than unity, that is, strong coupling occurs. As discussed in Sec. 9-2, this can lead to bistable operation. The coupling constant versus Δ is graphed in Fig. 10-5 for the special case of symmetric tuning, revealing a strong coupling region for mode spacings less than about 68 MHz. In Fig. 10-6, the intensities themselves are graphed versus detuning for a larger $\Delta(= 214$ MHz), but we see that a strong interaction still occurs for symmetric tuning. Note that in this weakly coupled regime one mode can inhibit oscillation of the other, but the reverse is not true (for the same set of parameters), as it is for strong coupling.

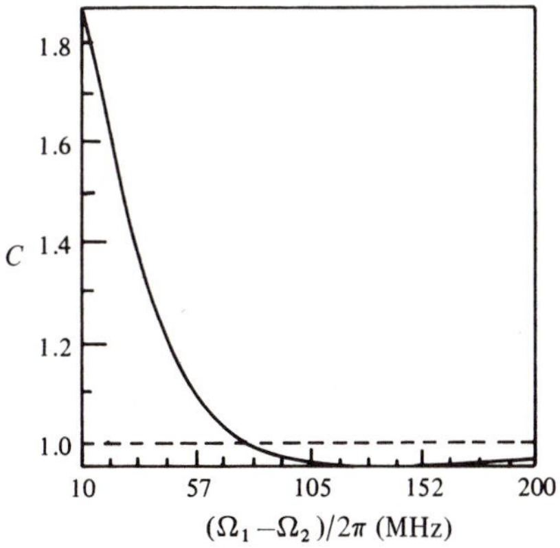

Figure 10-5. Graph of coupling parameter $C = \theta_{12}\theta_{21}/\beta_1\beta_2$ versus intermode spacing $\Omega_2 - \Omega_1$ for Doppler limit, symmetric tuning ($\Omega_2 - \omega = \omega - \Omega_1$), $\gamma = 2\pi \times 75$ MHz, $\gamma_a = \gamma_b = 2\pi \times 40$ MHz. We see that the mode coupling is strong ($C > 1$) for $\Omega_2 - \Omega_1 \simeq 2\pi \times 68$ MHz. The coupling is weak for larger mode spacings.

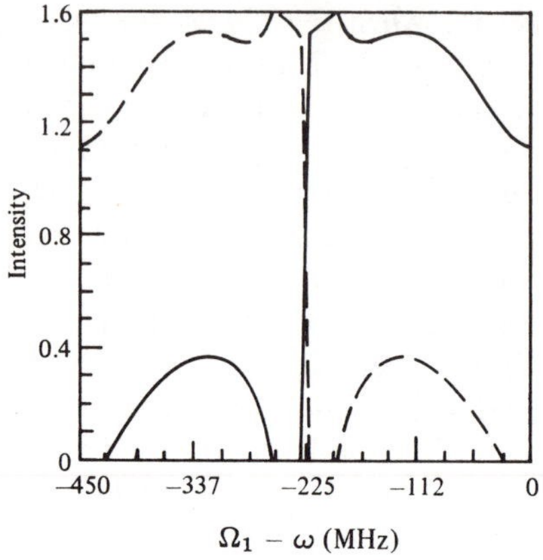

Figure 10-6. Graphs of two-mode dimensionless intensities versus detuning of the lower-frequency mode (solid line). Corresponding detuning of the higher-frequency mode (dashed line) is given by the abcissa value plus 450 MHz. Thus the modes straddle line center for $\Omega_1 - \omega = -225$ MHz. The Doppler width constant $Ku = 2\pi \times 1010$ MHz, the decay constants $\gamma = 2\pi \times 70$ MHz, $\gamma_a = 2\pi \times 20$ MHz, and $\gamma_b = 2\pi \times 40$ MHz, and the relative excitation $\mathfrak{N} = 1.2$.

Three-Mode Operation

In Sec. 9-3 the locking between beat frequencies for three-mode operation was discussed. After a general formulation of the problem, we obtained the result that for stationary atoms the two beat frequencies did, in fact, lock together for symmetric tuning, that is, when the second mode was tuned close to line center. The same situation occurs here for the gas medium, although there are quantitative differences.

The locking equation $\dot{\Psi} = d + l_s \sin\psi + l_c \cos\psi$ (9.54) simplifies here, as it did for stationary atoms, for central tuning of the middle mode, for the dispersion seen by the outer modes becomes equal and opposite and that for middle mode cancels out. Hence the d term goes to zero as central tuning is approached. Furthermore, the l_c term vanishes in this limit, and

$$l_s = \tfrac{1}{16}\pi^{1/2}\wp^4\bar{N}(\hbar^3 Ku\varepsilon_0)^{-1}(\gamma_a\gamma)^{-2}\mathscr{L}_a(\Delta)$$
$$\times\left[\frac{N_2}{\bar{N}}(\gamma_a\gamma - \tfrac{1}{2}\Delta^2)\mathscr{L}(\tfrac{1}{2}\Delta)\,(E_2^2 + 2E_1^2) + 2\left(\frac{\gamma}{\gamma_a}\right)E_1^2\right.$$
$$\left. + (\gamma_a\gamma - \Delta^2)\mathscr{L}(\Delta)E_2^2\right] + \text{same with } \gamma_a \to \gamma_b. \qquad (54)$$

Here we have assumed that the mode intensity $E_1^2 = E_3^2$, for which d can be positive or negative, as found for the homogeneous case of Sec. 9-2.

In Fig. 10-7 the three-mode intensities are graphed as the cavity tuning is varied. Because of mode coupling, particularly between E_1 and E_3, only two modes oscillate over the major portion of the frequency scan. In the region for which mode 2 is within about 12 MHz of line center, all three modes oscillate and the outer mode intensities cross over much as they did for two-mode operation. In a region for which mode 2 is within 6 MHz from line center, mode locking takes place, that is, the difference between beat frequencies vanishes. It is interesting to note that the outer intensities differ by more than a multiplicative factor of 3 at the edge of the locking region, and the fact that our assumption that the d term of Eq. (9.55) vanishes becomes invalid is not surprising. Experimental studies (Tomlinson, unpublished work, 1968) of three-mode operation indicate that the locking region is often considerably displaced from line center, presumably because of pressure effects, which we have ignored in our analysis. Because of the intricate interaction of modes, three-mode operation might be very sensitive to such omissions. A similar situation occurs in the Zeeman laser problem, where a small pressure effect is responsible for changing slightly weakly coupled competition into the strongly coupled type and hence for creating a dramatic change in the laser output (Wang and Tomlinson, 1969).

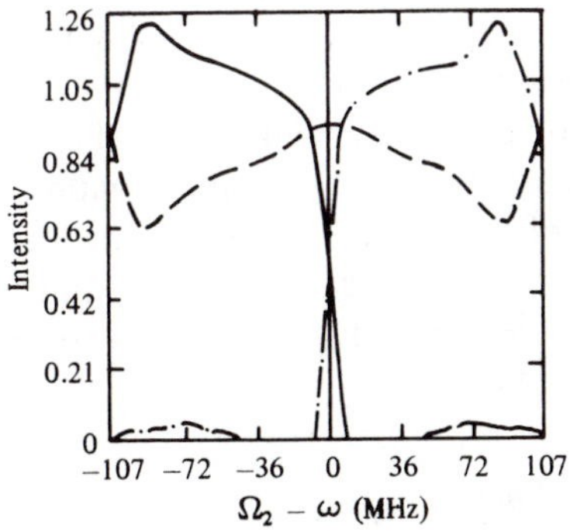

Figure 10-7. Graph of three mode intensities versus cavity detuning of the second mode. The detuning values for the first and third modes are given by subtracting and adding, respectively, the mode spacing, 214 MHz, The first mode intensity is given by the solid line, the second by the dashed line, and the third by the dash-dotted line. Laser parameters used are $Ku = 2\pi \times 1010$ *MHz,* $\gamma_a = 2\pi \times 15.5$ *MHz,* $\gamma_b = 2\pi \times 41$ *MHz,* $\gamma = 2\pi \times 128$ *MHz,* $\mathfrak{N} = 1.2$, $N_2/\bar{N}_T = -0.78$, *and* $N_4/\bar{N}_T = -0.042$.

Multimode Operation

The multimode problem can be treated as discussed in Sec. 9-4. One result from such analysis is that in free-running operation the mode pattern can be asymmetric with respect to line center and bistable, that is, more modes oscillate on one side of line center than on the other. Computer experiments attribute this asymmetry to the strong interaction between modes symmetrically placed with respect to line center, as discussed above for two-mode operation. The fact that the asymmetry disappears in the multimode unidirectional laser confirms this conclusion [see O'Bryan and Sargent (1973)].

Nevertheless, one might also expect the strong interaction to yield an alternation of modes on either side of line center. The bistable asymmetric mode distributions have been observed (Garside, 1968).

10-3. Single-Mode Strong-Signal Theory

The third-order theories described in detail in Secs. 10-1 and 10-2 and in Appendix D have much to commend them. Explicit expressions for the mode amplitude and frequencies near threshold are quite accurate provided effects neglected in the analysis, such as atomic collisions, do not become too important. Nevertheless, lasers are often operated far above threshold in regions where the third-order (and higher-order) theory fails not only quantitatively, but qualitatively as well. Hence it is worth investigating a more exact treatment even at the expense of transcendental solutions which can offer understanding only through numerical analysis. One such approach is quite feasible for the single-mode gas laser and is applicable near steady-state operation. In this section and in Appendix E, we modify the treatment of Sec. 10-1 along lines first proposed by Lax (1968) and derived by Stenholm and Lamb (1969). Some improvements due to Feld and Feldman (1970) are also incorporated. The lowest-order approximation of the theory gives the rate equation result of Sec. 10-1. In our discussion, we evaluate the validity of both the third-order and the rate equation approximations. For a recent treatment of this problem, the ring laser case and multimode operation, see Hambenne and Sargent (1976).

Polarization of Medium

Here, as in other strong-signal theories (Rabi flopping and the theory of nuclear magnetic resonance (See. 7-5)), it is convenient to transform the equations of motion for the density matrix into an interaction picture and eliminate thereby the optical frequency variations of the off-diagonal elements. Furthermore, it is advantageous to use the real and imaginary parts of the resulting off-diagonal elements in place of the latter.† Specifically, we introduce the complex polarization

$$\mathscr{P}_n(z, v, t) = 2\wp \exp[i(\nu_n t + \phi_n)]\, \rho_{ab}(z, v, t), \tag{55}$$

describing an ensemble of atoms moving with z component of velocity v located at position z at time t. We write (55) as the sum of real and imaginary parts:

$$\mathscr{P}_n(z, v, t) = C_n(z, v, t) + iS_n(z, v, t), \tag{56}$$

where C_n and S_n are the in-phase and in-quadrature components, respectively, of the ensemble. Comparing (55) with the complex polarization $\mathscr{P}_n(t)$ in (13), we see that

$$\mathscr{P}_n(t) = C_n(t) + iS_n(t) = \frac{1}{\mathscr{N}} \int_0^L dz\, U_n^*(z) \int_{-\infty}^{\infty} dv\, \mathscr{P}_n(z, v, t). \tag{57}$$

Here $S_n(t)$ and $C_n(t)$ are the imaginary and real parts required in the self-consistency equations (8.11) and (8.12). We introduce also the population difference

$$D(z, v, t) = \rho_{aa}(z, v, t) - \rho_{bb}(z, v, t) \tag{58}$$

and sum

$$M(z, v, t) = \rho_{aa}(z, v, t) + \rho_{bb}(z, v, t). \tag{59}$$

If the decay rates of the upper and lower levels are equal, this last quantity can be ignored.†

In Appendix E we derive integrodifferential equations for the in-quadrature component of the polarization $S_n(z, v, t)$ and the population difference $D(z, v, t)$ using the population matrix equations of motion (10)–(12). The solutions of these equations are given by Fourier series in terms of the position coordinate z. To understand this approach, recall that, to zeroth order in the electric-dipole interaction energy, the population difference was given simply by $N(z, v, t)$ [see (49)], which varies little in a wavelength or optical frequency period. The first-order contribution to the off-diagonal elements (and hence to S_n and C_n) contained the factor $\sin K_n z$ corresponding to one interaction with the electric field. The second-order term for the population difference contained two interactions and hence had $\sin^2 K_n z$ dependence, and the third-order term for ρ_{ab} had $\sin^3 K_n z$ dependence. In general, we see that the population difference D can be expressed in terms of even powers of exp $(iK_n z)$ and the off-diagonal elements in term of odd powers. Thus we expand S_n in the odd-term Fourier series:

†These changes parallel closely those given in Sec. 7-5 in the pictorial representation of the density matrix by a vector **R**.

†C_n, S_n of (56) and D of (58) correspond to the components R_1, $-R_2$, and R_3 of **R** discussed in Sec. 7-5, but we take $\gamma_a \neq \gamma_b$ here.

$$S_n(z, v, t) = -i\wp N(z, v, t) \sum_{j=-\infty}^{\infty} q_{2j+1}(v) \exp[(2j + 1)iK_nz| \tag{60}$$

and the population difference in the even-term Fourier series:

$$D(z, v, \dot{t}) = N(z, v, t) \sum_{j=-\infty}^{\infty} q_{2j}(v) \exp(2ijK_nz). \tag{61}$$

Here time dependence not due to excitation variation could be included in the expansion coefficients q_j. We are restricting our analysis to steady state.

In Appendix E expansions (60) and (61) are substituted into integro-differential equations for $S_n(z, v, t)$ and $D(z, v, t)$, and the Fourier coefficients q_n are determined in terms of a continued fraction $\mathfrak{F}_n$ (E.22). Subsequent use of (56) and (57) for the complex polarization $\mathscr{P}_n(t)$ with the self-consistency relation (8.11) yields the amplitude-determining equation (E.30):

$$\begin{aligned}\mathfrak{N}^{-1} = 2Ku[\gamma_{ab}Z_i(\gamma)]^{-1} \int_0^{\infty} dv\, W(v)\, \mathfrak{F}(v, \omega - \nu_n, I_n) \\ \times [1 + I_n\, \mathfrak{F}(v, \omega - \nu_n, I_n)]^{-1}.\end{aligned} \tag{62}$$

Here the Boltzman distribution (2) is assumed, the relative excitation $\mathfrak{N} = \bar{N}/\bar{N}_T$ with (39) is used, and I_n is the dimensionless intensity (8.45).

The lowest-order approximation to the continued fraction $\mathfrak{F}$ is given by (E. 32):

$$\mathfrak{F} \cong \frac{1}{2}\frac{\gamma_{ab}}{\gamma}[\mathscr{L}(\omega - \nu_n + Kv) + \mathscr{L}(\omega - \nu_n - Kv)]. \tag{63}$$

With this value, Eq. (62) coincides with the formula given by the rate equation polarization of (27). The complete equivalence of the lowest-order approximation and the rate equation approximation (REA) of Sec. 10-1 is shown in Appendix E. As we will see shortly, the approximation in (63) is remarkably accurate.

Equation (62) determines the electric field intensity I_n only implicitly, as contrasted with the third-order treatment, which resulted in the explicit equation (39). Nevertheless, it is possible to determine the intensity predicted by (62), given values of the relative excitation $\mathfrak{N}$ and detuning $\omega - \nu_n$, by use of a Newton-Raphson numerical method. If several curves of intensity versus detuning, as in Fig. 10-3, are desired, it is more efficient in regard to computing time to calculate the relative excitation from (62) for various values of the intensity and detuning and then to determine the tuning curves by linear interpolation. Figure 10-8 gives several intensity versus detuning curves, all exhibiting Lamb dips.

In Fig. 10-9 the REA and third-order series expansions are compared to the continued fraction theory. We see that the third-order theory noticeably overestimates the saturation for relative excitations as low as 1.1, whereas the REA theory remains reasonably accurate. For large $\mathfrak{N}$, the third-order expression (39) gives a maximum intensity of $I_n = 2\gamma/\gamma_{ab}$ (i.e., 2 for these graphs

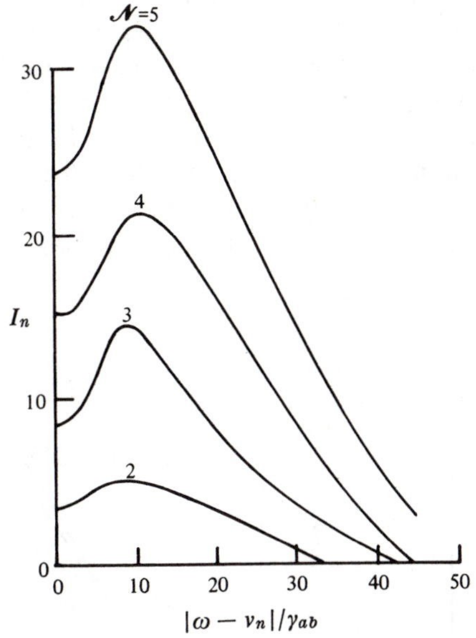

Figure 10-8. *Dimensionless intensity versus detuning for several values of relative excitation. Laser parameters are normalized with respect to the decay constant γ_{ab} as follows: $\gamma_a = 0.6\gamma_{ab}$, $\gamma_b = 1.4\gamma_{ab}$, $\gamma = \gamma_{ab}$, $Ku = 40\gamma_{ab}$ (extreme Doppler limit!). From Stenholm and Lamb (1969).*

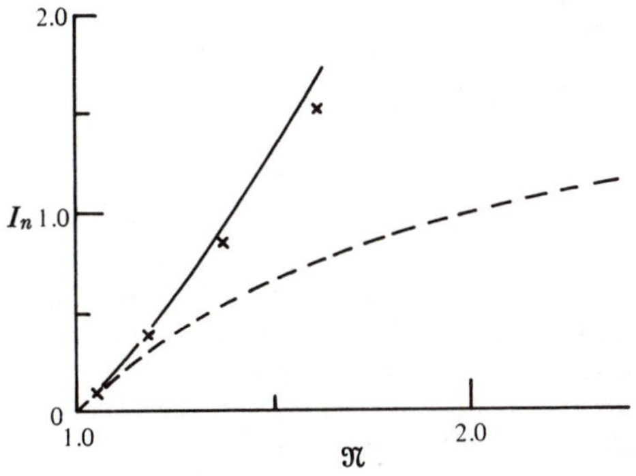

Figure 10-9. *Graph of laser output intensity for central tuning ($\nu_n = \omega$) as a function of relative excitation $\mathcal{N}$. Solid line gives exact calculation using continued fraction; dashed line gives third-order result; and crosses give values for the rate equation approximation (one term in continued fraction). The rate equation values are seen to be in good agreement with the exact values, while the third-order results differ appreciably for relative excitations as low as 1.1. From Stenholm and Lamb (1969).*

with $\gamma = \gamma_{ab}$). In contrast, Fig. 10-9 reveals that the REA continues to be remarkably accurate for considerably larger relative excitations, particularly for detuned operation. The error incurred on resonance is due to the occurrence of atoms with approximately zero axial velocity. These atoms are subject to spatial hole burning, which is assumed to average out in the REA [see Eq. (21)]. Since atoms at field nodes see no field, the actual average saturation for a given intensity is somewhat smaller than that predicted by the REA, leading to the underestimated intensities in Figs. 10-9 and 10-10.

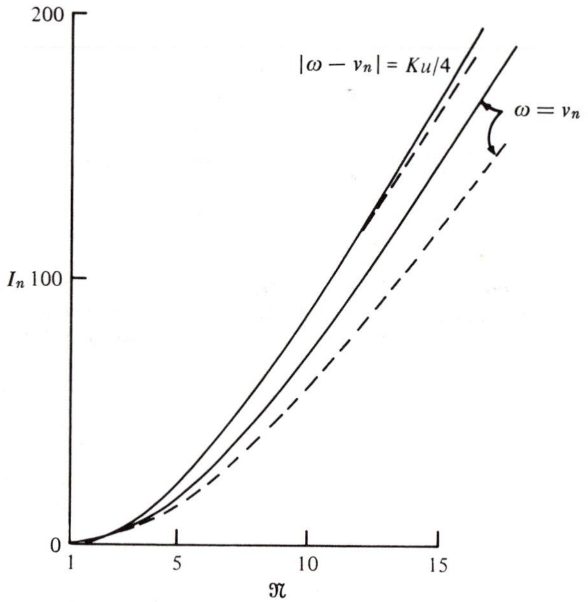

Figure 10-10. Laser output intensity I_n at resonance $\omega = \nu_n$ and for $|\omega - \nu_n| = \frac{1}{4}Ku$, as a function of relative excitation $\mathfrak{N}$ for large values of $\mathfrak{N}$. The results of the rate equation approximation are shown (broken line), and for the tuned laser $\omega = \nu_n$ they deviate appreciably from the exact results only for intensities larger than $I_n = 10$ (with $\mathfrak{N}$ larger than 3.5), whereas for the detuned laser the rate equation approximation gives good results even for the largest intensities considered. The third-order theory predicts an asymptote at $I_n = 2.0$ when $\mathfrak{N}$ goes to infinity. From Stenholm and Lamb (1969).

Further comparison between the series and continued fraction (exact) approaches is given in Fig. 10-11, in which the population difference $D(z, v, t)$ given by (61) is plotted versus a normalized velocity component, and the number of terms required for convergence is given. The series expansion

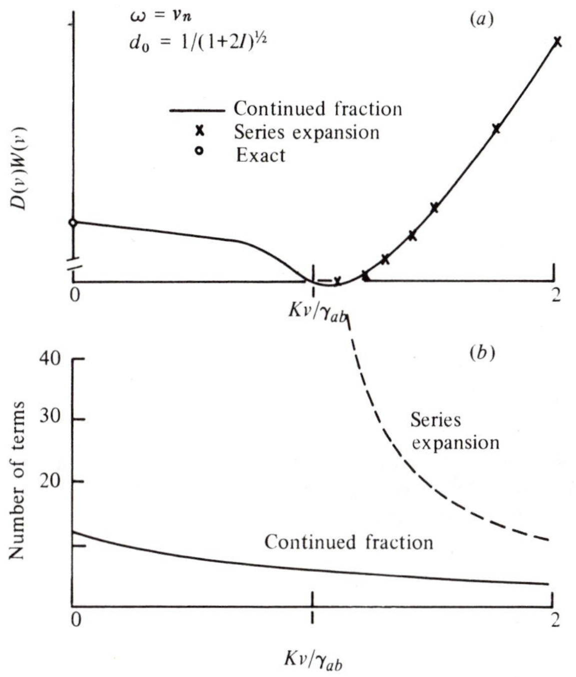

Figure 10-11. (a) Population difference D(v) for central tuning, depicting a bump not predicted by third-order and rate equation approximations. The circle for zero velocity is given by an exact calculation. (b) The number of terms required for convergence by continued fraction (solid line) and series expansion (dashed line) methods. The intensity $I_n = 3.66$ in this figure. From Stenholm and Lamb (1969).

fails to converge at all for this intensity and small velocity components, while the continued fraction gives the exact value even for $v = 0$, a value obtainable analytically as in Sec. 8-2. In Fig. 10-11*a*, the population difference is seen to have a bump for small v, in contrast to what would be expected from a velocity hole burning theory (REA). This is not surprising, however, since the bump is due to spatial hole burning, neglected in the REA, which reduces the average saturation below the REA value.

Finally, in Fig. 10-12, we see the population difference D plotted along the laser (z) axis for several values of v. For zero v, spatial holes are burned in accordance with the theory presented in Sec. 8-2 (see Fig. 8-4). For nonzero v, the z dependence gradually washes out, for the atoms move through more and more standing waves of the field, a fact which helps to justify the replacement of $U_n^2(z)$ by its average value $\frac{1}{2}$ in the rate constant of (21).

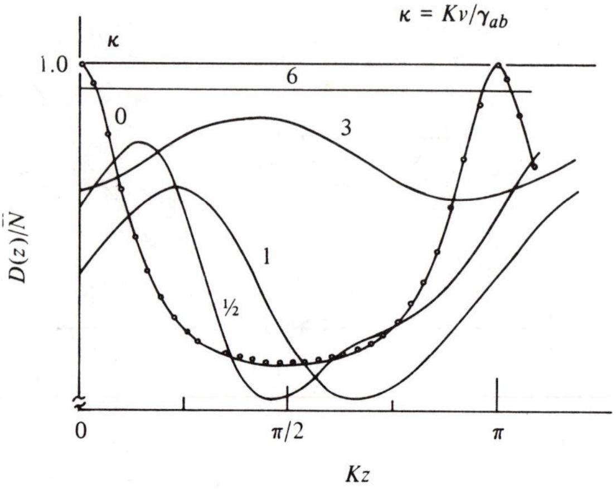

Figure 10-12. Normalized population difference versus spatial phase Kz along laser axis. Spatial holes burned by field intensity for nonmoving atoms are seen to wash out for rapidly moving atoms. From Stenholm and Lamb (1969).

Problems

10-1. Derive the population difference (22), the complex polarization (27), and the intensity (39) for the unidirectional ring laser, that is, for $U_n(z) = \exp(iK_n z)$. Why, physically, is there *no* Lamb dip?
10-2. Prove that Eq. (30) for the plasma dispersion function $Z(\upsilon)$ follows from (C.3) by completing the square in v.
10-3. Show the validity of the following relations for $Z(\upsilon)$:

$$(a)\ Z(\upsilon^*) = -Z^*(\upsilon),$$

$$(b)\ dZ/d\upsilon = (2/Ku)\ [(\upsilon/Ku)\ Z(\upsilon) - i].$$

Hint: For (*b*), use (30) and integrate by parts with a careful choice of parts.

$$(c)\ Z(\upsilon) = \frac{iK}{\sqrt{\pi}} \int_{-\infty}^{\infty} dv \exp(-v^2/u^2) \left[\frac{\upsilon}{\upsilon^2 + (Kv)^2}\right]$$

10-4. Equation (30) for $Z(\upsilon)$ is just the Fourier transform of the time-domain product

$$Z(t) = \begin{cases} 0, & t < 0, \\ iKu \exp(-\gamma t) \exp[-(Ku/2)^2 t^2], & t \geqslant 0. \end{cases}$$

Using the Fourier transform convolution theorem, obtain Eq. (29) directly. Why does the inhomogeneously broadened medium have this temporal response?

10-5. Prove that for relative excitations satisfying the inequality (41), a Lamb dip is observed. *Hint:* Set $\partial^2 I_n/\partial\nu^2|_{\nu=\omega} > 0$; calculate $\partial I_n/\partial\nu$ first and use $\nu = \omega$ while evaluating the second derivative.

10-6. Show that in the extreme Doppler limit $Ku \gg \gamma$ the half width at half "maximum" of the tuning dip is $\sqrt{2}\gamma$.

10-7. Show for two-mode operation with the modes symmetrically placed with respect to line center ($\omega - \nu_1 = \nu_2 - \omega$) and $\gamma_a = \gamma_b$ that the coupling constant C of (9.37) is given by

$$C = \left[1 + \frac{\frac{1}{2}\gamma_a\gamma(\gamma_a\gamma - \frac{1}{2}\Delta^2)(N_2/\bar{N} + 1)}{(\gamma^2 + \frac{1}{8}\Delta^2)(\gamma_a^2 + \Delta^2)}\right]^2.$$

When is this greater than 1 (strong coupling), and why is the coupling so large?

10-8. Reduce the product of sines

$$\sin K_n z \sin(K_\mu z - Kv\tau') \sin[K_\rho z - Kv(\tau' + \tau'')] \times \sin[K_\sigma z - Kv(\tau' + \tau'' + \tau''')]$$

appearing in the third-order polarization (D.2) by neglecting terms odd in v and rapidly varying in z.

Ans.: Equation (D.3).

10-9. Show that

$$\int_{-\infty}^{\infty} dv\, \mathscr{L}(\omega - \nu - Kv)f(v) = \int_{-\infty}^{\infty} dv\, \mathscr{L}(\omega - \nu + Kv)f(v),$$

where $f(-v) = f(v)$, that is, f is even in v. Also show that

$$\int_{-\infty}^{\infty} dv\, \exp[-(v/u)^2] \cos Kvx = \int_{-\infty}^{\infty} dv \exp[-(v/u)^2] \exp(-iKvx).$$

10-10. Derive the slowly varying equations of motion (E.1)–(E.4) from those for $\rho(z, v, t)$, Eqs. (10)–(12). How can the pictorial representation of the density matrix (Bloch equations) include $\gamma_a \neq \gamma_b$?

10-11. The complete third-order (not Doppler limit), single-mode coefficients are given by (D.19) for ϑ_{nnnn} as

$$\begin{aligned}\rho_n + i\beta_n = \frac{\gamma_a\gamma_b}{8\sqrt{\pi}}F_1 \Bigg\{ &\frac{1}{\gamma_a}\left[\frac{1}{\upsilon}Z(\upsilon) + \frac{i}{\gamma}Z_i(\upsilon) - \frac{\partial Z(\upsilon)}{\partial \upsilon} + \frac{i}{\omega - \nu_n}Z_r(\upsilon)\right] \\ &+ \frac{1}{2}\frac{1}{\upsilon - \gamma_a/2}\left[\frac{\partial Z(\upsilon)}{\partial \upsilon} - \frac{Z(\upsilon) - Z(\gamma_a/2)}{\upsilon - \gamma_a/2} \right. \\ &\left. - \frac{iZ_r(\upsilon)}{\omega - \nu_n} + \frac{Z^*(\upsilon) - Z^*(\gamma_a/2)}{\upsilon^* - \gamma_a/2}\right]\Bigg\} + \text{same with } \gamma_a \to \gamma_b. \qquad (64)\end{aligned}$$

Show that the Strong Doppler limit $[Z(\upsilon) = i\sqrt{\pi}]$ gives a better approximation to this value than it does to the T_{l1} integrals alone, the only integrals which survive in that limit. *Hint:* Expand the plasma dispersion function to first-order in υ/Ku.

Ans.: Value given in Table 10-1.

10-12. Show that the rate-equation-approximation polarization $\mathscr{P}_n(t)$ of (27) is given by

$$\mathscr{P}_n(t) = -\frac{1}{2}\left[\frac{\wp^2\bar{N}}{\hbar Ku}\right] E_n(\gamma + i\Delta)\ \{(1 + A)\,[Z(\upsilon_+)/\upsilon_+] + (1 - A)[Z(\upsilon_-)/\upsilon_-]\}\,, \tag{65}$$

where

$$\begin{aligned} \upsilon_\pm{}^2 &= \gamma'^2 - \Delta^2 \pm [(\gamma\gamma_{ab}I_n/2)^2 - 4\Delta^2\gamma'^2]^{1/2} \\ \gamma' &= \gamma\,[1 + \tfrac{1}{2}(\gamma_{ab}/\gamma)I_n]^{1/2} \\ A &= \frac{\gamma_{ab}I_n + 4i\Delta}{[(\gamma_{ab}I_n)^2 - 16\Delta^2(\gamma'/\gamma)^2]^{1/2}}. \end{aligned} \tag{66}$$

Hint: multiply out the Lorentzians in the denominator of (27), find the roots of the resulting quadratic, drop integrands with odd powers of v and use formula of Prob. 10–3d for $Z(\upsilon)$.

10-13. Show that for unidirectional operation, the polarization (13) becomes

$$\mathscr{P}_n(t) = -\frac{\wp^2}{\hbar Ku}\,\bar{N}\,E_n[Z_r(\gamma' + i\omega - i\nu_n) + i\frac{\gamma}{\gamma'}\,Z_i(\gamma' + i\omega - i\nu_n)], \tag{67}$$

where the power broadened decay constant γ' is given by (66).

References

The references of Chaps. 8 and 9 are useful here too. Some additional books and review articles are the following:

L. Allen and D. G. C. Jones, 1967, *Principles of Gas Lasers,* Plenum Publishing Corp., New York.

W. R. Bennett, Jr., 1971, in: *Brandeis University Summer Institute in Theoretical Physics, 1969*, Gordon and Breach, New York.

A. L. Bloom, 1968, *Gas Lasers*, John Wiley & Sons, New York.

C. G. B. Garrett, 1967, *Gas Lasers,* McGraw-Hill Book Co., New York.

O. P. McDuff, 1972, "Techniques of Gas Lasers," Chap. C5 in *Laser Handbook, op. cit.* in references for Chap. 8.

C. K. Rhodes and A. Szöke, 1972, "Gaseous Lasers: Atomic, Molecular and Ionic," Chap. B1 in *Laser Handbook, op. cit.* in references for Chap. 8.

M. Sargent III, 1974, "Strong Signal Laser Theory," in *High Energy Lasers and their Applications*, ed. by S. F. Jacobs, M. Sargent III, and M. O. Scully, Addison-Wesley Reading, Mass.

D. C. Sinclair and W. E. Bell, 1969, *Gas Laser Technology*, Holt, Rinehart and Winston, New York.

S. Stenholm, 1970, "The Semiclassical Theory of the Gas Laser," in: *Progress in Quantum Electronics*, ed. by J. H. Sanders and K. W. H. Stevens, Pergamon Press, Oxford.

Other pertinent references are as follows:

W. R. Bennett, Jr., 1962, *Phys. Rev.* **126**, 580.

R. G. Brewer, 1972, *Science* **178**, 247. See also articles in *High Resolution Laser Spectroscopy*, 1976, Ed. by K. Shimoda, Springer-Verlag, Heidelberg.

B. K. Garside, 1968, *IEEE J. Quantum Electronics* **QE-4**, 940.

B. J. Feldman and M. S. Feld, 1970, *Phys. Rev.* **A1,** 1375.

J. L. Hall, 1973, in *Atomic Physics 3*, ed. by S. J. Smith and D. K. Walters, Plenum, New York, p. 615.

A. Javan, W. R. Bennett, Jr., and D. R. Herriott, 1961, *Phys. Rev. Letters* **6**, 106.

S. Stenholm and W. E. Lamb, Jr., 1969, *Phys. Rev.* **181**, 618.

C. H. Wang and W. J. Tomlinson, 1969, *Phys. Rev.* **181**, 115.

XI

THE RING LASER

11. Ring Laser

In Chapters 8–10 we considered primarily the two-mirror, standing-wave laser. The analysis was written to cover also the theoretically simpler ring (or three mirror) laser depicted in Fig. 8-2, which is constructed to support only a *single*, clockwise running wave characterized by the normal-mode function $U_n(z) = \exp(iK_nz)$ of Eq. (8.7). The relative simplicity of this unidirectional ring laser stems from the fact that the two-mirror field is composed of *two* running waves traveling in opposite directions, as given by Eq. (10.25). Hence, for example, the standing-wave gas laser may exhibit a dip in intensity versus detuning (Lamb dip), as discussed in Sec. 10-1; the unidirectional case does not, for only one velocity ensemble contributes gain regardless of tuning. Similarly, the two-mirror case can be affected by spatial hole burning by the standing-wave intensity (Sec. 8-2); the unidirectional laser cannot. The simple operation of the latter is, however, a special case of operation more general than that of the two-mirror, standing-wave laser, for without nonreciprocal losses or suitable mode inhibitory effects the ring laser supports running waves in *both* directions, clockwise and counterclockwise. Unlike the standing-wave field of Eq. (10.25), the oppositely directed running waves in this bidirectional ring laser may have unrelated amplitudes, phases, and frequencies. In particular, a clockwise rotation of the laser about an axis perpendicular to the plane of the mirrors Doppler-downshifts the frequency of the clockwise running wave and upshifts the counterclockwise running wave. The beat note between the two waves provides a measure of the rotation rate, leading to the use of the ring laser as a gyroscope. A representation equivalent to the independent running waves is that with cos K_nz mode functions in addition to the sin K_nz functions of the two-mirror case.

In this chapter, we treat two-mode, bidirectional ring laser operation in which the atoms can have nonzero velocities as in Chap. 10. Specifically, in

Sec. 11-1 the appropriate amplitude- and frequency-determining equations are derived. In Sec. 11-2 some numerical results are given and the locking of the frequencies of oppositely directed running waves is considered. Both the two-mode analysis of Sec. 9-2 and the locking theory of Sec. 9-3 are used in this study. In Sec. 11-3 the multimode theory for unidirectional running waves derived in Appendix D is discussed and is used to show quantitatively that modes placed symmetrically with respect to line center do not compete strongly, as they did in the standing-wave case of Sec. 10-2.

11-1. Ring Laser Amplitude and Frequency Equations

The electric field in a ring laser $E(z, t)$ can be written as the superposition of sines *and* cosines, in contrast to the situation with the two-mirror, standing-wave laser. An equivalent and (for many problems) more convenient representation consists of oppositely directed running waves (Fig. 11-1), namely,

$$E(z, t) = \tfrac{1}{2}\sum_n \{E_{+n}(t)\exp[-i(\nu_{+n}t + \phi_{+n} - K_{+n}z)] + E_{-n}(t)\exp[-i(\nu_{-n}t + \phi_{-n} + K_{-n}z)]\} + \text{c.c.} \quad (1)$$

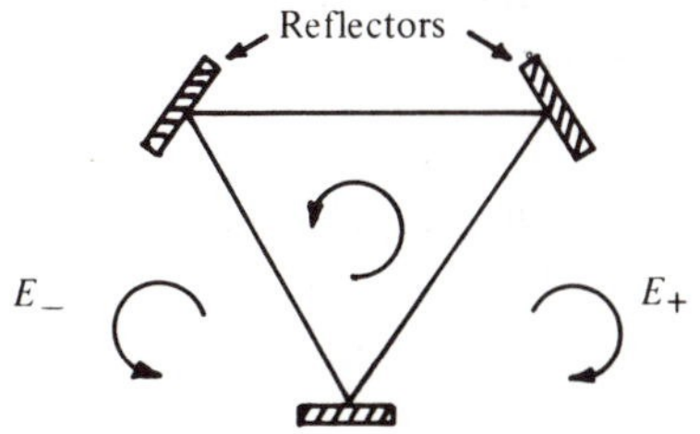

Figure 11-1. Diagram of possible ring laser cavity with three mirrors. The round-trip length is given by L; the running-wave mode amplitudes, by E_+ and E_-.

The amplitudes and phases are again assumed to be slowly varying. The wave numbers for oppositely directed waves and identical n can differ notably for a rotating cavity, for the round-trip length for one direction differs from that for the opposite direction. Accordingly the passive cavity eigenfrequencies differ ($\Omega_{+n} \neq \Omega_{-n}$).

For simplicity, we consider the oscillation of two oppositely directed running waves with approximately equal frequencies ($\nu_+ \simeq \nu_-$) and wave numbers ($K_+ \simeq K_-$). The electric field (1) for this is

$$E(z, t) = \tfrac{1}{2}\{E_+ \exp[-i(\nu_+ t + \phi_+ - K_+ z)] + E_- \exp[-i(\nu_- t + \phi_- + K_- z)]\} + \text{c.c.}, \tag{2}$$

with the corresponding induced polarization:

$$P(z, t) = \tfrac{1}{2}\{\mathscr{P}_+(t) \exp[-i(\nu_+ t + \phi_+ - K_+ z)] + \mathscr{P}_- \exp[-i(\nu_- t + \phi_- + K_- z)]\} + \text{c.c.} \tag{3}$$

In a fashion similar to that in Chap. 8, we find the self-consistency equations:

$$\dot{E}_+ + \frac{1}{2}\left(\frac{\nu}{Q_+}\right) E_+ = -\frac{1}{2}\frac{\nu}{\varepsilon_0}\,\text{Im}(\mathscr{P}_+), \tag{4}$$

$$\nu_+ + \dot{\phi}_+ = \Omega_+ - \frac{1}{2}\frac{\nu}{\varepsilon_0} E_+^{-1}\,\text{Re}(\mathscr{P}_+), \tag{5}$$

and similar equations in which the plus is replaced by the minus for the oppositely directed running wave. Here the possibility of a rotating frame is allowed for by taking $\Omega_+ \neq \Omega_-$.

In general, for moving atoms, the polarization is again given by (10.5), but the complex polarization involves different mode projections [compare with (10.13)]:

$$\mathscr{P}_+ = 2\wp \exp[i(\nu_+ t + \phi_+)] \frac{1}{L}\int_0^L dz \exp(-iK_+ z) \int_{-\infty}^{\infty} dv\, \rho_{ab}(z, v, t) \tag{6}$$

and a similar expression for $\mathscr{P}_-$ in which $\exp(iK_- z)$ appears.

To calculate this polarization, we substitute the electric-dipole interaction energy (in the rotating-wave approximation),

$$\mathscr{V}_{ab}(t) = -\tfrac{1}{2}\wp\{E_+ \exp[-i(\nu_+ t + \phi_+ - K_+ z)] + E_- \exp[-i(\nu_- + \phi_- + K_- z)]\}, \tag{7}$$

into the formal integrals (10.50) and (10.53) for the first- and third-order contributions to the population matrix element $\rho_{ab}(z, v, t)$. The first-order contribution is given by

$$\rho_{ab}^{(1)} = -\tfrac{1}{2} i \frac{\wp}{\hbar} N(z, v, t) \Big\{E_+ \exp[-i(\nu_+ t + \phi_+ - K_+ z)] \times \int_0^\infty d\tau' \exp[-i(\omega - \nu_+ + Kv)\tau' - \gamma\tau'] + E_- \exp[-i(\nu_- t + \phi_- + K_- z)] \int_0^\infty d\tau' \exp[-i(\omega - \nu_- - Kv)\tau' - \gamma\tau'] \Big\}.$$

Evaluating the integrals and using (6), we have the complex polarization

$$\mathscr{P}_+^{(1)}(t) = -\wp^2 \bar{N}(\hbar K u)^{-1} E_+ Z[\gamma + i(\omega - \nu_+)], \tag{8}$$

where the plasma dispersion fuction $Z(v)$ is given by (10.29).

The third-order contribution is given by (7) and (10.53):

$$\rho_{ab}^{(3)}(z, v, t) = \tfrac{1}{8}i\left[\frac{\wp}{\hbar}\right]^3 N(z, v, t)\int_0^\infty d\tau' \int_0^\infty d\tau'' \int_0^\infty d\tau''' \exp[-(i\omega + \gamma)\tau']$$
$$\times \{E_+\exp[-i(\nu_+t + \phi_+ - K_+z)] \exp[i(\nu_+ - Kv)\tau'] + E_-\exp[-i(\nu_-t + \phi_- + K_-z)] \exp[i(\nu_- + Kv)\tau']\}[\exp(-\gamma_a\tau'') + \exp(-\gamma_b\tau'')]$$
$$\times \{\exp[-(i\omega + \gamma)\tau'''][E_+ \exp[i(\nu_+t + \phi_+ - K_+z)] \exp[-i(\nu_+ - Kv)(\tau' + \tau'')] + E_-\exp[i(\nu_-t + \phi_- + K_-z)] \exp[-i(\nu_- + Kv)(\tau' + \tau'')]]$$
$$\times [E_+\exp[-i(\nu_+t + \phi_+ - K_+z)] \exp[i(\nu_+ - Kv)(\tau' + \tau'' + \tau''')] + E_-\exp[-i(\nu_-t + \phi_- + K_-z)] \exp[i(\nu_- + Kv)(\tau' + \tau'' + \tau''')]]$$
$$+ \exp[(i\omega - \gamma)\tau'''][E_+\exp[-i(\nu_+t + \phi_+ - K_+z)] \exp[i(\nu_+ - Kv)(\tau' + \tau'')] + E_-\exp[-i(\nu_-t + \phi_- + K_-z)] \exp[i(\nu_- + Kv)(\tau' + \tau'')]]$$
$$\times [E_+ \exp[i(\nu_+t + \phi_+ - K_+z) \exp[-i(\nu_+ - Kv)(\tau' + \tau'' + \tau''')] + E_- \exp[i(\nu_-t + \phi_- + K_-z)] \exp[-i(\nu_- + Kv)(\tau' + \tau'' + \tau''')]]\} \quad (9)$$

Keeping only terms with slowly varying z dependence (others vanish in spatial projection),† performing the time integrations, and using (6), we have the complex polarization:

$$\mathscr{P}_+^{(3)} = \tfrac{1}{4}\wp^4(\hbar^3Ku)^{-1}\bar{N}E_+(iKu)\int_{-\infty}^\infty dv\, W(v)\, \mathscr{D}(\omega - \nu_+ + Kv)$$
$$\times \{2\gamma_{ab}(\gamma_a\gamma_b)^{-1}[E_+^2\mathscr{D}(\omega - \nu_+ + Kv) + E_-^2\mathscr{D}(\omega - \nu_- - Kv) + \text{c.c.}] + E_-^2[\mathscr{D}_a(\nu_- - \nu_+ + 2Kv) + \mathscr{D}_b(\nu_- - \nu_+ + 2Kv)][\mathscr{D}(\omega - \nu_+ + Kv) + \mathscr{D}(\nu_- - \omega + Kv)]\}. \quad (10)$$

This can be written in terms of the plasma dispersion function (10.29) and its first derivative by separating the complex denominators by partial fractions as derived in general in Appendix D. We find

$$\mathscr{P}_+^{(3)} = \tfrac{1}{2}\wp^4\bar{N}(\hbar^3Ku)^{-1} E_+2\gamma_{ab}(\gamma_a\gamma_b)^{-1}\left\{E_+^2\left[\frac{-\partial Z(v_+)}{\partial v_+} + \frac{iZ_i(v_+)}{\gamma}\right] + \frac{\tfrac{1}{4}E_-^2[Z(v_+)+Z(v_-)]}{v_0} + \tfrac{1}{2}i\frac{E_-^2[Z(v_+) + Z^*(v_-)]}{\omega - \nu_0}\right\}$$
$$- \tfrac{1}{8}\wp^4\bar{N}(\hbar^3Ku)^{-1}E_+E_-^2\left\{\left(\frac{1}{v_0 - \tfrac{1}{2}\gamma_a}\right)\left[\frac{Z(v_+) - Z[\tfrac{1}{2}\gamma_a + \tfrac{1}{2}i(\nu_- - \nu_+)]}{v_G^* - \tfrac{1}{2}\gamma_a} - \frac{\partial Z(v_+)}{\partial v_+} + \tfrac{1}{2}i\frac{Z(v_+) + Z^*(v_-)}{\omega - \nu_0} - \frac{Z(\tfrac{1}{2}\gamma_a + \tfrac{1}{2}i(\nu_- - \nu_+)] + Z^*(v_-)}{v_0^* - \tfrac{1}{2}\gamma_a}\right]\right\}$$

† It is helpful to define $e_+ \equiv E_+ \exp[-i(\nu_+t + \phi_+ - K_+z)]$ and $e_- \equiv E_-\exp[-i(\nu_-t + \phi_- + K_-z)]$ and note that the six combinations in(9) which contribute to $\mathscr{P}_+$ are $e_+e_+^*e_+, e_+e_+e_+^*, e_+e_-^*e_-, e_+e_-e_-^*, e_-e_-^*e_+, e_-e_+e_-^*$, aside from time-varying exponentials. The combinations $e_+e_-^*e_+$, $e_+e_+e_-^*$, $e_-e_+^*e_-$, and $e_-e_-e_+^*$ have rapidly varying z dependence and vanish in the spatial projection. The remaining six combinations contribute to $\mathscr{P}_-$.

$$+ \text{ same with } \gamma_a \to \gamma_b \Big\}, \tag{11}$$

where the average frequency $\nu_0 = \frac{1}{2}(\nu_+ + \nu_-)$ and $\upsilon_0 = \gamma + i(\omega - \nu_0)$. Aside from a numerical factor, this reduces to the non-Doppler limit (10.64) for a single-mode standing wave ($E_- = E_+$, $\nu_- = \nu_+$). In the strong Doppler limit ($Ku \gg \gamma$, $\nu_\pm - \omega$), the dispersion function $Z(\upsilon) \simeq i\sqrt{\pi}$, for which (11) reduces to

$$\mathscr{P}_+^{(3)} = \tfrac{1}{2}i\sqrt{\pi}\,\wp^4\bar{N}\gamma_{ab}(\hbar^3Ku\gamma\gamma_a\gamma_b)^{-1}E_+[E_+^2 + E_-^2\gamma\,\mathscr{D}(\omega - \nu_0)]. \tag{12}$$

Combining the first- (8) and third-(12) order contributions to the complex polarization with the amplitude (4) and frequency (5) self-consistency equations, we have

$$\dot{I}_+ = 2I_+(a_+ - \beta_+I_+ - \theta_{+-}I_-), \tag{13}$$

$$\dot{I}_- = 2I_-[a_- - \beta_-I_- - \theta_{-+}I_+], \tag{14}$$

$$\nu_+ + \dot{\phi}_+ = \Omega_+ + \sigma_+ - \rho_+I_+ - \tau_{+-}I_-, \tag{15}$$

$$\nu_- + \dot{\phi}_- = \Omega_- + \sigma_- - \rho_-I_- - \tau_{-+}I_+, \tag{16}$$

where the coefficients are defined in Table 11-1.

TABLE 11-1 *Ring Laser Coefficients (for Gas Laser) in Intensity and Frequency-Determining Equations (13)–(16).*

The Doppler limit $\gamma \ll Ku$ has been assumed. Corresponding coefficients for the homogeneously broadened laser are easily derived.

Coefficient	Physical Context
$a_\pm = \exp[-(\omega - \nu_\pm)^2/(Ku)^2]F_1 - \frac{1}{2}\nu/Q_\pm$	Linear net gain
$\beta_\pm = \frac{1}{2}(\gamma_{ab}/\gamma)F_1 \exp[-(\nu_\pm - \omega^2)/(Ku)^2]$	Self-saturation
$\theta_{\pm\mp} = \beta_\pm\mathscr{L}(\omega - \nu_0)$	Cross-saturation
$\sigma_\pm = -2\pi^{-1}\exp(-\xi^2)\int_0^\xi F_1\exp(x^2)\,dx, \xi = (\nu_\pm - \omega)/Ku$	Mode pulling
$\rho_+ = \rho_- = 0$	Mode (self) pushing
$\tau_{\pm\mp} = \beta_\pm[(\omega - \nu_0)/\gamma]\mathscr{L}(\omega - \nu_0)$	Cross pushing
$\mathscr{L}(\omega - \nu_0) = \gamma^2/[\gamma^2 + (\omega - \nu_0)^2]$	Dimensionless Lorentzian
$\nu_0 = \frac{1}{2}(\nu_+ + \nu_-)$	Average frequency
$F_1 = \frac{1}{2}\nu\pi^{1/2}\wp^2(\hbar Ku\varepsilon_0)^{-1}\bar{N}$	First-order factor

11-2. Ring Laser Operation

The intensity equations (13) and (14) are just those for two-mode operation, described in Sec. 9-2. The values of the steady-state intensities $I_\pm \equiv E_\pm^2$ depend critically on the coupling parameter $C = \theta_{+-}\theta_{-+}/\beta_+\beta_-$, given by (see Table 11-1)

$$C = \mathscr{L}^2(\omega - \nu_0). \tag{17}$$

This is less than unity except for central tuning of both modes ($\nu_+ = \nu_- = \omega$), and falls off sharply for detuned operation, indicating that, except in the region near central tuning, the coupling is weak and the two running waves oscillate nearly independently of one another. This behavior is illustrated in Fig. 11-2 for typical gas laser parameters.

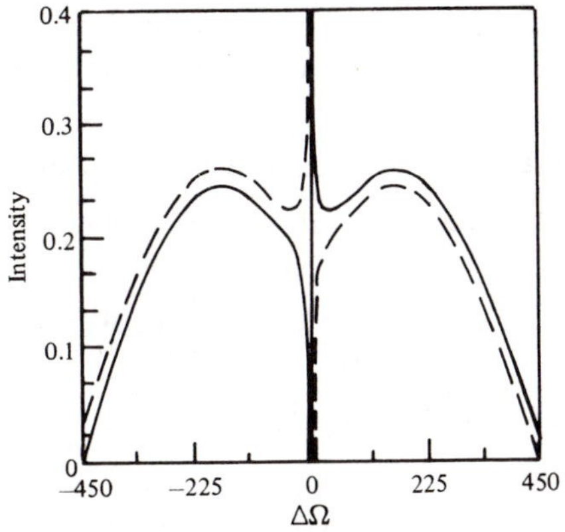

Figure 11-2. Intensities versus (average) detuning for oppositely directed running waves in a ring laser. The parameters $Ku = 2\pi \times 1010$ MHz, $\gamma = 2\pi \times 128$ MHz, $\gamma_a = 2\pi \times 15.5$ MHz, and $\gamma_b = 2\pi \times 41.0$ MHz, and the frequency splitting due to cavity rotation is $2\pi \times 15$ MHz.

As we have solved the problem, there is no locking of the running-wave frequencies, for no coupling terms with the relative phase angle

$$\Psi = (\nu_+ - \nu_-)t + \phi_+ - \phi_- \tag{18}$$

appear. Such a phase angle was responsible for the locking of beat frequencies in the three mode standing-wave laser (Sec. 9-3) We have neglected, however, the coupling introduced by back scattering from the mirrors and by the local mirror losses (see Aronowitz and Collins, 1966). The back scattering reflects part of a running wave into the direction of the oppositely directed running wave and contributes to the latter with the phase factor $\exp(\pm i\Psi)$. The mirror losses are probably even more effective in coupling the running waves. For simplicity, suppose that the mirror losses occur at a given point. A standing wave with node at this point experiences less loss and consequently has higher net gain than a mode with appreciable amplitude at the point. Inasmuch as a standing wave is a sum of two running waves with equal frequencies, the point loss tends to lock the running-wave frequencies.

Mathematically, the coupling due to either mechanism enters the self-consistency conditions through the terms containing the complex g matrix in the following equations:

$$\dot{E}_+ + \frac{1}{2}\frac{\nu}{Q_+} E_+ + \operatorname{Im}\,[ig_{+-}\exp(i\Psi)]\,E_- = -\frac{1}{2}\frac{\nu}{\varepsilon_0}\operatorname{Im}(\mathscr{P}_+), \tag{19}$$

$$\nu_+ + \dot{\phi}_+ + \operatorname{Re}[ig_{+-}\exp(i\Psi)]\left(\frac{E_-}{E_+}\right) = \Omega_+ - \frac{1}{2}\frac{\nu}{\varepsilon_0}\frac{\operatorname{Re}(\mathscr{P}_+)}{E_+}, \tag{20}$$

with similar equations for E_- and ν_-. Considering (20) and the counterpart for ν_-, we can derive with the polarization given in (8) and (12) and their counterparts for ν_- the equation of motion for the relative phase angle Ψ (18):

$$\dot{\Psi} = \nu_+ + \dot{\phi}_+ - \nu_- - \dot{\phi}_- = d + l\sin\Psi, \tag{21}$$

where the "constants" (they depend on the amplitudes, which may vary in time)

$$d = \Omega_+ - \Omega_- + \sigma_+ - \sigma_- - (\rho_+ - \tau_{-+})\,E_+^2 - (\tau_{+-} - \rho_-)\,E_-^2, \tag{22}$$

$$l = g_{+-}\left(\frac{E_-}{E_+} + \frac{E_+}{E_-}\right). \tag{23}$$

For small frequency splitting ($\Omega_+ \cong \Omega_-$) and near-central tuning, the constant d approaches zero, much as the corresponding constant for the three-mode standing-wave laser approaches zero as the central mode is tuned close to line center (Sec. 9-3). When the magnitude of d becomes smaller than that of l, the frequencies lock, as discussed in the three-mode problem. Here, the frequencies of oppositely directed running waves lock to the same value, whereas in the three-mode problem the beat frequency between the first and second modes locks with that between the second and third modes. As we will see in Chap. 12, the right and left circularly polarized components of a vector electric field lock to the same frequency because of an x-y Q anisotropy in a fashion entirely analogous to that considered here. It is primarily this locking phenomenon which limits the accuracy of the ring laser as a gyroscope, for at small rotation rates the running waves lock and their beat frequency, in principle proportional to the rotation rate, vanishes.

11-3. Multimode, Unidirectional Theory

In Sec. 10-2 on two-mode gas laser operation, a strong competition between modes placed symmetrically with respect to line center occurred, as shown in

Fig. 10-6. There we attributed this competition to the fact that both modes obtained gain from the same *pair* of velocity groups. This competition does not occur here for two modes traveling in the same direction and placed symmetrically about line center, for the modes are amplified by different velocity groups. We can see this quantitatively by using the unidirectional multimode field

$$E(z, t) = \tfrac{1}{2} \sum_n E_n \exp[-i(\nu_n t + \phi_n - K_n z)] + \text{c.c.} \tag{24}$$

in the first (10.50) and third-(10.53) order contributions to ρ_{ab}. This calculation is given in Appendix D. We find that the amplitude- and frequency-determining equations are (9.18) and (9.19) as for the standing wave laser, here with the third-order saturation coefficients

$$\vartheta_{n\mu\rho\sigma} = i \left(\frac{\frac{1}{2}\wp}{\hbar}\right)^2 [\mathscr{D}_a(\nu_\rho - \nu_\sigma) + \mathscr{D}_b(\nu_\rho - \nu_\sigma)]\, \mathscr{D}(\nu_\rho - \tfrac{1}{2}\nu_\mu - \tfrac{1}{2}\nu_\sigma) F_1, \tag{25}$$

where the factor F_1 is given in Table 11-1. This coefficient is the same as that in Table 10-2 for the standing-wave laser except for a factor of $\frac{1}{4}$ and the absence of the complex denominator $\mathscr{D}(\omega - \frac{1}{2}\nu_\mu + \frac{1}{2}\nu_\rho - \nu_\sigma)$. It was this denominator which resulted in the Lorentzian of β_n and hence in the Lamb dip. Here the Lamb dip does not occur, a fact which follows also from the self-saturation coefficients of Table 11-1.

Considering two-mode operation now, we find the cross-saturation coefficient:

$$\begin{aligned}\theta_{12} &= \frac{2\hbar^2 \gamma_a \gamma_b}{\wp^2} \operatorname{Im}\{\vartheta_{1122} + \vartheta_{1221}\} \\ &= \frac{1}{2}\frac{\gamma_b}{\gamma} F_1\, \mathscr{L}[\tfrac{1}{2}(\nu_2 - \nu_1)]\{1 + \mathscr{L}_a(\nu_2 - \nu_1) \\ &\qquad \times [1 - \frac{\frac{1}{2}(\nu_2 - \nu_1)^2}{\gamma\gamma_a}]\} + \text{same with } \gamma_a \to \gamma_b.\end{aligned} \tag{26}$$

Unlike the θ_{12} for the standing-wave gas laser (Table 10-2), this θ_{12} is independent of the detuning $(\omega - \nu_1)$ and hence shows no special value for symmetric tuning.

Problems

11-1 Draw the perturbation tree (see Fig. 10-4) for the third-order expression (10).

11-2. Show that Eq. (12) for the complex polarization $\mathscr{P}_+$ follows from (11) in the strong Doppler limit of $Z(v) = i\sqrt{\pi}$.

11-3. Show that Eq. (12) reduces to the standing-wave result of Prob. 10–11.

11-4. In terms of hole burning, discussed in Sec. 10-1 [see Eq. (10.25) and following text], explain the value of the coupling parameter (17).

11-5. Would you expect the three-mode locking region in the unidirectional laser to be larger or smaller than that for the standing-wave case? Support your answer with a physical argument.

11-6. Consider the operation of two modes traveling in opposite directions and differing in frequency by $c/2L$. Show by consideration of Eq. (9) that the population pulsation term averages to zero in the Doppler limit. Why is this true physically?

11-7. Show from Eq. (10) that for a homogeneously-broadened medium, the modes can be strong coupled; for $\nu_+ = \nu_-$, $\theta_{+-} = 2\beta_+$. See Hambenne and Sargent (1976) (op. cit. p. 142) for discussion.

References

F. Aronowitz, 1965, *Phys. Rev.* **139**, A635. See also his Ph. D. Thesis for New York University, 1969.

F. Aronowitz and R. J. Collins, 1966, *Appl. Phys. Letters* **9**, 55.

L. N. Menegozzi and W. E. Lamb, Jr., 1973, *Phys. Rev.* **A8**, 2103.

C. L. O'Bryan III and M. Sargent III, 1973, *Phys. Rev.* **A8**, 3071.

M. Sargent III, and M. O. Scully, 1972, "Theory of Laser Operation," Chap. A2 in *Laser Handbook*, *op. cit.* in references for Chap. 8.

S. G. Zeigler and E. E. Fradkin, 1966, *Opt. i Spektrosk*, **21**, 386 [Eng. trans.: *Opt. Spectrosc.* **21**, 217 (1966)].

XII

THE ZEEMAN LASER

12. Zeeman Laser

Up to this point we have assumed that the electric field is scalar, that is, polarized invariably in some particular direction, and that the medium responds along this direction. This is often a good approximation, but in general a more realistic assumption is that the field is vectorial with two transverse degrees of freedom. A particularly convenient choice of representation is that with circularly polarized components:

$$\mathbf{E}(z, t) = \tfrac{1}{2}\{\hat{\varepsilon}_+ E_+(t) \exp[-i(\nu_+ t + \phi_+)] + \hat{\varepsilon}_- E_-(t) \exp[-i(\nu_- t + \phi_-)]\} U(z) + \text{c.c.}, \tag{1}$$

where the complex circularly polarized unit vectors $\hat{\varepsilon}_\pm = 2^{-1/2}(\hat{x} \pm i\hat{y})$ (see Fig. 12-1), and amplitudes E_+, E_-, and phases ϕ_+, ϕ_- are slowly varying functions of time. A related generalization of the previous theory is the inclusion of cavity anisotropy, for example, different losses for different polarizations. A strong anisotropy favoring one linear polarization over another can be produced by a Brewster window. Similarly, different polarizations may see different cavity lengths.

In Sec. 12-1 we proceed in a fashion quite similar to that in Sec. 8-1 and obtain the same self-consistency equations (11.4) and (11.5) for the circularly polarized amplitudes, phases, and frequencies as were obtained in Chap. 11 for the ring laser running-wave components. Physically, of course, the equations are quite different, for scalar running waves are involved in one set and polarizations of a single standing wave in the other. The effects of cavity anisotropy are included in the derivation.

We assumed also in Chaps. 8–11 that the amplifying medium consists of two-level systems, with no field polarization preference. Real atoms have states with angular momenta and magnetic sublevels that contribute to transitions with quite definite polarization properties. A more complete atomic

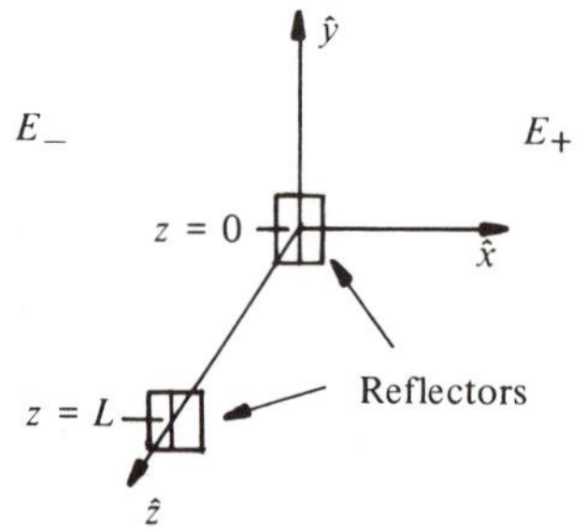

Figure 12-1. Here the unit vector $\hat{z}$ denotes the z (maser) axis. This is also chosen to be the atomic axis, for only axial magnetic fields are considered. The unit vectors $\hat{x}$ and $\hat{y}$ define the plane in which the electric vector **E** *oscillates.*

model consists of a system with $2J_a + 1$ upper states $|n_aJ_aa'\rangle$, where J_a is the angular momentum of the upper state, a' is the magnetic quantum number labeling a given sublevel, and n_a represents other quantum numbers. Similarly the system possesses $2J_b + 1$ lower states $|n_bJ_bb'\rangle$ (see Fig. 12-2). Hence the calculation for the polarization of the medium involves matrices with rank $2(J_a + J_b + 1)$ instead of 2. In the development, it is easy to include Zeeman splitting of the magnetic sublevels by defining, for example, $\omega_{a'} = \omega_a + \mu_B B g_a a'/\hbar$, where $\hbar\omega_a$ is the zero-field energy of the upper states, μ_B is the Bohr magneton, B is the applied magnetic induction and g_a is the Landé g

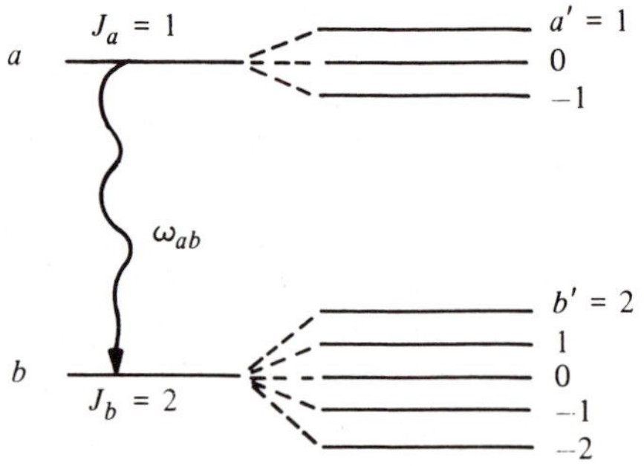

Figure 12-2. Possible level diagram showing how levels a and b might be split into $2J_a + 1 = 3$ and $2J_b + 1 = 5$ sublevels, respectively, by an applied dc magnetic field. Here J_a and J_b are the total angular momenta of levels a and b, a' and b' are corresponding magnetic quantum numbers, ω_{ab} is the zero-field optical frequency between levels a and b, and $\omega_{a'b'}$ is a representative optical frequency in the presence of a magnetic field.

factor for the upper states. In Sec. 12-2 the calculation of the polarization of a multilevel medium is motivated for the electric field (1) and a possible applied axial magnetic field. The details of the calculation for both the special J values, $J_a = 1$, $J_b = 0$, and general J values are deferred to Appendix F. In brief, the Zeeman laser theory is concerned with a vector electromagnetic field, a cavity with anisotropy, atoms whose levels may have nonzero angular momenta, and the influence of a dc magnetic field.

In Sec. 12-3 we find that the amplitude- and frequency-determining equations for an isotropic cavity and axial magnetic field are formally identical to those—(11.13)–(11.16)—for the ring laser. The simple two-mode form results from electric-dipole selection rules which require that E_+ cause transitions for which $a' - b' = +1$ and E_- cause those for which $a' - b' = -1$. The two-mode analysis of Sec. 9-2 applies, as for the ring laser, and yields some surprising results. Specifically, the solutions depend markedly on both the atomic angular momenta and the magnetic field strength, and strong coupling can occur. The locking effect of the $x - y$ Q anisotropy on the frequencies of the field is discussed. In this problem and others, there is a remarkable formal similarity between the orthogonal polarizations of the vector electric field and the oppositely directed running waves in the ring laser discussed in Chap. 11.

Finally, the two-level-atom, scalar field model is evaluated and found to be quite satisfactory provided one is not interested in the polarization of the electric field or in operation with nonzero magnetic fields. The primary values of the Zeeman theory are (1) the evaluation of the simpler two-level model in terms of a more rigorous model, and (2) the determination of the electric field polarization in the presence of applied magnetic fields and cavity anisotropy. The theory is also useful in determining atomic constants such as the decay rates and Landé g factors (Sargent, Lamb, and Fork, 1967b).

12-1. Electromagnetic Field Equations

When the electric field has two transverse degrees of freedom and the cavity has anisotropic loss, the wave equation (8.3) becomes

$$-\frac{\partial^2 \mathbf{E}(z, t)}{\partial z^2} + \mu_0 \overleftrightarrow{\sigma} \cdot \frac{\partial \mathbf{E}(z, t)}{\partial t} + \mu_0 \varepsilon_0 \frac{\partial^2 \mathbf{E}(z, t)}{\partial t^2} = -\mu_0 \frac{\partial^2 \mathbf{P}(z, t)}{\partial t^2} \tag{2}$$

where the losses are related to the electric field by the tensor constitutive relation

$$\mathbf{J} = \overleftrightarrow{\sigma} \cdot \mathbf{E}. \tag{3}$$

The induced polarization of the medium corresponding to field (1) has the form

$$\mathbf{P}(z,t) = \tfrac{1}{2}\{\hat{\varepsilon}_+\mathscr{P}_+(t)\exp[-i(\nu_+t+\phi_+)] + \hat{\varepsilon}_-\mathscr{P}_-(t)\exp[-i(\nu_-t+\phi_-)]\}\,U(z) + \text{c.c.}, \tag{4}$$

where the complex Fourier components of the polarization $\mathscr{P}_+$, $\mathscr{P}_-$ are slowly varying functions of time. Substituting field (1) and polarization (4) without complex conjugates into wave equation (2), projecting onto the unit vectors $\hat{\varepsilon}_\pm$ and onto $U(z)$, and neglecting small terms like $\ddot{\phi}_+$, $\sigma\dot{\phi}_+$, $\sigma\dot{E}_+$, $\ddot{E}_+$, and $\dot{\phi}_+\dot{E}_+$, we find the complex equation (Prob. 12-1):

$$(\nu_+ + \dot{\phi}_+ - \Omega)E_+ + i\,\{\dot{E}_+ + \tfrac{1}{2}\nu[g_{++}E_+ + g_{+-}E_-\exp(i\Psi)]\} = -\frac{1}{2}\frac{\nu}{\varepsilon_0}\mathscr{P}_+ \tag{5}$$

and a similar equation for E_- and ϕ_- in which the plus and minus of (5) are interchanged. Here the relative phase angle

$$\Psi = \nu_+t + \phi_+ - \nu_-t - \phi_-, \tag{6}$$

the conductivity matrix

$$G = \begin{pmatrix} g_{++} & g_{+-} \\ g_{-+} & g_{--} \end{pmatrix} = (\varepsilon_0\nu)^{-1}\sigma, \tag{7}$$

and the frequency $\nu \cong \nu_+ \cong \nu_-$.

In particular, if the losses are diagonal in this circularly polarized representation, the matrix elements, g_{++} and g_{--} are reciprocals of the cavity Q's:

$$g_{++} = Q_+^{-1}, \qquad g_{--} = Q_-^{-1}. \tag{8}$$

Equating the real and imaginary parts of (5) separately, we obtain the self-consistency equations:

$$\dot{E}_+ + \tfrac{1}{2}\nu\{g_{++}E_+ + \text{Im}[ig_{+-}E_-\exp(i\Psi)]\} = -\frac{1}{2}\frac{\nu}{\varepsilon_0}\text{Im}(\mathscr{P}_+), \tag{9}$$

$$(\nu_+ + \dot{\phi}_+ - \Omega)E_+ + \tfrac{1}{2}\nu\,\text{Re}\,[ig_{+-}E_-\exp(i\Psi)] = -\frac{1}{2}\frac{\nu}{\varepsilon_0}\text{Re}(\mathscr{P}_+), \tag{10}$$

with similar equations for E_- and ν_- in which the plus and minus of (9) and (10) are interchanged. For diagonal losses, the self-consistency equations reduce to

$$\dot{E}_+ + \frac{1}{2}\frac{\nu}{Q_+}E_+ = -\frac{1}{2}\frac{\nu}{\varepsilon_0}\text{Im}(\mathscr{P}_+), \tag{11}$$

$$\nu_+ + \dot{\phi}_+ = \Omega - \frac{1}{2}\frac{\nu}{\varepsilon_0}E_+^{-1}\,\text{Re}(\mathscr{P}_+), \tag{12}$$

with corresponding equations for E_- and ν_-.

Now suppose that the losses are diagonal in the $x-y$ representation for which the electric field has the decomposition

$$\mathbf{E}(z, t) = \tfrac{1}{2}\{\hat{x}E_x \exp[-i(\nu_x t + \phi_x)] + \hat{y}E_y \exp[-i(\nu_y t + \phi_y)]U(z) + \text{c.c.} \quad (13)$$

The corresponding conductivity matrix in the $+-$ representation, $(G)_\pm$, is determined by the similarity transformation from the matrix $(G)_{xy}$ in the $x-y$ representation:

$$(G)_\pm = S(G)_{xy}S^{-1}, \quad (14)$$

where the unitary matrix S is the composite of the $\hat{\varepsilon}_\pm$ unit vectors written in row vector form in the x-y basis:

$$S = \tfrac{1}{2}\sqrt{2}\begin{pmatrix} 1 & -i \\ 1 & +i \end{pmatrix}. \quad (15)$$

Hence

$$(G)_\pm = \frac{1}{2}\begin{pmatrix} 1/Q_x + 1/Q_y & 1/Q_x - 1/Q_y \\ 1/Q_x - 1/Q_y & 1/Q_x + 1/Q_y \end{pmatrix}. \quad (16)$$

If $Q_x \neq Q_y$, there are mixing terms with

$$g_{+-} = g_{-+} = \frac{1}{Q_x} - \frac{1}{Q_y} \quad (17)$$

in the self-consistency equations (9) and (10) containing the relative phase angle (7) which lead to the possibility of locking the frequencies ν_+ and ν_- together. We discuss this further in Sec. 12-3, using the theory developed in Sec. 9-3.

12-2. Polarization of the Atomic Medium

We consider a laser medium consisting of atoms with states $|n_aJ_aa'\rangle$ and $|n_bJ_bb'\rangle$, between which laser action takes place as illustrated in Fig. 12-2. Here J_a and J_b are the total angular momenta of the upper and lower laser levels; a' and b' are the corresponding magnetic sublevel indices, alternatively labeled m_{J_a} and m_{J_b}; and n_a and n_b are other quantum numbers. The existence of a nonzero magnetic field, which we take to be directed along the laser axis, splits the unperturbed Hamiltonian so that

$$\omega_{a'} = \omega_a + \mu_B B g_a a'/\hbar, \qquad \omega_{b'} = \omega_b + \mu_B B g_b b'/\hbar, \quad (18)$$

where ω_a and ω_b are the unperturbed energies of the states $|n_aJ_aa'\rangle$ and $|n_bJ_bb'\rangle$, respectively, μ_B is the Bohr magneton, B is the magnetic induction in gauss, and g_a and g_b are the Landé g factors for the levels.

The electric-dipole interaction energy has matrix elements

$$\mathscr{V}_{a'b'} = -\langle n_aJ_aa'|e\mathbf{E}\cdot\mathbf{r}|n_bJ_bb'\rangle. \quad (19)$$

Inasmuch as $\mathbf{E}$ is a vector not necessarily along the electron position vector $\mathbf{r}$ as assumed before, we decompose $\mathbf{r}$ as follows (see Fig. 1-4):

$$\begin{aligned}\mathbf{r} &= x\hat{x} + y\hat{y} + z\hat{z}\\ &= r\sin\theta(\cos\phi\hat{x} + \sin\phi\hat{y}) + r\cos\theta\hat{z}\\ &= \tfrac{1}{2}r\sin\theta[(\hat{x} - i\hat{y})\exp(i\phi) + (\hat{x} + i\hat{y})\exp-(i\phi)] + r\cos\theta\hat{z}.\end{aligned} \quad (20)$$

The electric-dipole selection rules yield the matrix elements (it is easy to show this for the hydrogen wave functions, Prob. 1–2)

$$(e\mathbf{r})_{a'b'} = \wp_{a'b'}[(\hat{x} - i\hat{y})\delta_{a',\,b'+1} + (\hat{x} + i\hat{y})\delta_{a',b'-1} + \hat{z}\delta_{a'b'}]. \quad (21)$$

Hence the matrix elements (19) of the electric-dipole interaction energy have the values

$$\mathscr{V}_{a'b'} = -\wp_{a'b'}[\mathbf{E}\cdot(\hat{x} - i\hat{y})\delta_{a',b'+1} + \mathbf{E}\cdot(\hat{x} + i\hat{y})\delta_{a',b'-1} + \mathbf{E}\cdot\hat{z}\delta_{a'b'}]. \quad (22)$$

Here, from Condon and Shortley (1935),

$$\begin{aligned}\wp_{a'b'} &= \mp\tfrac{1}{2}\wp[(J_b \pm a')(J_b \pm a' + 1)]^{1/2} & a' &= b' \mp 1 & &\multirow{2}{*}{}\\ &= \wp(J_b^2 - a'^2)^{1/2} & a' &= b' & &\Big\} J_a = J_b - 1\\ &= \tfrac{1}{2}\wp[(J_a \mp a')(J_a \pm a' + 1)]^{1/2} & a' &= b' \mp 1 & &\\ &= \wp a' & a' &= b' & &\Big\} J_a = J_b\\ &= \mp\tfrac{1}{2}\wp[(J_a \pm b')(J_a \pm b' + 1)]^{1/2} & a' &= b' \pm 1 & &\\ &= \wp(J_a^2 - b'^2)^{1/2} & a' &= b' & &\Big\} J_a = J_b + 1,\end{aligned} \quad (23)$$

where the reduced matrix element

$$\wp = (n_aJ_a|er|n_bJ_b). \quad (24)$$

With the electric field $\mathbf{E}(z, t)$ given by (1), we have (in the rotating-wave approximation),

$$\begin{aligned}\mathscr{V}_{a'b'} = -\tfrac{1}{2}\sqrt{2}\,\wp_{a'b'}\,U(z)\{&E_+\exp[-i(\nu_+t + \phi_+)]\delta_{a',b'+1}\\ &+ E_-\exp[-i(\nu_-t + \phi_-)]\delta_{a',b'-1}\}.\end{aligned} \quad (25)$$

Here we see that for an axial magnetic field (atomic axis along laser axis) and a circularly polarized decomposition (1), E_+ causes transitions for which the change in magnetic sublevel is $+1$ and E_- those for which this change is -1. This is an important and useful simplification, as will become clearer shortly.

The population matrix $\rho(z, v, t)$ is defined as before (10.6), but now has many elements corresponding to the increased number of levels. The macroscopic polarization is given by the trace

$$\mathbf{P}(z, t) = \int_{-\infty}^{\infty} dv\,\mathrm{Tr}(\rho e\mathbf{r}) \quad (26)$$

with projections

$$\begin{aligned}P_\pm(z, t) &= \int_{-\infty}^{\infty} dv\,\mathrm{Tr}(\rho e\hat{\varepsilon}_\pm{}^*\cdot\mathbf{r})\\ &= \sqrt{2}\int_{-\infty}^{\infty} dv\sum_{a'}\sum_{b'}\rho_{a'b'}\wp_{b'a'}\delta_{a',b'\pm1}.\end{aligned} \quad (27)$$

The complex polarizations of (4) become

$$\mathscr{P}_{\pm}(t) = 2\sqrt{2}\exp[i(\nu_{\pm}t + \phi_{\pm})]\frac{1}{\mathscr{N}}\int_0^L dz\, U^*(z)\int_{-\infty}^{\infty} dv \times \sum_{a'}\sum_{b'} \rho_{a'b'}\wp_{b'a'}\delta_{a',b'\pm 1}. \tag{28}$$

In Appendix F we calculate the polarization (28) for both arbitrary J values and for a $J = 1 \longleftrightarrow J = 0$ transition. It is instructive to motivate the latter here. For this the density matrix has the form

$$\rho = \begin{pmatrix} \rho_{++} & \rho_{+-} & \rho_{+b} \\ \rho_{-+} & \rho_{--} & \rho_{-b} \\ \rho_{b+} & \rho_{b-} & \rho_{bb} \end{pmatrix}, \tag{29}$$

in which we have written plus for the eigenstate $|n_aJ_a1\rangle$, minus for $|n_aJ_a - 1\rangle$, and b for $|n_bJ_b0\rangle$. The state $|n_aJ_a0\rangle$ is not important, for electric-dipole transitions are not allowed between this state and others for the axial magnetic field and transverse electromagnetic field that we are considering.

The dipole operator whose general matrix elements are given by (21) has the specific form

$$e\mathbf{r} = \tfrac{1}{2}\sqrt{2}\,\wp\begin{pmatrix} 0 & 0 & -(\hat{x} - i\hat{y}) \\ 0 & 0 & (\hat{x} + i\hat{y}) \\ -(\hat{x} + i\hat{y}) & (\hat{x} - i\hat{y}) & 0 \end{pmatrix}, \tag{30}$$

in which, according to (23),

$$\wp_{+b} = -\wp_{-b} = -\tfrac{1}{2}\sqrt{2}\,\wp. \tag{31}$$

The Hamiltonian

$$\mathscr{H} = \begin{pmatrix} \hbar\omega_+ & 0 & \mathscr{V}_{+b} \\ 0 & \hbar\omega_- & \mathscr{V}_{-b} \\ \mathscr{V}_{b+} & \mathscr{V}_{b-} & \hbar\omega_b \end{pmatrix}, \tag{32}$$

where the matrix elements [from (25)]

$$\mathscr{V}_{\pm b} = \pm\tfrac{1}{2}\wp E_{\pm}(t)\exp[-i(\nu_{\pm}t + \phi_{\pm})]U(z). \tag{33}$$

The complex polarization (28) reduces to

$$\mathscr{P}_{\pm}(t) = \mp 2\wp\exp[i(\nu_{\pm}t + \phi_{\pm})]\frac{1}{\mathscr{N}}\int_0^L dz\, U^*(z)\int_{-\infty}^{\infty} dv\, \rho_{\pm b}. \tag{34}$$

Here we see that E_+ causes transitions between the b level and the plus level; E_-, transitions between the b level and the minus level. Furthermore, the polarization so induced contributes to E_+ and E_-, respectively. At first glance, one might conclude that the two transitions operate independently of one another. However, the two field polarizations are coupled because they share the lower level b. Hence transitions between $|+\rangle$ and $|b\rangle$ caused by E_+

affect the population $\rho_{--} - \rho_{bb}$, which drives E_-. In addition, the electric quadrupole element ρ_{+-} induced by E_+ and E_- together couples the two field polarizations.

The equations of motion which must be integrated are as follows:

$$\dot{\rho}_{+b} = -(i\omega_{+b} + \gamma)\rho_{+b} + i\hbar^{-1}\mathscr{V}_{+b}(\rho_{++} - \rho_{bb}) + i\hbar^{-1}\rho_{+-}\mathscr{V}_{-b}, \quad (35)$$

$$\dot{\rho}_{++} = -\gamma_a\rho_{++} - (i\hbar^{-1}\mathscr{V}_{+b}\rho_{b+} + \text{c.c.}) + \lambda_+, \quad (36)$$

$$\dot{\rho}_{bb} = -\gamma_b\rho_{bb} + (i\hbar^{-1}\mathscr{V}_{+b}\rho_{b+} + \mathscr{V}_{-b}\rho_{b-} + \text{c.c.}) + \lambda_b, \quad (37)$$

$$\dot{\rho}_{+-} = -(i\omega_{+-} + \gamma_a)\rho_{+-} - i\hbar^{-1}(\mathscr{V}_{+b}\rho_{b-} - \mathscr{V}_{b-}\rho_{+b}). \quad (38)$$

Other equations are given by symmetry:

$$\left.\begin{array}{l}\rho_{-b} \longleftrightarrow \rho_{+b}\\ \rho_{--} \longleftrightarrow \rho_{++}\\ \rho_{-+} \longleftrightarrow \rho_{+-}\end{array}\right\} \text{ with plus and minus interchanged,} \quad (39)$$

and by Hermiticity of the density matrix:

$$\rho_{b+} = \rho_{+b}{}^*, \qquad \rho_{b-} = \rho_{-b}{}^*. \quad (40)$$

In Appendix F we show that the first-order contribution to the complex polarization $\mathscr{P}_+(t)$ is

$$\mathscr{P}_+{}^{(1)}(t) = i\frac{\wp^2}{\hbar Ku}\bar{N}E_+\, Z[\gamma + i(\omega_{+b} - \nu_+)] \quad (41)$$

and that the third-order contribution is

$$\begin{aligned}\mathscr{P}_+{}^{(3)}(t) = \tfrac{1}{16}\, i\wp^4\bar{N}(\hbar^3Ku)^{-1}E_+\{&E_+{}^2 2\gamma_{ab}(\gamma\gamma_a\gamma_b)^{-1}[1 + \gamma\mathscr{D}(\omega_{+b} - \nu_+)]\\ &+ E_-{}^2\gamma_b{}^{-1}[\mathscr{D}(\delta) + \mathscr{D}(\omega_0 - \nu_0)]\\ &+ E_-{}^2\mathscr{D}_a(2\delta)[\mathscr{D}(\delta) + \mathscr{D}(\omega_{+b} - \nu_+)]\},\end{aligned} \quad (42)$$

where the average atomic and field frequencies $\omega_0 = \frac{1}{2}(\omega_{+b} + \omega_{-b})$, $\nu_0 = \frac{1}{2}(\nu_+ + \nu_-)$, and the magnetic field splitting $\delta = \frac{1}{2}(\omega_{+-} - \nu_{+-}) \cong \mu_B gB/\hbar$. Corresponding contributions for $\mathscr{P}_-$ are given by an interchange of the plus and minus signs.

12-3. Steady-State Intensities

Combining the first- (41) and third- (42) order contributions to the complex polarization (34) with the self-consistency equations (11) and (12), we have the intensity- and frequency-determining equations:

$$\dot{I}_+ = 2I_+(\alpha_+ - \beta_+I_+ - \theta_{+-}I_-), \quad (43)$$

$$\nu_+ + \dot{\phi}_+ = \Omega + \sigma_+ - \rho_+I_+ - \tau_{+-}I_-, \quad (44)$$

where the coefficients are defined in Tables 12-1 and 12–2. Similar equations

TABLE 12-1 *Coefficients for a Gas Laser in Doppler Limit ($Ku \gg \gamma$) with a $J = 1 \leftrightarrow J = 0$, Transition According to Zeeman Laser Theory of Sec. 6-2*

More general J values involve summations over sublevels as given in Table 12-2. ω_{+b} is the atomic resonance frequency for the E_+ transition. Similar coefficients for a_-, etc., are given by the following with the plus subscript replaced by minus and the magnetic splitting $\delta \to -\delta$.

Coefficient	Physical Context
$\alpha_+ = \exp[-(\omega_{+b} - \nu_+)^2/(Ku)^2]F_1 - \frac{1}{2}\nu/Q_+$	Linear net gain
$\beta_+ = [1 + \mathscr{L}(\omega_{+b} - \nu_+)]F_3$	Self-saturation
$\theta_{+-} = \frac{1}{2}(\gamma_a/\gamma_{ab})[\mathscr{L}(\delta) + \mathscr{L}(\omega_0 - \nu_0)]F_3$ $+ \frac{1}{2}(\gamma_a/\gamma_{ab})\mathscr{L}(2\delta)\{\mathscr{L}(\delta)[1 - 2\delta^2(\gamma_a\gamma)^{-1}]$ $+ \mathscr{L}(\omega_{+b} - \nu_+)[1 - 2\delta(\omega_{+b} - \nu_+)(\gamma_a\gamma)^{-1}]\}F_3$	Cross saturation
$\sigma_+ = -2\exp(-\xi^2)\int_0^{\xi^2} dx \exp(x^2)F_1$	Linear pulling
$\rho_+ = [(\omega_{+b} - \nu_+)/\gamma]\mathscr{L}(\omega_{+b} - \nu_+)F_3$	Self-pushing
$\tau_{+-} = \{(\gamma_b\gamma)^{-1}\delta\mathscr{L}(\delta) + (\omega_0 - \nu_0)\mathscr{L}(\omega_0 - \nu_0)$ $+ \gamma_a^{-1}\mathscr{L}(2\delta)[\mathscr{L}(\delta)\,\delta(2\gamma_a^{-1} + \gamma^{-1})$ $+ \mathscr{L}(\omega_{+b} - \nu_+)[2\delta/\gamma_a + (\omega_{+b} - \nu_+)/\gamma]\}F_3$	Cross pushing
$\mathscr{L}_x(\Delta\Omega) = \gamma_x^2/[\gamma_x^2 + (\Delta\Omega)^2]$	Dimensionless Lorentzian
$\nu_0 = \frac{1}{2}(\nu_+ + \nu_-);\ \omega_0 = \frac{1}{2}(\omega_{+b} + \omega_{-b})$	Average field and atomic frequencies
$\delta = \mu_B Bg/\hbar - \frac{1}{2}(\nu_+ - \nu_-) \simeq \mu_B Bg/\hbar$	~ Zeeman shift
$F_1 = \frac{1}{2}\nu\sqrt{\pi}\,\wp\bar{N}^2(\hbar Ku\varepsilon_0)^{-1}$	First-order factor
$F_3 = \frac{1}{4}(\gamma_{ab}/\gamma)F_1$	Third-order factor

for I_- and ν_- are given by (43), (44), and Tables 12-1 and 12-2 with the pluses and minuses interchanged. Because the Q's of the cavity are high, the frequency difference $\nu_{+-} = \nu_+ - \nu_- \ll \omega_{+-}$, which is twice the Zeeman splitting $\mu_B gB/\hbar$. Hence the frequency $\delta \approx \mu_B gB/\hbar$, and the factors $\mathscr{L}(\delta)$ and $\mathscr{L}(2\delta)$ are essentially Lorentzians of the magnetic field splitting.

The intensity equations for I_+ and I_- have the form of the two-mode equations in Sec. 9-2, and hence that analysis applies here. Using the general coefficients in Table 12-2, we find that the coupling parameter $C = \theta_{+-}\theta_{-+}/(\beta_+\beta_-)$ depends markedly on both the angular momenta of the medium and on the magnetic field strength, as is depicted in Fig. 12-3. In zero field, the value of C can be given in a simple analytical form. From Table 12-2, the self- and cross-saturation coefficients reduce to

$$\beta_\pm = [1 + \mathscr{L}(\omega_0 - \nu_\pm)]F_3 \sum_{a'} \sum_{b'=a'\mp 1} |\wp_{a'b'}/\wp|^4, \tag{45}$$

$$\theta_{\pm\mp} = [1 + \mathscr{L}(\omega_0 - \nu_\pm)]F_3 \sum_{a'} \sum_{b'=a'\mp 1} |\wp_{a'b'}|^2[|\wp_{a'\mp 2,b'}|^2 + |\wp_{a',b'\pm 2}|^2]/\wp^4. \tag{46}$$

TABLE 12-2 *Coefficients in Eq. (43) for E_+ and their Counterparts for E_-*

Doppler limit $Z(v) = i\sqrt{\pi}$ is assumed for third-order coefficients. Laser levels have arbitrary angular momenta J_- and J_b. Coefficients for the special case $J = 1 \leftrightarrow J = 0$ are given in Table 12-1. The θ and τ coefficients are given by (F. 37).

Coefficent	Physical context
$\alpha_\pm = F_1 \sum_{a'} \sum_{b'} \delta_{a',b'\pm 1} \lvert \wp_{a'b'}/\wp \rvert^2 Z_i[\gamma + i(\omega_{a'b'} - \nu_\pm)] - \nu/2Q_\pm$	Linear net gain
$\beta_\pm = F_3 \sum_{a'} \sum_{b'} \delta_{a',b'\pm 1} \lvert \wp_{a'b'}/\wp \rvert^4 [1 + \mathscr{L}(\omega_{a'b'} - \nu_\pm)]$	Self-saturation
$\sigma_\pm = F_1 \sum_{a'} \sum_{b'} \delta_{a',b'\pm 1} \lvert \wp_{a'b'}/\wp \rvert^2 Z_r[\gamma + (i\omega_{a'b'} - \nu_\pm)]$	Mode pulling
$\rho_\pm = F_3 \sum_{a'} \sum_{b'} \delta_{a',b'\pm 1} \lvert \wp_{a'b'}/\wp \rvert^4 \gamma^{-1}(\omega_{a'b'} - \nu_\pm)\mathscr{L}(\omega_{a'b'} - \nu_\pm)$	Self-pushing
$\mathscr{L}_a(\Delta\omega) = (\gamma_a{}^2 + (\Delta\omega)^2]^{-1} \quad a = a, b$	Dimensionless Lorentzian
$\omega_{a'b'} = \omega_0 + (\mu_B/\hbar)B(g_a a' - g_b b')$	Zeeman frequency difference
$\delta_a = (\mu_B/\hbar)Bg_a + (\frac{1}{2}\nu_- - \nu_+)$	Approximate Zeeman splitting
$F_1 = \nu\bar{N}\wp^2(\varepsilon_0\hbar Ku)^1$	First-order factor
$F_3 = \frac{1}{2}\sqrt{\pi} F_1 \gamma_{ab} \gamma^{-1}$	Third-order factor

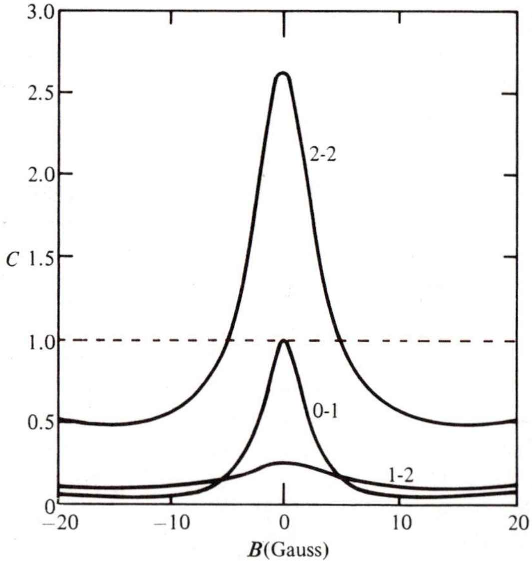

Figure 12-3. The coupling parameter $C = \theta_{+-}\theta_{-+}/(\beta_+\beta_-)$ is plotted versus axial magnetic field in gauss under central tuning for (in order of decreasing coupling) $J = 2 \leftrightarrow J = 2$, $0 \leftrightarrow 1$, $1 \leftrightarrow 2$. The other laser parameters are $\gamma_a = 2\pi \times 10$ MHz, $\gamma_b = 2\pi \times 40$ MHz, $\gamma = 2\pi \times 29$ MHz, $Ku = 2\pi \times 1010$ MHz, $\mathfrak{N}\pm = 1.20$, $g_a = g_b = 1.295$, $\Omega_+ = \Omega_-$, and $Q_+ = Q_-$. The dashed line for $C = 1$ represents neutral coupling and divides the graph into a strong coupling region ($C > 1$) and a weak-coupling region ($C < 1$). From Sargent, Lamb, and Fork (1967b).

Calculating C with these formulas and the sum rules in Sec. F-3, we have

$$C = \begin{cases} \left[\dfrac{(2J+3)(2J-1)}{2J^2+2J+1}\right]^2, & J \longleftrightarrow J, \\ \left(\dfrac{2J^2+4J+5}{6J^2+12J+5}\right)^2, & J \longleftrightarrow J+1. \end{cases} \tag{47}$$

Thus, for zero magnetic field, the $J = 2 \longleftrightarrow J = 2$ medium is strongly coupled $[C = (21/13)^2 = 2.6 > 1]$ and only one circular polarization oscillates at a given time in bistable fashion. The $J = 1 \longleftrightarrow J = 0$ transition is neutrally coupled ($C = 1$), leading to a variety of possible amplitudes, and the $J = 1 \longleftrightarrow J = 2$ transition (characteristic of the 6328 Å line in neon) is weakly coupled $[C = (11/23)^2 = 0.228 < 1]$, allowing both circular polarizations to oscillate simultaneously. In all cases, the coupling decreases as the magnetic field strength is increased, for two running waves of opposite polarization traveling in the same direction cease to interact with atoms in the same velocity range.

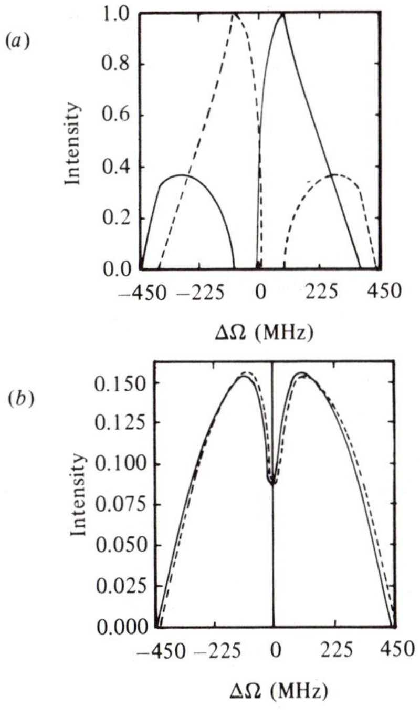

Figure 12-4. Intensities I_+ (dashed line) and I_- (solid line) versus cavity detuning in megahertz for (a) a $0 \leftrightarrow 1$ transition, (b) a $1 \leftrightarrow 2$ transition, axial field of 1 G and other parameters the same as in Fig. 12-3. The intensities are given in arbitrary units. From Sargent, Lamb, and Fork (1967b).

In Fig. 12-4 intensity versus detuning curves are given for $J = 1 \longleftrightarrow J = 0$ and $J = 1 \longleftrightarrow J = 2$ transitions.† Because of the small coupling, the $J = 1 \longleftrightarrow J = 2$ transition results in intensities which oscillate almost independently, each with its own Lamb dip structure. The $J = 1 \longleftrightarrow J = 0$ transition, however, has relatively stronger coupling and features tuning regions for which one polarization inhibits the oscillation of the other. Note that this is not strong coupling, for it is not bistable operation; regardless of the way in which a given tuning is reached, there is one and only one possible set of intensities.

In Fig. 12-5 intensities versus magnetic field are graphed for central tuning, revealing a magnetic tuning dip. The intensities for this case are equal and are given by

$$I_+ = I_- = \frac{a/\beta}{1 + \sqrt{C}}. \tag{48}$$

The dip results both from the quasi-Lorentzian dependence of $\sqrt{C}$ (see Fig. 12-3) and from a "Lamb dip" in the factor a/β. The second contribution differs from the ordinary Lamb dip in that the atomic gain is shifted by the magnetic field relative to a fixed cavity frequency, rather than that frequency being shifted by changing cavity length relative to fixed gain curves. These contributions can be resolved when more than one isotope is present (see Royce and Sargent, 1970).

In Fig. 12-6 the beat frequency $(\nu_+ - \nu_-)$ versus the magnetic field is graphed, exhibiting an initial variation opposite to the magnetic field splitting $\omega_+ - \omega_-$, and an ultimate slope several orders of magnitude less than that for $\omega_+ - \omega_-$. The initial variation is due to nonlinearities in the medium, and the small slope results from the success of the cavity in controlling the oscillation frequency (the cavity line width is much sharper than that of the medium). If the field splitting reaches the intermode spacing, oscillation occurs with one polarization on one cavity mode and the other on a different mode.

In general, the cavity exhibits some anisotropy, often in the form of different losses for x and y polarizations. This tends to lock the two circular polarizations to the same frequency in a fashion analogous to the way in which the point losses lock the two oppositely directed running waves in the ring laser. In fact the x-y Q anisotropy favors a linear polarization (two circular polarizations with the same frequency), just as the point losses favor a standing wave (two running waves with equal frequency). The mathematics of the

† These curves and others to follow agree quite well with corresponding experimental curves for the He-Ne laser given by Tomlinson and Fork (1967). One difference occurs for the $J=1 \leftrightarrow J = 0$ transition for zero magnetic field; here the theory predicts the borderline case of neutral coupling ($C = 1$), whereas experimentally strong coupling is observed because of pressure effects neglected in our analysis (see Wang and Tomlinson, 1969).

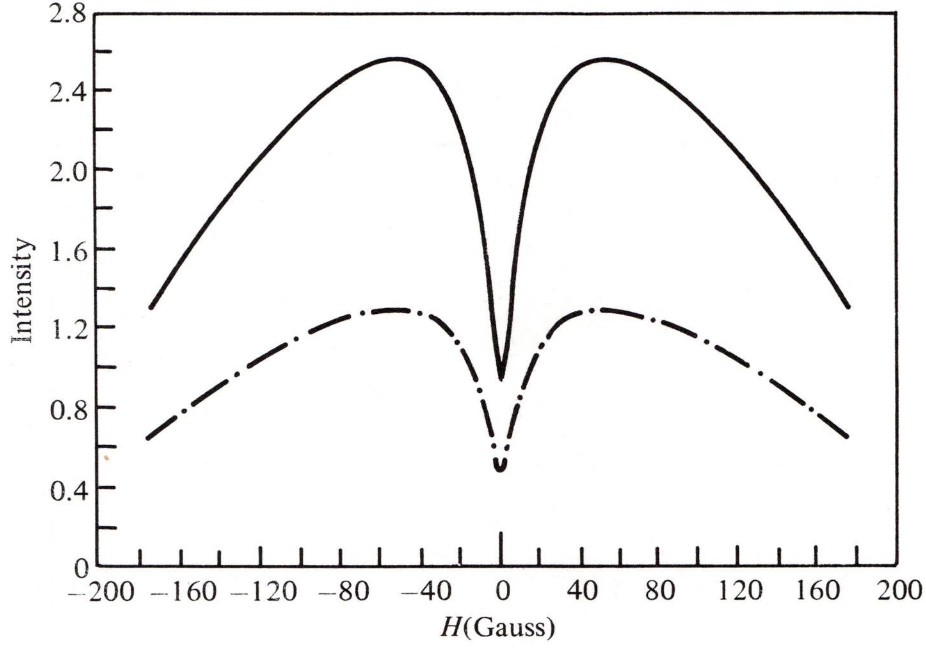

Figure 12-5. Mode intensities versus axial magnetic field for zero detuning. The solid line represents total intensity, and the dash-dot curve the equal intensities of the plus and minus polarizations as given by Eq. (48)· We have chosen parameters $J_a = 1$, $J_b = 0$, $g_a = 1.295$, $\mathfrak{N} = 1.20$, $\gamma_a = 2\pi \times 18.0$ MHz, $\gamma_a = 2\pi \times 40.0$ MHz, $\gamma_{ab} = 2\pi \times 29.0$ MHz, and $Ku = 2\pi \times 1010.0$ MHz. In this figure the intensities are given in arbitrary units equal to 1.64 times those in Fig. 12-4. From Sargent, Lamb, and Fork (1967b).

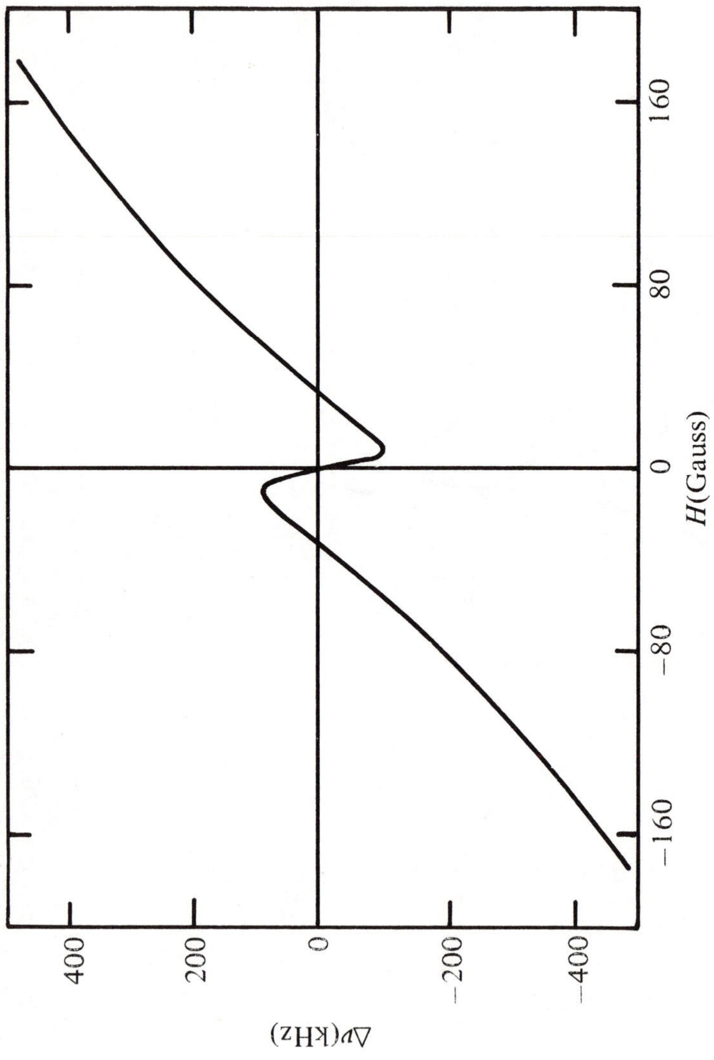

Figure 12-6. Beat frequency $\Delta\nu = \nu_+ - \nu_-$ in kilohertz versus magnetic field. Same parametes as in Fig. 12-5. From Sargent, Lamb, and Fork (1967b).

locking is just that described for the ring laser (Chap. 11). In fact, Fig. 12-7 reveals that, in regions where the frequency splitting $\nu_+ - \nu_-$ becomes small, locking occurs. The electric field polarization rotates with angular velocity $\dot{\Psi}$[see Eq. (6)] until the magnetic field reaches about 9 G. At this point locking occurs, and linear polarization ensues, rotated an angle $\frac{1}{2}\Psi$ from the x axis, where $\dot{\Psi}$ is given by Eq. (9.58) with the constants d and l chosen for the Zeeman laser.

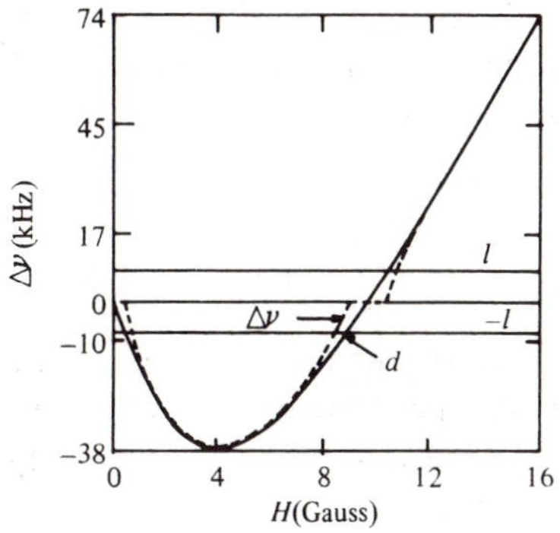

Figure 12-7. Plots of the coefficients d (solid curve) and l (horizontal solid lines) of Eq. (11.23) and the beat frequency $\overline{\Delta\nu}$ (dashed curve) of Eq. (9.61) versus magnetic field in gauss. The frequencies of circularly polarized modes are locked together for fields with $|d/l| < 1$, and the phase angle Ψ is given by Ψ_2 of Eq. (9.63). The x-y Q anisotropy $\frac{1}{4}\nu(Q_x^{-1} - Q_y^{-1}) = 2\pi \times 0.004$ MHz. Other laser parameters are $\gamma_a = 2\pi \times 15$ MHz, $\gamma_b = 2\pi \times 35$ MHz, $\gamma = 2\pi \times 25$MHz, $Ku = 2\pi \times 420.6$ MHz, $\mathfrak{N} = 1.10$, $\Omega_+ = \Omega_-$, $Q_+ = Q$. From Sargent, Lamb, and Fork (1967b).

We are now in a position to discuss the validity of the two-level, scalar field model used in preceding (and subsequent) chapters by considering the zero-field limit of the Zeeman theory. The two-level model assumes that the field is polarized in some definite direction and that the medium sustains this polarization alone. The primary limitation of the model (for zero magnetic field) is that it cannot specify what direction is involved. In this section we have seen that the polarization of the field is affected both by the angular momenta of the atomic levels and by the cavity anisotropy. Consider first a small cavity anisotropy. When the coupling between orthogonal circular polarizations is strong, as for the $J = 2 \longleftrightarrow J = 2$ transition, the medium is primarily responsible for choosing the polarization. For this transition a single circularly polarized field is sustained, and consequently the two-level theory will give correct intensities aside from a multiplicative factor (due to electric-dipole matrix elements). On the other hand, when the coupling is weak, as for the $J = 1 \longleftrightarrow J = 2$ transitions characteristic of popular neon lines, the polarization is determined by residual cavity anisotropy such as differences between the x and y Q values. The same conclusions are obtained in the stationary atom (homogeneously broadened) limit of the theory (see Prob. 12–8). A Brewster window provides a very large Q difference and generally assures linear polarization along the axis with higher Q (for weakly coupled media).

The two-level model is quite good provided one is not interested in the polarization of the electric field or in operation with an applied magnetic field. The incorrect overall multiplicative factors can be compensated for by proper choice of an effective electric-dipole matrix element. The limitations of using various incomplete schemes to incorporate level degeneracy have been discussed by Dienes (1968).

Problems

12-1. Fill in the steps in the derivation of (5) from the wave equation (2).

12-2. Using the perturbation energy (22) with the electric field (1), show that the rotating-wave approximation is not necessary for circularly polarized light, that is, E_+ or $E_- = 0$.

12-3. What helicity does E_+ have? Prove your answer.

12-4. Prove the summation formulas (F.39)–(F.41).

12-5. Solve for the polarization contribution $\mathscr{P}_\pm{}^{(3)}$ of (F.31) for the unidirectional Zeeman ring laser.

Ans.: Same as (F.31) without $\mathscr{D}\left(\frac{1}{2}\omega_{a'b'} + \frac{1}{2}\omega_{a'b''} - \cdots\right)$ and $\mathscr{D}\left(\frac{1}{2}\omega_{a'b'} + \omega_{a''b} - \cdots\right)$ terms. Why physically are these missing?

12-6. Using the summation formulas (F.38)–(F.41), show that the zero-field coupling constant C is given by (47).

12-7. Write the stationary atom expression for the complex polarization $\mathscr{P}_\pm{}^{(3)}$. *Hint*: Set $W(v) = \delta(v)$ in (F.30), take $U^4(z) = \frac{3}{8}$, and perform the time integrals.

Ans.:

$$\mathscr{P}_\pm{}^{(3)} = \tfrac{3}{4}i\bar{N}\hbar^{-3} \sum_{a''}\sum_{b''} \delta_{a',\,b'\pm 1}\, \wp_{b'a'}\wp_{a''b'}\wp_{b''a''}\wp_{a'b''} E_{a''b'}E_{b''a''}E_{a'b''}$$
$$\times\, \mathscr{D}(\omega_{a'b'} - \nu_{a''b'} + \nu_{a''b''} - \nu_{a'b''})$$
$$\times \{\mathscr{D}_a(\omega_{a'a''} + \nu_{a''b''} - \nu_{a'b''})\,[\mathscr{D}(\omega_{a'b''} - \nu_{a'b''}) + \mathscr{D}(\nu_{a''b''} - \omega_{a''b''})]$$
$$+\, \mathscr{D}_b(\omega_{b''b''} + \nu_{a''b''} - \nu_{a''b'})\,[\mathscr{D}(\omega_{a''b'} - \nu_{a''b'}) + \mathscr{D}(\nu_{a''b''} - \omega_{a''b''})]\}\,.$$

12-8. From Prob. 12–7, show that the stationary atom self-saturation coefficient is given by

$$\beta_\pm = F_3 \sum_{a'}\sum_{b'} \delta_{a',b'\pm 1} |\wp_{a'b'}|^4 \mathscr{L}^2(\omega_{a'b'} - \nu_+).$$

Also show in zero field and for $\nu_+ = \nu_-$ that the cross-saturation coefficient is given by

$$\theta_{+-} = F_3 \sum_{a'}\sum_{b'} \delta_{a',b'+1} |\wp_{a'b'}|^2 \{|\wp_{a'-2,b'}|^2 + |\wp_{a',b'+2}|^2\} \mathscr{L}^2(\omega - \nu_+).$$

Hence the zero-field coupling parameter is given by (47) for the stationary atom case as well.

References

E. U. Condon and G. H. Shortly, 1935, *The Theory of Atomic Spectra*, Cambridge University Press, New York, p. 63.

W. Culshaw and J. Kannelaud, 1966, *Phys. Rev.* **141**, 228, and **145**, 258.

W. Culshaw and J. Kannelaud, 1967, *Phys. Rev.* **156**, 308.

A. Dienes, 1968, *IEEE J. Quantum Electron.* **QE-4**, 260.

A. Dienes, 1968, *Phys. Rev.* **174**, 400 and 414.

C. V. Heer and R. D. Graft, 1965, *Phys. Rev.* **140**, A1088.

G. A. Royce and M. Sargent III, 1970, *Appl. Opt.* **9**, 2428.

M. Sargent III, W. E. Lamb, Jr., and R. L. Fork, 1967, *Phys. Rev.* **164**, (a) 436, (b) 450. See also *Phys. Rev.* **139**, A617 (1965) and D. R. Hanson and M. Sargent III, 1974, *Phys. Rev.* **A9**, 466.

W. J. Tomlinson and R. L. Fork, 1967, *Phys. Rev.* **164**, 466.

C. H. Wang and W. J. Tomlinson, 1969, *Phys. Rev.* **181**, 115.

M. I. D'Yakonov and V. E. Perel, 1966, *Opt. i Spektrosk.* **20**, 472 [Eng. trans.: *Opt. Spectrosc.* **20**, 257 (1966)].

Related three-level problems are discussed by

I. M. Beterov and V. P. Chebotaev, 1975, *Prog. in Quantum Electronics* **3**, 1.

T. Hänsch and P. E. Toschek, 1970, *Z. Physik* **236**, 213.

F. Najmabadi, M. Sargent III and F. A. Hopf, 1975, *Phys. Rev.* **A12**, 1553.

Application of Zeeman laser mode coupling to saturation spectroscopy is given by M. Sargent III, 1976, *Phys. Rev.* **A14**, 524.

XIII

COHERENT PULSE PROPAGATION

13. Pulse Propagation

In our semiclassical treatment of the laser in Chaps. 8–12, we "Fourier-analyzed" the electric field. This frequency-domain method has much to commend it: explicit expressions for the field are often obtained, the calculations generally do not require much computer time, and considerable insight can be gained into the nature of the atom-field interaction which takes place in the laser. There are also shortcomings, however, to the frequency-domain picture. The number of modes which can be treated easily is small, perhaps as few as ten. In general, the mode amplitudes and phases cannot vary appreciably in the atomic lifetimes (the total field, of course, can vary). Furthermore the strong-signal theories are rather complicated.

A time-domain treatment based on the slowly-varying envelope approximation (SVEA) complements the successes and shortcomings of the mode approaches. The SVEA derives its name from the form of the electric field:

$$E(z, t) = \tfrac{1}{2}\mathscr{E}(z, t) \exp[-i(\nu t - Kz)] + \text{c.c.}, \tag{1}$$

in which $\mathscr{E}(z, t)$ is a complex field envelope which varies little in an optical period or wavelength. Here ν is the optical frequency of the carrier, $K = \nu/c$ is the wave number, and c is the speed of light in the medium.

In this chapter we describe the development of the field (1) as it propagates through a resonant two-level atomic medium. We allow the medium to be inhomogeneously broadened. The theory is patterned after the work of Hopf and Scully (1969) and is applicable to a wide variety of problems. Inasmuch as the field (1) is a running wave, the Doppler-broadened gaseous medium acts as an ordinary inhomogeneously broadened medium (see discussion in Chap. 10 and in Icsevgi and Lamb (1969)).

The subject has benefited considerably from analogies with nuclear magnetic resonance. Hence the discussion in Sec. 7-5 on the vector model of the density matrix is quite valuable in the understanding of various phenomena, such as self-induced transparency and photon echo.

In Sec. 13-1 we derive the field equations of motion for homogeneously and inhomogeneously broadened media and discuss the important zero-phase ($\mathscr{E}$ real) solution. In Sec. 13-2 we discuss the rate equation approximation for a homogeneously broadened medium. Analytic results are obtained which reveal that the pulse envelope can propagate faster than the speed of light and can sharpen if the leading edge is truncated. The discussion provides a basis for comparison with propagation in inhomogeneously broadened media, as discussed in Sec. 13-3. In this third section we consider the area theorem of McCall and Hahn (1967) with regard to stable pulse propagation and self-induced transparency. The vector model is used to give a physical picture of the processes, and the theorem is proved rigorously by use of the formalism of Sec. 13-1. In particular self-induced transparency may occur when the atomic lifetimes are long compared to the pulse duration τ_p. In this limit atomic coherence plays an important role which is not readily perceived in our previous modal approaches. In Sec. 13-4 we discuss photon echo from three points of view: from a simple wave function approach, with the vector model, and in terms of the susceptibility integral (21). This section both introduces an interesting phenomenon (photon echo) and reveals some of the underlying nature of the field interaction with inhomogeneously broadened media.

In our treatment of pulse propagation, the size of the field envelope can be very large compared to the perturbation treatments of earlier chapters, although pulses of extreme intensity or of optical period duration render the slowly-varying envelope and rotating-wave approximations invalid (see Prob. 12-2). The major drawback of the approach is the often excessive amount of computer time required to integrate the coupled field-medium equations of motion. The time is justified, however, when simpler models fail to describe the problem in mind.

13-1. Field Envelope Equation of Motion

In this section, we derive the equation of motion for the electric field (1) propagating through both homogeneously and inhomogeneously broadened media. The electric field (1) induces the macroscopic polarization

$$P(z, t) = \tfrac{1}{2}\, \mathscr{P}(z, t) \exp[-i(\nu t - Kz)] + \text{c.c.}, \tag{2}$$

where $\mathscr{P}(z, t)$ is a complex function of z and t which varies little in an optical frequency or wavelength.

Substituting (1) and (2) into the wave equation (8.3), with (8.4) and neglecting second derivatives of the slowly varying quantities $\mathscr{E}$ and $\mathscr{P}$ as well as $\dot{\mathscr{P}}$

and $\sigma\dot{\mathscr{E}}$, we find the complex field self-consistency equation:

$$\frac{\partial \mathscr{E}}{\partial z} + c^{-1}\frac{\partial \mathscr{E}}{\partial t} + \kappa\mathscr{E} = \tfrac{1}{2}i\nu(c\varepsilon)^{-1}\mathscr{P}, \quad (3)$$

where κ is the loss per unit length, given by $\frac{1}{2}\sigma/(c\varepsilon)$.

We calculate the polarization of the medium using the population matrix $\rho(z, \omega, t)$, which has the component equations of motion (8.26)–(8.28). The electric-dipole perturbation energy which appears in these equations is given in the rotating-wave approximation as

$$\mathscr{V}_{ab} = -\tfrac{1}{2}\wp\,\mathscr{E}(z, t)\exp[-i(\nu t - Kz)]. \quad (4)$$

Combining this with Eq. (8.26) for the polarization matrix element ρ_{ab}, we have

$$\dot{\rho}_{ab}(z, \omega, t) = -(i\omega + \gamma)\,\rho_{ab} - \tfrac{1}{2}i\left(\frac{\wp}{\hbar}\right)\mathscr{E}(z, t)\exp[-i(\nu t - Kz)]D, \quad (5)$$

where D is the population difference (inversion):

$$D = \rho_{aa} - \rho_{bb}. \quad (6)$$

Further defining the population sum

$$M = \rho_{aa} + \rho_{bb}, \quad (7)$$

we find the equations of motion [from (8.27) and (8.28)]

$$\dot{D} = \lambda_a - \lambda_b - \gamma_{ab}D - \tfrac{1}{2}(\gamma_a - \gamma_b)M$$
$$+ \left[i\left(\frac{\wp}{\hbar}\right)\mathscr{E}(z, t)\exp[-i(\nu t - Kz)]\,\rho_{ba} + \text{c.c.}\right], \quad (8)$$

and

$$\dot{M} = \lambda_a + \lambda_b - \gamma_{ab}M - \tfrac{1}{2}(\gamma_a - \gamma_b)D. \quad (9)$$

We eliminate M from the equation of motion for D by substituting the formal integral of (9):

$$M = \frac{\lambda_a + \lambda_b}{\gamma_{ab}} - \tfrac{1}{2}(\gamma_a - \gamma_b)\int_{-\infty}^{t} dt' \exp[-\gamma_{ab}(t - t')]\,D(z, \omega, t')$$

into (8). We find

$$\dot{D} = \frac{\gamma_a\gamma_b}{\gamma_{ab}}N - \gamma_{ab}D + \tfrac{1}{4}(\gamma_a - \gamma_b)^2\int_{-\infty}^{t} dt' \exp[-\gamma_{ab}(t - t')]\,D(z, \omega, t')$$
$$+ \left\{i\left(\frac{\wp}{\hbar}\right)\mathscr{E}(z, t)\exp[-i(\nu t - Kz)]\,\rho_{ba} + \text{c.c.}\right\}, \quad (10)$$

where the (unperturbed) population inversion density

$$N = \lambda_a\gamma_a^{-1} - \lambda_b\gamma_b^{-1}. \quad (11)$$

We obtain an equation of motion for D in terms of $\mathscr{E}$ and D by substituting the formal integral of (5):

$$\rho_{ab}(z, \omega, t) = -\tfrac{1}{2}i\frac{\wp}{\hbar}\exp[-i(\nu t - Kz)]\int_{-\infty}^{t} dt'\,\mathscr{E}(z, t')D(z, \omega, t')$$

$$\times \exp[-(i\omega - i\nu + \gamma)(t - t')] \tag{12}$$

into (10). We find

$$\dot{D} = \frac{\gamma_a\gamma_b}{\gamma_{ab}}N - \gamma_{ab}D + \tfrac{1}{4}(\gamma_a - \gamma_b)^2\int_{-\infty}^{t} dt'\exp[-\gamma_{ab}(t - t')]\,D(z, \omega, t')$$

$$-\frac{1}{2}\left(\frac{\wp}{\hbar}\right)^2\int_{-\infty}^{t} dt'\,D(z, \omega, t')\,\{\exp[+(i\omega - i\nu - \gamma)(t - t')]$$

$$\times\,\mathscr{E}(z, t)\,\mathscr{E}^*(z, t') + \text{c.c.}\}. \tag{13}$$

The polarization of a homogeneously broadened medium is given by

$$P(z, t) = \wp\rho_{ab}(z, \omega, t) + \text{c.c.} \tag{14}$$

Identifying the positive frequency components of this equation and (2), we find the complex polarization component:

$$\mathscr{P}(z, t) = 2\,\wp\exp[i(\nu t - Kz)]\,\rho_{ab}. \tag{15}$$

With (12), this becomes

$$\mathscr{P}(z, t) = -i\frac{\wp^2}{\hbar}\int_{-\infty}^{t} dt'\,\mathscr{E}(z, t)D(z, \omega, t')\exp[-(i\omega - i\nu + \gamma)(t - t')]. \tag{16}$$

The frequency ν is arbitrary; for homogeneously broadened media, it is convenient to set it equal to ω. We then combine (16) with (3) to get the equation of motion for the field:

$$\left(\frac{\partial}{\partial z} + \frac{1}{c}\frac{\partial}{\partial t} + \kappa\right)\mathscr{E}(z,\ t) = a'\int_{-\infty}^{t} dt'\,\mathscr{E}(z, t')\exp[-\gamma(t - t')]$$

$$\times\,\frac{D(z, \omega, t')}{N}, \tag{17}$$

where the gain parameter a' is given by

$$a' = \frac{\wp^2\nu N}{2\hbar c\varepsilon}. \tag{18}$$

Equations (13) (with $\nu = \omega$) and (17) form a coupled set of integrodifferential equations which determine the development of the electric field in time and space. Inasmuch as the only approximations made are the slowly varying envelope and rotating wave, quite intense, short pulses of radiation can be treated. We discuss this further in Sec. 13-2 after extending the present treatment to the inhomogenous medium.

For this second kind of medium, the polarization (2) is given by

$$P(z, t) = \wp\int_{-\infty}^{\infty} d\omega\,W(\omega)\rho_{ab}(z, \omega, t) + \text{c.c.}$$

Hence the complex polarization component $\mathscr{P}(z, t)$ is given by

$$\mathscr{P}(z, t) = 2\wp\exp[i(\nu t - Kz)]\int_{-\infty}^{\infty} d\omega\, W(\omega)\,\rho_{ab}(z, \omega, t). \tag{19}$$

Combining this with (12), we have

$$\mathscr{P}(z, t) = -i\frac{\wp^2}{\hbar}\int_{-\infty}^{t} dt'\,\mathscr{E}(z, t')\exp[-\gamma(t - t')]$$
$$\times\int_{-\infty}^{\infty} d\omega\, W(\omega)\exp[-i(\omega - \nu)(t - t')]\,D(z, \omega, t'). \tag{20}$$

It is convenient to define the complex susceptibility integral:

$$\chi(z, T, t) = N^{-1}\int_{-\infty}^{\infty} d\omega\, W(\omega)\exp[-i(\omega - \nu)\mathrm{T}]D(z, \omega, t). \tag{21}$$

In terms of this, the polarization (20) is given by

$$\mathscr{P}(z, t) = -\,i\frac{\wp^2 N}{\hbar}\int_{-\infty}^{t} dt'\,\mathscr{E}(z, t')\exp[-\gamma(t - t')]\,\chi(z, t - t', t'). \tag{22}$$

With (3), this gives the field equation of motion:

$$\frac{\partial\mathscr{E}}{\partial z} + c^{-1}\frac{\partial\mathscr{E}}{\partial t} + \kappa\mathscr{E} = \alpha'\int_{-\infty}^{t} dt'\,\mathscr{E}(z, t')\exp[-\gamma(t - t')]\,\chi(z, t - t', t'). \tag{23}$$

We obtain the equation of motion for χ by taking the time derivative of (21). Using (13), we find

$$\frac{\partial\chi(z, T, t)}{\partial t} = N^{-1}\int_{-\infty}^{\infty} d\omega\, W(\omega)\exp[-i(\omega - \nu)T]\dot{D},$$

that is,

$$\frac{\partial}{\partial t}\chi(z, T, t) = \frac{\gamma_a\gamma_b}{\gamma_{ab}}\,\tilde{W}(T) - \gamma_{ab}\,\chi(z, T, t)$$
$$+\tfrac{1}{4}(\gamma_a - \gamma_b)^2\int_{-\infty}^{t} dt'\exp[-\gamma_{ab}(t - t')]\,\chi(z, T, t')$$
$$-\frac{1}{2}\left(\frac{\wp}{\hbar}\right)^2\int_{-\infty}^{t} dt'\exp[-\gamma(t - t')]$$
$$\times\,\{\mathscr{E}(z, t)\mathscr{E}^*(z, t')\chi(z, T - t + t', t')$$
$$+\,\mathscr{E}^*(z, t)\mathscr{E}(z, t')\chi(z, T + t - t', t')\}. \tag{24}$$

Here

$$\tilde{W}(T) = \int_{-\infty}^{\infty} d\omega\, W(\omega)\exp[-i(\omega - \nu)T] \tag{25}$$

is the Fourier transform of the inhomogeneous distribution $W(\omega)$. Equations (24) and (23) form a coupled set of integrodifferential equations which determine the time and space dependence of the electric field (1) as it propagates through an inhomogeneously broadened medium.

We will be concerned primarily with the important special case of (23) and (24), in which the field envelope $\mathscr{E}(z, t)$ is real, the distribution $W(\omega)$ is symmetrical about its center frequency ω_0, and the field carrier frequency $\nu = \omega_0$ (central tuning). That this is a solution to the equations is evident, for the population difference D is then symmetric in frequency [see (13)], the susceptibility χ of (21) is real (the imaginary part vanishes in the integration over ω), and hence Eq. (23) is real. The solution is, furthermore, a useful metastable one. A partial proof of this is the content of Prob. 13-11. Equation (23) and (24) with both $\mathscr{E}$ and χ real are our principal working equations. This real solution is often called the zero-phase solution, for the complex $\mathscr{E} = |\mathscr{E}| \exp(-i\phi)$ is real when $\phi = 0$ (or an integral multiple of π). Negative solutions for "zero phase" are possible and correspond to a (positive) $|\mathscr{E}|$ with the phase $\phi = \pi$.

13-2. Rate Equation Approximation

In this section we illustrate some of the basic phenomena of pulse propagation in the simple situation of a resonant, homogeneously broadened medium subject to a pulse whose duration τ_p satisfies the condition

$$\frac{1}{\gamma} \equiv T_2 \ll \tau_p \ll T_1 \simeq \frac{1}{\gamma_a}, \frac{1}{\gamma_b}. \tag{26}$$

This is a rate equation approximation differing from that in Chap. 8, which is characterized by T_2, $T_1 \ll \tau_p$. For (26), the population difference D satisfies the equation of motion [from (13)]

$$\dot{D}(z, t) = -T_2 I(z, t) D(z, t), \tag{27}$$

where the "intensity" $I(z, t)$ is defined by

$$I(z, t) = \left| \frac{\wp \mathscr{E}(z, t)}{\hbar} \right|^2. \tag{28}$$

From (17), we see that this intensity has the equation of motion

$$\left(\frac{\partial}{\partial z} + \frac{1}{c}\frac{\partial}{\partial t}\right) I(z, t) = a \frac{D}{N} I(z, t), \tag{29}$$

in which the gain coefficient a is given by

$$a = 2T_2 a' = \frac{\wp^2 \nu N}{\hbar c \varepsilon} T_2 \tag{30}$$

for this homogeneous case.

Equation (27) has the formal integral

$$D(z, t) = N \exp[-T_2 \mathscr{I}(z, t)], \tag{31}$$

where the partial energy integral $\mathscr{I}(z, t)$ is defined by

$$\mathscr{I}(z, t) = \int_{-\infty}^{t} dt' \, I(z, t'). \tag{32}$$

With use of (29), we find that $I(z, t)$, in turn, has the equation of motion

$$\left(\frac{\partial}{\partial z} + \frac{1}{c}\frac{\partial}{\partial t}\right) \mathscr{I}(z, t) = \int_{-\infty}^{t} dt' \left(\frac{\partial}{\partial z} + \frac{1}{c}\frac{\partial}{\partial t'}\right) I(z, t')$$

$$= a\int_{-\infty}^{t} dt' \, I(z, t') \exp[-T_2\mathscr{I}(z, t')]. \tag{33}$$

Since $d\mathscr{I}' = I\,dt'$, this reduces to

$$\left(\frac{\partial}{\partial z} + \frac{1}{c}\frac{\partial}{\partial t}\right)\mathscr{I}(z, t) = a\int_{0}^{\mathscr{I}(z,t)} d\mathscr{I}' \exp(-T_2\mathscr{I}')$$

$$= \frac{a}{T_2}[1 - \exp(-T_2\mathscr{I})]. \tag{34}$$

In particular, we define the total pulse energy passing the point z as

$$\mathscr{T}(z) \equiv \mathscr{I}(z, \infty) = \int_{-\infty}^{\infty} dt' \; I(z, t') = \int_{-\infty}^{\infty} dt' \left|\frac{\wp\mathscr{E}(z, t')}{\hbar}\right|^2. \tag{35}$$

According to (34), this has the equation of motion

$$\frac{\partial}{\partial z}\mathscr{T}(z) = \frac{a}{T_2}\{1 - \exp[-T_2\mathscr{T}(z)]\}. \tag{36}$$

For small energy, the exponential can be expanded to first order in $\mathscr{T}(z)$ to give

$$\frac{\partial}{\partial z}\mathscr{T}(z) \simeq a\mathscr{T}(z), \tag{37}$$

that is, exponential gain (or loss if $a < 0$). For attenuating media ($a < 0$), this is just the classical Beer's law for absorption.

For larger $\mathscr{T}(z)$, saturation sets in and Eq. (36) reduces to

$$\frac{\partial}{\partial z}\mathscr{T}(z) \simeq \frac{a}{T_2}, \tag{38}$$

that is, linear gain (or loss). For attenuating media, this saturation effect is called bleaching.

We can integrate (34) as for Prob. 13–2, thereby obtaining the wave form

$$I(z, \tau) = I(0, \tau)\exp(az)\left\{\frac{\exp\left[T_2\int_{-\infty}^{\tau} d\tau' \, I(0, \tau')\right]}{1 + \exp(az)\left\{\exp\left[T_2\int_{-\infty}^{\tau} d\tau' \, I(0, \tau')\right] - 1\right\}}\right\}, \tag{39}$$

in which $\tau = t - z/c$. We can gain some insight into the propagation through this kind of medium by considering the initial wave form modification and that at considerably later times. For the former (leading edge of the pulse), the integrals in (39) are approximately zero, giving unit exponentials and hence the following value of (39):

$$I(z, \tau) \simeq I(0, t) \exp(az)$$

This is just exponential gain (or loss) as expected from Beer's law. For large τ (trailing edge), the integrals achieve nonzero values, giving the exponential $e^{\int}$. For sufficiently large z and an amplifying medium, we obtain

$$I(z, \tau) \simeq I(0, t) \frac{e^{\int}}{e^{\int} - 1},$$

that is,

$$I(0, t) < I(z, \tau) \ll I(0, t) \exp(az)$$

In particular, we see that the trailing edges (as well as leading edges) of pulses at large z lie above those at smaller z (see Fig. 13-1). This is a basic characteristic of rate equation solutions with zero loss and is illustrated in Fig. 13-1 for a hyperbolic secant pulse (see Fig. 13-6 for a comparison of this shape with the Gaussian and Lorentzian).

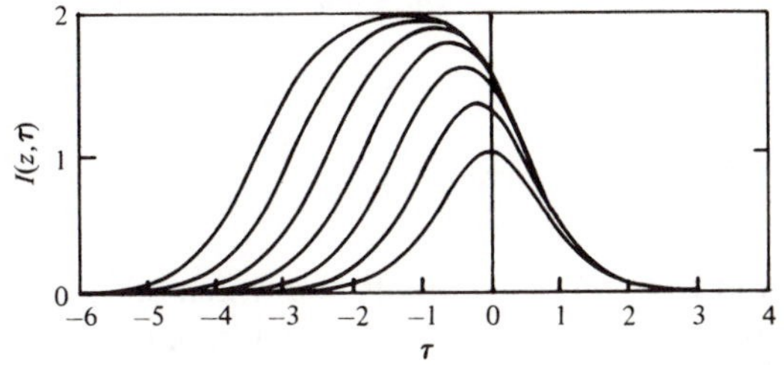

Figure 13-1. Evolution of a hyperbolic secant (h.s.) pulse as given by Eq. (39) (see Prob. 13-9 for the specific formula) in an amplifying medium. The distances z are 0, $1/a$, $2/a, \ldots, 6/a$ in the order of increasing intensity. The time interval $\tau = t - z/c$ is given in units of τ_s, the h.s. parameter. The input intensity constant $I_0 = 1/T_2\tau_s$, $I(z, \tau)$ is given in units of I_0. Note that the front edge (negative τ) is amplified substantially, whereas the back edge remains relatively unchanged. This is due to the change in saturation of the medium during passage of the pulse. The net effect is to cause the pulse center to propagate faster than the speed of light in the medium.

The net effect of the linear amplification of the leading edge and the saturated amplification of the trailing edge is to cause the center of the pulse to propagate *faster than the speed of light* in the medium. This phenomenon (which has been observed by Basov et al, 1966) does not, of course, violate special relativity, since information cannot be propagated on the pulse. In fact if the leading edge goes to zero for small τ, the pulse bunches up at the start, that is, becomes *sharper*. This phenomenon is illustrated in Fig. 13-2 for a h.s. pulse whose baseline (zero) was shifted up slightly. Also included in the calculations leading to Fig. 13-2 is some loss which causes the trailing edge of larger z values to dip below those of earlier z values. We see in the next section how an inhomogeneously broadened medium can greatly accentuate this dip effect.

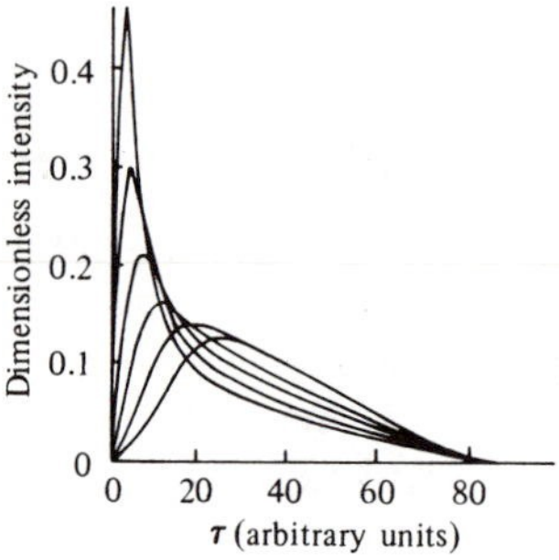

Figure 13-2. Evolution of hyperbolic secant whose baseline is shifted upward slightly for distances $z = 120a$ to $180a$ in steps of $12a$. A loss of $0.15a$ was included in the calculations. From Icsevgi and Lamb (1969).

13-3. The Pulse Area Theorem

As mentioned in Sec. 2-4, a π pulse is one which takes a two-level system from one level to the other. In particular, a resonant atom initially in the lower level acquires an upper-level probability $|C_a(\tau_p)|^2 = \sin^2(\wp E_0\tau_p/2\hbar)$ in a field E_0 with duration τ_p. This probability is unity for $\wp E_0\tau_p/\hbar = \pi$. In Sec. 7-5 we noted that such a pulse rotates the **R** vector from pointing down ($\mathbf{R} = -\hat{e}_3$) to up ($\mathbf{R} = \hat{e}_3$), that is, through an angle π. To treat more complicated pulses and general angles of rotation, we follow the lead of nuclear magnetic resonance and define the pulse "area"

$$\theta(z) = \frac{\wp}{\hbar}\int_{-\infty}^{\infty} dt\, \mathscr{E}(z, t). \tag{40}$$

For a square pulse of duration τ_p and amplitude E_0, this reduces to $\wp E_0\tau_p/\hbar$ as before. The area (40) does not determine the total energy $\mathscr{T}(z)$ of (35). In particular, the amplitude $\mathscr{E}(z, t)$ can go *negative*, yielding zero for the area (40), while the energy (35) is arbitrarily large.

In this section we discuss this area for large pulses propagating through media whose atomic lifetimes greatly exceed the pulse duration. In contrast to the situation met in classical optics, the area develops according to a remarkable theorem due to McCall and Hahn (1969), which predicts that some pulses can propagate at reduced speed and with unchanging shape through attenuating media. The theorem is a consequence of atomic coherence. After physical motivation, the theorem is discussed for various pulses, on and off resonance. The section closes with a proof of the theorem based on the formalism of Sec. 13-1.

In classical linear optics, the area obeys the equation of motion

$$\frac{d}{dz}\theta(z) = -\tfrac{1}{2}|a|\theta(z), \tag{41}$$

where the "gain" coefficient a is negative since the media absorb. This follows for simple media and long pulses from Beer's law (37). It gives an exponential decay of the area [as does (37) for $\mathcal{T}(z)$]. McCall and Hahn (1967) discovered a remarkable generalization of this law for large pulses which drive two-level systems into *nonlinear* regimes. They found for atomic lifetimes long compared to τ_p that $\theta(z)$ has the equation of motion

$$\frac{d}{dz}\theta(z) = \tfrac{1}{2}\, a \sin[\theta(z)]. \tag{42}$$

For small θ and attenuating media ($a < 0$), this reduces to the classical value (41). More generally it predicts no change in θ for increasing z when $\theta(z) = n\pi$. The stability of these solutions is easily determined by the small vibrations analysis of Sec. 9-2 (see Prob. 13-8). A more complete analysis is provided by the solution of (42). For this, we rewrite (42) as we did the mode-locking equation (9.58), that is,

$$\frac{d\theta(z)}{\sin[\theta(z)]} = \tfrac{1}{2}\, a\, dz.$$

Integrating the left-hand side from $\theta_0 = \theta(0)$ to $\theta(z)$ and z from 0 to z, we find

$$\begin{aligned}\tfrac{1}{2}\, az = \frac{a}{2}\int_0^z dz &= \int_{\theta_0}^{\theta}\frac{d\theta'}{\sin\theta'} = \int_{\theta_0}^{\theta}\frac{d\theta'}{2\sin(\theta'/2)\cos(\theta'/2)}\\ &= \int_{\theta_0}^{\theta}\frac{\sec^2(\theta'/2)\, d\theta'}{2\tan(\theta'/2)} = \int_{\tan(\theta_0/2)}^{\tan(\theta/2)}\frac{du}{u}\\ &= \ln\left[\frac{\tan(\theta/2)}{\tan(\theta_0/2)}\right],\end{aligned}$$

that is,

$$\theta(z) = 2\tan^{-1}[\tan(\theta_0/2)\exp(az/2)]. \tag{43}$$

Hence, as illustrated in Fig. 13-3, the area has the limit

$$\lim_{z\to\infty}\theta(z) = \left\{\begin{matrix} 2n\pi, & a < 0 \\ (2n+1)\,\pi, & a > 0\end{matrix}\right\} \text{ with } n \text{ determined by } \theta_0. \tag{44}$$

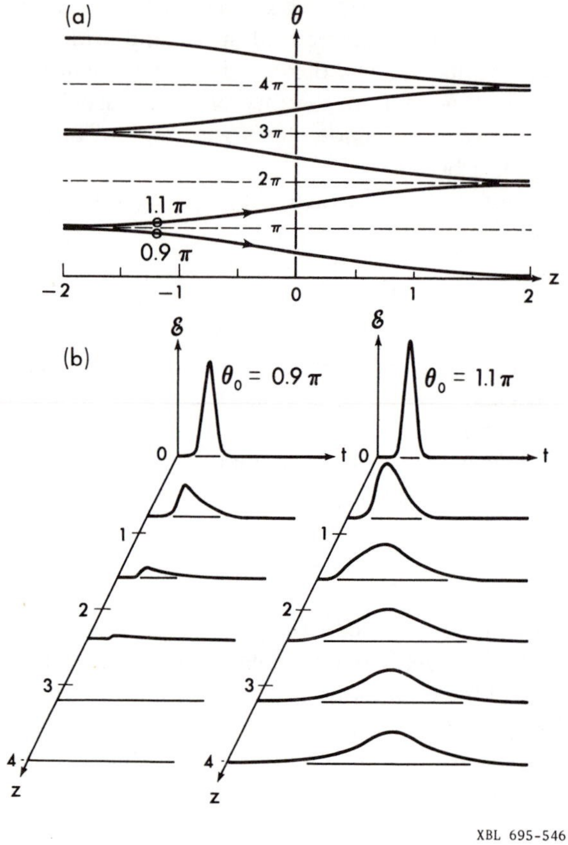

Figure 13-3. (a) Area solutions given by (43) are plotted versus z. For an absorbing (amplifying) medium with $a < 0$ ($a > 0$), the pulse area evolves in the direction of increasing (decreasing) distance z toward the nearest even (odd) multiple of π. The z axis is given in units of π/a cm for both (a) and (b). (b) Computer plots of pulse evolution for input values $\theta_0 = 0.9\pi$ and $\theta_0 = 1.1\pi$. The time scale is given in arbitrary units. From McCall and Hahn (1969).

McCall and Hahn discovered the area theorem from analysis of computer data. They explain it with the effective analogy depicted in Fig. 13-4. There a slowly rolling ball hits a sequence of pendulums and slows down. This is like Beer's law. A ball rolling sufficiently fast, however, can cause the pendulums to swing all the way around and to hit the ball on its back side, thereby returning all the energy initially absorbed from the ball. In this case, the ball continues past the pendulums, moving with the same speed as before the collisions, although somewhat retarded with respect to a collisionless path.

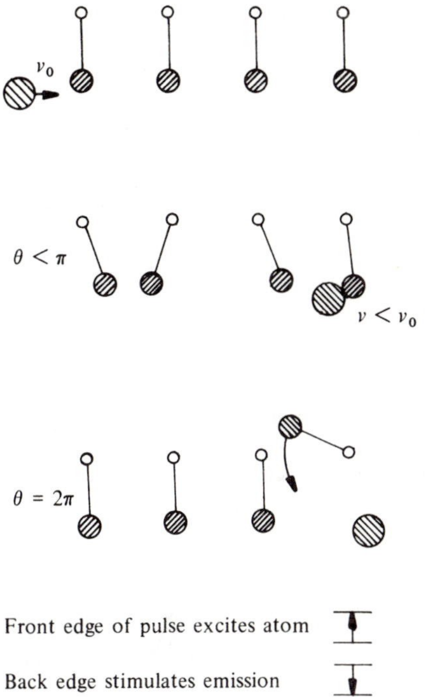

Figure 13-4. Bowling ball analogy for self-induced transparency.

Similarly the **R** vector for an ensemble of lower-state atoms rotates upward slightly for a weak incident pulse, absorbing pulse energy. A strong pulse, say 2π, causes the **R** vector to swing up (absorption) *and* come back down (stimulated emission), returning all the energy initially absorbed somewhat after a freely propagating pulse would have passed. According to (42), the area of the pulse stays 2π as it propagates. The phenomenon is called "self-induced transparency," a marvelous example of a coherent, nonlinear interaction between radiation and matter. We hasten to add that the circumstances must be special: a sharp (compared to atomic lifetimes), strong pulse interacts with a medium accurately described by two-level atoms. In particular, the area depends on both the field and the electric-dipole matrix element $\wp$. Hence, for many-level transitions such as those encountered in the Zeeman laser discussion of Chap. 12, a π pulse for one set of levels is not a π pulse for another set. This leads to averaging effects, which can be important in the interpretation of self-induced transparency data. We restrict ourselves to the two-level problem and refer the reader to the work by Hopf, Rhodes and Szöke (1970) for further discussion of the many-level system.

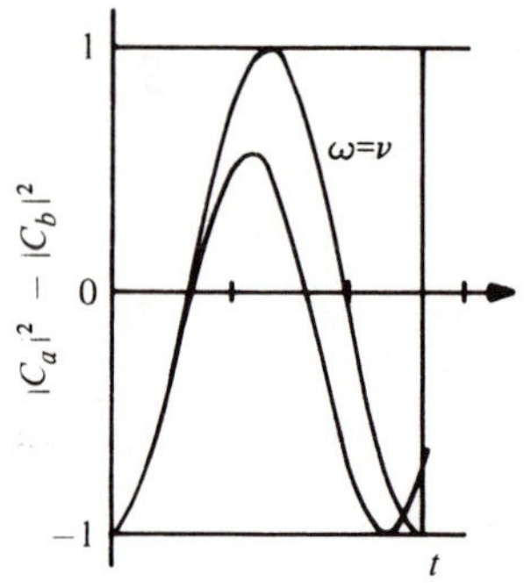

Figure 13-5. The development of atomic probability difference for resonant 2π square pulse and one with the detuning $(\omega - \nu) = 0.4/\tau_s$, where τ_s is the secant parameter used in Fig. 13-7. The time axis is given in units of τ_s.

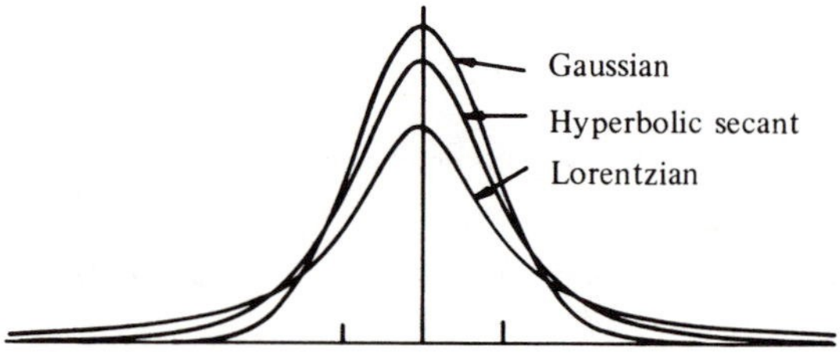

Figure 13-6. Comparison of hyperbolic secant function with Lorentzian and Gaussian functions all having the same half widths and areas. Abscissa values for half-width points are indicated.

The question of how a pulse which is 2π for resonant atoms ($\omega = \nu$) can also be transparent for nonresonant atoms may be raised. Specifically the square pulse defined by $\wp E_0 \tau_p/\hbar = 2\pi$ lacks this property, as can be seen from the lower-level probability, given by (2.63) as

$$|C_b(\tau_p)|^2 = 1 - |C_a(\tau_p)|^2$$

$$= 1 - \left(\frac{\wp E_0}{2\hbar}\right)^2 \left[\frac{\sin^2(\mu\tau_p/2)}{(\mu/2)^2}\right], \tag{45}$$

where the frequency $\mu = [(\omega - \nu)^2 + (\wp E_0/\hbar)^2]^{1/2}$ exceeds $\wp E_0/\hbar$ off resonance. The corresponding probability difference $|C_a|^2 - |C_b|^2$ is depicted in Fig. 13-5, in which we see that the off-resonance pulse overshoots the 2π mark. McCall and Hahn have shown, nevertheless, that the pulse shape (Fig. 13-6)

$$\mathscr{E}(z, t) = \frac{2\hbar}{\wp\tau_s} \operatorname{sech}\left(\frac{t - z/v_p}{\tau_s}\right), \tag{46}$$

for some retarded pulse speed v_p, can induce transparency for *all* frequencies and simultaneously satisfies Maxwell's equation. In fact, they showed that the normalized population difference $D/N \equiv R_3$ (this corresponds to the probability difference of Fig. 13-5) has the time dependence (see Prob. 13-10)

$$R_3(t) = -1 + \left(\frac{2}{1 + \tau_s^2(\omega - \nu)^2}\right) \operatorname{sech}^2(t/\tau_s). \tag{47}$$

This value is plotted versus t for several values of detuning $\omega - \nu$ in Fig. 13-7*b* for the hyperbolic secant pulse graphed in Fig. 13-7*a*. The **R** vector diagram for a number of detunings is given in Fig. 13-8.

The coherence effects represented by the area theorem are most pronounced in the pulse delay τ_d, illustrated in Fig. 13-9*a*. This delay is shown for a number of input pulse strengths and values of T_2 in Fig. 13-9*b*. We see that the π pulse suffers considerably more delay than the 2π. Furthermore, we see that, as the atomic coherence time T_2 decreases, the pulse delay becomes less pronounced. This is what would be expected, for, as T_2 decreases, the atomic coherence is washed out, ultimately leading to a valid rate-equation regime.

Propagation is more complicated through amplifying media than through attenuating media. Although the "stable" area remains $(2n + 1)\pi$, the energy can increase by both the sharpening and the negative swings of the pulse

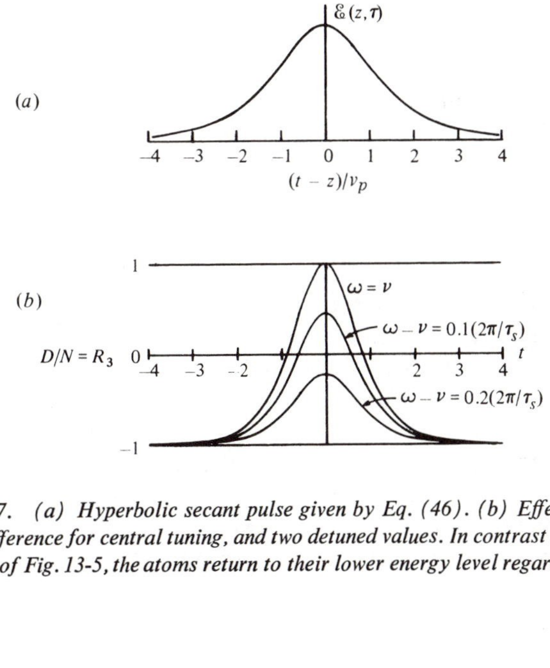

Figure 13-7. (a) Hyperbolic secant pulse given by Eq. (46). (b) Effect on normalized population difference for central tuning, and two detuned values. In contrast to the Rabi flopping response of Fig. 13-5, the atoms return to their lower energy level regardless of detuning.

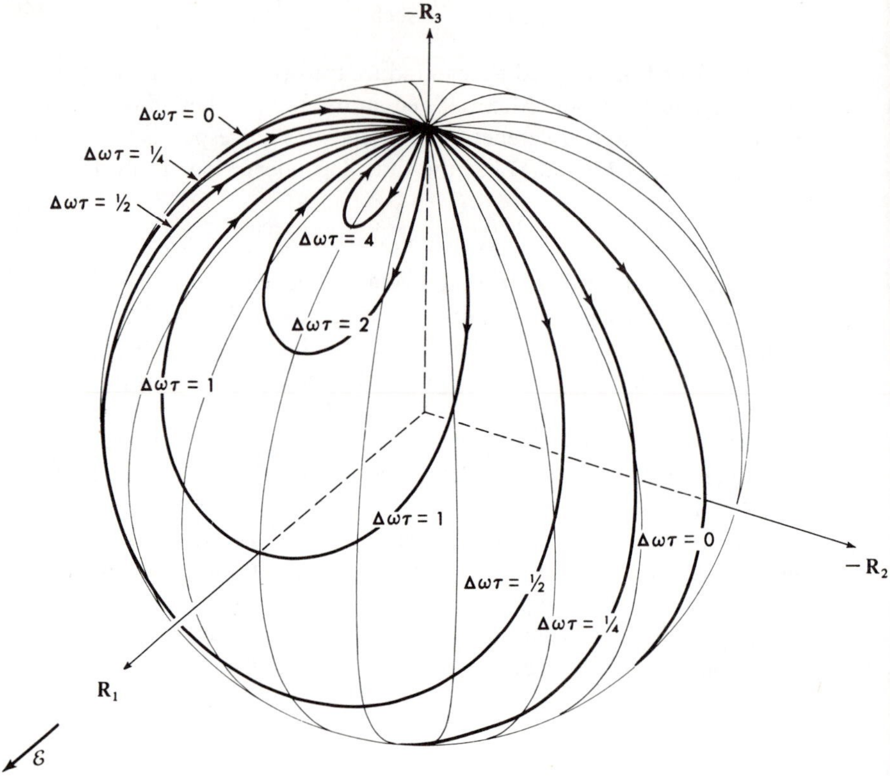

Figure 13-8. Trajectories of **R** *vector driven by a* 2π *hyperbolic secant field pulse for various values of* $\Delta\omega\tau \equiv (\omega - \nu)\tau_s$. *From McCall and Hahn (1969).*

amplitude. In general, quantitative effects of the inhomogeneously broadened amplifying medium require numerical analysis of the pulse equations of motion (23) and (24). Nevertheless, we can gain an intuitive feel for some of the qualitative features by considering the polarization (3.54) for a medium with long lifetimes and large inhomogeneous broadening. In this case the polarization envelope is proportional to $J_0(\wp E_0 t/\hbar)$, a Bessel function which *attenuates* at some times while amplifying at others. This behavior represents the combined contributions of atoms undergoing Rabi flopping at a distribution of frequencies. It contrasts with the rate equation treatment of the homogeneously broadened medium (Sec. 13-2), for which the polarization of Eq. (29) invariably amplifies. The attenuating capability of the inhomogeneously broadened medium can even cause negative swings of the pulse am-

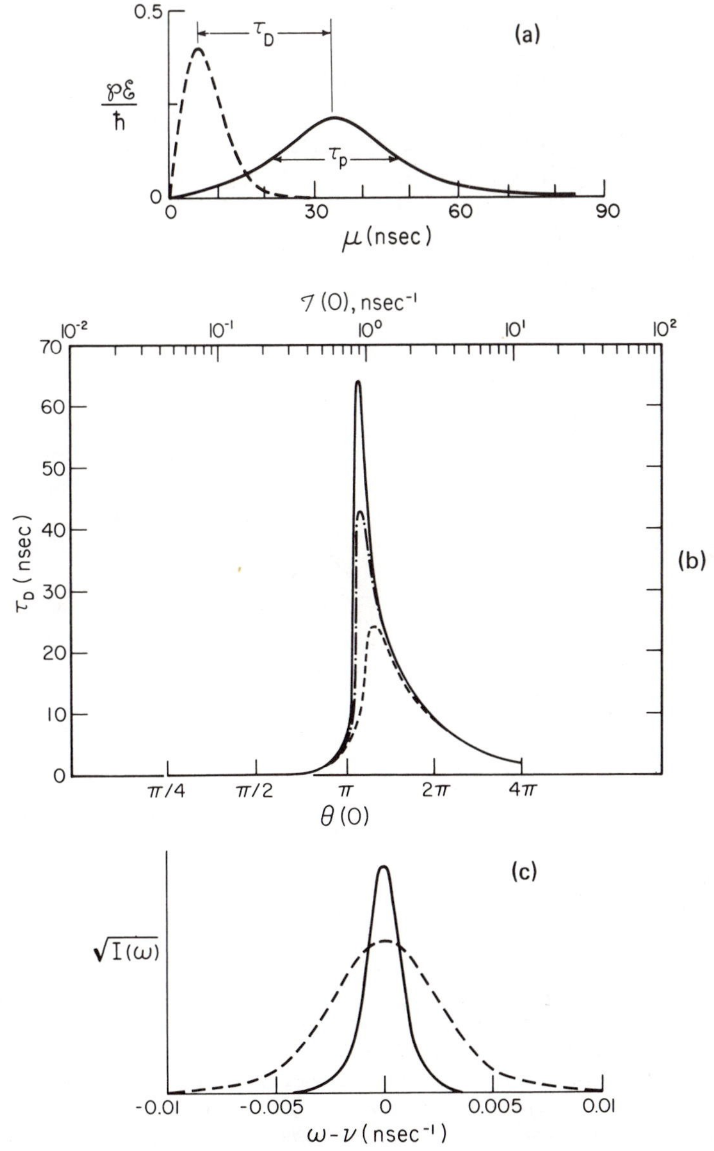

Figure 13-9. (a) Illustration of pulse delay and compression. (b) Delay time τ_d versus input area; corresponding value of input energy is given at the top scale. The solid curve is for the case $T_2 \gg \tau_p$, $T_1 \gg \tau_p$, no loss, and zero detuning. The dot-dash curve is for $T_2 = 150$ nsec, and the dashed curve for $T_2 = 50$ nsec. Other parameters are the same as for the solid curve. (c) Spectrum of pulse envelope for initial value of $\theta(z) = 1.3\pi$ (dashed line) and for evolved 1.9π pulse (solid line). Here $\theta(z)$ gives the center value of this spectrum and has increased as predicted.

plitude, as shown in Fig. 13-10*c*. We further see in Fig. 13-10*b* that pulse sharpening can occur even without the leading edge cutoff that led to the sharpening in the homogeneously broadened case of Sec. 13-2. The effect of inhomogeneously broadened media on pulse energy varies greatly depending on the input pulse area, as illustrated in Fig. 13-10*a*. This, too, contrasts with the rate equation treatment, which amplifies the energy equally for all initial pulse areas having the same energy. For further discussion of the amplifier, we refer the reader to the literature (see the list at the end of the chapter) and turn now to a proof of the area theorem based on the theory of Sec. 13-1.

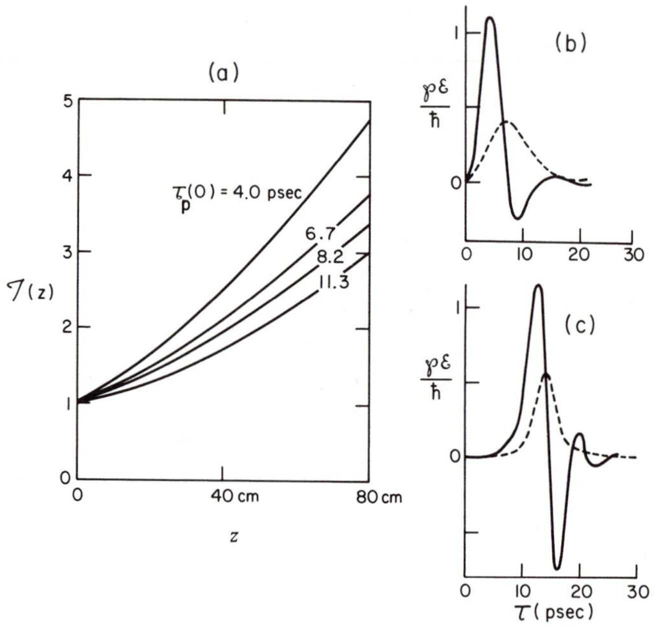

Figure 13-10. Four pulse configurations are considered here, all having the same input integrated intensity $\mathcal{T}(z)$. The pulses are defined by the initial values: (1) $\tau_p = 11.3$ psec, $\theta(0) = 3.99$; (2) $\tau_p(0) = 8.2$ psec, $\theta(0) = 3.44$; (3) $\tau_p(0) = 6.7$ psec, $\theta(0) = 3.13$: and (4) $\tau_p(0) = 4$ psec, $\theta(0) = 2.72$. (a) Pulse energy $\mathcal{T}(z)$ versus z for four pulses. (b) Output pulse shape (solid line) for case 2 (broken line). (c) Same as (b) for pulse 4(a hyperbolic secant pulse). From Hopf and Scully (1969).

The analysis is based on the real versions of the equations of motion for the field $\mathcal{E}(z, t)$ of (23) and the susceptibility $\chi(z, T, t)$ of (24), in which we set the atomic lifetimes to infinity, for they are assumed to be long compared to the pulse duration. We furthermore turn the pump off at time t_0, just before the pulse interacts with the medium. The equations of motion reduce to

$$\frac{\partial \mathscr{E}}{\partial z} + c^{-1}\frac{\partial \mathscr{E}}{\partial t} + \kappa \mathscr{E} = \alpha' \int_{t_0}^{t} dt'\, \mathscr{E}(z, t')\, \chi(z, t - t', t'), \tag{48}$$

$$\frac{\partial \chi(z, T, t)}{\partial t} = -\frac{1}{2}\left(\frac{\wp}{\hbar}\right)^2 \int_{t_0}^{t} dt'\, \mathscr{E}(z, t)\mathscr{E}(z, t') \times [\chi(z, T - t + t', t') + \chi(z, T + t - t', t')]. \tag{49}$$

Integrating both sides of (48) with respect to time, multiplying by $\wp/\hbar$, and extending the lower limit of integration to $-\infty$ $[\mathscr{E}(z, t < t_0) = 0]$, we find

$$\frac{d\theta}{dz} + \kappa\theta = \frac{\alpha'\wp}{\hbar}\int_{-\infty}^{\infty} dt \int_{-\infty}^{t} dt'\, \mathscr{E}(z, t')\, \chi(z, t - t', t').$$

Here the time integral of $\partial\mathscr{E}(z, t)/\partial t$ vanishes, inasmuch as the field itself vanishes at $\pm\infty$. Interchanging the order of time integrations with appropriate changes in limits for t, we have

$$\frac{d\theta}{dz} + \kappa\theta = \frac{\alpha'\wp}{\hbar}\int_{-\infty}^{\infty} dt'\, \mathscr{E}(z, t') \int_{t'}^{\infty} \chi(z, t - t', t')\, dt.$$

It is convenient to change the variable t to $T = t - t'$, with limits 0 to ∞. We can use the symmetry of the (real) susceptibility, $\chi(z, -T, t) = \chi(z, T, t)$, to extend the lower limit of integration to $-\infty$, provided that we include a multiplicative factor of $\frac{1}{2}$:

$$\frac{d\theta}{dz} + \kappa\theta = \frac{1}{2}\frac{\alpha'\wp}{\hbar}\int_{-\infty}^{\infty} dt'\, \mathscr{E}(z, t') \int_{-\infty}^{\infty} dT\, \chi(z, T, t'). \tag{50}$$

We now integrate (49) with respect to T. The finite additive factors $t - t'$ and $-t + t'$ appearing with T in the arguments of χ on the right-hand side of (49) can be dropped in the integration from $-\infty$ to ∞ over T. Hence we find

$$\frac{\partial}{\partial t}\int_{-\infty}^{\infty} dT\, \chi(z, T, t) = -\left(\frac{\wp}{\hbar}\right)^2 \int_{t_0}^{t} dt'\, \mathscr{E}(z, t)\mathscr{E}(z, t') \int_{-\infty}^{\infty} dT\, \chi(z, T, t'). \tag{51}$$

We have defined the susceptibility χ so that at time $t = t_0$ it has the value $\tilde{W}(T)$, which is normalized such that

$$\int_{-\infty}^{\infty} dT\, \chi(z, T, t_0) = \int_{-\infty}^{\infty} dT\, \tilde{W}(T) = 2\pi\, W(\nu). \tag{52}$$

The results of Prob. 13–5 show that (51) has the solution which, with the initial condition (52), is given by

$$\int_{-\infty}^{\infty} dT\, \chi(z, T, t) = 2\pi\, W(\nu) \cos\left[(\wp/\hbar)\int_{-\infty}^{t} dt'\, \mathscr{E}(z, t')\right]. \tag{53}$$

Here we have again extended the lower limit of integration over $\mathscr{E}(z, t)$ to $-\infty$ since $\mathscr{E}(z, t < t_0) = 0$. Substituting (53) into the equation of motion for the area (50), we have

$$\frac{d}{dz}[\theta(z)] + \kappa\theta(z) = \frac{\alpha}{2}\int_{-\infty}^{\infty} dt'\, \frac{\wp}{\hbar}\mathscr{E}(z, t') \cos\left[\int_{-\infty}^{t'} dt''(\wp/\hbar)\, \mathscr{E}(z, t'')\right],$$

in which the gain coefficient $a = 2\pi\, W(\nu)\, a'$ for this inhomogeneously broadened case. Now, changing the variable of integration t' to

$$\vartheta(z, t) = \int_{-\infty}^{t} dt' \frac{\wp}{\hbar} \mathscr{E}(z, t'),$$

and using the definition of $\theta(z)$, we see that $\vartheta(z, \infty) = \theta(z)$ and therefore that

$$\frac{d}{dz}[\theta(z)] + \kappa\theta(z) = \frac{a}{2}\int_{0}^{\theta(z)} d\vartheta \cos \vartheta$$

$$= \frac{a}{2} \sin [\theta(z)],$$

which proves (42) as asserted.

13-4. Photon Echo

Photon echo is an optical frequency analog of the spin echo found in nuclear magnetic resonance (Hahn, 1950). Inasmuch as the phenomenon can be treated completely without appeal to the quantum theory of radiation, it might better be called "optical spin echo"; however, Abella, Kurnit, and Hartmann (1966), its discoverers (and namers), consider "photon echo" a catchier designation. The echo occurs in the polarization of an inhomogeneously broadened medium after the passage of two sharp pulses, as depicted in Fig. 13-11. Typically the medium initially in its ground state is subjected to a 90° pulse, which leaves the atoms in a 50–50 superposition of states. After a time τ, the medium is subjected to a π pulse, which effectively interchanges the probability amplitudes for being in the upper and lower states. A time τ still later, the polarization acquires an appreciable value for a

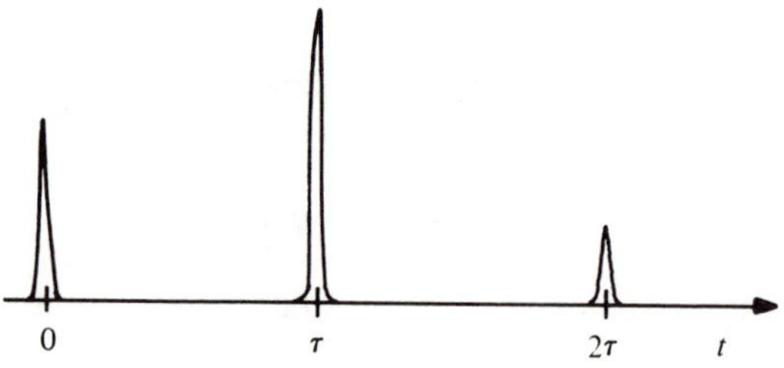

Figure 13-11. Description of simple photon echo: at time $t = 0$, the medium is subjected to a 90° pulse; at $t = \tau$, it is subjected to a 180° pulse; at $t = 2\tau$, the polarization acquires a nonzero value, producing an echo pulse in accord with Maxwell's equations.

time inversely proportional to the width of the inhomogeneous broadening, provided this broadening is large compared to other widths. In practice the broadening is not so large, but the results of the simple approach still occur. We refer the reader to the literature for a detailed explanation and give an intuitive discussion here.

A simple derivation of photon echo involves the two-level wave function

$$\psi(\mathbf{r}, t) = C_a(t)\exp(-i\omega_a t)u_a(\mathbf{r}) + C_b(t)\exp(-i\omega_b t)u_b(\mathbf{r}) \tag{54}$$

of Eq. (2.14). We suppose that, at time $t = -0$ (just before 0), all of the atoms are in their ground states, that is, $\psi(\mathbf{r}, -0) = u_b(\mathbf{r})$. At $t = 0$, the 90° pulse stimulates them into the coherent superposition

$$\psi(\mathbf{r}, 0) = \frac{1}{\sqrt{2}}(u_a + u_b).$$

Here we suppose for the sake of simplicity that the pulse occurs instantaneously. After a time τ_- (just before the arrival of the second pulse), the wave function has changed to

$$\psi(\mathbf{r}, \tau_-) = \frac{1}{\sqrt{2}}[\exp(-i\omega_a\tau)u_a + \exp(-i\omega_b\tau)u_b].$$

The 180° pulse interchanges the probability amplitudes, yielding

$$\psi(\mathbf{r}, \tau) = \frac{1}{\sqrt{2}}[\exp(-i\omega_b\tau)u_a + \exp(-i\omega_a\tau)u_b].$$

After an additonal time τ', the wave function becomes

$$\psi(\mathbf{r}, \tau + \tau') = \frac{1}{\sqrt{2}}[\exp(-i\omega_b\tau - i\omega_a\tau')u_a + \exp(-i\omega_a\tau - i\omega_b\tau')u_b].$$

The polarization of the medium is given by integration of (1.5) over the inhomogeneous distribution $W(\omega)$, that is, by (ignoring vector character)

$$\begin{aligned}\langle er\rangle &= \int d\omega\, W(\omega) \int d^3 r\,|\psi(\mathbf{r}, \tau + \tau')|^2\, er \\ &= \tfrac{1}{2}\wp \int d\omega\, W(\omega)\exp[-i\omega(\tau' - \tau)] + \text{c.c.}\end{aligned} \tag{55}$$

For a Gaussian frequency distribution

$$W(\omega) = \frac{1}{\sqrt{\pi}\,\Delta\omega}\exp[-(\omega - \omega_0)^2/(\Delta\omega)^2] \tag{56}$$

($\Delta\omega$ is often referred to as $2/T_2{}^*$), this gives

$$\langle er\rangle = \wp\exp[-(\tau' - \tau)^2(\Delta\omega)^2]\cos\omega_0 t, \tag{57}$$

that is, for equal time intervals, $\tau' = \tau$, a polarization evolves, yielding the "echo." Essentially the second pulse subtracts (in the exponentials) the time development during the interval $0 \to \tau$ with the result that the atoms arrive back at their time $t = 0$ states at $t = \tau + \tau'$.

We can visualize this process with the help of the **R** vector described in Sec. 7-5. Before the first pulse, **R** points down, that is, $\mathbf{R}(\omega, -0) = -\hat{e}_3$. After the 90° pulse, **R** has rotated up to $\hat{e}_2$:$\mathbf{R}(\omega, 0) = \hat{e}_2$. The component equations of motion (7.72) and (7.73) [or (7.76)] show that the **R** vectors for different resonant frequencies now "fan out" in the $\hat{e}_2 - \hat{e}_1$ plane (Fig. 13-12*a*). At

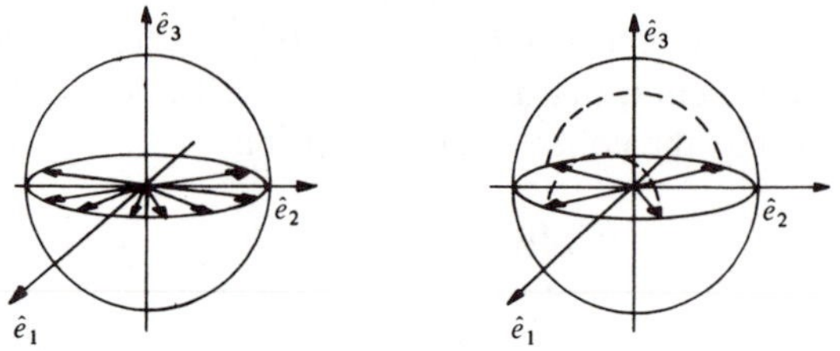

Figure 13-12. (a) ***R*** *vectors for atoms of varying frequency spread out in the $\hat{e}_1 - \hat{e}_2$ plane. (b) A π pulse rotates vectors about $\hat{e}_1$ axis. Vectors then continue to rotate in the same direction, arriving back at $\hat{e}_2$ at time 2τ.*

time $t = \tau$, the 180° pulse arrives, yielding an effective field **B** of Eq. (7.77) pointed along $\hat{e}_1$ (the field is taken sufficiently large and short that $\wp E_0/\hbar$ greatly exceeds the atomic detuning $\omega - \nu$). Hence **R** rotates 180° about $\hat{e}_1$, causing R_2 to go to $-R_2$ (Fig. 13-12*b*). In a time τ, all atoms get back to $\hat{e}_2$, where they started. The situation is analogous to a set of hounds loosed from their halters at time $t = 0$ on a racetrack. They start to spread out along the track because of their unequal speeds. At time $t = \tau$ they are miraculously forced to about-face and return to the starting line, maintaining the same speed with which they departed. At time $t = 2\tau$ they arrive back simultaneously. We note that these treatments are only approximate; a complete discussion involves a careful analysis of the pulse area.

Using the formalism of Sec. 13-1, we can predict not only the simple photon echo described above, but also multiple echoes occurring for three of more pulses. We illustrate the technique for the two-pulse case and assign a three-pulse version as Prob. 13–6. Since the echo phenomenon is not a propagation problem, we consider a medium at $z = 0$. At times less than $t = 0$, the initial susceptibility $\chi(0, T, t < 0)$ of Eq. (21) is given by the Fourier transform of the frequency distribution [since $(\rho_{aa} - \rho_{bb})/N = -1$]

$$\chi(0,\ T, t < 0) = [2\pi W(\nu)]^{-1} \int_{-\infty}^{\infty} d\omega\ W(\omega) \cos[(\omega - \nu)T]. \tag{58}$$

For the Gaussian distribution (56), this gives values of $\chi(0, T, t < 0)$ over a range of T points inversely proportional to the width $\Delta\omega$, as indicated

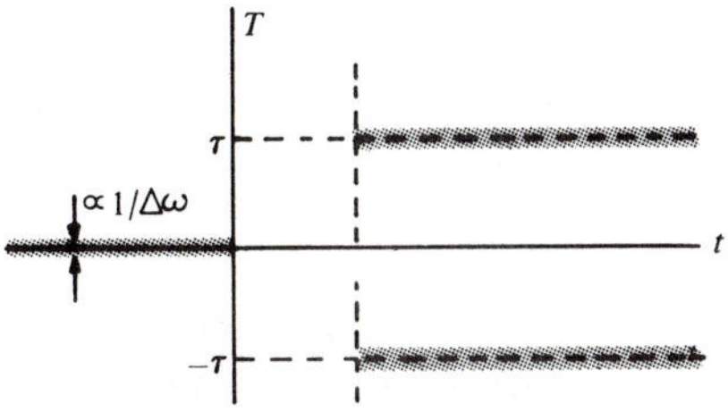

Figure 13-13. Diagram of coordinate values (shaded areas), for which the susceptibility $\chi(0, T, t)$ achieves appreciable magnitude for the pulses of Eq. (60) and initial condition (58).

schematically in Fig. 13-13. At time $t = 0$, the 90° pulse equilibrates the level populations, causing $\chi(0, T, 0)$ to vanish. To find the susceptibility at a later time, we integrate the equation of motion (49), finding

$$\chi(0, T, t) = -\frac{1}{2}\left(\frac{\wp}{\hbar}\right)^2 \int_{-\infty}^{t} dt' \, \mathscr{E}(0, t') \int_{-\infty}^{t'} dt'' \, \mathscr{E}(0, t'') \times [\chi(0, T + t' - t'', t'') + \chi(0, T - t' + t'', t'')]. \quad (59)$$

We define our pulses to be simple delta functions, yielding the areas $\pi/2$ and π:

$$\mathscr{E}(0, t'') = \frac{\pi}{2}\frac{\hbar}{\wp}\,\delta(t''), \qquad \mathscr{E}(0, t') = \pi\frac{\hbar}{\wp}\,\delta(t' - \tau). \quad (60)$$

Equation (59) yields for these pulses

$$\chi(0, T, \tau) = -\tfrac{1}{4}\pi^2[\chi(0, T + \tau, 0) + \chi(0, T - \tau, 0)]. \quad (61)$$

Since $\chi(0, 0, 0)$ has the value given by (58), we find new values for $T = \pm\tau$, as suggested in Fig. 13-13. These values continue for times $t > \tau$. The field integral (48) gives for the pulses (60) the two contributions

$$a\int_{t_0}^{t} dt' \mathscr{E}(0, t')\,\chi(0, t - t', t') = a\frac{\pi}{2}\frac{\hbar}{\wp}\,\chi(0, t, 0) + a\pi\frac{\hbar}{\wp}\,\chi(0, t - \tau, \tau). \quad (62)$$

The first contribution represents absorption of the $\pi/2$ pulse. The second term has a nonzero value for $t - \tau = \tau$, that is, for $t = 2\tau$. For the Gaussian distribution (56) this gives (57). This procedure can be continued, extending the range in Fig. 13-13 for more pulses.

Our treatment is for phase diffusion times T_2 and pulse separation times τ long compared to the pulse duration. When T_2 becomes important with respect to the latter time, the echo is found to decay as $\exp(-t/T_2)$. This can be shown in a straightforward way from Eps. (23) and (24) with the simple pulses of Eq. (60) (Prob. 13-7). Hence the photon echo provides a means to measure T_2. For pulse sequences other than the 90°–180°, echos can still occur, but with reduced peaks.

Problems

13-1. Show in the rate equation approximation for which $\tau_p \gg$ all other times that the population difference (6) is given by

$$D(z, t) = \frac{N(z, t)}{1 + |\wp \mathscr{E}/\hbar|^2 \gamma_{ab}/(\gamma_a \gamma_b \gamma)}.$$

13-2. Show from Eqs. (32) and (34) that the intensity $I(z, t)$ is given in the rate equation approximation of (26) by (39).

13-3. Show for $\gamma_a \simeq 0$(i.e., compared to other constants) that the population difference D of (6) becomes

$$D(z, t) = \frac{2\lambda_a \gamma}{|\wp\mathscr{E}/\hbar|^2},$$

i.e., independent of λ_b; excitation to the relatively undecaying upper level and induced transitions must balance out.

13-4. Show that the pulse width (full width at half maximum) τ_p of the hyperbolic secant function

$$\mathscr{S}(t) = \frac{1}{\pi\tau_s} \operatorname{sech}(t/\tau_s) = \frac{2}{\pi\tau_s}\left[\frac{1}{\exp(t/\tau_s) + \exp(-t/\tau_s)}\right]$$

is

$$\tau_p = 2 \ln (2 + \sqrt{3})\, \tau_s \simeq 2.63\tau_s.$$

13-5. Show that the equations

$$\dot{C}_a = i\left[\frac{\wp\mathscr{E}(t)}{2\hbar}\right] C_b, \qquad \dot{C}_b = i\left[\frac{\wp\mathscr{E}(t)}{2\hbar}\right] C_a$$

have the solution

$$C_a(t) = A \sin [\tfrac{1}{2}\vartheta(t)] + B \cos [\tfrac{1}{2}\vartheta(t)]$$

with a corresponding expression for $C_b(t)$, where the phase angle

$$\vartheta(t) = \frac{\wp}{\hbar}\int_{-\infty}^{t} dt'\, \mathscr{E}(t'),$$

Hint: Use the method of Prob. 2–4b.

13-6. Carry out the discussion of Fig. 13-13 on photon echoes, using three pulses instead of two.

13-7. Show for the pulses of Eq. (60) that the echo decays according to the multiplicative factor $\exp(-t/T_2)$.
13-8. Show that the stationary values (44) for the area $\theta(z)$ are stable, as indicated by using the small vibrations analysis of Sec. 9-2.
13-9. Show for the hyperbolic secant pulse of (46) that the intensity (39) becomes

$$I(z,\tau) = I(0,t)\exp(az)\left\{\frac{\exp\{T_2\tau_s(\wp\mathscr{E}_0/\hbar)^2[\tanh(\tau/\tau_s)+1]\}}{1+\exp(az)[\exp\{T_2\tau_s(\wp\mathscr{E}_0/\hbar)^2[\tanh(\tau/\tau_s)+1]\}-1]}\right\}.$$

Give a qualitative description of the intensity development in time for several z. Your description should match Fig. 13-1.
13-10. Show that Eq. (47) satisfies the Bloch equations (7.72), (7.73) and (7. 75) with $\gamma = 1/T_1 = 0$ and E_0 given by the hyperbolic secant (46).

References

Review Articles

F. T. Arecchi, G. L. Masserini, and P. Schwendimann, 1969, *Nuovo Cimento 1*, 181.
E. Courtens, 1972,"Nonlinear Coherent Resonant Phenomena," Chap. E5 in *Laser Handbook, op. cit.* in references for Chap. 8.
G. L. Lamb, Jr., 1971, *Rev. Mod. Phys.* **43**, 99.
F. A. Hopf, 1974. "Amplifier Theory," in *High Energy Lasers and Their Applications*, ed. by S. F. Jacobs, M. Sargent III, and M. O. Scully, Addison-Wesley, Reading, Mass.

Other References

I. D. Abella, N. A. Kurnit, and S. R. Hartmann, 1966, *Phys. Rev.* **141**, 391.
N. G. Basov, R. V. Ambartsumyan, V. S. Zuev, P. G. Kryukov, and V. S. Letokhov, 1966, *Zh. Eksp. i Teor. Fiz.* **50**, 23 [Eng. trans.: *Sov. Phys. JETP* **23**, 16 (1966)].
E. L. Hahn, 1950, *Phys. Rev.* **80**, 580.
F. A. Hopf and M. O. Scully, 1969, *Phys. Rev.* **179**, 399.
F. A. Hopf and M. O. Scully, 1970, *Phys. Rev. B* **1**, 50.
F. A. Hopf, C.K. Rhodes and A. Szöke, 1970, *Phys. Rev.* **B1**, 2833.
A. Icsevgi and W. E. Lamb, Jr., 1969, *Phys. Rev.* **185**, 517.
S. L. McCall and E. L. Hahn, 1967, *Phys. Rev. Letters* **18**, 908.
S. L. McCall and E. L. Hahn, 1969, *Phys. Rev.* **183**, 457.

XIV

QUANTUM THEORY OF RADIATION

14. Quantum Theory of Radiation

In our preceding treatments of the interaction of radiation with matter, we assumed that the electric field is classical. In most cases, this assumption is validated by experiment. There are several instances, however, for which a classical field fails to give experimentally observed results, whereas a quantized field succeeds. This is true, for example, of spontaneous emission, a phenomenon which was described phenomenologically in our earlier work (see Chaps. 2 and 7). Another case involves the $2s\frac{1}{2}$ and $2p\frac{1}{2}$ states in the hydrogen atom, which are predicted to have the same energy not only by the simple theory of Chap. 1, but also by the more sophisticated Dirac theory. In both cases, a classical electric field was assumed. Experimentally, the two levels differ by approximately the Lamb shift, 1057 MHz. A fully quantized treatment of the field and atom system gives impressive agreement with the experimentally observed shift. Derivation of the fluctuations in intensity of a laser near threshold requires the quantum theory of radiation. In addition there are other problems for which the concept of a "particle," the photon, is either necessary or convenient. Despite a popular impression to the contrary, however the explanation of the photoelectric effect is not one of these (see Prob. 2-10).

Believing, then, that a quantized field approach leads to physically observed results, we benefit from understanding the method both in evaluating the semiclassical theory and in deriving results beyond the scope of the latter. In Sec. 14-1 we show how the electric and magnetic fields of a single-mode, cavity resonator correspond to the position and momentum of a simple harmonic oscillator and thus can be quantized along the lines mentioned in Chap. 1. This leads naturally to a discussion of the photon, the quantum of the electromagnetic field. The treatment is then generalized to a multimode (nonmonochromatic) field in Sec. 14-2. In Sec. 14-3 we discuss the atom-field interaction in the electric-dipole approximation and give a fully quantum-mechanical treatment of Rabi flopping. This phenomenon was treated in the semiclassical approximation in Sec. 2-4. In Sec. 14-4 the Weisskopf-Wigner theory of spontaneous emission is presented. This theory justifies the inclusion of phenomenological decay rates γ_a and γ_b in the Schrödinger equations for the atomic probability amplitudes of Sec. 2-3. In Appendix G we consider radiation from a collection of atoms, a "superradiant" process.

14-1. Quantization of the Electromagnetic Field

In most treatments of the quantum theory of radiation,† unbounded regions are considered and the vector potential is used. Inasmuch as we are primarily concerned with the interaction between laser radiation and decaying atoms in the electric-dipole approximation, we prefer to develop the theory in a gauge-invariant form more appropriate for quantum electronics, emphasizing the electric and magnetic fields. Specifically, we start with Maxwell's equations for a free field in mks units, as introduced in Chap. 8:

$$\nabla \times \mathbf{H} = \frac{\partial \mathbf{D}}{\partial t}, \tag{1}$$

$$\nabla \times \mathbf{E} = -\frac{\partial \mathbf{B}}{\partial t}, \tag{2}$$

$$\nabla \cdot \mathbf{B} = 0, \tag{3}$$

$$\nabla \cdot \mathbf{E} = 0, \tag{4}$$

with constitutive relations

$$\mathbf{B} = \mu_0 \mathbf{H}, \qquad \mathbf{D} = \varepsilon_0 \mathbf{E}. \tag{5}$$

Furthermore, we take the electric field to have the space dependence appropriate for the cavity resonator of Fig. 14-1.

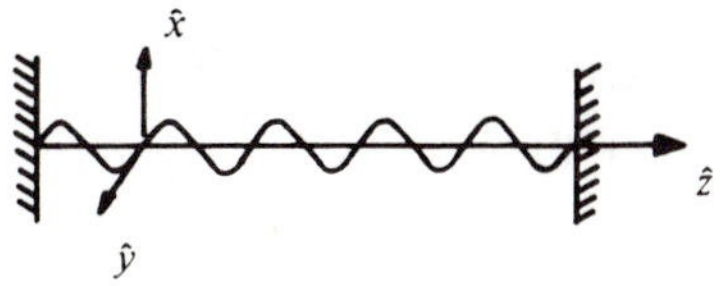

Figure 14-1. Cavity with resonance frequency Ω. The electromagnetic field is assumed to be transverse with the electric field polarized in the x direction.

† See, for example, Heitler (1956) or Louisell (1964). The treatment given here follows that in Section II of Scully and Lamb (1967).

Specifically we write the nonvanishing component of the electric field as

$$E_x(z, t) = q(t)\left(\frac{2\Omega^2 M}{V\varepsilon_0}\right)^{1/2} \sin Kz, \tag{6}$$

where V is the cavity volume, M is a constant with dimensions of a mass introduced to simplify subsequent equations, K is the wave number Ω/c, and $q(t)$ is a variable with dimensions of a length. Equations (1), (5) and (6) yield the nonvanishing magnetic field component:

$$H_y(z, t) = \dot{q}(t)\frac{\varepsilon_0}{K}\left(\frac{2\Omega^2 M}{V\varepsilon_0}\right)^{1/2} \cos Kz. \tag{7}$$

The classical Hamiltonian for the field is

$$\mathscr{H} = \frac{1}{2}\int d\tau\,(\varepsilon_0 E_x{}^2 + \mu_0 H_y{}^2), \tag{8}$$

where the integration is over the volume of the cavity. Using (6) and (7), we can show (Prob. 14-1) that

$$\mathscr{H} = \tfrac{1}{2}(M\Omega^2 q^2 + M\dot{q}^2) = \frac{1}{2}\left(M\Omega^2 q^2 + \frac{p^2}{M}\right), \tag{9}$$

where the momentum

$$p \equiv M\dot{q}.$$

Equation (9) is just the Hamiltonian for a simple harmonic oscillator of mass M, frequency Ω, and positional coordinate $q(t)$. This dynamical problem can be quantized by identifying q and p as operators obeying the commutation relation

$$[q, p] \equiv qp - pq = i\hbar, \tag{10}$$

which leads to the energy eigenfunctions and eigenvalues given in Sec. 1-3.

We now define the operator combinations

$$a = (2M\hbar\Omega)^{-1/2}(M\Omega q + ip), \tag{11}$$

$$a^\dagger = (2M\hbar\Omega)^{-1/2}(M\Omega q - ip), \tag{12}$$

in terms of which the Hamiltonian (9) becomes (Prob. 14-2)

$$\mathscr{H}(a, a^\dagger) = \hbar\Omega(a^\dagger a + \tfrac{1}{2}), \tag{13}$$

the commutation relation (10) becomes

$$[a, a^\dagger] = 1, \tag{14}$$

and, in addition, we have the relations

$$[a, a] = [a^\dagger, a^\dagger] = 0. \tag{15}$$

The electric field operator becomes

$$E_x(z, t) = \mathscr{E}\,(a + a^\dagger) \sin Kz, \tag{16}$$

where the factor

$$\mathscr{E} = \left(\frac{\hbar\Omega}{V\varepsilon_0}\right)^{1/2} \tag{17}$$

has the dimensions of an electric field and corresponds to the electric field "per photon." It is straightforward to show the additional commutation relations

$$[\mathscr{H}, a] = -\hbar\Omega a, \qquad [\mathscr{H}, a^\dagger] = \hbar\Omega a^\dagger. \tag{18}$$

Hence a obeys the (Heisenberg) equation of motion:

$$\dot{a}(t) = \frac{i}{\hbar}[\mathscr{H}, a] = -i\Omega a \tag{19}$$

with solution $a(t) = a(0)\exp(-i\Omega t)$. Similarly $a^\dagger(t) = a^\dagger(0)\exp(i\Omega t)$, and the electric field (16) has a similar monochromatic time dependence. We will have occasion to consider these operators in both the Schrödinger and the Heisenberg pictures. In Chaps. 14–18 we use the Schrödinger and interaction pictures; in Chaps. 19 and 20, the Heisenberg picture.

Because of the fundamental role, that the simple harmonic oscillator plays in quantum optics, we now solve its eigenvalue equation:

$$\mathscr{H}|\mathscr{H}\rangle = \hbar\omega|\mathscr{H}\rangle \tag{20}$$

for the energy eigenstates $|n\rangle$ and eigenvalues $\hbar\omega_n$. Using the first commutation relation of (18), we have

$$\mathscr{H}a|\mathscr{H}\rangle = [a\mathscr{H} - \hbar\Omega a]|\mathscr{H}\rangle = \hbar(\omega - \Omega)\,a|\mathscr{H}\rangle. \tag{21}$$

Hence $a|\mathscr{H}\rangle$ is an energy eigenstate with the eigenvalue $\hbar(\omega - \Omega)$. Because the operator a lowers the energy, it is often called an annihilation, destruction, or absorption operator. By repeatedly applying the annihilation operator, we can find eigenstates with increasingly smaller eigenvalues. We can see that the lowest of these eigenvalues is positive as follows. For an arbitrary vector $|\phi\rangle$, the expectation value of the Hamiltonian operator is

$$\langle\phi|\hbar\Omega(a^\dagger a + \tfrac{1}{2})|\phi\rangle = \hbar\Omega\langle\phi'|\phi'\rangle + \tfrac{1}{2}\hbar\Omega,$$

where $|\phi'\rangle = a|\phi\rangle$. Calling the lowest eigenvalue $\hbar\omega_0$ with eigenstate $|0\rangle$, we have

$$a|0\rangle = 0 \tag{22}$$

and, from (20),

$$\mathscr{H}|0\rangle = \hbar\Omega(a^\dagger a + \tfrac{1}{2})|0\rangle = \hbar\omega_0|0\rangle, \tag{23}$$

that is, the lowest-energy eigenvalue $\hbar\omega_0 = \frac{1}{2}\hbar\Omega$.

Using the second commutation relation of (18), we find higher eigenvalues and eigenstates:

$$\mathscr{H}a^\dagger|0\rangle = [a^\dagger\mathscr{H} + \hbar\Omega a^\dagger]|0\rangle$$

$$= \hbar\Omega(1 + \tfrac{1}{2})\, a^\dagger|0\rangle. \tag{24}$$

Similarly,

$$\mathscr{H}(a^\dagger)^n|0\rangle = \hbar\Omega(n + \tfrac{1}{2})\,(a^\dagger)^n|0\rangle. \tag{25}$$

Hence we have generated the energy eigenvectors which we call $|n\rangle$ and have found the eigenvalues

$$\hbar\omega_n = \hbar\Omega(n + \tfrac{1}{2}). \tag{26}$$

We can determine the normalization of the eigenvectors in (25) by noting from (21) and (26) that

$$a|n\rangle = s_n|n - 1\rangle,$$

for some scalar s_n.

Taking the magnitude of this expression, we have

$$|s_n|^2\langle n - 1\ |n - 1\rangle = \langle n|a^\dagger a|n\rangle = n\langle n|n\rangle,$$

that its, $s_n = \sqrt{n}$, and

$$a|n\rangle = \sqrt{n}\ |n - 1\rangle. \tag{27}$$

Furthermore, from (25), we find that

$$a^\dagger|n\rangle = s_{n+1}|n + 1\rangle$$

with magnitude

$$|s_{n+1}|^2\langle n + 1|n + 1\rangle = \langle n|aa^\dagger|n\rangle = \langle n|a^\dagger a + 1|n\rangle = n + 1,$$

that is, $s_{n+1} = \sqrt{n + 1}$ in agreement with (27). Thus,

$$a^\dagger|n\rangle = \sqrt{n + 1}\ |n + 1\rangle. \tag{28}$$

With (25) this yields the normalized eigenstates

$$|n\rangle = \frac{1}{\sqrt{n!}}(a^\dagger)^n|0\rangle. \tag{29}$$

Finally, we consider the coordinate representation of $|n\rangle$,

$$\phi_n(q) = \langle q|n\rangle. \tag{30}$$

Using Eq. (22), we have

$$(M\Omega q + ip)\phi_0(q) = 0, \qquad \frac{d}{dq}\phi_0(q) = -\left(\frac{M\Omega}{\hbar}\right)q\phi_0(q).$$

These have the solution

$$\phi_0(q) = \mathscr{N}_0 \exp[-\tfrac{1}{2}(M\Omega/\hbar)q^2].$$

Setting $\int dq|\phi_0(q)|^2 = 1$, we find the normalization constant $\mathscr{N}_0$ such that

$$\phi_0(q) = \left(\frac{M\Omega}{\pi\hbar}\right)^{1/4} \exp[-\tfrac{1}{2}(M\Omega/\hbar)q^2]. \tag{31}$$

The higher-energy eigenfunctions are then given by the coordinate representation of (30):

$$\phi_n(q) = \frac{1}{\sqrt{n!}}(a^\dagger)^n\phi_0(q) = (n!)^{-1/2}(2M\hbar\Omega)^{-n/2}\left(M\Omega q - \hbar\frac{d}{dq}\right)^n\phi_0(q)$$
$$= (2^n n!)^{-1/2}\, H_n\!\left[\left(\frac{M\Omega}{\hbar}\right)^{1/2} q\right]\phi_0(q), \tag{32}$$

where H_n are the Hermite polynomials.

The various important results of this section are collected in Table 14-1.

An interesting property of the number state $|n\rangle$ is that the corresponding expectation value of the electric field operator (16) vanishes:

TABLE 14-1. *Harmonic Oscillator Relationships*

$\mathscr{H} = \hbar\Omega\,(a^\dagger a + \frac{1}{2})$	$a\|n\rangle = \sqrt{n}\,\|n-1\rangle$
$[a, a^\dagger] = aa^\dagger - a^\dagger a = 1$	$a^\dagger\|n\rangle = \sqrt{n+1}\,\|n+1\rangle$
$[a, a] = [a^\dagger, a^\dagger] = 0$	$a^\dagger a\|n\rangle = n\|n\rangle$
$E_x\,(z, t) = \mathscr{E}\,(a + a^\dagger)\sin Kz$	$\mathscr{H}\|n\rangle = \hbar\Omega\,(n + \frac{1}{2})\,\|n\rangle$
$\phi_0\,(q) = (M\Omega/\pi\hbar)^{1/4}\exp[-\frac{1}{2}(M\Omega/\hbar)q^2]$	$\phi_n(q) = (2^n n!)^{-1/2}\,H_n[(M\Omega/\hbar)^{1/2}q]\phi_0(q)$

$$\langle n|E|n\rangle = \mathscr{E}\sin Kz\langle n|a(t)|n\rangle + \text{c.c.} = 0. \tag{33}$$

However, the expectation value for the "intensity" operator E^2 is

$$\langle n|E^2|n\rangle = \mathscr{E}^2\sin^2(Kz)\,\langle n|a^\dagger a^\dagger + aa^\dagger + a^\dagger a + aa|n\rangle$$
$$= 2\mathscr{E}^2\sin^2(Kz)(n + \tfrac{1}{2}), \tag{34}$$

that is, there are fluctuations in the field about its zero ensemble average. This does not mean that no electric field exists, that no deflection will occur if an electron passes through the cavity. Rather, the average deflection of many such "measurements" on identically prepared systems vanishes. We wish to emphasize that Eqs. (33) and (34) involve quantum-mechanical averages, not time averages.

It is useful to interpret the eigenvalues (26) as corresponding to the presence in the cavity of n quanta or "photons" of energy $\hbar\Omega$. The eigênstates $|n\rangle$ are often called the photon number states. The energy eigenvalues are discrete in contrast to the classical expression (9), which can assume any value. However, the energy expectation value can also take on any value, for the state vector is, in general, an arbitrary superposition of energy eigenstates. The residual energy $\frac{1}{2}\hbar\Omega$ is called the zero-point energy, and the associated field fluctuations of (34) can be considered to "stimulate" an excited atom to emit spontaneously.

The operators a and $a^\dagger$ annihilate and create photons, respectively, for they change a state with n photons into one with $n - 1$ or $n + 1$ photons. These operators are not themselves Hermitean ($a \neq a^\dagger$) and do not represent observable quantities such as the electric field amplitude and intensity. However, some combinations of the operators are Hermitean, for example, the very useful representations E_x in (16) and $\mathscr{H}$ in (13).

Photons are quanta of a single (monochromatic) mode of the radiation field and are not localized at any particular position and time within the cavity like fuzzy balls; rather, they are spread out over the entire cavity. In fact, no satisfactory quantum theory of photons as particles has ever been given. On the other hand, the quantum theory of radiation seems to offer amazingly satisfactory accounts of a very wide range of radiative problems, and there is no real need to have a corpuscular theory of photons.

Dirac has written in Chap. 1 of his book *Quantum Mechanics*, "Each photon then only interferes with itself. Interference between two different photons never occurs." This statement has caused a lot of confusion. It is known that two separated radio transmitters can produce interference effects, and, for that matter, two lasers can as well. If one thinks that each transmitter sends out its own photons, there is an apparent contradiction with Dirac's statement. The difficulty disappears when one remembers that both transmitters are coupled to the modes of the universal radiation field. A photon is simply a particular energy eigenstate of one such radiation mode.

The fields encountered in most problems are not single $|n\rangle$ states, but superpositions:

$$|\psi\rangle = \sum_n c_n |n\rangle. \tag{35}$$

In fact, the state vector most nearly corresponding to a coherent classical field is such a superposition and is called the coherent state $|\alpha\rangle$. This has a Poisson distribution among the $|n\rangle$ states, that is, there exists some number of photons n_p which is most probable and other numbers are increasingly less probable the more they differ from n_p. We return to the coherent state in Chap. 15.

14-2. Multimode Electromagnetic Field

In the preceding section we considered a single-mode field and found that in general the wave function could be written as a linear superposition of photon number states $|n\rangle$. In this section we extend that formalism to deal with multimode fields, that is, fields with many frequencies. This more general formalism is required to treat, for example, spontaneous emission which is nonmonochromatic with a Lorentzian frequency spread inversely proportional to the lifetime of the radiating systems.

The multimode electric field has the form

$$\mathbf{E}(z, t) = \hat{x} \sum_s q_s(t) \left(\frac{2\Omega_s{}^2 M_s}{V\varepsilon_0}\right)^{1/2} \sin K_s z, \tag{36}$$

$$\mathbf{H}(z, t) = \hat{y} \sum_s \dot{q}_s(t)\varepsilon_0 \, K_s^{-1}\left(\frac{2\Omega_s^2 M_s}{V\varepsilon_0}\right)^{1/2} \cos K_s z, \tag{37}$$

where the eigenfrequencies and wave numbers

$$\Omega_s = \frac{s\pi c}{L}, \qquad K_s = \frac{s\pi}{L}. \tag{38}$$

The Hamiltonian (8) becomes

$$\mathscr{H} = \frac{1}{2} \sum_s \left(M_s \Omega_s^2 q_s^2 + \frac{p_s^2}{M_s}\right) = \sum_s \mathscr{H}_s, \tag{39}$$

where $\mathscr{H}_s$ are the Hamiltonians for individual modes. To quantize this many-particle harmonic oscillator, we introduce the generalized commutation relations

$$[q_s, p_{s'}] = i\hbar\delta_{ss'}, \tag{40}$$

$$[q_s, q_{s'}] = [p_s, p_{s'}] = 0.$$

We further introduce the annihilation and creation operators

$$a_s = (2M_s\hbar\Omega_s)^{-1/2} \, [\Omega_s M_s q_s + ip_s], \tag{41}$$

$$a_s^\dagger = (2M_s\hbar\Omega_s)^{-1/2} \, [\Omega_s M_s q_s - ip_s], \tag{42}$$

in terms of which the Hamiltonian

$$\mathscr{H} = \sum_s \hbar\Omega_s \, [a_s^\dagger a_s + \tfrac{1}{2}] \tag{43}$$

and the electric field operator

$$E_x(z, t) = \sum_s \mathscr{E}_s(a_s + a_s^\dagger) \sin K_s z. \tag{44}$$

In particular, for the single-mode field in the preceding section we write the photon number state $|n\rangle$ as $|n_s\rangle$ and the energy eigenvalue equation as

$$\mathscr{H}_s|n_s\rangle = \hbar\Omega_s(n_s + \tfrac{1}{2})|n_s\rangle. \tag{45}$$

The general eigenstate can have n_1 photons in the first mode, n_2 in the second, n_s in the sth and so forth, and can be written as

$$|n_1\rangle \, |n_2\rangle \ldots |n_s\rangle \ldots$$

or, more conveniently,

$$|n_1 n_2 \ldots n_s \ldots\rangle \equiv |\{n_s\}\rangle. \tag{46}$$

Note that the annihilation operator a_s lowers the n_s entry alone:

$$a_s|n_1 n_2 \ldots n_s \ldots\rangle = \sqrt{n_s}\,|n_1 n_2 \ldots n_s - 1 \ldots\rangle. \tag{47}$$

The general state vector for the field is a linear superposition of these eigenstates:

$$|\psi\rangle = \sum_{n_1}\sum_{n_2} \ldots \sum_{n_s} \ldots c_{n_1 n_2 \ldots n_s \ldots} |n_1 n_2 \ldots n_s \ldots\rangle$$

$$\equiv \sum_{\{n_s\}} c_{\{n_s\}} \,|\, \{n_s\}\rangle. \tag{48}$$

This is a more general superposition than

$$|\psi\rangle = |\psi_1\rangle|\psi_2\rangle \cdots |\psi_s\rangle \cdots, \tag{49}$$

where the $|\psi_s\rangle$ are state vectors for individual modes. Equation (48) includes state vectors of type (49) as well as more general states having correlations between the field modes which can result from perturbations. Such perturbations occur, for example, when the field is coupled to a set of atoms by an electric-dipole interaction. It is to such a coupled field-atom system that we now turn.

14-3. Atom-Field Interaction

We now consider the interaction between a single-mode radiation field and a two-level atom (see Fig. 14-2) located at the point z in a cavity. Our treatment differs from that in Chap. 2 in that here both the atoms and field are treated quantum mechanically. Assuming an initial atomic state vector†

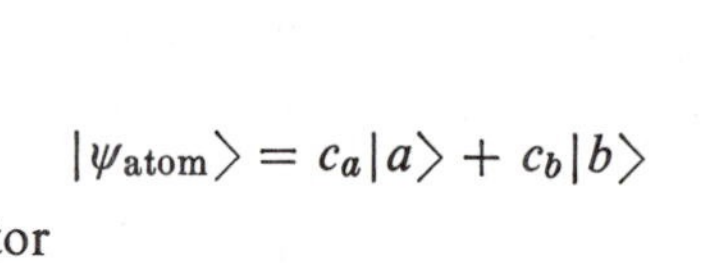

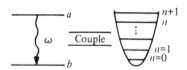

Figure 14-2. Diagram depicting two-level atomic system coupled to single-mode radiation field.

$$|\psi_{\text{atom}}\rangle = c_a|a\rangle + c_b|b\rangle \tag{50}$$

and a field state vector

$$|\psi_{\text{field}}\rangle = \sum_n c_n|n\rangle, \tag{51}$$

we then have the atom-field state vector:

$$|\psi_{\text{a-f}}\rangle = \sum_n [c_{a,n}|a\rangle|n\rangle + c_{b,n}|b\rangle|n\rangle]. \tag{52}$$

Here, for example, $c_{a,n}$ is the probability amplitude that the atom is in the upper state $|a\rangle$ and the field has n photons. Because of the interaction

† The ket $|a\rangle$ is the upper atomic state (not to be confused with a state of the radiation annihilation operator a).

between the two systems, the probability amplitudes $c_{a,n}$ and $c_{b,n}$ change in time. For example, if at time $t = 0$,

$$|\psi(0)_{\text{a-f}}\rangle = |a\rangle|n\rangle, \tag{53}$$

then at some later time t

$$|\psi(t)_{\text{a-f}}\rangle = c_{a,n}|a\rangle|n\rangle + c_{b,n+1}|b\rangle|n+1\rangle, \tag{54}$$

that is, there is a finite probability that the atom has made a transition to the lower (b) level and emitted a photon.

The total Hamiltonian for the system is the sum of the atomic and field energies and of the interaction energy:

$$\mathscr{H} = \mathscr{H}_0{}^{\text{atom}} + \mathscr{H}_0{}^{\text{field}} + \mathscr{V}, \tag{55}$$

where the unperturbed Hamiltonians are given by

$$\mathscr{H}_0{}^{\text{atom}} = \hbar\begin{pmatrix}\omega_a & 0\\ 0 & \omega_b\end{pmatrix} \tag{56}$$

$$\mathscr{H}_0{}^{\text{field}} = \hbar\Omega(a^\dagger a + \tfrac{1}{2}). \tag{57}$$

In order to obtain $\mathscr{V}$ we consider the situation in the semiclassical theory, where electric-dipole interaction energy for an atom at z

$$\mathscr{V}^{sc} = -\wp\, E_x(z,t)\begin{pmatrix}0 & 1\\ 1 & 0\end{pmatrix} = -\wp\, E_x(z,t)(\sigma + \sigma^\dagger),$$

in which $E_x(z, t)$ is a number (not an operator), and the σ operators defined by Eqs. (1.44) and (1.45) flip an atom from one level to the other. The fully quantum-mechanical interaction energy contains the electric field operator (16):

$$\mathscr{V} = -\wp\,\mathscr{E}\sin(Kz)(a + a^\dagger)(\sigma + \sigma^\dagger). \tag{58}$$

Hence the total Hamiltonian (55) has the explicit form

$$\mathscr{H} = \hbar\begin{pmatrix}\omega_a & 0\\ 0 & \omega_b\end{pmatrix} + \hbar\Omega(a^\dagger a + \tfrac{1}{2}) + \hbar g(a + a^\dagger)(\sigma + \sigma^\dagger), \tag{59}$$

where the atom-field coupling constant

$$g = -\frac{\wp}{\hbar}\mathscr{E}\sin Kz. \tag{60}$$

The interaction energy (58) consists of four terms, two of which do not conserve energy. In particular, the σ operator takes an atom in the upper state into the lower state. In order to conserve energy, a photon should be created (stimulated emission), as depicted in Fig. 14-3c. The combination $a^\dagger\sigma$

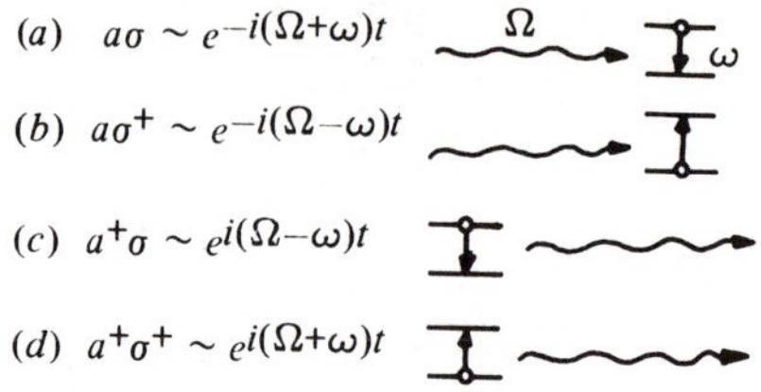

Figure 14-3. Diagram depicting processes corresponding to operator combinations appearing in interaction Hamiltonian (58). Cases (a) and (d) do not conserve energy, whereas (b) and (c) do. The exponential dependences are explained in the text [*see Eqs. (62)–(65)*].

describes the process, while $a\sigma$ annihilates a photon, resulting in the loss of approximately $2\hbar\Omega$ in energy. Similarly, $a\sigma^\dagger$ flips an atom from the lower into the upper state and annihilates a photon (stimulated absorbtion), but $a^\dagger\sigma^\dagger$ creates a photon, resulting in the gain of $2\hbar\Omega$. Dropping the energy non-conserving terms $a\sigma$ and $a^\dagger\sigma^\dagger$ corresponds to the rotating-wave approximation made for Eq. (2.23). To see this and to facilitate later calculations, we now transform our state vector (52) and Hamiltonian (59) from the Schrödinger picture into the interaction picture (see Sec. 6-4).

Inasmuch as the field and atom operators are independent, the interaction picture perturbation energy for the equation of motion (6.63) is

$$\mathscr{V}_I = \hbar g \exp[i\Omega(a^\dagger a + \tfrac{1}{2})t](a + a^\dagger) \exp[-i\Omega(a^\dagger a + \tfrac{1}{2})t]$$
$$\times \exp\left[i\begin{pmatrix}\omega_a & 0\\ 0 & \omega_b\end{pmatrix}t\right]\begin{pmatrix}0 & 1\\ 1 & 0\end{pmatrix}\exp\left[-i\begin{pmatrix}\omega_a & 0\\ 0 & \omega_b\end{pmatrix}t\right]. \tag{61}$$

We can show that (Prob. 14-5)

$$\exp[i\Omega(a^\dagger a + \tfrac{1}{2})t](a + a^\dagger) \exp[-i\Omega(a^\dagger a + \tfrac{1}{2})t]$$
$$= a\exp(-i\Omega t) + a^\dagger \exp(i\Omega t). \tag{62}$$

by using the relation

$$(a^\dagger a)^n\, a = a(a^\dagger a - 1)^n \tag{63}$$

or by taking the time rate of change of (62) [as in (6.61)] and integrating the result. Furthermore, from Prob. 6-4 we have

$$\exp\left[i\begin{pmatrix}\omega_a & 0\\ 0 & \omega_b\end{pmatrix}t\right]\begin{pmatrix}0 & 1\\ 1 & 0\end{pmatrix}\exp\left[-i\begin{pmatrix}\omega_a & 0\\ 0 & \omega_b\end{pmatrix}t\right]$$
$$= \sigma^\dagger \exp(i\omega t) + \sigma\exp(-i\omega t), \tag{64}$$

where the frequency difference $\omega = \omega_a - \omega_b$. Combining (61), (62), and (64), we have

$$\mathscr{V}_I = \hbar g[a \exp(-i\Omega t) + a^\dagger \exp(i\Omega t)]\,[\sigma \exp(-i\omega t) + \sigma^\dagger \exp(i\omega t)]$$
$$= \hbar g\{a\sigma \exp[-i(\Omega + \omega)t] + a\sigma^\dagger \exp[-i(\Omega - \omega)t]$$
$$+ a^\dagger\sigma \exp[i(\Omega - \omega)t] + a^\dagger\sigma^\dagger \exp[i(\Omega + \omega)t]\}\,. \quad (65)$$

Here, as noted earlier, we see that the energy-conserving terms $a\sigma^\dagger$ and $a^\dagger\sigma$ are multiplied by slowly varying factors $\exp[\pm i(\Omega - \omega)t]$, whereas the energy-nonconserving terms $a\sigma$ and $a^\dagger\sigma^\dagger$ are multiplied by the rapidly varying terms $\exp[\pm i(\Omega + \omega)t]$. A time integration of the Schrödinger equation (6.63) leads to a resonant denominator for the former and to an antiresonant denominator for the latter, which we neglect in the rotating-wave approximation as discussed in Chap. 2. Hence our interaction Hamiltonian reduces to

$$\mathscr{V}_I(t) = \hbar g\{a\sigma^\dagger \exp[-i(\Omega - \omega)t] + a^\dagger\sigma \exp[i(\Omega - \omega)t]\}\,. \quad (66)$$

We now proceed to solve the equation of motion (6.63) with (66) for $|\psi\rangle$, starting from two initial conditions: (53), for which the field has n photons and the atom is in the upper state $|a\rangle$, and

$$|\psi(0)\rangle = |b, n + 1\rangle, \quad (67)$$

for which there are $n + 1$ photons and the atom is in the lower state. These two cases give probabilities for emission and absorption, respectively. Inasmuch as we are using the interaction picture, we use the slowly varying probability amplitudes $C_{a,n}$ and $C_{b,n+1}$ in place of $c_{a,n}$ and $c_{b,n+1}$. Because the interaction energy (66) can only cause transitions between the states $|a\,n\rangle$ and $|b\,n + 1\rangle$, the state vector (52) reduces to

$$|\psi(t)\rangle = C_{a,n}|a\,n\rangle + C_{b,n+1}|b\,n + 1\rangle, \quad (68)$$

and the equation of motion (6.63) becomes

$$|\dot{\psi}\rangle = \dot{C}_{a,n}(t)|a\,n\rangle + \dot{C}_{b,n+1}(t)|b\,n + 1\rangle$$
$$= -\,ig\{a\sigma^\dagger \exp[-i(\Omega - \omega)t] + \text{adjoint}\}$$
$$\times\,[C_{a,n}|a\,n\rangle + C_{b,n+1}|b\,n + 1\rangle]. \quad (69)$$

Projecting (69) onto $\langle a\,n|$ we have the equation of motion†

$$\dot{C}_{an}(t) = -ig\sqrt{n + 1}\,\exp[-i(\Omega - \omega)t]\,C_{b,n+1}(t). \quad (70)$$

Similarly,

$$\dot{C}_{b,n+1}(t) = -ig\sqrt{n + 1}\,\exp[i(\Omega - \omega)t]\,C_{an}(t). \quad (71)$$

For an atom initially in the lower state $|b\,n + 1\rangle$[see (68) and Fig. 14-4*a*], the amplitudes are given by

†Here, for typographical simplicity, we omit the comma in $C_{a,n}$ since there is no ambiguity.

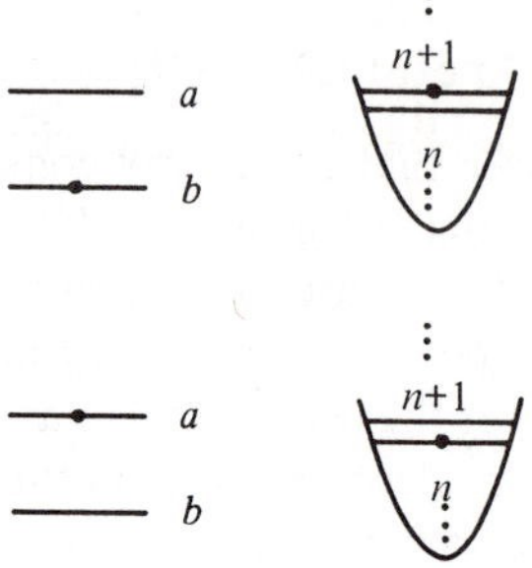

Figure 14-4. Initial conditions of atom-field system for (a) absorption and (b) emission.

$$C_{an}(0) = 0 \qquad \text{and} \qquad C_{b,n+1}(0) = 1. \tag{72}$$

A formal integral of (70) yields

$$C_{an}(t) = -ig\sqrt{n+1}\int_0^t dt'\, C_{b,n+1}(t') \exp[-i(\Omega - \omega)t']. \tag{73}$$

with $C_{b,n+1}(t') \simeq 1$(first-order time-dependent perturbation theory), this is

$$C_{an}(t) \simeq -ig\sqrt{n+1}\left\{\frac{\exp[-i(\Omega-\omega)t] - 1}{-i(\Omega-\omega)}\right\}, \tag{74}$$

which gives the probability of absorption

$$|C_{an}(t)|^2 = g^2(n+1)t^2\left\{\frac{\sin^2[(\Omega-\omega)t/2]}{[(\Omega-\omega)t/2]^2}\right\}. \tag{75}$$

Similarly, for an atom initially in the upper state (53), the amplitudes are given by

$$C_{an}(0) = 1, \qquad C_{b,n+1}(0) = 0 \tag{76}$$

(see Fig. 14-4*b*), and we find the approximate probability for emission:

$$|C_{b,n+1}(t)|^2 \simeq g^2(n+1)t^2\left\{\frac{\sin^2[(\Omega-\omega)t/2]}{[(\Omega-\omega)t/2]^2}\right\}. \tag{77}$$

Here the factor n can be attributed to stimulated emission and the 1 to spontaneous emission, for, in contrast with the absorption in a field of $n + 1$ photons, we have emission stimulated by only n photons. In fact, even if no photons are present, there is still a (resonant) probability for emission, namely,

$$|C_{b,1}(t)|^2 = g^2t^2. \tag{78}$$

This is the short-term probability for spontaneous emission into a single-mode field. In Sec. 14-4 we consider spontaneous emission into a many-mode field, a situation which corresponds more closely to physical reality.

The reader will recognize the approximation in (74) and (77) as first-order perturbation theory. Let us now solve the equations of motion (70) and (71) exactly in extension of the Rabi solution of Sec. 2-4. For simplicity we consider resonance($\Omega = \omega$). Taking the derivative of (70) and using (71), we have

$$\begin{aligned} \ddot{C}_{an}(t) &= -ig\sqrt{n+1}\,\dot{C}_{b,n+1}(t) \\ &= -g^2(n+1)\,C_{an}(t). \end{aligned} \tag{79}$$

This has the general solution

$$C_{an}(t) = A\sin(g\sqrt{n+1}\,t) + B\cos(g\sqrt{n+1}\,t). \tag{80}$$

Combining this with (70), we have the amplitude

$$\begin{aligned} C_{b,n+1}(t) &= i(g\sqrt{n+1})^{-1}g\sqrt{n+1}\,[A\cos(g\sqrt{n+1}\,t) \\ &\qquad - B\sin(g\sqrt{n+1}\,t)] \\ &= iA\cos(g\sqrt{n+1}\,t) - iB\sin(g\sqrt{n+1}\,t), \end{aligned} \tag{81}$$

For absorption, we have the initial condition (72), which with (81) gives $A = -i$ and with (80) gives $B = 0$. Thus

$$C_{a,n}(t) = -i\sin(g\sqrt{n+1}\,t), \tag{82}$$

$$C_{b,n+1}(t) = \cos(g\sqrt{n+1}\,t). \tag{83}$$

The corresponding probabilities are plotted in Fig. 14-5. The factor $g\sqrt{n+1}$ is the Rabi flopping frequency, Similarly, for emission, the initial condition (77) implies $B = 1$ and $A = 0$, for which the amplitudes

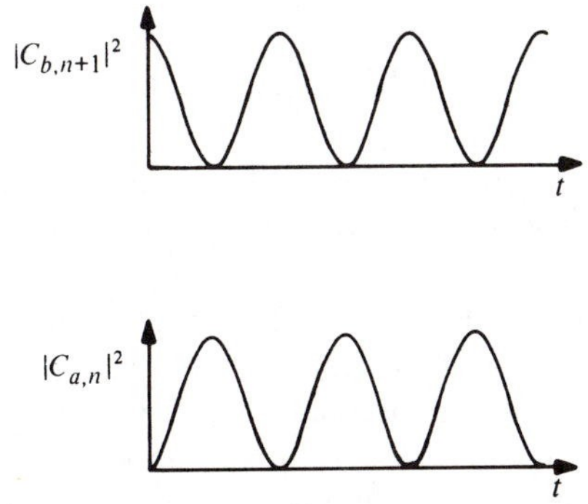

Figure 14-5. Probabilities [from (82) and (83)] for being in the upper (a) and lower (b) states as a function of time when initially in the lower state. The probabilities for the initial condition of being in the upper state are interchanged.

$$C_{a,n}(t) = \cos(g\sqrt{n+1}\,t), \tag{84}$$

$$C_{b,n+1}(t) = -i\sin(g\sqrt{n+1}\,t), \tag{85}$$

and the probabilities given in Fig. 14-5 are interchanged.

Recall now the semiclassical treatment of Rabi flopping (Sec. 2-4), according to which the flopping frequency (2.61) had the resonance value $\wp E_0/\hbar$. Comparing this to $g\sqrt{n+1}$, we see that E_0 corresponds to $\mathscr{E}\sqrt{n+1}$. For large fields, the factor 1 can be neglected compared to n and the correspondence $E_0 \longleftrightarrow \mathscr{E}\sqrt{n}$ made. For small fields and specifically for zero fields, however, this correspondence and, hence, the semiclassical theory, are invalid.

14-4. Weisskopf-Wigner Theory of Spontaneous Emission

We have found that even in the absence of an applied field an atom initially in the upper state makes transitions back and forth to the lower state in time. Experimentally, it is found that an excited atom does make a transition to a lower state and ultimately to the ground state, but does not flop back and forth. In physical reality our model of a single-mode field is often insufficient since in many cases there is a continuum of modes corresponding to a quantization cavity which is infinite in extent. Consequently our discussion of spontaneous emission must include many modes of the field. Usually the atom decays in a characteristic lifetime which, according to the uncertainty principle, limits the precision with which we can measure the energy. Specifically, the longest time during which energy measurements can be made is approximately the atomic lifetime τ, and hence the energy difference of the levels is uncertain by the amount

$$\Delta E = \frac{\hbar}{\tau}. \tag{86}$$

This relation implies that the spontaneously emitted radiation is not perfectly monochromatic, but instead has a frequency spectrum with width inversely proportional to τ.

A more detailed explanation of this line broadening lies in the fact that, when an atom emits, it creates a radiation field which acts back on the atom. For a single-mode field, this interaction produces two frequencies, one slightly above and one below the uncoupled resonance frequencies of either oscillator, just as two weakly coupled pendulums with identical resonant frequencies have shifted normal modes of oscillations. For a multimode field, this interaction produces a band of shifted frequencies which in the simple two-level atom case considered here has a Lorentzian profile.

We will show shortly that an atom does decay with a characteristic lifetime by integrating the Schrödinger equation in the Weisskopf-Wigner approximation. As for the single-mode case, it is convenient first to find equations of motion for the probability amplitudes of a general state vector for the atom-field system.

The Hamiltonian for this system

$$\mathscr{H} = \sum_s \hbar\Omega_s(a_s{}^\dagger a_s + \tfrac{1}{2}) + \hbar\begin{pmatrix}\omega_a & 0\\ 0 & \omega_b\end{pmatrix} + \hbar\sum_s g_s(\sigma a_s{}^\dagger + \sigma^\dagger a_s). \qquad (87)$$

In the interaction picture, the perturbation energy becomes

$$\mathscr{V}(\mathrm{t}) = \hbar\sum_s g_s\{\sigma a_s{}^\dagger \exp[i(\Omega_s - \omega)t] + \sigma^\dagger a_s \exp[-i(\Omega_s - \omega)t]\}, \qquad (88)$$

and the general state vector in the interaction picture is†

$$|\psi\rangle = \sum_{\alpha=a,b}\sum_{n_1}\sum_{n_2}\cdots\sum_{n_s=0}^{\infty}\cdots C_{\alpha,n_1n_2\cdots n_s\cdots}|\alpha\, n_1n_2\ldots n_s\ldots\rangle. \qquad (89)$$

The eigenkets in (89) represent the state for which the atom is in the αth level (a or b), and the first mode of the field has n_1 photons, the second n_2 photons, the third n_3 photons, and so on, for an infinite number of modes. Consequenctly there are an infinite number of summations over photon numbers, and each individually runs from 0 to∞. Projecting the Schrödinger equation (6.63) onto $\langle\alpha, n_1\ldots n_s\ldots|$ and inserting a complete set of states between $\mathscr{V}$ and $|\psi\rangle$, we find ($\alpha = a$ or b)

$$\begin{aligned}\dot{C}_{\alpha,n_1\cdots n_s\cdots} &= -\frac{i}{\hbar}\sum_{m_1}\sum_{m_2}\cdots\sum_{m_r}\cdots\sum_{\beta}\\ &\times\langle\alpha, n_1\ldots n_s\ldots|\sum_r\mathscr{V}_r|\beta, m_1\ldots m_r\ldots\rangle\\ &\times C_{\beta,m_1\cdots m_r\cdots},\end{aligned} \qquad (90)$$

where m_r represents m_r photons in the rth mode. The single-mode energy $\mathscr{V}_r$ affects only the rth mode, and hence the matrix element in (90) reduces to

$$\begin{aligned}&\sum_r\langle\alpha, n_r|\mathscr{V}_r|\beta, m_r\rangle\langle n_1\ldots n_{r-1}, n_{r+1}\ldots|m_1\ldots m_{r-1},\ldots m_{r+1}\ldots\rangle\\ &= \sum_r\langle\alpha, n_r|\mathscr{V}_r|\beta, m_r\rangle\delta_{n_1m_1}\cdots\delta_{n_{r-1}m_{r-1}}\delta_{n_{r+1}m_{r+1}}\cdots.\end{aligned} \qquad (91)$$

The delta symbols in (91) reduce all summations over photon numbers in (90) to single terms except that for the rth mode. Hence (90) becomes

$$\dot{C}_{\alpha,n_1\cdots n_s\cdots} = -\frac{i}{\hbar}\sum_\beta\sum_r\sum_{m_r}\langle\alpha,n_r|\mathscr{V}_r|\beta,m_r\rangle C_{\beta,m_1\cdots m_r\cdots}. \qquad (92)$$

The matrix elements for single modes are evaluated as in Sec. 14-3. Here the more general matrix elements

†The use of α in the eigenvectors should not be confused with the coherent state $|\alpha\rangle$ of Chap. 15.

$$\langle a|\sigma|\beta\rangle = \delta_{ab}\delta_{\beta a}, \tag{93}$$

$$\langle a|\sigma^\dagger|\beta\rangle = \delta_{aa}\delta_{\beta b}, \tag{94}$$

are useful. Finally we find the general equation of motion:

$$\dot{C}_{a,n_1}\ldots{}_{n_s}\ldots(t) = -i\sum_r g_r\{\sqrt{n_r}\exp[i(\Omega_r-\omega)t]\,\delta_{ab}\,C_{a,n_1}\ldots{}_{n_r-1}\ldots$$
$$+\sqrt{n_r+1}\exp[-i(\Omega_r-\omega)t]\,\delta_{aa}\,C_{b,n_1}\ldots{}_{n_r+1}\ldots\}. \tag{95}$$

Now suppose that at time $t = 0$ the atom is in the upper state $|a\rangle$ and there are no photons in any mode of the field. For this, the probability amplitude

$$C_{a,0_10_2}\cdots{}_{0_s}\cdots(0) \equiv C_{a,\{0\}}(0) = 1 \tag{96}$$

and all others are zero. In time the probability amplitude

$$C_{b,0_1}\cdots{}_{1_s}\cdots(t) \equiv C_{b,\{1_s\}}(t) \tag{97}$$

for each s may acquire a nonzero value. The equation of motion (95) for the amplitude in (96) reduces to

$$\dot{C}_{a,\{0\}} = -i\sum_r g_r\exp[-i(\Omega_r-\omega)t]\,C_{b,\{1_r\}}. \tag{98}$$

Furthermore the equations for the amplitudes (97) reduce to

$$\dot{C}_{b,\{1_s\}} = -ig_s\exp[i(\Omega_s-\omega)t]\,C_{a,\{0\}}. \tag{99}$$

Substituting a formal integral for (99) into (98), we obtain the integro-differential equation

$$\dot{C}_{a,\{0\}} = -\sum_r g_r^2\int_0^t dt'\exp[-i(\Omega_r-\omega)(t-t')]\,C_{a,\{0\}}(t'). \tag{100}$$

Now, assuming that the modes of the field are closely spaced in frequency, we replace the summation over r by an integral over Ω:

$$\sum_r \to \int d\Omega\,\mathfrak{D}(\Omega), \tag{101}$$

where $\mathfrak{D}(\Omega)$ is the density of final states, $V\Omega^2/(\pi^2c^3)$. This density and the coupling constant $g(\Omega)$ are proportional to powers of Ω and therefore vary little in the frequency interval τ^{-1} about ω, for which the time integral in (100) is nonnegligible.† This fact makes it possible to evaluate the product $\mathfrak{D}(\Omega)g(\Omega)$ for $\Omega = \omega$ and to factor it outside the integral over Ω. We find

$$\dot{C}_{a,\{0\}} = -g^2(\omega)\mathfrak{D}(\omega)\int d\Omega\int_0^t dt'\exp[-i(\Omega-\omega)(t-t')]\,C_{a,\{0\}}(t'). \tag{102}$$

The integral

$$\int_0^t dt'\exp[-i(\Omega-\omega)(t-t')] = \pi\delta(\Omega-\omega) - \mathscr{P}\left(\frac{i}{\Omega-\omega}\right), \tag{103}$$

†A number $\tau^{-1}\sim 10^8$ Hz is being compared to an optical frequency like 3×10^{14} Hz.

which with (102) yields the equation (without the imaginary part†)

$$\dot{C}_{a,\{0\}}(t) = -\tfrac{1}{2}\gamma_a C_{a,\{0\}}(t), \tag{104}$$

where the decay rate

$$\gamma_a = 2\pi g^2(\omega)\,\mathfrak{D}(\omega). \tag{105}$$

Equation (104) has the integral $|C_{a,\{0\}}(t)|^2 = \exp(-\gamma_a t)$, which in the excited state $|a\rangle$ in zero field decays exponentially in time with the lifetime $\tau = 1/\gamma_a$. This is what we set out to prove.

We use the density operator in much of our work because we deal with open systems which cause the system of interest to be described by a mixture of state vectors. In particular, the density operator makes it possible to ignore certain parts of a problem. For example, suppose that we desire to know the equation of motion for the probability that an atom is in the ground state $|b\rangle$ because of spontaneous emission. The state vector for the problem is

$$|\psi(t)\rangle = C_{a,\{0\}}(t)|a\rangle|\{0\}\rangle + \sum_r C_{b,\{1\}_r}(t)|b\rangle|\{1_r\}\rangle. \tag{106}$$

Here we mean by $\{1_r\}$ one photon in the rth mode and none in the other modes. Hence the probability that the atom is in $|b\rangle$, regardless of what mode the emitted photon is in, is given by the sum

$$\sum_r |C_{b,\{1_r\}}(t)|^2. \tag{107}$$

Since $\langle\psi|\psi\rangle = 1$, we have

$$|C_{a,\{0\}}(t)|^2 = 1 - \sum_r |C_{b,\{1_r\}}(t)|^2,$$

which with (104) gives the equation of motion,

$$\frac{d}{dt}\left\{\sum_r |C_{b,\{1_r\}}(t)|^2\right\} = -\frac{d}{dt}|C_{a,\{0\}}(t)|^2 = \gamma_a |C_{a,\{0\}}(t)|^2. \tag{108}$$

We can represent this dissipation easily by using the density operator. For the complete system of atom and radiation, we have for (7.17)

$$\begin{aligned}\rho_{\text{a-f}}(t) &= \sum_\psi P_\psi |\psi\rangle\langle\psi| \\ &= \sum_\psi P_\psi \{|C_{a,\{0\}}(t)|^2 |a\rangle\langle a| \otimes |\{0\}\rangle\langle\{0\}| \\ &\quad + \sum_r |C_{b,\{1_r\}}(t)|^2 |b\rangle\langle b| \otimes |\{1_r\}\rangle\langle\{1_r\}|\} \\ &\quad + \text{terms off field diagonal.}\end{aligned} \tag{109}$$

The "atom only" density operator is given by the trace over the field states:

$$\begin{aligned}\rho_{\text{atom}} &= \mathrm{Tr}_{\text{field}}\,\rho_{\text{a-f}}(t) \\ &= \rho_{aa}|a\rangle\langle a| + \rho_{bb}|b\rangle\langle b|,\end{aligned} \tag{110}$$

†The imaginary part can lead to a level shift. See Sakurai (1967), Sec. 2-8. In principle one should consider 3 dimensions. See Sakurai for complete development.

where the matrix elements

$$\rho_{aa} = \sum_{\psi} P_{\psi} |C_{a,\{0\}}|^2, \tag{111}$$

$$\rho_{bb} = \sum_{\psi} P_{\psi} \sum_{r} |C_{b,\{1_r\}}|^2. \tag{112}$$

Note that the coherence terms like ρ_{ab} are lost in this process, that is, no single state vector can represent the atom-only system. The term ρ_{atom} is called a reduced density operator, an operator of considerable importance in Chap. 16. For now, we simply note that Eq. (108) becomes

$$\dot{\rho}_{bb} = -\dot{\rho}_{aa} = \gamma_a \rho_{aa}. \tag{113}$$

Problems

14-1. Show for the electric (6) and magnetic (7) fields that the Hamiltonian (8) has the form for a simple harmonic oscillator (9).

14-2. Show that the Hamiltonian (9) when taken to be an operator is given by (13) in terms of the annihilation and creation operators (11) and (12).

14-3. Show that the root-mean-square deviation of the electric field operator (16) $\Delta E = [\langle E^2\rangle - \langle E\rangle^2]^{1/2}$ for the state

$$|\psi\rangle = \frac{1}{\sqrt{2}}[|n\rangle + \exp(-i\Omega t)|n+1\rangle] \tag{114}$$

is given by

$$\Delta E = \mathscr{E}\sqrt{n+1} \sin Kz \,[1 + \sin^2 \Omega t]^{1/2}. \tag{115}$$

14-4. Calculate the electric field expectation value $\langle E\rangle$ for (*a*) a single-mode field (16) in state $|\psi\rangle$ of (35) and (*b*) the multimode field (44) in state (48).

Ans. to (a):

$$\langle E\rangle = \mathscr{E} \sin Kz \sum_{n} \sqrt{n+1}\, c_{n+1}{}^{*} c_n + \text{c.c.} \tag{116}$$

14-5. Show that

$$\exp(i\Omega a^{\dagger}at)\, a \exp(-i\Omega a^{\dagger}at) = a \exp(-i\Omega t) \tag{117}$$

by proving first, using mathematical induction, the validity of (63).

14-6. Show for the unperturbed Hamiltonian

$$\mathscr{H}_0 = \sum_{n} \hbar\omega_n |n\rangle\langle n|$$

that

$$\exp(i\mathscr{H}_0 t/\hbar) = \sum_{n} \exp(i\omega_n t)|n\rangle\langle n|. \tag{118}$$

References

See references for Chap. 6 and the following:

N. N. Bogoliubov and D. V. Shirkov, 1959, *Introduction to the Theory of Quantized Fields,* Academic Press, New York.

R. H. Dicke, 1954, *Phys. Rev.* **93**, 121.

K. Gottfried, 1966, *Quantum Mechanics,* W. A. Benjamin, New York.

W. Heitler, 1956, *The Quantum Theory of Radiation,* Oxford University Press, New York.

F. A. Kaempfer, 1965, *Concepts in Quantum Mechanics,* Academic Press, New York.

G. Källen, 1972, *Quantum Electrodynamics,* Springer-Verlag, New York (trans. by C. Iddings and M. Mizushima).

R. Loudon, 1974, *The Quantum Theory of Light,* Oxford University Press, New York.

W. H. Louisell, 1964, *Radiation and Noise in Quantum Electronics,* McGraw-Hill Book Co., New York.

E. A. Power, 1970, *Introduction to Quantum Electrodynamics and Radiation,* McGraw-Hill Book Co., New York.

J. J. Sakurai, 1967, *Advanced Quantum Mechanics,* Addison-Wesley Publishing Co., Reading, Mass.

S. S. Schweber, 1962, *An Introduction to Relativistic Quantum Field Theory,* Harper and Row, New York.

M. O. Scully and W. E. Lamb, Jr., 1967, *Phys. Rev.* **159**, 208.

M. O. Scully and M. Sargent III, 1972, *Phys. Today,* **25**, 38.

V. F. Weisskopf and E. Wigner, 1930, *Z. Phys.* **63**, 54.

XV

COHERENT STATES

15. Coherent States of Radiation Field

Having developed a formalism for the quantum theory of radiation, we will find it both interesting and profitable to determine what wave function most nearly describes the classical electromagnetic field given by $E_0 \cos(\nu t + \phi)$ and $H_0 \sin(\nu t + \phi)$. This field is supposedly known with absolute certainty, whereas the quantum operators for the electric and magnetic fields are associated with nonzero uncertainties obeying the relation

$$\Delta E \, \Delta H \geq \tfrac{1}{2}\hbar \left(\frac{2\Omega^2}{VK}\right) \tag{1}$$

(symbols defined in Sec. 14-1), just as the position q and momentum p obey the relation $\Delta q \, \Delta p \geqslant \frac{1}{2}\hbar$. We propose that the wave function which corresponds most closely to the classical field must have minimum uncertainty [equality in (1)] for all time when subject to the appropriate simple harmonic potential. We will find in Appendix H that minimum uncertainty merely at one time is not sufficient, for if the width of the wave packet is initially chosen too small, it becomes too large at a later time, as illustrated in Fig. 15-1*b*. Furthermore, it fails to satisfy the conditions for minimum uncertainty except at isolated points in time. If the right width is chosen for the potential, however, the packet bounces back and forth sinusoidally without changing width, as shown in Fig. 15-2*b*. This latter packet "coheres," always has minimum uncertainty, and resembles (Fig. 15-2*a*) the classical field as closely as quantum mechanics permits.

As Schrödinger first showed, this "coherent" wave function is just that of a displaced ground state of the simple harmonic oscillator. The corresponding state vector is the "coherent state" $|\alpha\rangle$ of Glauber (1963), which is the eigenstate of the positive frequency part of the electric field operator. The eigenvalue $\mathscr{E}\alpha$ can be identified with the classical $\frac{1}{2}E_0 \exp(-i\phi)$. The coherent state

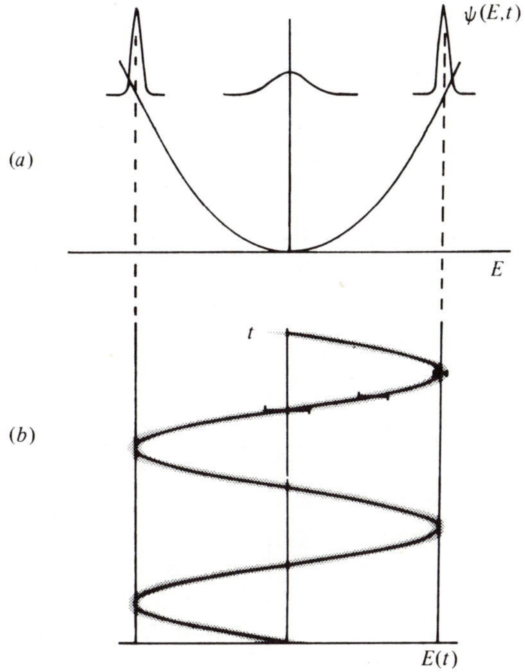

Figure 15-1. (a) Non-minimum-uncertainty wave packet $\psi^*\psi$ given by Eq. (H. 42). (b) Corresponding electric field with limits of 1/e points of wave packet.

is not simply a photon number state, a state yielding vanishing expectation value for the field. Rather, it is given by an appropriate superposition of these energy eigenstates. In fact, as shown in Sec. 3-1 for a two-level atom, a superposition of at least two energy eigenstates is required to give an oscillating probability distribution in that case.

In Appendix H, we derive the coherent wave function by requiring that the function have minimum uncertainty for all time. Section 15-1 and the problems develop various properties of the coherent state $|\alpha\rangle$. In Sec. 15-2 we introduce the multimode extension of the coherent state and show that the radiation from a classical current is described by such a many-mode state. In Sec. 15-3 we give the $R(\alpha^*, \beta)$ and $P(\alpha)$ representatives of the density operator ρ for the radiation field. The P representation is the more useful and is defined by

$$\rho = \int d^2\alpha \, P(\alpha) \, |\alpha\rangle\langle\alpha| , \tag{2}$$

in which the integration extends over the complex plane. In particular, the

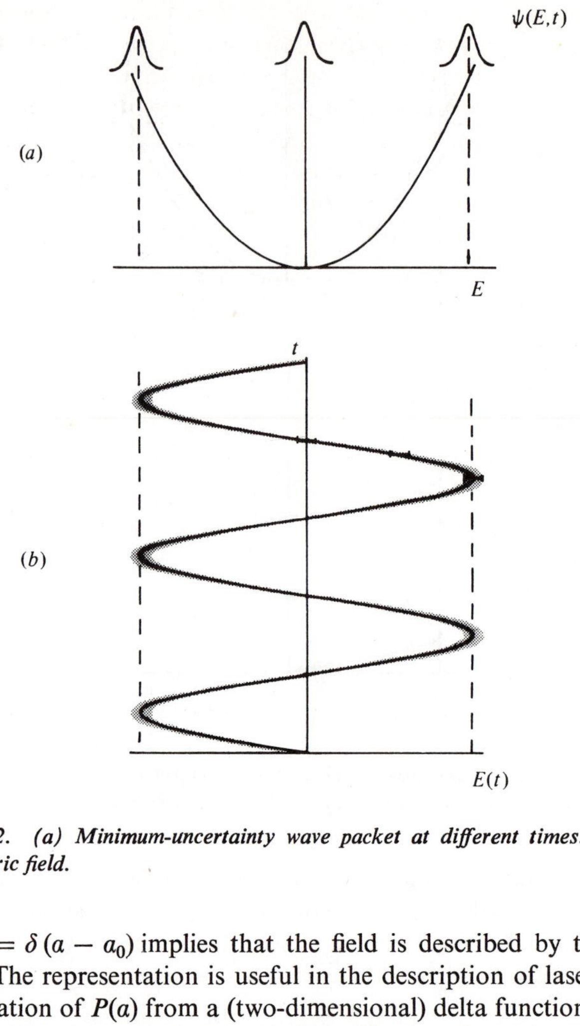

Figure 15-2. (a) Minimum-uncertainty wave packet at different times. (b) Corresponding electric field.

value $P(\alpha) = \delta(\alpha - \alpha_0)$ implies that the field is described by the coherent state $|\alpha_0\rangle$. The representation is useful in the description of laser radiation, for the deviation of $P(\alpha)$ from a (two-dimensional) delta function determines the degree of incoherence of the radiation. The time dependence of the $P(\alpha)$ distribution is discussed in Sec. 16-3. The present chapter closes with a comparison between the photon statistics of thermal and of coherent radiation.

15-1. Properties of Coherent States

The coherent state for the radiation field is defined as that whose uncertainty product $\Delta E \, \Delta H$ is minimum for all time when subject to the simple harmonic potential characteristic of the field. The corresponding wave packet coheres (does not spread) in time. The derivation of the explicit value [Eq.

(11)] for this state and of its time development is given in Appendix H. We summarize that treatment here for convenience. In our discussion we use the coordinate q and momentum p operators instead of the electric and magnetic field operators. They are related, of course, by Eqs. (14.6) and (14.7).

The derivation begins with a review of the uncertainty principle, which provides the conditions for minimum uncertainty. The principle states that the product of the mean-squared deviations $\Delta q\,\Delta p$ is restricted by the inequality

$$\Delta q\,\Delta p \geq \tfrac{1}{2}\hbar, \tag{3}$$

where the deviations

$$(\Delta q)^2 = \langle(q - \langle q\rangle)^2\rangle, \qquad (\Delta p)^2 = \langle(p - \langle p\rangle)^2\rangle. \tag{4}$$

The corresponding product (1) for the electric and magnetic field operators results from inserting relations (14.6) and (14.7) into (3). In physical terms, the positions and momenta of many identically prepared systems, that is, of systems described by the same state vector, are measured. One multitude of measurements may be used to obtain the expectation values for $(\Delta q)^2$ in (4) and another for $(\Delta p)^2$. We wish to emphasize that the uncertainty relation does not apply to a single measurement, for it involves expectation values, that is, ensemble averages resulting from many measurements. Quantum mechanics provides no way of treating "simultaneous" measurements of q and p except in terms of the ensemble averages described above.

The conditions for minimum uncertainty [equality in (3)] yield a simple differential equation for the wave function $\psi(q)$. The solution is a Gaussian with arbitrary width and center. The time development of this function subject to a simple harmonic potential can be obtained by using the appropriate Green's function (See. H-2). Choosing the width so that the function maintains minimum uncertainty in time, we find (Sec. H-2) that

$$\psi(q, t) = \exp[\tfrac{1}{2}(ae^{-i\Omega t})^2 - |a|^2)]\,\phi_0(\xi - \sqrt{2}ae^{-i\Omega t}), \tag{5}$$

where the complex number

$$a = (2M\hbar\Omega)^{-1/2}\,(M\Omega\langle q\rangle + i\langle p\rangle), \tag{6}$$

the dimensionless coordinate

$$\xi = \left(\frac{M\Omega}{\hbar}\right)^{1/2} q, \tag{7}$$

and ϕ_0 is the lowest simple harmonic oscillator eigenfunction (1.22), which the reader will recall is Gaussian. Equation (5) yields the probability density

$$\psi^*\psi = \{\phi_0[\xi - \sqrt{2}\,|a|\cos(\Omega t + \phi)]\}^2, \tag{8}$$

where we have written

$$a = |a|\exp(-i\,\phi)$$

In terms of the electric field, (8) is written as

$$|\psi(E,t)|^2 = (\sqrt{2\pi}\mathscr{E})^{-1} \exp\{-[(\sqrt{2}\,\mathscr{E})^{-1} E - \sqrt{2}\,|a| \cos(\Omega t + \phi)]^2\}. \quad (9)$$

Thus the electric field oscillates back and forth with constant spread $\sqrt{2}\mathscr{E}$ and maximum extent $2\mathscr{E}|a|$. This has been depicted in Fig. 15-2. It is particularly interesting that the spread depends on the photon energy $\hbar\Omega$ and the volume of the cavity, but not on the magnitude of the displacement a.

The oscillation is analogous to that of a bob on a pendulum initially displaced as depicted in Fig. 15-3. The bob is released, and the pendulum oscillates back and forth in simple harmonic motion.

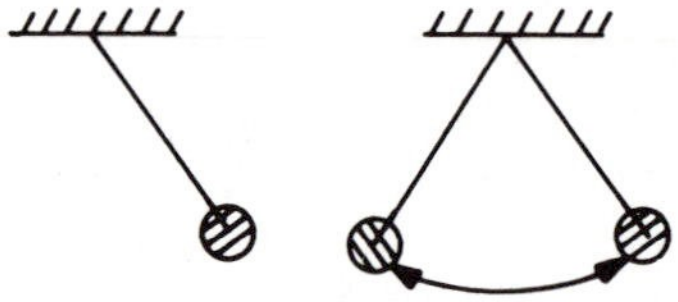

Figure 15-3. Oscillation of pendulum analogous to coherent state.

In the number representation, the coherent wave function is found to be (Sec. H-3)

$$\psi(q) = \sum_{n=0}^{\infty} \left(\frac{a^n}{\sqrt{n!}}\right) \exp(-\tfrac{1}{2}|a|^2)\, \phi_n(\xi), \quad (10)$$

which is the coordinate representative of the coherent state

$$|a\rangle = \sum_{n=0}^{\infty} \left[\frac{a^n}{\sqrt{n!}}\right] \exp(-\tfrac{1}{2}|a|^2)|n\rangle. \quad (11)$$

This state is the eigenstate of the annihilation operator with eigenvalue a, for

$$\begin{aligned} a|a\rangle &= a \sum_{n=1}^{\infty} \left[\frac{a^{n-1}}{\sqrt{n!}}\right] \exp(-\tfrac{1}{2}|a|^2)\,\sqrt{n}\,|n-1\rangle \\ &= a|a\rangle. \end{aligned} \quad (12)$$

Noting that the positive frequency part of the electric field operator (14.16) is $\mathscr{E}a$, we see that the state $\mathscr{E}|a\rangle$ is the eigenstate of the complex electric field. It is this state which most closely corresponds to the classical electric field, $E_0 \cos(\Omega t + \phi)$.

We now discuss characteristics of the coherent states $|a\rangle$, such as their completeness, degree of orthogonality, and relationship to the photon number states $|n\rangle$. Starting with this last property first, we see that the probability for finding n photons in a coherent state $|a\rangle$ is given by

$$P_n(a) = |\langle n|a\rangle|^2 = \left[\frac{(a^*a)^n}{n!}\right] \exp(-a^*a), \tag{13}$$

which is a Poisson distribution with average number

$$\langle n\rangle = a^*a. \tag{14}$$

As we shall see in Sec. 17-2, the corresponding probability for the laser approaches this distribution for sufficiently high excitations but is in general broader (less peaked), as illustrated in Fig. 17-3. In thermal equilibrium, the radiation in the single-mode cavity has a Boltzmann distribution (7.25) with the ambient temperature. The coherent and thermal distributions are compared in Fig. 15-4.

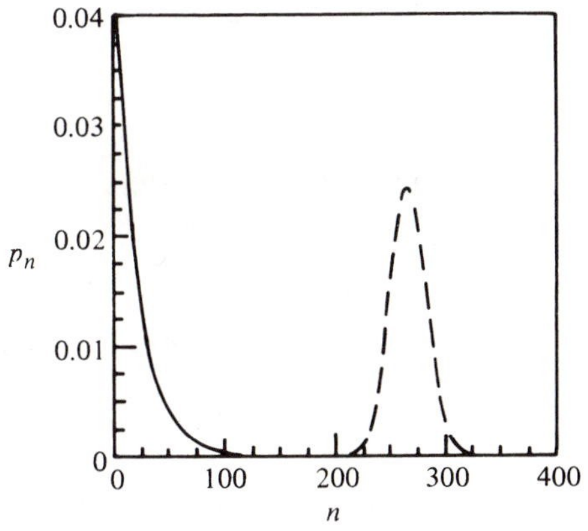

Figure 15-4. Comparison of blackbody (thermal) distribution, given by solid line, and the coherent-state distribution (dashed line) of (13).

A valuable formula for the coherent state used in the derivation in Appendix H is

$$|a\rangle = \exp(aa^\dagger - \tfrac{1}{2}a^*a)|0\rangle, \tag{15}$$

which follows immediately from the relation

$$|n\rangle = \left[\frac{(a^\dagger)^n}{\sqrt{n!}}\right]|0\rangle,$$

used in the derivation of Eq. (11) and the Maclaurin series $\exp(x) = \sum_n x^n/n!$ Using formula (15) and the derivative

$$\frac{d}{da}[\exp(aa^\dagger - \tfrac{1}{2}a^*a)] = [a^\dagger - \tfrac{1}{2}a^*]\exp(aa^\dagger - \tfrac{1}{2}a^*a),$$

we find

$$a^\dagger|a\rangle = \left(\frac{\partial}{\partial a} + \tfrac{1}{2}a^*\right)|a\rangle. \tag{16}$$

The coherent states satisfy the completeness relation:

$$\pi^{-1}\int d^2a|a\rangle\langle a| = 1. \tag{17}$$

In fact, the differential

$$d^2a = d[\mathrm{Re}(a)]\,d[\mathrm{Im}(a)] = d(|a|)|a|\,d\phi \tag{18}$$

which leads to

$$\int d^2a|a\rangle\langle a| = \int_0^\infty d|a|\int_0^{2\pi}|a|\,d\phi\sum_m\sum_n\left[\frac{|a|^m}{\sqrt{m!}}\right]\exp(im\phi)$$
$$\times\left[\frac{|a|^n}{\sqrt{n!}}\right]\exp(-in\phi)\exp(-|a|^2)|m\rangle\langle n|. \tag{19}$$

Inasmuch as

$$\int_0^{2\pi} d\phi\exp[i(m-n)\phi] = 2\pi\delta_{nm},$$

Eq. (19) reduces to

$$\begin{aligned}\int d^2a\;|a\rangle\langle a| &= \pi\sum_n(n!)^{-1}\int_0^\infty 2|a|\,d(|a|)\;|a|^{2n}\exp(-|a|^2)|n\rangle\langle n|\\ &= \pi\sum_n(n!)^{-1}\int_0^\infty dx\,x^n\exp(-x)|n\rangle\langle n|\\ &= \pi\sum_n|n\rangle\langle n| = \pi,\end{aligned} \tag{20}$$

which yields (17).

The completeness property is essential for the utility of a set of states. A very useful, although nonessential, property is orthogonality. The coherent states are not orthogonal, for

$$\begin{aligned}\langle\beta|a\rangle &= \sum_n\sum_m\left[\frac{\beta^{*m}}{\sqrt{m!}}\right]\left[\frac{a^n}{\sqrt{n!}}\right]\exp[-\tfrac{1}{2}(|a|^2+|\beta|^2)]\langle m|n\rangle\\ &= \exp[-\tfrac{1}{2}(|a|^2+|\beta|^2)+\beta^*a]\end{aligned} \tag{21}$$

and

$$|\langle\beta|a\rangle|^2 = \exp(-|\beta-a|^2). \tag{22}$$

Here we see that, if the magnitude of $a-\beta$ is much greater than unity, the states $|a\rangle$ and $|\beta\rangle$ are nearly orthogonal to one another. The degree to which these wave functions overlap one another determines the size of the inner product $\langle\beta|a\rangle$ (see Fig. 15-5). A consequence of the lack of orthogo-

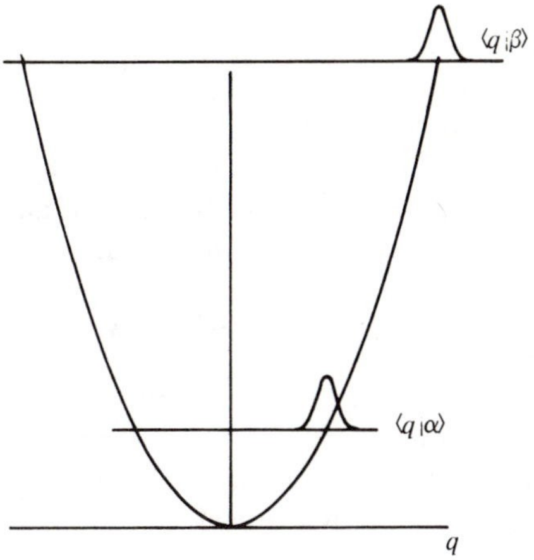

Figure 15-5. Graph showing degree of orthogonality of two states $|\alpha\rangle$ and $|\beta\rangle$ having substantially different magnitudes.

nality given by (22) is the fact that any coherent state can be expanded in terms of the other states:

$$|\alpha\rangle = \pi^{-1} \int |\beta\rangle \exp(-\tfrac{1}{2}|\alpha|^2 - \tfrac{1}{2}|\beta|^2 + \beta^*\alpha)\, d^2\beta. \tag{23}$$

This indicates that the coherent states are "overcomplete." Additional properties of the coherent states are discussed in the paper by Glauber (1963) and in the problems.

15-2. Radiation from a Classical Current

In this section we define the multimode coherent state and show that the radiation emitted by a classical current distribution is such a state. By "classical," we mean that the current can be described by a prescribed vector $\mathbf{J}(\mathbf{r}, t)$, that is, not an operator.

In Sec. 14-2 we found that the multimode energy eigenstates (14.46) are outer products of the single-mode eigenstates. Similarly we define the multimode coherent states as the products of single-mode coherent states:

$$|\{\alpha_k\}\rangle \equiv \prod_k |\alpha_k\rangle, \tag{24}$$

where for notational convenience we have used the braces to stand for the set of all α_k. The eigenvalue equation (12) then becomes the system of equations

$$a_s|\{\alpha_k\}\rangle = \alpha_s|\{\alpha_k\}\rangle, \tag{25}$$

and the completeness relation is

$$\int |\{\alpha_k\}\rangle\langle\{\alpha_k\}| \prod_k \frac{d^2\alpha_k}{\pi} = 1. \tag{26}$$

Let us now consider radiation by a system with a current specified by the classical vector $\mathbf{J}(\mathbf{r}, t)$. Complications introduced by operator algebra (commutation relations, etc.) do not oecur for $\mathbf{J}(\mathbf{r}, t)$, although they will for the field operators. We also consider the multimode vector potential operator

$$\mathbf{A}(\mathbf{r}, t) = c\sum_n \left(\frac{\frac{1}{2}\hbar}{\Omega_k}\right)^{1/2} [a_k\mathbf{U}_k(\mathbf{r}) \exp(-i\Omega_k t) + \text{adjoint}] \tag{27}$$

instead of the electric field operator, for here the current distribution may not be confined to a region which satisfies the electric-dipole approximation. The Hamiltonian which describes the interaction between the field and the current is then given by (in first order)

$$\mathscr{V}(t) = -\frac{1}{c}\int \mathbf{J}(\mathbf{r}, t) \cdot \mathbf{A}(\mathbf{r}, t)\, d^3r, \tag{28}$$

and the state vector for the combined system obeys the interaction picture Schrödinger equation (6.63).

The vector function $\mathbf{J}(\mathbf{r}, t)$ commutes with itself at different times, but the operator $\mathbf{A}(\mathbf{r}, t)$ does not. Hence the interaction energy $\mathscr{V}(t)$ does not either, and ordinarily the Schrödinger equation cannot be integrated as

$$|\psi(t)\rangle = \exp\left[-i\hbar^{-1}\int_{-\infty}^{t} dt'\mathscr{V}(t')\right]|\psi(0)\rangle. \tag{29}$$

However, as discussed by Glauber, the various commutators introduced in obtaining the correct integration yield (29) multiplied by an overall phase factor which we discard. With (27) and (28), the exponential in (29) becomes

$$\exp\left[-\frac{i}{\hbar}\int_{-\infty}^{t} dt'\mathscr{V}(t')\right] = \prod_k \exp(\alpha_k a_k^\dagger - \alpha_k^* a_k), \tag{30}$$

where the complex time-dependent amplitudes

$$\alpha_k = i(2\hbar\Omega_k)^{1/2}\int_{-\infty}^{t} dt' \int d^3r\, U_k^*(\mathbf{r})\cdot\mathbf{J}(\mathbf{r}, t)\exp(i\Omega_k t'). \tag{31}$$

We choose the initial state $|\psi(0)\rangle$ to be the vacuum $|0\rangle$, and note that

$$e^{A+B} = e^A e^B e^{-[A,B]/2} \tag{32}$$

when the commutator $[A, B]$ commutes with the operators A and B. With (30) and (32), the state vector (29) becomes

$$\begin{aligned}|\psi(t)\rangle &= \prod_k \exp(\alpha_k a_k^\dagger - \alpha_k^* a_k)|0\rangle \\ &= \prod_k \exp(\alpha_k a_k^\dagger)\exp(-\alpha_k^* a_k)\exp(-\tfrac{1}{2}\alpha_k\alpha_k^*)|0\rangle_k\end{aligned}$$

$$= \prod_k \exp(-\tfrac{1}{2}a_k a_k^*) \exp(a_k a_k^\dagger)|0\rangle_k$$

$$= \prod_k |a_k\rangle_k. \tag{33}$$

Thus we see that a classically oscillating current distribution radiates a coherent state. This confirms our identification of the coherent state with the classical field.

15-3. Density Operator Expansions—$P(a)$, $R(a^*, \beta)$

The radiation field is, in general, described by the density operator defined in Chap. 7. As stated earlier, this operator is just the outer product (7.16)

$$\rho = |\psi\rangle\langle\psi|$$

when the complete state $|\psi\rangle$ of the system (here the field) is known. However, when one has only the statistical knowledge that the system has probability P_ψ of being in state $|\psi\rangle$, the density operator is given by (7.17)

$$\rho = \sum_\psi P_\psi |\psi\rangle\langle\psi|.$$

The density operator often is more easily applicable than the state vector, and it is useful to expand ρ in terms of the coherent states. The values of the expansion coefficients express the degree to which the field is actually coherent in the correct quantum-mechanical sense.

Note first the more obvious expansion in terms of the photon number states:

$$\rho = \mathscr{I} \cdot \rho \cdot \mathscr{I} = \sum_n \sum_m |n\rangle\langle n|\rho|m\rangle\langle m| = \sum_n \sum_m \rho_{nm}|n\rangle\langle m|, \tag{34}$$

where $\mathscr{I}$ is the identity operator (6.13). Similarly,

$$\rho = \mathscr{I} \cdot \rho \cdot \mathscr{I} = \pi^{-2}\int d^2a \int d^2\beta \; |a\rangle\langle a|\rho|\beta\rangle\langle\beta|$$

$$= \pi^{-2}\int d^2a \int d^2\beta \, R(a^*, \beta) \exp(-\tfrac{1}{2}|a|^2 - \tfrac{1}{2}|\beta|^2)|a\rangle\langle\beta|, \tag{35}$$

where the expansion function

$$R(a^*, \beta) = \langle a|\rho|\beta\rangle \exp(\tfrac{1}{2}|a|^2 + \tfrac{1}{2}|\beta|^2). \tag{36}$$

Here $R(a^*, \beta)$ is just $\rho_{a\beta}$ apart from a normalization factor included to make the relation between $R(a^*, \beta)$ and ρ_{nm} a simple one. In fact, writing $|a\rangle$ and $|\beta\rangle$ in terms of number states, we find for (36)

$$R(a^*, \beta) = \sum_n \sum_m \frac{(a^*)^n \beta^m}{\sqrt{n!\, m!}} \rho_{nm}. \tag{37}$$

Thus, when the matrix elements ρ_{nm} are known, $R(a^*,\beta)$ follows immediately from Eq. (37).

For example, the density operator for single-mode thermal radiation is given by

$$\rho = \frac{\exp(-\hbar\Omega a^\dagger a/k_B T)}{Tr\{\exp(-\hbar\Omega a^\dagger a/k_B T)\}}, \tag{38}$$

which has matrix elements

$$\begin{aligned}\rho_{nm} &= \exp(-n\hbar\Omega/k_B T)\,[\sum_n \exp(-n\hbar\Omega/k_B T)]^{-1}\delta_{nm}\\ &= \exp(-n\hbar\Omega/k_B T)\,[1-\exp(-\hbar\Omega/k_B T)]\delta_{nm}.\end{aligned} \tag{39}$$

Combining this with (37), we have

$$\begin{aligned}R(a^*,\beta) &= [1-\exp(-\hbar\Omega/k_B T)]\sum_n (a^*\beta)^n \exp(-n\hbar\Omega/k_B T)/n!\\ &= [1-\exp(-\hbar\Omega/k_B T)]\exp[a^*\beta\exp(-\hbar\Omega/k_B T)].\end{aligned} \tag{40}$$

Although this expansion of the density operator is straightforward, it is somewhat messy. We can often obtain a simpler expansion with real coefficents $P(a)$:

$$\rho = \int d^2a\, P(a)\,|a\rangle\langle a|. \tag{41}$$

If the coherent states were orthogonal, this would be a diagonal expansion with limited application. The overlap of the coherent states, however, results in a rather general expansion.

The expectation value of an operator $\mathcal{O}$ is given by $\mathrm{Tr}(\rho\mathcal{O})$, as discussed in Sec. 7-2. We can find the expectation value in terms of $P(a)$ as follows. Using (41), we find

$$\begin{aligned}\langle\mathcal{O}\rangle = \mathrm{Tr}(\rho\mathcal{O}) &= \sum_n \langle n|\int d^2a\, P(a)\,|a\rangle\langle a|\mathcal{O}\,|n\rangle\\ &= \int d^2a\, P(a)\sum_n \langle a|\mathcal{O}\,|n\rangle\langle n|a\rangle\\ &= \int d^2a\, P(a)\,\mathcal{O}(a),\end{aligned} \tag{42}$$

where $\mathcal{O}(a) = \langle a|\,\mathcal{O}\,|a\rangle$. Here we have assumed the validity of (41). To show the result independently, we expand ρ and $\mathcal{O}$ in terms of the annihilation and creation operators. Specifically we write the density operator in antinormal order (annihilation operators precede creation operators):

$$\rho^{(a)}(a, a^\dagger) = \sum_i\sum_j \rho_{ij}{}^{(a)} a^i (a^\dagger)^j, \tag{43}$$

and the operator $\mathcal{O}$ in normal order (creation operators precede annihilation operators):

$$\mathcal{O} = \sum_l \sum_m \mathcal{O}_{lm}{}^{(n)} (a^\dagger)^l a^m, \tag{44}$$

where the superscripts (a) and (n) have been used to indicate the particular ordering. Using (43) and (44), we find for the expectation value in (42)

$$\begin{aligned}\langle\mathcal{O}\rangle &= \mathrm{Tr}\{\sum_i\sum_j\sum_l\sum_m \rho_{ij}{}^{(a)}\mathcal{O}_{lm}{}^{(n)}a^i(a^\dagger)^{j+l}a^m\} \\ &= \int d^2\alpha \sum \rho_{ij}{}^{(a)}\mathcal{O}_{lm}{}^{(n)} \langle\alpha|(a^\dagger)^{j+l}a^{i+m}|\alpha\rangle \\ &= \int d^2\alpha \sum_i\sum_j\sum_l\sum_m \rho_{ij}{}^{(a)}\alpha^i\alpha^{*j}\mathcal{O}_{lm}{}^{(n)}\alpha^{*l}\alpha^m \\ &= \int d^2\alpha P(\alpha)\mathcal{O}(\alpha).\end{aligned} \tag{45}$$

Here we have used the fact that operators can be cyclically interchanged underneath the trace operation. We see that the expectation value in (42) can be written in terms of a function $P(\alpha)$, given by

$$P(\alpha) = \sum_i\sum_j \rho_{ij}{}^{(a)}\alpha^i(\alpha^*)^j = \rho^{(a)}(\alpha, \alpha^*), \tag{46}$$

and consequently we identify the density operator as in (41). In this derivation, we have assumed that the power series (43) and (44) exist, an assumption which leads sometimes to weird coefficients like the nth derivative of the Dirac delta function (see Prob. 15-6).

As an example, we derive $P(\alpha)$ for the thermal field. This is most easily accomplished by noting from (41) and (22) that

$$\begin{aligned}\langle\alpha|\rho|\alpha\rangle &= \int d^2\beta\, P(\beta)\langle\alpha|\beta\rangle\langle\beta|\alpha\rangle \\ &= \int d^2\beta\, P(\beta)\exp(-|\alpha-\beta|^2),\end{aligned} \tag{47}$$

that is, $\langle\alpha|\rho|\alpha\rangle$ is a convolution of $P(\alpha)$ with a Gaussian. A simple derivation. (Prob. 15–5) reveals that in the general case $\langle\alpha|\rho|\alpha\rangle$ can be used to calculate the expectation values of antinormally ordered operators in the same fashion that $P(\alpha)$ is used to calculate those for normally ordered operators. We can determine $P(\alpha)$ by taking the Fourier transform† of (47), which by the inversion theorem for convolutions is the product

$$\mathcal{F}[\langle\alpha|\rho|\alpha\rangle] = \mathcal{F}[P(\beta)]\mathcal{F}[\exp(-|\alpha|^2)]. \tag{48}$$

Dividing through by $\mathcal{F}[\exp(-|\alpha|^2)]$ and taking the inverse transform, we have

---

†Our notation is illustrated by

$$g(k_x, k_y) = \mathcal{F}[f(\alpha)] = \int_{-\infty}^{\infty} dx \int_{-\infty}^{\infty} dy\, f(x, y)\exp(-ik_x x - ik_y y),$$

where $\alpha = x + iy$.

$$P(\beta) = \mathscr{F}^{-1}\left\{\frac{\mathscr{F}[\langle a|\rho|a\rangle]}{\mathscr{F}[\exp(-|a|^2)]}\right\}. \tag{49}$$

In fact, from (34) and (39), we have, setting $x = \hbar\omega/k_{\rm B}T$,

$$\begin{aligned}\langle a|\rho|a\rangle &= \sum_n\sum_m \langle a|n\rangle\rho_{nm}\langle m|a\rangle \\ &= [1-\exp(-x)]\sum_n \exp(-nx)\langle a|n\rangle\langle n|a\rangle \\ &= [1-\exp(-x)]\sum_n \exp(-nx)\left[\frac{(a^*a)^n}{n!}\right]\exp(-a^*a) \\ &= [1-\exp(-x)]\exp(-|a|^2)\exp(|a|^2e^{-x}) \\ &= [1-\exp(-x)]\exp[-|a|^2(1-e^{-x})].\end{aligned}$$

The factor $[1-\exp(-x)]$ can be conveniently expressed in terms of the average number of photons given by (7.81) as

$$\langle n\rangle = \frac{1}{\exp(x)-1} = \frac{1}{\exp(\hbar\Omega/k_{\rm B}T)-1}. \tag{51}$$

Inverting this expression, we have

$$1-\exp(-x) = (\langle n\rangle+1)^{-1}$$

and hence

$$\langle a|\rho|a\rangle = (\langle n\rangle+1)^{-1}\exp[-|a|^2/(\langle n\rangle+1)]. \tag{52}$$

Following the procedure given in (48) and (49), we take the Fourier transform of (52). This is a two-dimensional transform which separates into the product of two one-dimensional transforms, that is,

$$\begin{aligned}\mathscr{F}(\langle a|\rho|a\rangle) &= (\langle n\rangle+1)^{-1}\mathscr{F}\{\exp[-x^2/(\langle n\rangle+1)]\} \\ &\quad\times \mathscr{F}\{\exp[-y^2/(\langle n\rangle+1)]\} \\ &= \pi\exp[-\tfrac{1}{4}k^2(\langle n\rangle+1)],\end{aligned} \tag{53}$$

where we have set $|a|^2 = \mathrm{Re}(a)^2 + \mathrm{Im}(a)^2 = x^2+y^2$. Similarly, we have

$$\mathscr{F}[\exp(-|a|^2)] = \pi\exp(-\tfrac{1}{4}k^2). \tag{54}$$

Substituting this and Eq. (53) into (49), we find

$$\begin{aligned}P(a) &= \mathscr{F}^{-1}[\exp(-\tfrac{1}{4}k^2/\langle n\rangle)] \\ &= (\pi\langle n\rangle)^{-1}\exp(-|a|^2/\langle n\rangle).\end{aligned} \tag{55}$$

It is interesting to note that, in the classical limit of large $\langle n\rangle$, $\langle a|\rho|a\rangle$ given by (52) and $P(a)$ coincide. Also, in the classical limit, distinctions depending on the ordering of operators vanish. This point is discussed further in Glauber's *Les Houches Lectures* (1965).

The probability for finding n photons is given by the photon statistical distribution ρ_{nn}. As shown in Fig. 15-4, the distributions for thermal light given by (39) and for coherent light (13) are dramatically different. The difference between the $P(a)$ for the two cases is perhaps even more striking, for thermal light has a Gaussian dependence (55) on the magnitude of a, whereas the coherent state is a delta function:

$$P(a) = \delta(a - a_0). \tag{56}$$

We have interpreted $P(a)$ as a probability distribution for finding the coherent state $|a\rangle$. This is not always correct, however, for $P(a)$ may take on negative values or singularities worse than the delta function. In particular, whenever the photon distribution ρ_{nn} is narrower than the Poisson distribution for the coherent state, $P(a)$ becomes badly behaved. Such a case is the photon number state $|n\rangle$ (see Prob. 15-6).

Problems

15-1. The electric field at a point $\mathbf{r}$ at time t can be decomposed into positive and negative frequency components as follows:

$$E(\mathbf{r}, t) = E^{(+)}(\mathbf{r}, t) + E^{(-)}(\mathbf{r}, t), \tag{57}$$

where for single mode the positive frequency term is given by

$$E^{(+)}(\mathbf{r}, t) = i\,(\tfrac{1}{2}\hbar\Omega)^{1/2}\, au\,(\mathbf{r})\exp(-i\Omega t) \tag{58}$$

and the negative frequency part is given by the adjoint. A useful quantity in the description of the coherence properties of a field with density operator is the nth-order correlation function

$$G^{(n)}(x_1, \ldots, x_n, x_{n+1}, \ldots, x_{2n})$$
$$= \mathrm{Tr}\{\rho E^{(-)}(x_1) \cdots E^{(-)}(x_n)\, E^{(+)}(x_{n+1}) \cdots E^{(+)}(x_{2n})\}, \tag{59}$$

where for simplicity we have used x_j for the combination $\mathbf{r}_j$, t_j. Show that, when the field is described by a coherent state $|a\rangle$, this correlation function factors in the form

$$G^{(n)} = \mathscr{E} a^*(x_1) \cdots a^*(x_n)\, a(x_{n+1}) \cdots a(x_{2n}),$$

where $a(x_j) = a\,u(r_j)\exp(-i\Omega t_j)$.

15-2. Show that

$$\frac{1}{\pi}\int \exp(\beta^* a - |a|^2)\, f(a^*)\, d^2a = f(\beta^*), \tag{60}$$

where $f(\beta)$ is a function which has a power series expansion.

15-3. Calculate the expectation value of the displacement operator

$$D(a) = \exp(aa^\dagger - a^* a) \tag{61}$$

for a thermal field. *Hint*: Show first that

$$\langle D(\alpha)\rangle = \mathrm{Tr}\,[\rho D(\alpha)] = \frac{1}{\pi\langle n\rangle}\int d^2\beta \exp(-|\beta|^2/\langle n\rangle)\langle\beta|D(\alpha)|\beta\rangle \quad (62)$$

and then use Eq. (32) to evaluate $\langle\beta|D(\alpha)|\beta\rangle$.

Ans.:

$$\langle D(\alpha)\rangle = \exp[-|\alpha|^2(\langle n\rangle + \tfrac{1}{2})]. \quad (63)$$

15-4. Determine the expansion coefficients c_n for the coherent state written in terms of the photon number states $|\alpha\rangle = \sum_n c_n|n\rangle$, by projecting the eigenvalue equation $a|\alpha\rangle = \alpha|\alpha\rangle$ onto the eigenbra $\langle n|$.

15-5. Show that the distribution $\langle\alpha|\rho|\alpha\rangle$ can be used to calculate the expectation values of antinormally ordered operators when ρ itself is normally ordered. The proof parallels that for $P(\alpha)$, given in (45).

15-6. Show that

$$P(\alpha) = \left[\frac{n!}{2\pi r(2n)!}\right]\exp(r^2)\left(-\frac{\partial}{\partial r}\right)^{2n}\delta(r), \quad (64)$$

where $\alpha = r\exp(i\theta)$, yields the photon number state $\rho = |n\rangle\langle n|$.

15-7. Show for the thermal distribution (39) that the simple harmonic oscillator density matrix in the coordinate representation is

$$\rho(x, x_0) = \left[\frac{M\omega}{2\pi\hbar\sinh(\hbar\Omega/k_\mathrm{B}T)}\right]^{1/2}\exp\left\{-\frac{M\omega}{2\hbar\sinh(\hbar\Omega/k_\mathrm{B}T)}\right.$$
$$\left.\times\,[(x^2 + x_0^2)\cosh(\hbar\omega/k_\mathrm{B}T) - 2xx_0]\right\}. \quad (65)$$

Hint: Equate the two values of the Green functions given in Appendix H [(Eqs. (H. 21) and (H. 29)].

References

R. J. Glauber, 1963, *Phys. Rev.* **130**, 2529, and **131**, 2766.

R. J. Glauber, 1965, in: *Quantum Optics and Electronics*, Les Houches, Edited by C. DeWitt, A. Blandin and C. Cohen-Tannoudji, Gordon and Breach, New York, pp. 331–381.

R. J. Glauber, ed., 1969, *Quantum Optics*, Proceedings of the International School of Physics, "Enrico Fermi," Course XLII, Academic Press, New York.

S. M. Kay and A. Maitland, eds., 1970, *Quantum Optics*, Proceedings of the 10th Session of Scottish Universities Summer School in Physics 1969, Academic Press, London.

J. R. Klauder and E. C. G. Sudarshan, 1970, *Fundamentals of Quantum Optics*, W. A. Benjamin, New York.

W. H. Louisell, 1964, *Radiation and Noise in Quantum Electronics*, McGraw-Hill Book Co., New York.

L. Mandel and E. Wolf, 1970, *Selected Papers on Coherence and Fluctuations of Light*, Dover Publications, New York.

H. M. Nussenzveig, 1974, *Introduction to Quantum Optics*, Gordon and Breach, New York.

XVI

RESERVOIR THEORY – DENSITY OPERATOR METHOD

16. Reservoir Theory — Density Operator Method

Our treatment of the quantum theory of radiation so far has predominantly involved entire systems, such as the electromagnetic field in a cavity. In most areas of quantum optics, however, we are interested in only part of the entire system. For example, in the laser problem, we want to know the field but are not particularly interested in what happens to the atoms. We find it convenient to separate the system of primary interest from that (or those) of secondary interest and to call the former simply the "system" and the latter the "reservoir." We can eliminate the reservoir by using the *reduced* density operator method in the Schrödinger (or interaction) picture or the noise operator method in the Heisenberg picture. In this chapter we discuss the density operator method, which provides computational convenience while stressing the statistical aspects of the problem. In Chap. 19 we use the noise operator approach, which often requires more calculation but offers a direct physical appeal in its resemblance to the classical Brownian motion problem.

In Sec. 16-1 we introduce the reservoir concept by considering a system consisting of a simple harmonic oscillator (e.g., radiation field) interacting with a reservoir of resonant, two-level atoms. The field is taken to vary little during the lifetime of an atom (as in the semiclassical theory of Chap. 8), allowing the use of a "coarse-grained" time rate of change for the field density matrix. The atom-field probability amplitudes derived in Chap. 14 are used in the derivation. With a good knowledge of this section, the reader is prepared for the quantum theory of the laser in Chap. 17.

In Sec. 16-2 we carry out a more general approach in which the density operator is used for the entire caculation. The formalism is illustrated by the atomic beam reservoir problem of Sec. 16-1 and allows an easy calculation of the coherent-state representation. This latter equation of motion has the form of a two-dimensional Fokker-Planck equation. In Sec. 16-3 we develop the classical Fokker-Planck equation for the probability density $P(v, t)$ for a particle with velocity v in Brownian motion and for a random walk problem. The equations are discussed pictorially and are related to the quantum version of Sec. 16-2. In Sec. 16-4 a general density operator approach is used which places emphasis on the correlation developed between two interacting systems. The theory is applied to the damping of a field and of a two-level atom by a reservoir of simple harmonic oscillators. These cases are also discussed in the problems from the point of view of Sec. 16-2.

16-1. Light Field Damping by Atomic Beam Reservoir

To illustrate the reservoir concept in simple terms, we couple a simple harmonic oscillator representing a time-dependent electric field to a beam of two-level atoms, some of which are initially excited. The atomic energy distribution is characterized by a temperature T given by the Boltzmann distribution:

$$\frac{r_a}{r_b} = \exp(-\hbar\omega/k_B T), \tag{1}$$

where r_a and r_b are the numbers of atoms per second in the upper and lower levels which pass through the cavity (see Fig. 16-1). The effect of the beam on the simple harmonic oscillator is to bring it to the same temperature T, as we shall see through the derivation in this section.

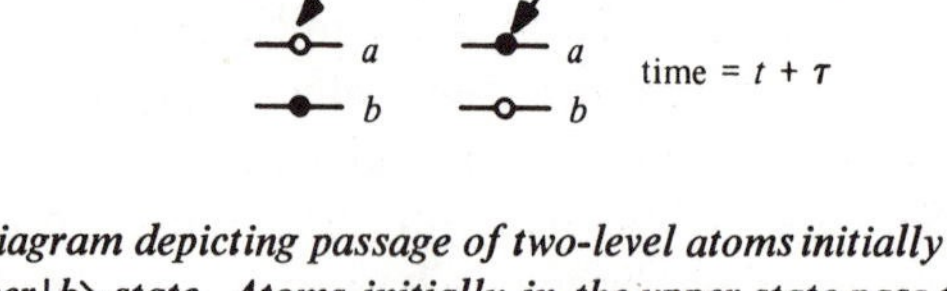

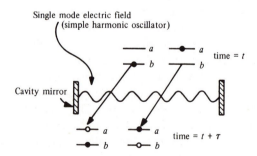

Figure 16-1. Diagram depicting passage of two-level atoms initially (time = t) in either the upper |a⟩ or lower |b⟩ state. Atoms initially in the upper state pass through the cavity at the rate r_a per second, and r_b atoms initially in the lower state pass through per second. The atoms interact with the field, bringing it to an equilibrium temperature T defined by (1), a fact ultimately demonstrated in Eq. (30).

In general, the field is described by a mixture of states conveniently represented by the field density operator:

$$\rho(t) = \sum_{\psi} P_{\psi}(t) |\psi(t)\rangle\langle\psi(t)| . \tag{2}$$

Here P_{ψ} is the probability that the field has the state vector $|\psi\rangle$.

To derive the equation of motion for this density operator, we consider first the small change produced in the transit of a single atom, namely,

$$\delta\rho(t) = \rho(t + \tau) - \rho(t), \tag{3}$$

for which $\rho(t + \tau)$ is determined from an integration of a Schrödinger equation for the coupled atom-field system, as will be discussed shortly. A "coarse-grained" time rate of change for $\rho(t)$ is then given by the sum

$$\dot{\rho}(t) = r_a(\delta\rho)_a + r_b(\delta\rho)_b, \tag{4}$$

where $(\delta\rho)_a$ and $(\delta\rho)_b$ are the changes caused by the interaction with atoms injected in the upper and lower states, respectively. Equation (4) is an average time rate of change and, as such, does not describe fluctuations about this average which are responsible for the atomic contribution to shot noise in, for example, a laser. To treat these latter effects, we could include in (4) a random noise operator of the sort discussed in Chap. 19.

Two methods of integration for the coupled atom-field system are worth mentioning. For the first, the combined atom-field density operator $\rho_{\text{a-f}}(t)$ is found at time $t + \tau$ by means of an equation of motion (7.29) with the initial condition

$$\rho_{\text{a-f}}(t) = \rho(t) \otimes |\beta\rangle\langle\beta| . \tag{5}$$

Here an "outer product" (as distinguished from an inner product) of the initial field density operator $\rho(t)$ is made with the initial density operator $|\beta\rangle\langle\beta|$ for an atom injected into the energy eigenstate $|\beta\rangle$. The field operator $\rho(t + \tau)$ is then given by the trace over atomic coordinates:

$$\rho(t + \tau) = \text{Tr}_{\text{atom}} \{\rho_{\text{a-f}}(t + \tau)\} . \tag{6}$$

This field density operator can be used to compute the expectation value of an arbitrary operator $\mathcal{O}_{\text{f}}$ which depends on field coordinates alone, for

$$\mathcal{O}_{\text{f}} = \text{Tr}_{\text{a-f}} \{\mathcal{O}_{\text{f}}\rho_{\text{a-f}}\} = \text{Tr}_{\text{field}} \{\mathcal{O}_{\text{f}}\rho\} .$$

The tracing operation in (6) automatically leaves $\rho(t + \tau)$ in a mixture of states even if $\rho(t)$ described a pure case (single state vector). Hence an equation of the form (2) for $\rho(t + \tau)$ would involve different P_{ψ}'s from those in (2). This process of contraction (tracing) provides a fundamental, quantum-mechanical source of noise in the field. We will see more of this in Sec. 16-2 (and in Chap. 17) when field diffusion is considered.

A second integration method consists of integrating the *state vector* $|\psi_{\text{atom-field}}(t)\rangle$ for the atom coupled to a *single* state of the field. This problem was solved in Sec. 14-3 by the use of ordinary time-dependent perturbation theory. The resulting atom-field density operator is given by

$$\rho_{\text{a-f}}(t+\tau) = \sum_{\psi} P_{\psi}(t)\,|\psi_{\text{a-f}}(t+\tau)\rangle\langle\psi_{\text{a-f}}(t+\tau)|, \tag{7}$$

which is then used in (5) for $\rho(t+\tau)$. The initial condition for the combined atom-field state vector is the outer product

$$|\psi_{\text{a-f}}(t)\rangle = |\psi(t)\rangle|\beta\rangle, \tag{8}$$

where the state vector $|\psi(t)\rangle$ is any one of those appearing in (2). Inasmuch as this second method is considerably easier for the simple reservoir interaction at hand, we use it here and postpone further discussion of the straight density operator method to the general theory of Sec. 16-2. It is interesting to note that a very realistic model of the laser consists of simple extensions of the field-beam problem, and we will need in Chap. 17 no more than our present method to give a fairly general quantum theory of the laser. In that development we use the strong-signal results of Sec. 14-3. Here we need consider only second-order perturbation theory solutions on resonance.

Our initial field state vector (at time t of atom injection) is given in the number representation and interaction picture as the superposition (note that C_n depends on the ψ in question)

$$|\psi(t)\rangle = \sum_n C_n(t)|n\rangle. \tag{9}$$

Corresponding to this is the initial field density operator:

$$\begin{aligned}\rho(t) = \sum_{\psi} P_{\psi}|\psi\rangle\langle\psi| &= \sum_{\psi} P_{\psi}\sum_n\sum_m C_n C_m^*|n\rangle\langle m|\\ &\equiv \sum_n\sum_m \rho_{nm}(t)|n\rangle\langle m|.\end{aligned} \tag{10}$$

The initial atom-field state vector is just the (outer) product:

$$\begin{aligned}|\psi_{\text{a-f}}(t)\rangle &= |\psi_{\text{atom}}(t)\rangle|\psi(t)\rangle\\ &= |\psi_{\text{atom}}(t)\rangle\sum_n C_n(t)|n\rangle.\end{aligned} \tag{11}$$

From (14.70) and (14.71), we know that the atom-field probability amplitudes $C_{a,n}(t)$ and $C_{b,n+1}(t)$ are coupled together in time according to the (on resonance) equations of motion

$$\dot{C}_{a,n}(t) = -\,ig(n+1)^{1/2}\,C_{b,n+1}(t), \tag{12}$$

$$\dot{C}_{b,n+1}(t) = -\,ig(n+1)^{1/2}\,C_{a,n}(t). \tag{13}$$

Hence we write the atom-field state vector at the later time $t+\tau$ as

$$|\psi_{\text{a-f}}(t+\tau)\rangle = \sum_n [C_{a,n}(t+\tau)|a\rangle|n\rangle + C_{b,n+1}(t+\tau)|b\rangle|n+1\rangle]. \tag{14}$$

Suppose first that the atoms are injected into the lower state, that is, $|\psi_{\text{atom}}(t)\rangle = |b\rangle$. Then the initial state vector for the combined atom-field system is

$$|\psi_{a-f}(t)\rangle = \sum_n C_{n+1}(t)|n+1\rangle|b\rangle, \tag{15}$$

where we have used $n + 1$ instead of n to interface with the results in Sec. 14-3. Hence the probability amplitudes $C_{a,n}(t)$ and $C_{b,n+1}(t)$ in that section have the initial conditions $C_{an}(t) = 0$, $C_{b,n+1}(t) = C_{n+1}(t)$. Equation (14.73) then gives, to first-order,

$$C_{a,n}(t+\tau) \approx -ig\sqrt{n+1}\,\tau C_{b,n+1}(t) = -ig\tau\sqrt{n+1}\,C_{n+1}(t). \tag{16}$$

The corresponding amplitude for remaining in the lower state is, to second order in the interaction,

$$\begin{aligned} C_{b,n+1}(t+\tau) &\approx [1 - \tfrac{1}{2}g^2\tau^2(n+1)]C_{b,n+1}(t) \\ &= [1 - \tfrac{1}{2}g^2\tau^2(n+1)]C_{n+1}(t). \end{aligned} \tag{17}$$

These solutions could have been obtained, of course, by expansion of the sines and cosines in the exact resonant solution (14.82 and 14.83). The total density matrix elements $\rho_{a,n;a,m}(t+\tau)$ required in the trace of (6) are given by

$$\rho_{a,n;a,m}(t+\tau) = \sum_\psi P_\psi C_{a,n}(t+\tau)C_{a,m}{}^*(t+\tau), \tag{18}$$

in which the sum over state vectors refers to the field mixture. Hence, with (16) and (17), we have

$$\rho_{a,n;a,m}(t+\tau) \simeq g^2\tau^2\sqrt{(n+1)(m+1)}\,\rho_{n+1,m+1}(t) \tag{19}$$

$$\begin{aligned} \rho_{b,n;b,m}(t+\tau) &\simeq (1 - \tfrac{1}{2}g^2\tau^2 n)(1 - \tfrac{1}{2}g^2\tau^2 m)\,\rho_{nm}(t) \\ &\simeq [1 - \tfrac{1}{2}g^2\tau^2(n+m)]\rho_{nm}(t), \end{aligned} \tag{20}$$

that is,

$$\begin{aligned} \rho_{nm}(t+\tau) = {} & g^2\tau^2\sqrt{(n+1)(m+1)}\,\rho_{n+1,m+1}(t) \\ & + [1 - \tfrac{1}{2}g^2\tau^2(n+m)]\rho_{nm}(t). \end{aligned}$$

Hence the coarse-grained time rate of change of $\rho_{nm}(t)$ due to atoms initially in the lower state $|b\rangle$ is

$$\begin{aligned} \dot\rho_{nm}(t)|_{|b\rangle\,\text{atoms}} &= r_b[\rho_{nm}(t+\tau) - \rho_{nm}(t)]_{|b\rangle\,\text{atoms}} \\ &= -\tfrac{1}{2}\mathscr{R}_b(n+m)\rho_{nm} \\ &\quad + \mathscr{R}_b[(n+1)(m+1)]^{1/2}\rho_{n+1,m+1}, \end{aligned} \tag{21}$$

where the rate coefficient

$$\mathscr{R}_b = r_b g^2\tau^2. \tag{22}$$

Similarly, for atoms injected into the $|a\rangle$ eigenstate, we have

$$\begin{aligned} C_{a,n}(t+\tau) &= [1 - \tfrac{1}{2}g^2\tau^2(n+1)]C_n(t), \\ C_{b,n+1}(t+\tau) &= ig\tau(n+1)^{1/2}\,C_n(t), \end{aligned} \tag{23}$$

which gives the contribution to the equation of motion for $\rho_{nm}(t)$:

$$\dot{\rho}_{nm}(t)|_{|a\rangle \text{ atoms}} = -\tfrac{1}{2}\mathscr{R}_a(n+1+m+1)\,\rho_{nm} + \mathscr{R}_a\sqrt{nm}\,\rho_{n-1,m-1}, \tag{24}$$

in which the rate coefficient

$$\mathscr{R}_a = r_a g^2 \tau^2. \tag{25}$$

Hence the total (coarse-grained) time rate of change of the density matrix element ρ_{nm} is given by

$$\begin{aligned}\dot{\rho}_{nm}(t) &= \dot{\rho}_{nm}|_{|a\rangle \text{ atoms}} + \dot{\rho}_{nm}|_{|b\rangle \text{ atoms}} \\ &= -\tfrac{1}{2}[\mathscr{R}_a(n+1+m+1) + \mathscr{R}_b(n+m)]\,\rho_{nm} + \mathscr{R}_a\sqrt{nm}\,\rho_{n-1,m-1} \\ &\quad + \mathscr{R}_b\,[(n+1)(m+1)]^{1/2}\,\rho_{n+1,m+1}.\end{aligned} \tag{26}$$

In particular, this gives the photon rate equation

$$\begin{aligned}\dot{\rho}_{nn} = &-[\mathscr{R}_a(n+1) + \mathscr{R}_b n]\rho_{nn} + \mathscr{R}_a n\rho_{n-1,\,n-1} \\ &+ \mathscr{R}_b(n+1)\rho_{n+1,n+1}.\end{aligned} \tag{27}$$

Here each term is simply understood in terms of the probabilities that atoms do (or do not) make transitions in the presence of a given number of photons. For example,

$$\begin{aligned}\mathscr{R}_a n\rho_{n-1,n-1} = &\ (\text{rate at which atoms are injected into state } |a\rangle) \\ &\times (\text{probability of stimulated emission by an } n-1 \text{ photon field}) \\ &\times (\text{probability of } n-1 \text{ photons}).\end{aligned} \tag{28}$$

We can further understand (27) in terms of Fig. 16-2, in which the flow of photon number probability is depicted. The $\mathscr{R}_a$ terms involve emissions and hence correspond to arrows pointing up in the diagram, while the $\mathscr{R}_b$ arrows

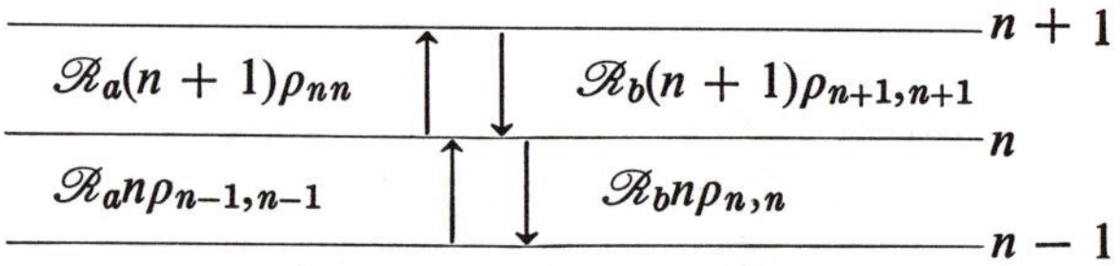

Figure 16-2. Diagram of photon number probability, ρ_{nn}. The $\mathscr{R}_a$ terms are due to emission, and the $\mathscr{R}_b$ to absorption. Each term is simply understood in terms of rates and transition probilities, as discussed in connection with Eq. (28).

point down. Equilibrium is obtained when the net flow between all pairs of levels vanishes, for example,

$$\mathscr{R}_a n \rho_{n-1,n-1} = \mathscr{R}_b n \rho_{nn}. \tag{29}$$

This condition for arbitrary n is referred to as detailed balancing. The solution of (29) with (1) is

$$\begin{aligned}\rho_{nn} &= \frac{\mathscr{R}_a}{\mathscr{R}_b} \rho_{n-1,n-1} \\ &= \frac{r_a}{r_b} \rho_{n-1,n-1} \\ &= \exp(-\hbar\omega/k_\mathrm{B}T)\, \rho_{n-1,n-1}.\end{aligned}$$

Iteration of this result for successively lower n yields the Planck distribution for single mode:

$$\begin{aligned}\rho_{nn} &= \rho_{00} \exp(-n\hbar\omega/k_\mathrm{B}T) \\ &= [1 - \exp(-\hbar\omega/k_\mathrm{B}T)] \exp(-n\hbar\omega/k_\mathrm{B}T), \end{aligned} \tag{30}$$

where we have obtained ρ_{00} by imposing the normalization condition $\sum_n \rho_{nn} = 1$. Thus we see that in the steady state the field comes to the temperature of the atomic beam reservoir, as it should.

Another quantity of interest is the average photon number:

$$\langle n(t) \rangle = \sum_n n \rho_{nn}(t). \tag{31}$$

As shown in Prob. 7-1, the steady-state value of this number is given by the Bose-Einstein expression:

$$\langle n(\infty) \rangle \equiv \bar{n} = \sum_n n \rho_{nn} = \frac{1}{\exp(\hbar\omega/k_\mathrm{B}T) - 1}, \tag{32}$$

in which ρ_{nn} is the steady-state distribution (30). Inserting the equation of motion (27) into (31), we find

$$\begin{aligned}\frac{d}{dt}\langle n(t)\rangle = \sum_n n\dot{\rho}_{nn}(t) = \sum_n [-\mathscr{R}_a(n^2 + n)\, \rho_{nn} - \mathscr{R}_b n^2\, \rho_{nn}(t) \\ + \mathscr{R}_b(n^2 + n)\, \rho_{n+1,n+1}(t) + \mathscr{R}_a n^2\, \rho_{n-1,n-1}].\end{aligned}$$

Letting $m = n + 1$ in the summation over $\rho_{n+1,n+1}$ and $m = n - 1$ in the summation over $\rho_{n-1,n-1}$, we find

$$\begin{aligned}\frac{d}{dt}\langle n(t)\rangle &= -\mathscr{R}_a[\langle n^2\rangle + \langle n\rangle] - \mathscr{R}_b\langle n^2\rangle \\ &\quad + \mathscr{R}_b \sum_{m=1}^{\infty} (m^2 - m)\rho_{m,m} \\ &\quad + \mathscr{R}_a \sum_{m=0}^{\infty} (m^2 + 2m + 1)\, \rho_{mm} \\ &= (\mathscr{R}_a - \mathscr{R}_b)\langle n\rangle + \mathscr{R}_a. \end{aligned} \tag{33}$$

Here the lone $\mathscr{R}_a$ term derives from a 1 in an emission factor $n + 1$ and hence can be regarded as the result of spontaneous emission. This implies a buildup of the radiation until an equilibrium is reached. In steady state, (33) yields

$$\langle n(\infty)\rangle = \frac{\mathscr{R}_a}{\mathscr{R}_b - \mathscr{R}_a} = \bar{n} \tag{34}$$

in agreement (of course) with (32).

Another quantity of particular interest is the electric field expectation value defined by [see (14.16)]

$$\begin{aligned}\langle E(t)\rangle &= \mathscr{E} \sin Kz \operatorname{Tr}\{\rho(t)(a + a^\dagger)\}\\ &= \mathscr{E} \sin Kz \operatorname{Tr}\{\rho(t)a\} + \text{c.c.}\\ &= \mathscr{E} \sin Kz \sum_{n=0}^{\infty} \langle n|\rho(t)a|n\rangle + \text{c.c.}\\ &= \mathscr{E} \sin Kz \sum_{n=0}^{\infty} \sqrt{n}\, \rho_{n,n-1}(t) + \text{c.c.}\end{aligned} \tag{35}$$

Inserting the equation of motion (26) for $\rho_{n,n-1}$ and using the techniques leading to (33), we can show (Prob. 16-4) that the positive frequency part of the field expectation value (35) has the equation of motion

$$\frac{d}{dt}\langle \mathscr{E}a\rangle = -\tfrac{1}{2}\mathscr{R}_b(\bar{n} + 1)^{-1}\langle \mathscr{E}a\rangle - i\nu\langle \mathscr{E}a\rangle. \tag{36}$$

apart from the sin Kz. Thus the average field decays to zero (dephases with zero ensemble average) in time with decreasing rate for larger temperatures. This does not imply that the field intensity vanishes in time; this quantity is proportional to the average photon number $\bar{n}$.

In the laser theory of Chap. 17, we use an atomic beam reservoir to simulate the finite cavity Q. For this, we suppose that the average electric field decays according to

$$\frac{d}{dt}\langle \mathscr{E}(t)\rangle = -\frac{1}{2}\frac{\nu}{Q}\langle \mathscr{E}(t)\rangle - i\nu\langle \mathscr{E}(t)\rangle \tag{37}$$

or, equivalently, that the average photon number has the decay term $-(\nu/Q)$ $\langle n(t)\rangle$. Comparing (37) with (36), we see that the Q is given by

$$\frac{\nu}{Q} = \frac{\mathscr{R}_b}{\bar{n} + 1} = \frac{\mathscr{R}_a}{\bar{n}} = \mathscr{R}_b - \mathscr{R}_a, \tag{38}$$

that is,

$$\mathscr{R}_a = \frac{\nu}{Q}\bar{n}, \qquad \mathscr{R}_b = \frac{\nu}{Q}(\bar{n} + 1). \tag{39}$$

In these terms, the density matrix element ρ_{nm} has the equation of motion

$$\begin{aligned}\dot{\rho}_{nm} = -\frac{1}{2}\frac{\nu}{Q}[2\bar{n}(n + m + 1) + n + m]\,\rho_{nm} + \frac{\nu}{Q}\bar{n}\sqrt{nm}\,\rho_{n-1,m-1}\\ + \frac{\nu}{Q}(\bar{n} + 1)\,[(n + 1)(m + 1)]^{1/2}\rho_{n+1,m+1}.\end{aligned} \tag{40}$$

16-2. Density Operator and Quantum Fokker-Planck Equations of Motion

We now consider the system-reservoir interaction from a totally operator point of view. This method is often more convenient in treating more complicated problems than the two-level atomic beam reservoir interacting with a simple harmonic oscillator. We illustrate the formalism with this simple problem and then derive a corresponding Fokker-Planck equation for the diagonal distribution $P(a)$ of Sec. 15-3. The reader will recall that $P(a)$ is ordinarily the probability density for the coherent state $|a\rangle$, although there exist pathological examples for which $P(a)$ has negative and highly nonsingular values. The Fokker-Planck equation is fairly easy to derive with the operator formalism of this section, although it is relatively difficult to obtain from the photon number representation (ρ_{nm}) found earlier by use of the state vector. The theory can also be used to treat the damping of a field or an atom by a reservoir of oscillators, as discussed in Prob. 16-14.

We consider in general a system denoted by A interacting with a reservoir denoted by B. The combined density operator is denoted by $\rho_{AB}(t)$, and the system reduced operator is given in extension of (6) by the trace over the reservoir coordinates:

$$\rho_A(t) = \mathrm{Tr}_B\{\rho_{AB}(t)\} = \sum_B \langle B|\rho_{AB}(t)|B\rangle. \tag{41}$$

At the initial time t, the system and reservoir are taken to be uncorrelated, and hence the initial value of $\rho_{AB}(t)$ is given by the outer product [like(5)]:

$$\rho_{AB}(t) = \rho_A(t) \otimes \rho_B(t). \tag{42}$$

For the atomic beam problem of Sec. 16-1, we derived a coarse time rate of change for the field-atom combination by writing an appropriate ρ_{AB} in terms of a superposition of state vectors whose time evolution was known from calculations in Sec. 14-3. Here we determine this time development for ρ_{AB} itself. Specially, in the interaction picture, the equation of motion for $\rho_{AB}(t)$ is

$$\dot{\rho}_{AB}(t) = -\frac{i}{\hbar}[\mathscr{V}_{AB}(t), \rho_{AB}(t)]. \tag{43}$$

Solving this through iteration as indicated in Sec. 7-4, we obtain $\rho_A(t+\tau)$ by tracing over the B subsystems:

$$\begin{aligned}\rho_A(t+\tau) = \mathrm{Tr}_B\{\rho_{AB}(t+\tau)\} &= \mathrm{Tr}_B\{\rho_A(t)\otimes\rho_B(t) \\ &+ \sum_s\left(-\frac{i}{\hbar}\right)^s\int_t^{t+\tau}dt_1\int_t^{t_1}dt_2\ldots\int_t^{t_{s-1}}dt_s \\ &\times[\mathscr{V}(t_1),[\mathscr{V}(t_2),\ldots,[\mathscr{V}(t_s),\rho_A(t)\otimes\rho_B(t)]\ldots]]\}.\end{aligned} \tag{44}$$

It is instructive to write the lowest-order terms explicitly:

$$\rho_A(t+\tau) = \mathrm{Tr}_B\{\rho_A(t)\otimes\rho_B(t) - \frac{i}{\hbar}\int_t^{t+\tau}dt'[\mathscr{V}(t'),\rho_A(t)\otimes\rho_B(t)] + \cdots\}$$

$$= \underbrace{\rho_A(t)}_{\substack{\text{(initial } A \text{ system}\\ \text{density matrix)}}} - \underbrace{\frac{i}{\hbar}\int_t^{t+\tau} dt'\, \mathrm{Tr}_B\, [\mathscr{V}(t'), \rho_A(t) \otimes \rho_B(t)] + \cdots}_{\substack{\text{(correlations produced by}\\ \text{interactions with } B \text{ system)}}}.$$

It is important to remember that $\rho_A(t + \tau)$ will be in general an impure case (statistical mixture) even though $\rho_A(t)$ may be a pure case (single state vector). The process of tracing over the unobserved states leads to an irreversible dynamics for the A system, that is, damping and "noise." We see this for the laser in Chap. 17.

Let us now illustrate this formalism by considering again the simple harmonic oscillator (e.g., electric field) damped by atomic beam reservoirs as depicted in Fig. 16-1. The initial atomic density matrix for a Boltzmann mixture of atoms in upper and lower levels is given by

$$\rho_{\mathrm{atom}}(t) = Z^{-1}\begin{pmatrix} \exp(-\hbar\omega_a/k_{\mathrm{B}}T) & 0 \\ 0 & \exp(-\hbar\omega_b/k_{\mathrm{B}}T) \end{pmatrix}$$

$$= \begin{pmatrix} \rho_{aa} & 0 \\ 0 & \rho_{bb} \end{pmatrix}, \tag{45}$$

where the normalization factor

$$Z = \exp(-\hbar\omega_a/k_{\mathrm{B}}T) + \exp(-\hbar\omega_b/k_{\mathrm{B}}T).$$

The frequency ω of Eq. (1) is given by the difference $\omega_a - \omega_b$. The on-resonance interaction energy required for (44) is given by (14.66) as

$$\mathscr{V} = \hbar g \sigma a^\dagger + \text{adjoint} = \hbar g \begin{pmatrix} 0 & a \\ a^\dagger & 0 \end{pmatrix}. \tag{46}$$

Keeping terms up to second order in $\mathscr{V}$ and capitalizing on the time independence of $\mathscr{V}$ (field on resonance), we have from (44) the fairly simple result

$$\rho(t + \tau) \equiv \rho_{\mathrm{field}}(t + \tau) = \rho(t) - \frac{i}{\hbar}\tau\, \mathrm{Tr}_{\mathrm{atom}}\, [\mathscr{V}, \rho_{\mathrm{a-f}}(t)]$$

$$- \frac{1}{2}\left(\frac{\tau}{\hbar}\right)^2 \mathrm{Tr}_{\mathrm{atom}}[\mathscr{V}, [\mathscr{V}, \rho_{\mathrm{a-f}}(t)]]. \tag{47}$$

With (45), the initial condition for $\rho_{\mathrm{a-f}}(t)$ is

$$\rho_{\mathrm{a-f}}(t) = \rho(t) \otimes \begin{pmatrix} \rho_{aa} & 0 \\ 0 & \rho_{bb} \end{pmatrix} = \begin{pmatrix} \rho_{aa}\rho(t) & 0 \\ 0 & \rho_{bb}\rho(t) \end{pmatrix}.$$

The first-order contribution in (47) has only the off-diagonal elements

$$[\mathscr{V}, \rho_{\mathrm{a-f}}(t)] = \hbar g \begin{pmatrix} 0 & a\rho_{bb}\rho(t) \\ a^\dagger \rho_{aa}\rho(t) & 0 \end{pmatrix} - \begin{pmatrix} 0 & \rho_{aa}\rho(t)a \\ \rho_{bb}\rho(t)a^\dagger & 0 \end{pmatrix} \hbar g$$

$$= \begin{pmatrix} 0 & a\rho_{bb}\rho(t) - \rho_{aa}\rho(t)a \\ a^\dagger \rho_{aa}\rho(t) - \rho_{bb}\rho(t)a^\dagger & 0 \end{pmatrix}. \tag{48}$$

The second-order commutator is given with the use of (46) and (48) as

$$[\mathscr{V}, [\mathscr{V}, \rho_{a-f}(t)]] = \hbar^2 g^2 \begin{pmatrix} aa^\dagger \rho_{aa}\rho(t) - a\rho_{bb}\rho(t)a^\dagger & 0 \\ 0 & a^\dagger a\rho_{bb}\rho(t) - a^\dagger \rho_{aa}\rho(t)a \end{pmatrix} + \text{adjoint}. \tag{49}$$

Substitution of (48) and (49) into (47) yields

$$\rho(t+\tau) \simeq \rho(t) - \tfrac{1}{2}g^2\tau^2[(aa^\dagger\rho - a^\dagger\rho a)\rho_{aa} + (a^\dagger a\rho - a\rho a^\dagger)\rho_{bb} + \text{adjoint}].$$

The coarse-grained time rate of change of $\rho(t)$ is then given by

$$\dot{\rho}(t) = r[\rho(t+\tau) - \rho(t)], \tag{50}$$

where r is the rate of injection of atoms per second. Noting here that $r_a = r\rho_{aa}$ and $r_b = r\rho_{bb}$, we obtain

$$\dot{\rho}(t) = -\tfrac{1}{2}\mathscr{R}_a[aa^\dagger\rho - a^\dagger\rho a] - \tfrac{1}{2}\mathscr{R}_b[a^\dagger a\rho - a\rho a^\dagger] + \text{adjoint}, \tag{51}$$

where $\mathscr{R}_a = r_a g^2\tau^2$ as in (25) and (22). The reader can verify quite easily (Prob. 16-7) that the equation of motion for the photon number matrix element ρ_{nm} is just that (26) obtained in Sec. 16-1 and hence that the equations of motion for the average photon number $\langle n(t)\rangle$ and the field $\langle E(t)\rangle$ are just (33) and (36), respectively. In terms of the cavity Q of (38), (51) reads

$$\dot{\rho}(t) = -\frac{1}{2}\frac{\nu}{Q}\left\{\bar{n}[aa^\dagger\rho - a^\dagger\rho a] + (\bar{n}+1)[a^\dagger a\rho - a\rho a^\dagger]\right\} + \text{adjoint}. \tag{52}$$

A particularly interesting representation of the density operator equation of motion (51) is the coherent state. To find this, we substitute the coherent-state expansion (15.2)

$$\rho(t) = \int d^2\alpha P(\alpha, t)|\alpha\rangle\langle\alpha|$$

into (51), obtaining

$$\int d^2\alpha\, \dot{P}(\alpha, t)|\alpha\rangle\langle\alpha| = -\tfrac{1}{2}\int d^2\alpha P(\alpha, t)\,\{\mathscr{R}_a[aa^\dagger|\alpha\rangle\langle\alpha| - a^\dagger|\alpha\rangle\langle\alpha|a] + \mathscr{R}_b[a^\dagger a|\alpha\rangle\langle\alpha| - a|\alpha\rangle\langle\alpha|a^\dagger]\} + \text{adjoint}. \tag{53}$$

The expression in curly braces can be simplified through the use of the relation

$$a^\dagger|\alpha\rangle\langle\alpha| = (\partial_\alpha + \alpha^*)|\alpha\rangle\langle\alpha| \tag{54}$$

and its adjoint

$$|\alpha\rangle\langle\alpha|a = (\partial_{\alpha^*} + \alpha)|\alpha\rangle\langle\alpha|, \tag{55}$$

in which we have set

$$\partial_a = \frac{\partial}{\partial a} \tag{56}$$

for typographical simplicity. These relations follow (Prob. 16–8) from the formula

$$|a\rangle\langle a| = \exp(-aa^*)\exp(aa^\dagger)|0\rangle\langle 0|\exp(a^*a), \tag{57}$$

which, in turn, follows from Eq. (15.15) for the coherent state $|a\rangle$. In particular, we find in Eq. (53) that

$$\begin{aligned} a^\dagger a|a\rangle\langle a| - a|a\rangle\langle a|a^\dagger &+ \text{adjoint} \\ &= a^\dagger a|a\rangle\langle a| - a|a\rangle\langle a|a^\dagger + \text{adjoint} \\ &= [a(\partial_a + a^*) - aa^*]|a\rangle\langle a| + \text{adjoint} \\ &= a\partial_a|a\rangle\langle a| + \text{adjoint} \end{aligned} \tag{58}$$

and similarly that

$$\begin{aligned} aa^\dagger|a\rangle\langle a| - a^\dagger|a\rangle\langle a|a &+ \text{adjoint} \\ &= -[a\partial_a + \partial_a\partial_{a^*}]|a\rangle\langle a| + \text{adjoint}. \end{aligned} \tag{59}$$

We plan to substitute (58) and (59) into (53) and integrate the result by parts. In doing so, we encounter integrals like

$$\int d^2a\, Pa\partial_a|a\rangle\langle a| = aP|a\rangle\langle a|\Big|_{-\infty}^{\infty}\Big|_{-\infty}^{\infty} - \int d^2a\, \partial_a(aP)|a\rangle\langle a|. \tag{60}$$

The distribution $P(a, t)$ vanishes at the infinite limits, and therefore (60) becomes

$$\int d^2a\, Pa\partial_a|a\rangle\langle a| = -\int d^2a\, \partial_a(aP)|a\rangle\langle a|.$$

The $\partial_a\partial_{a^*}$ term requires two such integrations (yielding two minus signs). With these considerations, Eq. (53) becomes

$$\int d^2a\, \dot{P}|a\rangle\langle a| = -\int d^2a\{\tfrac{1}{2}(\mathscr{R}_a - \mathscr{R}_b)[\partial_a(aP) + \text{c.c.}] - \mathscr{R}_a\partial_a\partial_{a^*}(P)\}|a\rangle\langle a|,$$

where we obtain a complex conjugate from the adjoints since P is real. Identifying coefficients of $|a\rangle\langle a|$ in the integrands, we have the equation of motion for $P(a, t)$:

$$\dot{P}(a,t) = -\tfrac{1}{2}(\mathscr{R}_a - \mathscr{R}_b)\left\{\frac{\partial}{\partial a}[aP(a,t)] + \text{c.c.}\right\} + \mathscr{R}_a\frac{\partial^2 P(a,t)}{\partial a\partial a^*}. \tag{61}$$

In the next section we discuss a classical version of such an equation, which is called the Fokker-Planck equation. For the present we simply note that (61) can be used to calculate the time rates of change of various expectation values, much as (26) for $\rho_{nm}(t)$ was used. In particular, the rate of change for the complex electric field value $\mathscr{E}a$ is given by

$$\frac{d}{dt}\langle\mathscr{E}a\rangle = \int d^2a\mathscr{E}a\dot{P}(a, t) = -\tfrac{1}{2}(\mathscr{R}_a - \mathscr{R}_b)\mathscr{E}\int d^2aa\left[\frac{\partial}{\partial a}(aP) + \text{c.c.}\right]$$

$$= -\tfrac{1}{2}(\mathscr{R}_b - \mathscr{R}_a)\ \mathscr{E}\int d^2a\, aP$$

$$= -\tfrac{1}{2}\ \mathscr{R}_b(\bar{n} + 1)^{-1}\langle\mathscr{E}a\rangle \tag{62}$$

in agreement with the density matrix calculation of Eq. (36).

In terms of a cavity Q (38), the Fokker-Planck equation (61) reads

$$\dot{P}(a, t) = \frac{1}{2}\frac{\nu}{Q}\left\{\frac{\partial}{\partial a}[aP(a, t)] + \text{c.c.}\right\} + \frac{\nu}{Q}\bar{n}\frac{\partial^2 P(a, t)}{\partial a\partial a^*}. \tag{63}$$

16-3. The Fokker-Planck Equation

Having seen the Fokker-Planck equation in the preceding section on the damped oscillator, let us consider here how this equation arises naturally in stochastic problems and interpret its time evolution. The simplest problem is that of random walk along a line. The coordinate on the line can be, for example, a position or a velocity. We consider two possible methods, one involving discrete, small position changes, and the other involving (one-dimensional) velocity changes.

For the first method, we suppose that there are many positions of a line (Fig. 16-3) which are occupied with the probability $P(x_n, t)$. We desire the

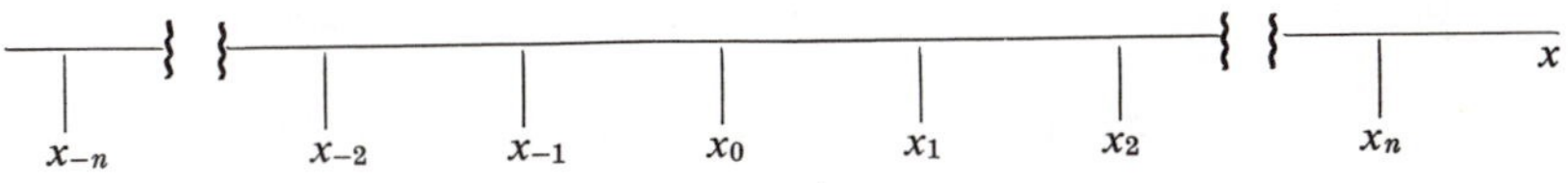

Figure 16-3. Discrete positions on a line which might be occupied by a particle.

equation of motion for this probability. We further suppose that the particle hops with a probability $p_\pm$ one position at a time, that is, $x_n \to x_{n\pm 1}$. The change in probability of being at x_n in the time interval Δt is then given by

$$P(x_n, t + \Delta t) - P(x_n, t) = -(p_+ + p_-)\,P(x_n, t)$$
$$+ p_- P(x_{n+1}, t) + p_+ P(x_{n-1}, t). \tag{64}$$

The reader will note that this equation is quite reminiscent of the photon rate equations in Sec. 16-1 and hence that a similar analysis can be applied to that problem to obtain a Fokker-Planck equation. With the definition $x_{n\pm 1} = x_n \pm \Delta x$ and appropriate first-order expansions of the P's, Eq. (64) yields

$$\begin{aligned}\frac{\partial}{\partial t} P(x_n, t)\,\Delta t = &-(p_+ + p_-)P(x_n, t)\\ &+ p_-\left\{P(x_n, t) + \Delta x\frac{\partial}{\partial x_n}P(x_n, t) + \frac{(\Delta x)^2}{2}\frac{\partial^2}{\partial x_n^2}P(x_n, t)\right\}\\ &+ p_+\left\{P(x_n, t) - \Delta x\frac{\partial}{\partial x_n}P(x_n, t) + \frac{(\Delta x)^2}{2}\frac{\partial^2}{\partial x_n^2}P(x_n, t)\right\}. \qquad (65)\end{aligned}$$

This reduces to the Fokker-Planck equation:

$$\frac{\partial}{\partial t} P(x, t) = -M_1 \frac{\partial}{\partial x} P(x, t) + \tfrac{1}{2}M_2 \frac{\partial^2}{\partial x^2} P(x, t), \qquad (66)$$

where we have taken Δx to be small and x_n to be x, and defined the moments

$$M_1 = (p_+ - p_-)\frac{\Delta x}{\Delta t}, \qquad (67)$$

$$M_2 = (p_- + p_+)\frac{(\Delta x)^2}{\Delta t}. \qquad (68)$$

We will interpret this equation after deriving it again in a slightly different context.

For this second method, the probability density of a particle scattering from a velocity v_0 at time t_0 to velocity v at time t is denoted by $P(v, t|\, v_0, t_0)$. We consider processes which are stationary in time, that is, which depend only on the time difference $s = t - t_0$. We then wish to find an equation of motion for the probability density $P(v, s|\, v_0)$. To this end, we note that the probability density of the particle's scattering into v in the time interval $s + \Delta s$ is

$$P(v, s + \Delta s|\, v_0) = \int dv_1\, P(v, s + \Delta s|\, v_1)\, P(v_1, s|\, v_0), \qquad (69)$$

that is, for an initial v_0, the particle scatters in the interval s with probability $P(v, s|\, v_0)$ to the intermediate velocity v_1, and then at a time Δs later it scatters with probability $P(v, s + \Delta s|\, v_1)$ to v. The total probability density of scattering to v in the time $s + \Delta s$ is then given by summation over all intermediate velocities, v_1.

In the limit $\Delta s \to 0$, we may expand $P(v, s + \Delta s|\, v_0)$ to first order in Δs, obtaining

$$P(v, s + \Delta s|\, v_0) \approx P(v, s|\, v_0) + \frac{\partial}{\partial s} P(v, s|\, v_0)\,\Delta s. \qquad (70)$$

With the definition $\Delta v = v - v_1$, the integrand of Eq. (69) is given by

$$P(v, s + \Delta s|\, v - \Delta v)\, P(v - \Delta v, s|\, v_0)$$

$$= \sum_{n=0}^{\infty} \frac{(-\Delta v)^n}{n!} \frac{\partial^n}{\partial v^n} P(v + \Delta v, s + \Delta s | v) P(v, s | v_0). \tag{71}$$

Inverting (70) and inserting (69) with (71), we find the equation of motion:

$$\begin{aligned}
\frac{\partial}{\partial s} P(v, s \,|\, v_0) &= \frac{1}{\Delta s} [P(v, s + \Delta s \,|\, v_0) - P(v, s \,|\, v_0)] \\
&= \frac{1}{\Delta s} \left[\int d(\Delta v)\, P(v + \Delta v, s + \Delta s | v) - 1 \right] P(v, s | v_0) \\
&= \frac{1}{\Delta s} \sum_{n=1}^{\infty} \frac{\partial^n}{\partial v^n} \int d(\Delta v) \frac{(-\Delta v)^n}{n!} P(v + \Delta v, s + \Delta s | v) P(v, s | v_0) \\
&= \sum_{n=1}^{\infty} (-1)^n \frac{\partial^n}{\partial v^n} [M_n P(v, s | v_0)],
\end{aligned} \tag{72}$$

where the moments (note that $\Delta t = \Delta s$)

$$\begin{aligned}
M_n &= \frac{1}{\Delta t} \int_{-\infty}^{\infty} d(\Delta v)\, \frac{(\Delta v)^n}{n!} P(v + \Delta v, s + \Delta s | v) \\
&= \frac{\langle (\Delta v)^n \rangle}{\Delta t}.
\end{aligned} \tag{73}$$

For sufficiently small Δt and $n > 2$, $\langle (\Delta v)^n \rangle \to 0$ faster than Δt. Hence we can truncate (72) to two terms. We further integrate over all initial velocities to obtain the total probability density for a particle's having velocity v. We thus obtain the Fokker-Planck equation:

$$\frac{\partial}{\partial t} P(v, t) = -\frac{\partial}{\partial v} (M_1 P) + \frac{1}{2} \frac{\partial^2}{\partial v^2} (M_2 P). \tag{74}$$

In Fig. 16-4*a* we show how the first derivative term causes the center of the

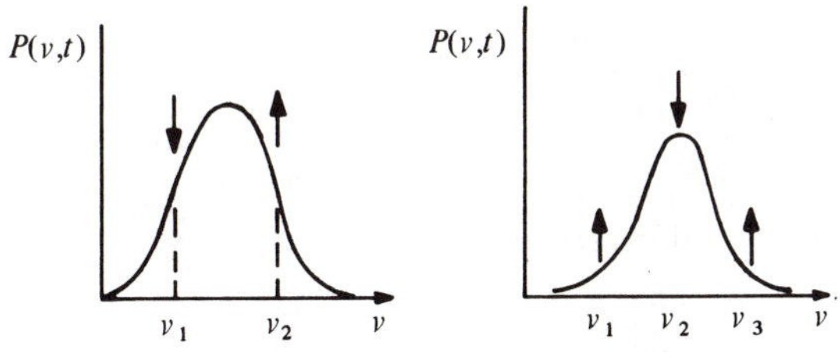

Figure 16-4. (a) For $M_1 > 0$, the first derivative $\partial(M_1 P)/\partial v$ at $v = v_1$ is positive, causing $P(v_1, t)$ to decrease in time. For $v = v_2$, the derivative is negative, causing $P(v_2, t)$ to increase in time. Thus the center of gravity of $P(v, t)$ moves or "drifts" toward larger v. (b) Here the second derivative $\partial^2(M_2 P)/\partial v^2$ is positive for $v = v_1$ and $v = v_3$, leading to increases in $P(v, t)$ in time, while for $v = v_2$ a decrease occurs, that is, the distribution spreads or diffuses.

distribution to shift or "drift" toward larger values of v if M_1 is positive and toward smaller values for M_1 negative. Mathematically, this is represented by the fact that the average v value has the equation of motion

$$\frac{d}{dt}\langle v(t)\rangle = M_1, \tag{75}$$

which is the content of Prob. 16-9. The first moment M_1 is consequently called the "drift term." In Fig. 16-4*b* we show how the distribution spreads or diffuses. In fact, the mean-square deviation

$$\langle v^2(t)\rangle - \langle v(t)\rangle^2 \tag{76}$$

has the equation of motion $[M_1 = -\Gamma v]$

$$\frac{1}{2}\frac{d}{dt}[\langle v^2(t)\rangle - \langle v(t)\rangle^2] = -\Gamma[\langle v^2(t)\rangle - \langle v(t)\rangle^2] + M_2/2, \tag{77}$$

as the reader can show by solving Prob. 16–10. For $M_1 = 0\,(\Gamma = 0)$, as may occur in (66), the distribution diffuses with the values

$$\langle x^2\rangle - \langle x\rangle^2 = M_2 t.$$

Hence M_2 is called the diffusion term. For $M_1 < 0\,(\Gamma > 0)$, (77) has the steady-state value

$$\langle v^2(t)\rangle - \langle v(t)\rangle^2 = \frac{M_2}{\Gamma},$$

in which the spreading speed is limited by the damping. We return to a discussion of this in Sec. 19-1 on Brownian motion.

The Fokker-Planck equation (63) for our atomic beam reservoir problem is a two-dimensional extension of (66). In it the complex variables a and a^* are independent. It is often convenient to use instead Cartesian or polar coordinates. We consider the latter here and leave the Cartesian representation to Prob. 16-11. With

$$a = r\exp(i\theta), \tag{78}$$

we transform to r and θ by writing

$$r^2 = aa^*, \qquad \theta = \tfrac{1}{2}i\ln(a^*/a). \tag{79}$$

These give the partial derivatives

$$\frac{\partial r}{\partial a} = \frac{1}{2}\frac{a^*}{r} = \tfrac{1}{2}\exp(-i\theta), \tag{80}$$

$$\frac{\partial\theta}{\partial a} = -\tfrac{1}{2}i\frac{\partial\ln(a)}{\partial a} = -\frac{1}{2}\frac{i}{a} = -\frac{1}{2}\frac{i}{r}\exp(-i\theta), \tag{81}$$

and hence by the chain rule

$$\frac{\partial}{\partial a} = \frac{\partial r}{\partial a}\frac{\partial}{\partial r} + \frac{\partial\theta}{\partial a}\frac{\partial}{\partial\theta}$$

$$= \tfrac{1}{2}\exp(-i\theta)\left(\frac{\partial}{\partial r} - \frac{i}{r}\frac{\partial}{\partial \theta}\right). \tag{82}$$

The partial $\partial/\partial a^*$ is given by the complex conjugate of (82):

$$\frac{\partial}{\partial a^*} = \tfrac{1}{2}\exp(i\theta)\left(\frac{\partial}{\partial r} + \frac{i}{r}\frac{\partial}{\partial \theta}\right), \tag{83}$$

which with (82), gives the second derivative (note that the purely imaginary terms have to cancel)

$$\frac{\partial^2}{\partial a\,\partial a^*} = \frac{1}{4}\left(\frac{\partial^2}{\partial r^2} + \frac{1}{r}\frac{\partial}{\partial r} + \frac{\partial^2}{\partial \theta^2}\right)$$
$$= \frac{1}{4r^2}\left(r\frac{\partial}{\partial r}r\frac{\partial}{\partial r} + \frac{\partial^2}{\partial \theta^2}\right). \tag{84}$$

Noting that

$$\frac{\partial}{\partial a}a = 1 + a\frac{\partial}{\partial a} = 1 + \tfrac{1}{2}r\frac{\partial}{\partial r} - \tfrac{1}{2}i\frac{\partial}{\partial \theta} = \frac{1}{2}\frac{1}{r}\frac{\partial}{\partial r}r^2 - \tfrac{1}{2}i\frac{\partial}{\partial \theta}, \tag{85}$$

we obtain the transformed Fokker-Planck equation:

$$\frac{d}{dt}P(r,\theta,t) = \frac{1}{2}\frac{\nu}{Q}\frac{\partial}{\partial r}\left[r^2P(r,\theta,t)\right] + \frac{1}{4}\frac{(\nu/Q)\bar{n}}{r^2}\left(r\frac{\partial}{\partial r}r\frac{\partial}{\partial r} + \frac{\partial^2}{\partial \theta^2}\right)P(r,\theta,t). \tag{86}$$

16-4. Generalized Reservoir Theory

We have dealt with the atomic beam reservoir problem at some length, since it illustrates the reservoir concept and the coarse-grained time derivative in a simple manner. We have considerable use for these concepts in Chap. 17 on the quantum theory of the laser. We now turn to the problem of two interacting systems in a more general context in which both systems are treated on an equal footing and emphasis is placed on the correlation developed between the systems. We apply the theory to the spontaneous decay of an excited two-level atom.

Consider two systems interacting with energy $\mathscr{V}(t)$. As in Sec. 16-2, the reduced density matrices ρ_A and ρ_B for the A and B systems, respectively, are given by traces over the complete set of states of the other system, that is,

$$\rho_A(t) = \mathrm{Tr}_B\{\rho_{AB}(t)\} = \sum_B \langle B|\rho_{AB}(t)|B\rangle, \tag{87}$$

$$\rho_B(t) = \mathrm{Tr}_A\{\rho_{AB}(t)\} = \sum_A \langle A|\rho_{AB}(t)|A\rangle. \tag{88}$$

Equations of motion for ρ_A and ρ_B are then, from (44) (in the interaction picture),

$$i\hbar\frac{d}{dt}\rho_A(t) = \mathrm{Tr}_B[\mathscr{V}(t), \rho_{AB}(t)], \tag{89}$$

$$i\hbar \frac{d}{dt} \rho_B(t) = \text{Tr}_A[\mathscr{V}(t), \rho_{AB}(t)]. \tag{90}$$

Our procedure (following Lax) for solving these equations is to write the total

$$\rho_{AB}(t) = \rho_A(t) \otimes \rho_B(t) + \rho_c(t). \tag{91}$$

Here $\rho_c(t)$ represents the correlation which develops in time between the systems because of their interaction. Next we write the interaction term $[\mathscr{V}(t), \rho_A(t) \otimes \rho_B(t)]$ in the form

$$\begin{aligned}[\mathscr{V}(t), \rho_A(t) \otimes \rho_B(t)] &= \mathscr{V}(t)\rho_A(t) \otimes \rho_B(t) - \rho_A(t) \otimes \rho_B(t)\mathscr{V}(t) \\ &= \mathscr{V}(t)\rho_A(t) \otimes \rho_B(t) - \rho_B(t)\mathscr{V}(t)\rho_A(t) \\ &\quad + \rho_B(t)\mathscr{V}(t)\rho_A(t) - \rho_A(t) \otimes \rho_B(t)\mathscr{V}(t). \end{aligned} \tag{92}$$

Here $\mathscr{V}(t)$ includes both system and reservoir operators and thus implicitly contains an outer product. Using (92) and the fact that operators commute under a trace, we find

$$\begin{aligned}\text{Tr}_B[\mathscr{V}(t), \rho_A(t) \otimes \rho_B(t)] &= \text{Tr}_B[\mathscr{V}(t), \rho_B(t)]\rho_A(t) + [\text{Tr}_B\{\rho_B(t)\mathscr{V}(t)\}, \rho_A(t)] \\ &= [\text{Tr}_B\{\mathscr{V}(t)\rho_B(t)\}, \rho_A(t)] \\ &= [\mathscr{V}_A(t), \rho_A(t)], \end{aligned} \tag{93}$$

where

$$\mathscr{V}_A(t) = \text{Tr}_B\{\mathscr{V}(t)\rho_B(t)\} \tag{94}$$

is a Hartree-type self-consistent energy for the A system. Similarly we find

$$\text{Tr}_A[\mathscr{V}(t), \rho_A(t) \otimes \rho_B(t)] = [\mathscr{V}_B(t), \rho_B(t)], \tag{95}$$

where

$$\mathscr{V}_B(t) = \text{Tr}_A\{\mathscr{V}(t)\, \rho_A(t)\} \tag{96}$$

is a Hartree energy for the B system. With (93)–(96) and (91), the equations of motion (89) and (90) become

$$i\hbar \frac{d}{dt} \rho_A(t) = [\mathscr{V}_A(t), \rho_A(t)] + \text{Tr}_B[\mathscr{V}(t), \rho_c(t)], \tag{97}$$

$$i\hbar \frac{d}{dt} \rho_B(t) = [\mathscr{V}_B(t), \rho_B(t)] + \text{Tr}_A[\mathscr{V}(t), \rho_c(t)]. \tag{98}$$

In order to integrate these expressions, we need an equation for the correlation operator $\rho_c(t)$. From (91), (97), and (98), this is

$$\begin{aligned} i\hbar\, \dot\rho_c(t) &= i\hbar\{\dot\rho_{AB}(t) - \dot\rho_A \otimes \rho_B - \rho_A \otimes \dot\rho_B\} \\ &= [\mathscr{V}(t), \rho_A \otimes \rho_B + \rho_c] - [\mathscr{V}_A(t), \rho_A] \otimes \rho_B - \text{Tr}_B[\mathscr{V}(t), \rho_c]\rho_B \\ &\quad - \rho_A \otimes [\mathscr{V}_B(t), \rho_B] - \rho_A \otimes \text{Tr}_A[\mathscr{V}(t), \rho_c]. \end{aligned} \tag{99}$$

We consider problems for which the Hartree energies $\mathscr{V}_A$ and $\mathscr{V}_B$ are zero(see

Prob. 16-13), and we neglect the higher-order terms containing the commutator $[\mathscr{V}, \rho_c]$. In these approximations, we have the equation of motion

$$i\hbar \frac{d}{dt} \rho_c(t) = [\mathscr{V}(t), \rho_A(t) \otimes \rho_B(t)], \tag{100}$$

with the formal solution

$$\rho_c(t) = -\frac{i}{\hbar} \int_{t_0}^{t} dt' [\mathscr{V}(t'), \rho_A(t') \otimes \rho_B(t')], \tag{101}$$

where we have assumed that initially the two systems are uncorrelated, so that $\rho_c(t_0) = 0$. Equation (101) can be thought of as the beginning of an iterative solution for ρ_c in terms of ρ_A and ρ_B.

The operator equations of motion (97), (98), and (101) treat both systems on an equal footing, making it possible to study how each system reacts back on the other in the course of time. In general, the equations have to be solved in a self-consistent fashion. Their solution is simplified in the study of reservoir interactions for which one of the systems (the reservoir) is too large to respond significantly to energy inputs from the other. Good examples of this situation are provided by an ensemble of oscillators (e.g., photons or phonons) regarded collectively as a reservoir in interaction with a mode of the radiation field or an atomic system.

Suppose that $\rho_A(t)$ is the density matrix for the small system and $\rho_B(t)$ is the density matrix for the reservoir. The reservoir assumption allows us to take

$$\frac{d}{dt} \rho_B(t) \approx 0. \tag{102}$$

Hence we can set $\rho_B(t') = \rho_B(t_0)$ in Eq. (101) for $\rho_c(t)$. Substituting this modified value of $\rho_c(t)$ into (97) for $\rho_A(t)$ in which we drop the Hartree term, we find the equation of motion

$$\begin{aligned}\dot{\rho}_A(t) &= -\frac{1}{\hbar^2} \int_{t_0}^{t} dt' \, \mathrm{Tr}_B[\mathscr{V}(t), [\mathscr{V}(t'), \rho_A(t') \otimes \rho_B(t_0)]]. \\ &= -\hbar^{-2} \int_{t_0}^{t} dt' \, \mathrm{Tr}_B\{\mathscr{V}(t)\mathscr{V}(t')\rho_A(t') \otimes \rho_B(t_0) \\ &\quad -\mathscr{V}(t)\rho_A(t') \otimes \rho_B(t_0)\mathscr{V}(t')\} + \text{adjoint}. \end{aligned} \tag{103}$$

This is a valid equation for a system represented by ρ_A interacting with a reservoir represented by ρ_B. Note that, if the reservoir produces a shift in the frequency of the system (e.g., the Hartree term is not zero), this can be included in the unperturbed frequency of that system.

Because the t' integration is over correlation functions of the reservoir which are characterized by times short compared to those (decay times, etc.) of the system, we can often set

$$\rho_A(t') = \rho_A(t) \tag{104}$$

in the integrand of (103). This is called the "Markoff approximation." The models we treat immediately below allow us to work out the interaction-induced correlations explicitly and hence to note to what degree the Markoff approximation is valid. Note that (103) with (104), is obtained from the second-order approximation of (44) by letting the difference $\tau = t - t_0$ be small (Prob. 16-14).

16-5. Field and Atom Damping by Simple Harmonic Oscillator Reservoir

In this section we apply the generalized reservoir theory of Sec. 16-4 to the damping of a simple harmonic oscillator and of a two-level atom by a reservoir of simple harmonic oscillators. The field damping provides an alternative model for a laser cavity Q and is, in fact, the model used later in the Langevin treatment of laser oscillation (Sec. 20-2). The atom damping is just the Weisskopf-Wigner problem of Sec. 14-4, here treated with density operator techniques. In Prob. 16-14 this phenomenon is analysed with the perturbation method of Sec. 16-2.

We describe the single-mode field by the annihilation operator a, creation operator $a^\dagger$, and frequency Ω. We describe the simple harmonic oscillators of the reservoir by annihilation operators b_k, creation operators $b_k{}^\dagger$, and densely distributed frequencies ω_k. The interaction energy is then given by

$$\mathscr{V}(t) = \hbar \sum_k g_k a b_k{}^\dagger \exp[-i(\Omega - \omega_k)t] + \text{adjoint}, \tag{105}$$

where we have made the rotating-wave approximation as discussed in Sec. 14-3. We assume that the reservoir variables are distributed in uncorrelated thermal equilibrium mixtures of states. Hence the reservoir reduced density operator is the multimode extension of the thermal operator (15.38), namely,

$$\rho_B = \prod_k \exp(-\hbar\omega_k b_k{}^\dagger b_k/k_B T)\{1 - \exp(-\hbar\omega_k/k_B T)\}. \tag{106}$$

In Prob. 16-13, we show, using (106), that

$$\mathrm{Tr}_B(b_k{}^\dagger b_j \rho_B) = \bar{n}_k\, \delta_{k,j}, \tag{107}$$

where the thermal average boson number

$$\bar{n}_k = \frac{1}{\exp(\hbar\omega_k/k_B T) - 1}. \tag{108}$$

We further show that

$$\mathrm{Tr}_B\{b_k b_j{}^\dagger \rho_B\} = (\bar{n}_k + 1)\delta_{k,j}, \tag{109}$$

$$\mathrm{Tr}_B\{b_k b_j \rho_B\} = \mathrm{Tr}_B\{b_k{}^\dagger b_j{}^\dagger \rho_B\} = 0. \tag{110}$$

Hence insertion of the interaction energy (105) into the equation of motion (103) for $\rho_A(t)$ yields

$$\begin{aligned}\dot{\rho}_A(t) = -\int_{t_0}^{t} dt' \sum_k g_k^2 \{ & a^\dagger a \rho_A(t')(\bar{n}_k + 1) \exp[i(\Omega - \omega_k)(t - t') \\ & + a a^\dagger \rho_A(t')\bar{n}_k \exp[-i(\Omega - \omega_k)(t - t') \\ & - a^\dagger \rho_A(t')\, a\, (\bar{n}_k + 1) \exp[i(\Omega - \omega_k)(t - t')] \\ & - a \rho_A(t') a^\dagger \bar{n}_k \exp[-i(\Omega - \omega_k)(t - t')]\} + \text{adjoint}. \qquad (111)\end{aligned}$$

We now carry out the same procedure as was used in the Weisskopf-Wigner theory of spontaneous emission (Sec. 14-4). Specifically, we change the summation over k in (111) to an integration over frequency

$$\sum_k \to \int \mathfrak{D}(\omega) d\omega, \qquad (112)$$

where $\mathfrak{D}(\omega)$ is the density of reservoir oscillator states. Furthermore, we replace the resulting $\mathfrak{D}(\omega)\ \bar{n}(\omega)\ g^2(\omega)$ by $\mathfrak{D}(\Omega)\ \bar{n}(\Omega)\ g^2(\Omega)$, inasmuch as this product varies little in the slowly varying region of the exponentials. Finally we set

$$\int_{t_0}^{t} dt' \exp[\pm i(\Omega - \omega)(t - t')] = \pi\delta\,(\Omega - \omega), \qquad (113)$$

where we neglect the principal part, as discussed in Sec. 14-4. In these approximations, Eq. (111) becomes

$$\begin{aligned}\dot{\rho}_A(t) = -\tfrac{1}{2}\gamma_a \{ & \bar{n}[a a^\dagger \rho_A(t) - a^\dagger \rho_A(t) a] + (\bar{n} + 1)[a^\dagger a \rho_A(t) - a \rho_A(t) a^\dagger]\} \\ & + \text{adjoint}, \qquad (114)\end{aligned}$$

where $\gamma_a = 2\pi\mathfrak{D}(\Omega)g^2(\Omega)$ is the Fermi golden rule rate constant derived in Eq. (14-105) and $\bar{n} = \bar{n}(\Omega)$. Equation (114) with γ_a replaced by ν/Q is the same as (52), obtained in Sec. 16-2 for a two-level atom reservoir. Thus the explicit model of a reservoir in thermal equilibrium does not affect our current results.

The equation of motion for a two-level atom damped by a reservoir of simple harmonic oscillators (spontaneous emission problem) can literally be obtained by the replacement of a and $a^\dagger$ by σ and $\sigma^\dagger$, respectively, in the interaction energy (105), in the intermediate result(111), and hence in Eq.(114). The explanations for the derivation are identical. We thus find for the two-level atom density operator $\rho_{\text{atom}}(t)$

$$\begin{aligned}\dot{\rho}_{\text{atom}} = -\tfrac{1}{2}\gamma_a \{ & \bar{n}[\sigma\sigma^\dagger \rho_{\text{atom}} - \sigma^\dagger \rho_{\text{atom}}\sigma] \\ & + (\bar{n} + 1)\, [\sigma^\dagger\sigma\rho_{\text{atom}} - \sigma\rho_{\text{atom}}\sigma^\dagger]\} + \text{adjoint}. \qquad (115)\end{aligned}$$

In particular, for zero temperature ($\bar{n} = 0$) and an initially excited atom [$\rho_{aa}\,(0) = 1$], Eq. (115) implies

$$\dot{\rho}_{aa}(t) = -\ \gamma_a \rho_{aa}, \qquad (116)$$

which is just the Weisskopf-Wigner result derived in Sec. 14-4 with the use of the state vector. We return to this important problem in the Heisenberg picture in Chap. 19.

In conclusion, let us consider more carefully what reservoir characteristics are required for the damping to be truly Markoffian. The double integrals involved in (111) with (112) have the form

$$\int_0^{t-t_0} d\tau \int d\omega\, g^2(\omega)\mathfrak{D}(\omega) \exp[i(\Omega - \omega)\tau]. \tag{117}$$

As the interval $t - t_0$ increases, the integration over τ acts as a frequency filter. Only reservoir modes with frequencies close to $\Omega(|\omega - \Omega| \lesssim 1/\tau)$ have significant effect on the motion of the system of interest. If the reservoir spectrum is densely and broadly distributed around Ω, the frequency integral becomes negligible as τ increases because of destructive interference. We call the time interval τ for which the frequency integral has appreciable value the correlation time τ_c. If little change occurs in the A system during this time (i.e., $\tau_c \ll 1/\gamma_a$), the Markoff approximation of Eq. (104) is valid. This approximation is also very useful in the noise operator approach of Chaps. 19 and 20.

Problems

16-1. What is the average energy of an atom described by the density matrix of (19) and (20) to order g^2?

Ans. $\langle \mathscr{H}_{\text{atom}}(t + \tau)\rangle = \hbar\omega_b + \hbar(\omega_a - \omega_b)g^2\tau^2\langle n\rangle$.

16-2. To order g, calculate the expectation value of the dipole moment operator er, using the state vector (14).

Ans. $\langle er\rangle = \wp g\tau \sum_n (n + 1)^{1/2}\, iC_{n+1}{}^*(t)\, C_n(t) + \text{c.c.}$

16-3. Derive Eq. (24) for $\dot{\rho}_{nm}|_{|a\rangle \text{ atoms}}$ by filling in the steps for Eqs. (23) and so forth.

16-4. Using Eq. (26) for $\dot{\rho}_{nm}$, calculate the equation of motion for the ensemble average electric field $\langle E(t)\rangle$.

Ans. Equation (36).

16-5. Along the lines leading to (33) for $(d/dt)\langle n(t)\rangle$, calculate the corresponding equation for $\langle n^2(t)\rangle$ and its steady-state value.

16-6. What is the average photon number $\bar{n}$ of (32) for (*a*) $\sim$ zero temperature, (*b*) room temperature ($k_B T \approx \frac{1}{40}$ eV) and optical frequencies (1μ is 1.24 eV), (*c*) microwave frequencies (10,000 MHz)?

16-7. Derive the equation of motion (26) for ρ_{nm} by projecting the operator equation (51) into the number representation.

16-8. Show that Eqs. (54) and (55) for $a^\dagger|a\rangle\langle a|$ and $|a\rangle\langle a|a$ are true as outlined around Eq. (57).

16-9. Using the Fokker-Planck equation (74), show that the average velocity $\langle v(t)\rangle$ has the equation of motion (75).
16-10. Again using the Fokker-Planck equation (74), show that the mean-square deviation (76) has the equation of motion (77).
16-11. Show that, in the Cartesian representation defined by the transformation

$$a = x + iy, \qquad a^* = x - iy, \qquad x = \tfrac{1}{2}(a + a^*), \qquad y = -\tfrac{1}{2}i(a - a^*), \quad (118)$$

the Fokker-Planck equation (63) is given by

$$\frac{\partial}{\partial t}P(x, y, t) = \frac{1}{2}\frac{\nu}{Q}\left(\frac{\partial}{\partial x}x + \frac{\partial}{\partial y}y\right)P(x, y, t) + \frac{1}{4}\frac{\nu}{Q}\bar{n}\left(\frac{\partial^2}{\partial x^2} + \frac{\partial^2}{\partial y^2}\right)P(x, y, t). \quad (119)$$

16-12. Show that, for a two-level atom or a simple harmonic oscillator (SHO) interacting with a reservoir of SHO's, the Hartree energies vanish.
16-13. Show that the reservoir average

$$\langle b_j{}^\dagger b_k\rangle_B = \mathrm{Tr}_B[b_j{}^\dagger b_k\rho_B] = \bar{n}_k\delta_{jk},$$

with $\bar{n}_k$ given by (108), in which the reduced density operator ρ_B is the many-mode thermal operator (106). Further show that

$$\langle b_j b_k{}^\dagger\rangle_B = (\bar{n}_k + 1)\delta_{jk}, \qquad \langle b_j b_k\rangle_B = \langle b_j{}^\dagger b_k{}^\dagger\rangle_B = 0.$$

16-14. Using the iterative expansion (44) to second order, show that both the two-level atom and the SHO systems damped by reservoirs of SHO's, are governed by

$$\dot{\rho}_A(t) \approx \frac{1}{\tau}[\rho_A(t + \tau) - \rho_A(t)] = -(\tau\hbar^2)^{-1}\int_t^{t+\tau} dt'\int_t^{t'} dt'' \times \mathrm{Tr}_B[\mathscr{V}(t')\mathscr{V}(t'')\,\rho_A(t)\otimes\rho_B(t) - \mathscr{V}(t'')\rho_A(t)\otimes\rho_B(t)\mathscr{V}(t'')] + \text{adjoint},$$

which also yields the equations of motion (114) and (115) in the Markoff approximation. The analysis here is highly analogous to that in Sec. 19-3, where the time development is determined by equations of motion for the annihilation and creation (or spin-flip) operators.

16–15. In Sec. 2–2, the energy density laws for the Planck and Wien laws are compared and shown to be fairly close in value. The difference between the two laws is that the Wien neglects stimulated emission. Calculate the photon distribution ρ_{nn} for this case from Eq. (27) by dropping the stimulated emission terms (those proportional to n). Then show that the second moment $\langle n^2\rangle$ is half the Planck value of Prob. 16-5.

References

H. B. Callen and T. A. Welton, 1951, *Phys. Rev.* **83**, 34.

S. Fujita, 1966, *Introduction to Non-Equilibrium Quantum Statistical Mechanics,* W. B. Saunders, Co., Philadelphia.

R. Kubo, 1957, *J. Phys. Soc. Japan* **12**, 570.

W. H. Louisell, 1973, *Quantum Statistical Properties of Radiation in Pure and Applied Optics,* John Wiley & Sons, New York.

D. ter Haar, 1961, *Rept. Progr. Phys* **24**, 304.

R. K. Wangsness and F. Bloch, 1956, *Phy. Rev.* **89**, 728.

V. F. Weisskopf and E. Wigner, 1930, *Z. Physik* **63**, 54.

R. Zwanzig, 1960, *J. Chem. Phys.* **33**, 1338.

XVII

QUANTUM THEORY OF THE LASER

17. Quantum Theory of the Laser

In this chapter we treat single-mode laser operation in which both the atoms and the field obey the laws of quantum mechanics. For simplicity, we suppose for most of the discussion that the field is tuned to atomic line center and that the active atoms are two-level systems which pass through the cavity in a time τ, as in Sec. 16-1. To simulate a homogeneously broadened medium, we suppose that the transit times are distributed with the probability $\gamma \exp(-\gamma\tau)$. The more realistic (and complicated) calculation underlying the theory of Chap. 8 is given in Appendix I. We also suppose that excitation occurs only to the upper level; Probs. 17-13 and 17-14 consider excitation to the lower level as well. We describe cavity losses by an ensemble of nonresonant atoms injected into the lower level, as discussed in Sec. 16-1. The response of these "loss" atoms does not saturate (i.e., small coupling constant), as does that of the active atoms. The loss atoms here play the same role as the fictitious conductivity in the semiclassical laser theory of Chap. 8. As shown in Sec. 16-5, an equivalent description is obtained by a reservoir of simple harmonic oscillators with a spread of frequencies. Our discussion depends on the reservoir theory of Sec. 16-1, in which the reduced density operator represents the radiation field. In Chap. 20 we give a related development in which the Langevin (Heisenberg picture) approach is used.

In Sec. 17-1 we derive the equation of motion for the laser field reduced density matrix (photon number representation), using the probability amplitudes of Sec. 14-3 as discussed in Sec. 16-1. The equations are given a simple and intuitively reasonable interpretation [see Eq. (10)]. We compare a fourth-order approximation to the theory with the semiclassical treatment of Chap. 8. The corresponding density operator and Fokker-Planck equations are derived in Probs. 17-1 and 17-4.

In Sec. 17-2 the equations are solved for the steady-state photon statistics and integrated to give the field buildup from vacuum. In Sec. 17-3 the off-diagonal elements (n representation) and the Fokker-Planck equation (coherent-state representation) are used separately to yield the laser linewidth. Chapter 18 discusses the philosophy and measurement of the results.

17-1. Field Equation of Motion

We wish to calculate the effect that an atom has on the laser electric field in its passage through the cavity. The calculation we give is essentially the same as that for the atomic beam reservoir in Sec. 16-1 with the exceptions of strong-signal results and simulation of ordinary atomic decay by a distribution of transit times. Central to our approach here (as in Sec. 16-1) is the assumption that the field photon number probability amplitude $C_n(t)$ in a representative field state vector:

$$|\psi(t)\rangle = \sum_n C_n(t)|n\rangle$$

does not change appreciably because of the interaction with a single atom (or even several atoms). This is physically reasonable since the effect of the radiation emitted by a single atom is distributed over *all* of the many $C_n(t)$'s, not just one or two (for which a larger change would occur). Our assumption is the quantum field counterpart of the semiclassical approximation that the electric field phasor $E_n(t)\exp[-i\phi_n(t)]$ does not vary appreciably in atomic lifetimes and hence can be factored outside atomic integrations over time. Here we suppose that the $C_n(t)$ do not change much in the coarse-grained time interval Δt.

As in Sec. 16-1, we calculate a coarse-grained time derivative of the field density matrix $\rho_{nm}(t)$. For this, recall that the change in ρ_{nm} due to one atom interacting for a time τ is given by

$$\delta\rho_{nm}^{(\tau)}(t) = \sum_a \rho_{an;am}(t+\tau) - \rho_{nm}(t) \tag{1}$$

and the change due to N atoms acting in a small time Δt is

$$\Delta\rho_{nm}^{(\tau)}(t) = N(\Delta t)\,\delta\rho_{nm}^{(\tau)}(t).$$

If we define the rate of atomic injection as r, then $N(\Delta t) = r\,\Delta t$, and we can write the coarse-grained derivative for $\rho_{nm}(t)$ as

$$\frac{\Delta\rho_{nm}^{(\tau)}(t)}{\Delta t} = r\,\delta\rho_{nm}^{(\tau)}(t). \tag{2}$$

This assumes that each atom lives for a time τ and is then removed.

Closer to the real physical situation is a model which removes atoms after a time τ (by collisions or spontaneous emission to a nonlasing level), described by a probability distribution $P(\tau)$. The average change in the radiation field per atom is then given by

$$\delta\rho_{nm}(t) = \int_0^{''\infty''} d\tau\, P(\tau)\, \delta\rho_{nm}{}^{(\tau)}(t), \tag{3}$$

where the upper limit $''\infty''$ is a time long compared to the average τ, given by $P(\tau)$, but short compared to Δt. The choice

$$P(\tau) = \gamma \exp(-\gamma\tau)$$

simulates spontaneous emission with decay constants $\gamma_a = \gamma_b = \gamma$, a fact proved in Appendix I. With this, Eqs. (1), (2), and (3) yield the average coarse-grained time rate of change:

$$\dot\rho_{nm} \simeq \frac{\Delta\rho_{nm}(t)}{\Delta t} = r\int_0^{\infty} d\tau\, \gamma \exp(-\gamma\tau)\, \{\sum_a \rho_{an;am}(t+\tau) - \rho_{nm}(t)\}. \tag{4}$$

We now calculate the matrix elements $\rho_{an;am}$ using the state vector results of Sec. 14-3. Our initial field state vector $|\psi(t)\rangle$ is given by (16.9). For an initially excited atom, the combined atom-field state vector is given by

$$|\psi_{a-f}(t)\rangle = |a\rangle|\psi(t)\rangle. \tag{5}$$

This vector develops in time according to Eqs. (16.12) and (16.13) to the value

$$|\psi_{a-f}(t+\tau)\rangle = \sum_n \{C_{an}(t+\tau)|a\rangle|n\rangle + C_{b,n+1}(t+\tau)|b\rangle|n+1\rangle\}$$

at time $t + \tau$, for which the initial condition (5) reads

$$C_{an}(t) = C_n(t), \qquad C_{b,n+1}(t) = 0. \tag{6}$$

The resonant solutions of (16.12) and (16.13) are [as for (14.84) and (14.85)]

$$C_{an}(t+\tau) = C_n(t)\cos(g\tau\sqrt{n+1}), \tag{7}$$

$$C_{b,n+1}(t+\tau) = -iC_n(t)\sin(g\tau\sqrt{n+1}). \tag{8}$$

The matrix element $\rho_{an;am}(t+\tau)$ required by (4) is then given by

$$\begin{aligned}\rho_{an;am}\,(t+\tau) &= \sum_\psi P_\psi C_{an}(t+\tau)\, C_{am}{}^*(t+\tau)\\ &= \sum_\psi P_\psi C_n(t) C_m{}^*(t)\cos(g\tau\sqrt{n+1})\cos(g\tau\sqrt{m+1})\\ &= \rho_{nm}(t)\cos(g\tau\sqrt{n+1})\cos(g\tau\sqrt{m+1}).\end{aligned}$$

Here the sum over ψ applies to the possible probability amplitude products $C_nC_m{}^*$ in the field mixture described by ρ_{nm}. Similarly,

$$\rho_{bn;bm}(t+\tau) = \rho_{n-1,m-1}(t)\sin(g\tau\sqrt{n})\sin(g\tau\sqrt{m}).$$

Inserting these matrix elements into (4), we find the average coarse-grained time derivative:

$$\begin{aligned}\dot\rho_{nm}(t) \simeq &-r_a\rho_{nm}(t)\,[1-\gamma\int_0^\infty d\tau \exp(-\gamma\tau)\cos(g\tau\sqrt{n+1})\cos(g\tau\sqrt{m+1})]\\ &+ r_a\rho_{n-1,m-1}(t)\,\gamma\int_0^\infty d\tau\exp(-\gamma\tau)\sin(g\tau\sqrt{n})\sin(g\tau\sqrt{m}).\end{aligned} \tag{9}$$

In particular, the diagonal elements ρ_{nn} (probabilities of n photons) have the equation of motion

$$\dot{\rho}_{nn} = -r_a\rho_{nn}(t)[1 - \gamma\int_0^\infty d\tau \exp(-\gamma\tau)\cos^2(g\tau\sqrt{n+1})]$$

$$+ r_a\rho_{n-1,n-1}(t)\,\gamma\int_0^\infty d\tau \exp(-\gamma\tau)\sin^2(g\tau\sqrt{n}).$$

This equation has the simple interpretation

$\dot{\rho}_{nn} = -$(no. of atoms injected/second) $\times$ (probability of n photons)
$\times$(average probability of stimulated emission by n-photon field)
$+$(no. of atoms injected/second) $\times$ (probability of $n - 1$ photons)
$\times$(average probability of stimulated emission by $(n - 1)$-photon field). (10)

Noting that

$$\gamma\int_0^\infty d\tau \exp(-\gamma\tau)\begin{pmatrix}\cos(g\tau\sqrt{n+1})\cos(g\tau\sqrt{m+1})\\ \sin(g\tau\sqrt{n+1})\sin(g\tau\sqrt{m+1})\end{pmatrix}$$

$$= \frac{1}{4}\gamma\int_0^\infty d\tau\{\exp[-\gamma\tau - ig\tau(\sqrt{n+1} - \sqrt{m+1})]$$

$$\pm\exp[-\gamma\tau - ig\tau(\sqrt{n+1} + \sqrt{m+1})]\} + \text{c.c.}$$

$$= \frac{\begin{pmatrix}1 + (g/\gamma)^2(n+1+m+1)\\ 2(g/\gamma)^2[(m+1)(n+1)]^{1/2}\end{pmatrix}}{1 + 2(g/\gamma)^2(n+1+m+1) + (g/\gamma)^4(n-m)^2}, \tag{11}$$

we find for (9)

$$\dot{\rho}_{nm} = -\left(\frac{\mathcal{N}_{nm}'\mathscr{A}}{1+\mathcal{N}_{nm}\mathscr{B}/\mathscr{A}}\right)\rho_{nm} + \left(\frac{\sqrt{nm}\,\mathscr{A}}{1 + \mathcal{N}_{n-1,m-1}\mathscr{B}/\mathscr{A}}\right)\rho_{n-1,m-1}, \tag{12}$$

where the linear gain coefficient

$$\mathscr{A} = 2r_a\left(\frac{g}{\gamma}\right)^2, \tag{13}$$

the self-saturation coefficient

$$\mathscr{B} = 4\left[\frac{g}{\gamma}\right]^2\mathscr{A}, \tag{14}$$

and the dimensionless factors:

$$\mathcal{N}_{nm}' = \tfrac{1}{2}(n+1+m+1) + \frac{\frac{1}{8}(n-m)^2\mathscr{B}}{\mathscr{A}}, \tag{15a}$$

$$\mathcal{N}_{nm} = \tfrac{1}{2}(n+1+m+1) + \frac{\frac{1}{16}(n-m)^2\mathscr{B}}{\mathscr{A}}. \tag{15b}$$

Equation (12) determines the time evolution of the density matrix due to the gain medium. We describe the cavity losses by an atomic beam of two-level atoms initially in their lower level, as discussed in Sec. 16-1. Specifically adding (12) to the contribution given by Eq. (16-40) with $\mathscr{R}_a = \nu\bar{n}/Q = 0$, we find

$$\dot{\rho}_{nm} = -\left(\frac{\mathscr{N}_{nm}'\,\mathscr{A}}{1+\mathscr{N}_{nm}\mathscr{B}/\mathscr{A}}\right)\rho_{nm} + \left(\frac{\sqrt{nm}\,\mathscr{A}}{1+\mathscr{N}_{n-1,m-1}\mathscr{B}/\mathscr{A}}\right)\rho_{n-1,m-1}$$

$$-\frac{1}{2}\frac{\nu}{Q}(n+m)\,\rho_{nm} + \frac{\nu}{Q}\Big[(n+1)\,(m+1)\Big]^{1/2}\rho_{n+1,m+1}, \qquad (16)$$

which constitutes one of our basic results. In particular, the diagonal element ρ_{nn} (probability of n photons) has the equation of motion ($\mathscr{N}_{nn}' = \mathscr{N}_{nn} = n+1$)

$$\dot{\rho}_{nn}(t) = -\left(\frac{(n+1)\,\mathscr{A}}{1+(n+1)\,\mathscr{B}/\mathscr{A}}\right)\rho_{nn}(t) + \left(\frac{n\mathscr{A}}{1+n\mathscr{B}/\mathscr{A}}\right)\rho_{n-1,n-1}(t)$$

$$-\frac{\nu}{Q}n\,\rho_{nn} + \frac{\nu}{Q}(n+1)\,\rho_{n+1,\,n+1}(t). \qquad (17)$$

It is interesting to note at this point that the diagonal elements are coupled only to diagonal elements and that, more generally, only off-diagonal elements with the same difference $n - m$ are coupled. Examination of (7) and (8) reveals that this simplification is due to (*a*) injection of the atoms in an eigenstate, and (*b*) the trace over atomic coordinates. The trace eliminates the dependence in the field density matrix on the *coherence* induced in the interaction. Here we are interested only in the *probabilities* that the atoms make transitions. It is possible, of course, to ask other questions, such as what the polarization of the emitting atoms is, a question central to the semiclassical treatment but only peripheral in the present approach.

In our work to follow, we consider the fourth-order approximation to (17) in comparing the development of this section both to the operator discussion in the problems and to the third-order semiclassical theory of Chap. 8. The approximation is given by expansion of the denominators in (17) to two terms, that is,

$$\dot{\rho}_{nn}(t) \approx -[\mathscr{A} - (n+1)\mathscr{B}](n+1)\,\rho_{nn}(t) + (\mathscr{A} - n\mathscr{B})n\,\rho_{n-1,\,n-1}(t)$$

$$-\frac{\nu}{Q}n\,\rho_{nn}(t) + \frac{\nu}{Q}(n+1)\,\rho_{n+1,\,n+1}(t). \qquad (18)$$

We summarize the physical meaning of this equation in Fig. 17-1. There we see that probability is "flowing" in and out of the $|n\rangle$ state from and to the neighboring $|n+1\rangle$ and $|n-1\rangle$ states. The $(\nu/Q)\,n\,\rho_{nn}$ term, for example, represents the "flow" from the $|n\rangle$ state to the $|n-1\rangle$ state due to the absorption of photons by loss atoms and is equal to the rate ν/Q times the number of photons n times the probability of having n photons, ρ_{nn}. The term

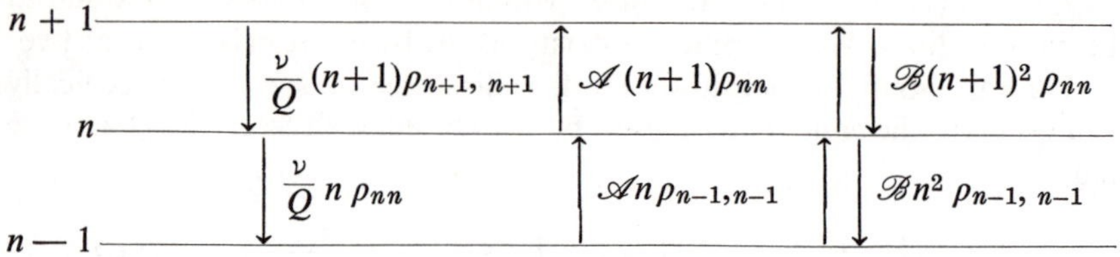

Figure 17-1. Flow of probability in photon rate equation (18).

$[\mathscr{A} - (n+1)\,\mathscr{B}]\,(n+1)\,\rho_{nn}$ represents the "flow" of probability from the $|n\rangle$ state to the $|n+1\rangle$ state due to the emission of photons by lasing atoms initially in their upper states. As summarized in Eq. (9), this term is the (rate of atomic injection into the upper state $|a\rangle$) times the (average probability of stimulated emission by n photons) times the probability of n photons).

To gain further insight into Eq. (18), let us consider the average photon number

$$\langle n(t)\rangle = \sum_n n\,\rho_{nn}(t). \tag{19}$$

In Prob. 17-3, the reader shows, using (18), that

$$\frac{d}{dt}\langle n(t)\rangle = \left(\mathscr{A} - \frac{\nu}{Q}\right)\langle n\rangle + \mathscr{A} - \mathscr{B}[\langle n^2\rangle + 2\langle n\rangle + 1]. \tag{20}$$

Aside from spatial considerations, this corresponds to the classical single-mode intensity equation of motion given by (8.51) for the dimensionless intensity I_n of (8.45):

$$\dot{I}_n = 2I_n(a_n - \beta_n I_n). \tag{21}$$

Noting that the energy in the cavity is given by $\hbar\nu\langle n\rangle$ or semiclassically by $\frac{1}{2}\varepsilon_0 E^2 V$, where V is the volume of the cavity, we see that the net-gain coefficient a_n (Table 8-1) has the correspondence

$$a_n \longleftrightarrow \frac{1}{2}\left(\mathscr{A} - \frac{\nu}{Q}\right), \tag{22}$$

and the self-saturation coefficient β_n

$$\beta_n \longleftrightarrow \mathscr{B}\left(\frac{8\hbar\nu}{\varepsilon_0 V}\right). \tag{23}$$

The second $\mathscr{A}$ in (20) gives the spontaneous emission into the laser field mode and does not appear in the classical equation (21). We see that for $I_n = 0$ the semiclassical field remains zero for all time. In contrast, the quantum equation (20) can build up from zero because of this spontaneous emission term, $\mathscr{A}$. We discuss the details of the buildup in Sec. 17-2.

In Prob. 17-1 the reader shows that the density operator equation of motion to fourth order in the interaction energy (16.46) is given by

$$\dot{\rho}(t) = -\tfrac{1}{2}\mathscr{A}(\rho a a^\dagger - a^\dagger \rho a) - \frac{1}{2}\frac{\nu}{Q}(\rho a^\dagger a - a \rho a^\dagger)$$
$$+ \tfrac{1}{8}\mathscr{B}[\rho(a a^\dagger)^2 + 3 a a^\dagger \rho a a^\dagger - 4 a^\dagger \rho a a^\dagger a] + \text{adjoint}, \qquad (24)$$

where the atoms interact for a time τ, and where $\mathscr{A} = 2r_a g^2 \tau^2$. In Prob. 17-2 the reader shows, using (24), that the diagonal matrix element ρ_{nn} has the equation of motion (18), previously derived with the use of probability amplitudes. Finally he shows in Prob. 17-4 that the coherent-state representation of (24) (see Chap. 15) has the form of a Fokker-Planck equation (see Sec. 16-3), namely,

$$\frac{\partial}{\partial t} P(a, t) = -\frac{1}{2}\left\{\frac{\partial}{\partial a}\left[\left(\mathscr{A} - \frac{\nu}{Q} - \mathscr{B}|a|^2\right) a P\right] + \text{c.c.}\right\} + \mathscr{A}\frac{\partial^2 P}{\partial a \, \partial a^*}. \qquad (25)$$

The terms in curly braces here represent the drift or net amplification of the field, whereas the last term describes the diffusion of the electric field. As mentioned earlier, this diffusion is responsible for the laser linewidth, a fact discussed in Sec. 17-3.

17-2. Laser Photon Statistics

In Sec. 17-1 we saw that the probability ρ_{nn} of an n-photon laser field was changing in time because of (*a*) gain produced by stimulated emission, and (*b*) losses. In this section, we consider steady-state and time-varying solutions of the equation of motion (18) for ρ_{nn} and the photon number statistics implied by the steady-state solution.

The steady-state condition $\dot{\rho}_{nn} = 0$ is satisfied if the flow of probability between, for example, the $|n+1\rangle$ and $|n\rangle$ states is exactly zero, that is,

$$\frac{\nu}{Q}(n+1)\,\rho_{n+1,\,n+1} - [\mathscr{A} - (n+1)\mathscr{B}]\,(n+1)\,\rho_{nn} = 0. \qquad (26)$$

An equivalent equation for the flow of probability between the $|n\rangle$ and $|n-1\rangle$ states can also be written. The solution of either yields the recursion relation

$$\rho_{n+1,\,n+1} = \left[\frac{\mathscr{A} - (n+1)\mathscr{B}}{\nu/Q}\right]\rho_{nn}, \qquad (27)$$

which determines ρ_{nn} for any n in terms of ρ_{00}. Specifically, the probability of having one photon is

$$\rho_{11} = \left(\frac{\mathscr{A} - \mathscr{B}}{\nu/Q}\right)\rho_{00}; \qquad (28)$$

that for two photons is

$$\rho_{22} = \left(\frac{\mathscr{A} - 2\mathscr{B}}{\nu/Q}\right)\rho_{11} = \left(\frac{\mathscr{A} - 2\mathscr{B}}{\nu/Q}\right)\left(\frac{\mathscr{A} - \mathscr{B}}{\nu/Q}\right)\rho_{00}; \tag{29}$$

and thus, by induction, that for n photons is

$$\rho_{nn} = \rho_{00} \prod_{k=1}^{n} \frac{\mathscr{A} - \mathscr{B}k}{\nu/Q}. \tag{30}$$

Inasmuch as probability is conserved, we have

$$\begin{aligned} 1 &= \sum_{n=0}^{\infty} \rho_{nn} \\ &= \rho_{00} + \rho_{00} \sum_{n=1}^{\infty} \prod_{k=1}^{n} \frac{\mathscr{A} - \mathscr{B}k}{\nu/Q}, \end{aligned}$$

which gives

$$\rho_{00} = \left(1 + \sum_{n=1}^{\infty} \prod_{k=1}^{n} \frac{\mathscr{A} - \mathscr{B}k}{\nu/Q}\right)^{-1}. \tag{31}$$

Thus ρ_{00} is just some number which serves as a normalization constant. Calling it $1/\mathscr{N}_p$, we have

$$\rho_{nn} = \mathscr{N}_p^{-1} \prod_{k=1}^{n} \frac{\mathscr{A} - \mathscr{B}k}{\nu/Q}. \tag{32}$$

Let us now consider the form of ρ_{nn} as a function of n. Above threshold, $\mathscr{A} > \nu/Q$. At steady state, the gain equals the losses, so that $\mathscr{A} - \mathscr{B}\bar{n}_{ss} = \nu/Q$, that is, the steady-state average photon number $\bar{n}_{ss}$ is given by

$$\bar{n}_{ss} = \frac{\mathscr{A} - (\nu/Q)}{\mathscr{B}}. \tag{33}$$

For $n \ll \bar{n}_{ss}$, $\mathscr{B}n \ll \mathscr{A}$ and $\rho_{nn} \sim (\mathscr{A}Q/\nu)^n$, an increasing function of n. For larger n, the fraction $(\mathscr{A} - \mathscr{B}k)/(\nu/Q)$ in (32) approaches unity as k approaches $\bar{n}_{ss}$ and then becomes smaller as k increases beyond $\bar{n}_{ss}$. We see that, for $n \approx \bar{n}_{ss}$, the curve for ρ_{nn} versus n "tops out," and that, for large n, ρ_{nn} falls to zero. This behavior is shown by the dashed graph in Fig. 17-2. Note that ρ_{nn} is peaked at approximately $\bar{n}_{ss}$. This is quite different from a thermal distribution, which has a most probable photon number of zero, like that for the solid line graph in Fig. 17-2.

At threshold ($\mathscr{A} = \nu/Q$), the fraction $(\mathscr{A} - \mathscr{B}k)/(\nu/Q) = 1 - (\mathscr{B}/\mathscr{A})k$ is less than unity for all $k > 0$. Hence the largest photon number probability is ρ_{00} and others decrease monotonically, as shown in the dot-dashed graph of Fig. 17-2. Finally, for operation below threshold ($\mathscr{A} < \nu/Q$), the fraction $(\mathscr{A} - \mathscr{B}k)/(\nu/Q) \approx \mathscr{A}Q/\nu$, for which ρ_{nn} has the exponentially decreasing value $(\mathscr{A}Q/\nu)^n$, as shown in the solid line graph of Fig. 17-2.

There is a difficulty with expression (32) for ρ_{nn} when the index k becomes greater than $\mathscr{A}/\mathscr{B}$, for then ρ_{nn} goes alternately positive and negative as n

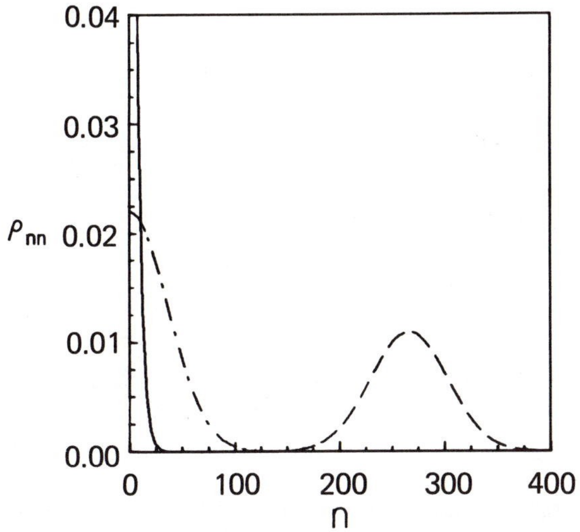

Figure 17-2. Graph of steady-state solution of Eq. (18) for three excitation levels: below (solid line), at (dot-dashed line), and above (dashed line) threshold.

increases. The problem arises from the use of the fourth-order approximation (18) to the strong-signal equation (17). To resolve this problem and to determine the photon statistics of strong-signal operation, we apply the principle of detailed balance [see Eq. (26)] to (17). This gives the distribution

$$p_{nn} = \mathcal{N}_s^{-1}\left\{\frac{\left[\dfrac{\mathcal{A}^2}{\mathcal{B}(\nu/Q)}\right]^{n+\mathcal{A}/\mathcal{B}}}{\left[n+\dfrac{\mathcal{A}}{\mathcal{B}}\right]!}\right\}, \tag{34}$$

where $\mathcal{N}_s$ is a normalization factor.

The average photon number for this distribution is (Prob. 17-8)

$$\bar{n}_{ss} = \frac{\mathcal{A}}{\nu/Q}\,\frac{\mathcal{A}-(\nu/Q)}{\mathcal{B}}. \tag{35}$$

The statistics of the strong-signal distribution (34) are compared to those (Poisson) for light in a coherent state $|\alpha\rangle$ (15.13) in Fig. 17-3. The operation considered is only 20% above threshold, and the laser distribution is substantially wider than the coherent. However, for very high excitation ($\mathcal{A} \gg \nu/Q$), the average photon number

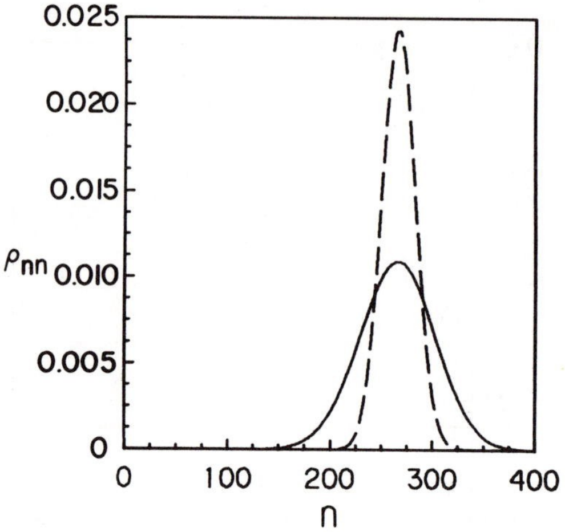

Figure 17-3. Graphs of photon number probabilities ρ_{nn} versus photon number for a laser 20% above threshold (solid line) from Eq. (34), and for a Poisson distribution (dashed line) with the same average number of photons from Eq. 13.

$$\langle n \rangle \to \frac{\mathscr{A}^2}{\mathscr{B}(\nu/Q)} \tag{36}$$

and photon numbers in the vicinity of $\langle n \rangle$ exceed $\mathscr{A}/\mathscr{B}$ by the (large) ratio $\mathscr{A}Q/\nu$. Hence (34) approaches the value

$$\rho_{nn} \approx \frac{\mathscr{N}_s^{-1}\langle n \rangle^n}{n!} = \frac{e^{-\langle n \rangle}\langle n \rangle^n}{n!}, \tag{37}$$

that is, the statistics approach those of a coherent state. The laser statistics given here are not those measured directly by a photodetector. The latter involve a convolution of laser and detector statistics and are discussed in Sec. 18-2.

We conclude this section with a discussion of laser buildup from vacuum fluctuations. As determined in connection with Eq. (20), the quantum equations have the spontaneous emission source for building up in the absence of initial radiation. We can integrate a set of $\dot{\rho}_{nn}$ equations (18) suitably truncated for large n values as illustrated in Fig. 17-4 for an average photon number of 267. Larger numbers can be treated by either the density matrix method (Wang and Lamb, 1973; see Fig. 17-5) or by the Fokker-Planck method (Risken, 1967). One advantage of the density matrix method is that large excitation can be treated with the strong-signal version given by (17).

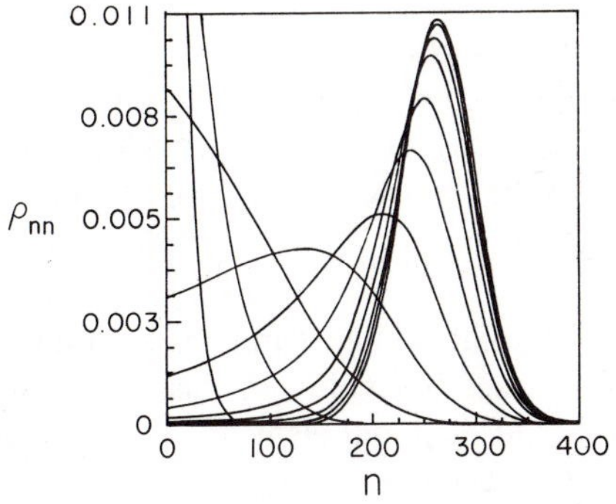

Figure 17-4. Graphs of photon number probabilities ρ_{nn} versus photon number n at times (in order of increasing $\langle n \rangle$) 5, 10, 15, 20, 25, 30, 35, 40, 45, 50 and 55 μsec. Laser parameters used in the integration of Eq. (18) are $\mathscr{A} = 1.2$, $\nu/Q = 1.0$, and $\mathscr{B} = \mathscr{A}/1600 = 0.00075$, yielding $\langle n \rangle = 267$. The choice $\mathscr{A}/\mathscr{B} = 1600$ causes ρ_{nn} to equal zero when $n = 1600$ [see Eq. (32)]. The time development shown here is depicted more completely by the computer movie printed in the upper right-hand corners of Sargent, Scully, and Lamb (1970), where it can be seen by thumbing through the pages.

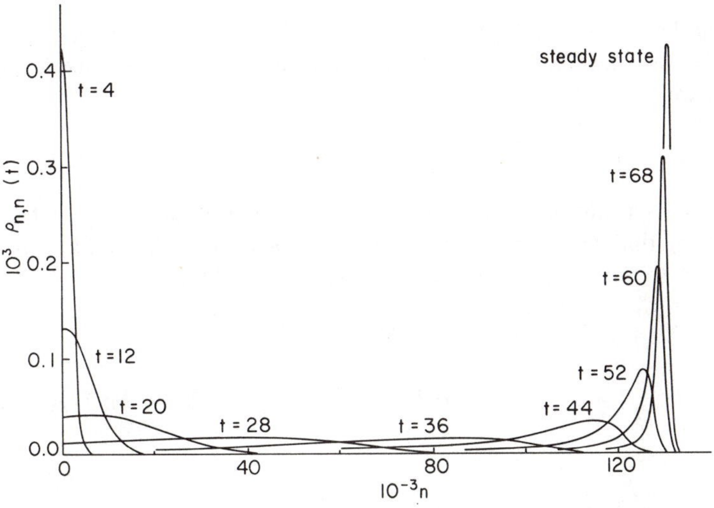

Fig. 17-5. Plots of ρ_{nn} (t) at times indicated for $\mathscr{A} = 1.15\mu\text{sec}^{-1}$, $\mathscr{B} = \mathscr{A} \times 10^{-6}$ and $\nu/Q = 1.0\mu\text{sec}^{-1}$. From Wang and Lamb (1973).

17-3. Off-Diagonal Elements and the Laser Linewidth

The diagonal elements ρ_{nn} of the density operator come to nonzero, steady-state values as given by (32). We see in this section that, in contrast, the off-diagonal element $\rho_{n,\, n+1}$ decays to zero. Inasmuch as this element determines the ensemble average of the electric field (14.16) according to Eq. (16.35):

$$\langle E(t)\rangle = \tfrac{1}{2}\,\mathscr{E} \sum_n \rho_{n,\, n+1}(t)\, \sqrt{n+1}\, \exp(-i\nu t) + \text{c.c.}, \tag{38}$$

the decay of the off-diagonal elements implies a simultaneous decay of the field, thus producing the laser linewidth. In this section we consider the equations of motion for the off-diagonal element $\rho_{n,\, n+1}(t)$, derive the laser linewidth, and interpret the results.

For the sake of simplicity, we write (16) for $\dot{\rho}_{n,\, n+1}$ to fourth order in the interaction energy, that is, we set

$$\frac{1}{1+\mathscr{N}_{nm}\mathscr{B}/\mathscr{A}} \simeq 1 - \frac{\mathscr{N}_{nm}\mathscr{B}}{\mathscr{A}}, \tag{39}$$

and discard terms of order $\mathscr{B}^2/\mathscr{A}$. We find the equation of motion

$$\begin{aligned}\dot{\rho}_{n,\, n+1} = &-\left\{[\mathscr{A} - \mathscr{B}(n+\tfrac{3}{2})]\,(n+\tfrac{3}{2}) - \tfrac{1}{8}\mathscr{B} - \frac{\nu}{Q}(n+\tfrac{1}{2})\right\}\rho_{n,\, n+1}\\ &+ [\mathscr{A} - \mathscr{B}(n+\tfrac{1}{2})]\,[n\,(n+1)]^{1/2}\,\rho_{n-1,n}\\ &+ \frac{\nu}{Q}\,[(n+1)\,(n+2)]^{1/2}\,\rho_{n+1,\, n+2}\,.\end{aligned} \tag{40}$$

If the laser is far enough above threshold, we expect that $\rho_{n,\, n+1}(t)$ will resemble the diagonal solution (32) with the inclusion of an exponential decay. Hence we try a solution of the form

$$\rho_{n,\, n+1}(t) = \mathscr{N}_1 \left(\prod_{l=1}^{n} \frac{\mathscr{A} - \mathscr{B}l}{\nu/Q} \prod_{m=1}^{n+1} \frac{\mathscr{A} - \mathscr{B}m}{\nu/Q}\right)^{1/2} \exp(-\mu_1 t), \tag{41}$$

where $\mathscr{N}_1$ is a constant determined by the initial conditions and μ_1 is the decay parameter that we desire to find. With (41), the elements $\rho_{n-1,n}$ and $\rho_{n+1,n+2}$ are given to very good approximation (since $n \gg 1$) as

$$\begin{aligned}\rho_{n-1,n} &= \frac{\nu}{Q}\{(\mathscr{A} - \mathscr{B}n)[\mathscr{A} - \mathscr{B}(n+1)]\}^{-1/2}\,\rho_{n,n+1}\\ &\approx \frac{\nu}{Q}\,[\mathscr{A} - \mathscr{B}(n+\tfrac{1}{2})]^{-1}\,\{1 + \tfrac{1}{8}\,\mathscr{B}^2[\mathscr{A} - \mathscr{B}(n+\tfrac{1}{2})]^{-2}\}\,\rho_{n,n+1}\end{aligned} \tag{42}$$

and

$$\begin{aligned}\rho_{n+1,n+2} &= \{[\mathscr{A} - \mathscr{B}(n+1)]\,[\mathscr{A} - \mathscr{B}(n+2)]\}^{1/2}\,\frac{Q}{\nu}\,\rho_{n,n+1}\\ &\approx [\mathscr{A} - \mathscr{B}(n+\tfrac{3}{2})]\,\{1 - \tfrac{1}{8}\,\mathscr{B}^2\,[\mathscr{A} - \mathscr{B}(n+\tfrac{3}{2})]^{-2}\}\,\frac{Q}{\nu}\,\rho_{n,n+1}.\end{aligned} \tag{43}$$

We further expand the remaining square roots in (40) as

$$[n(n+1)]^{1/2} \approx n + \tfrac{1}{2} - \tfrac{1}{8}n^{-1} \tag{44}$$

and

$$[(n+1)(n+2)]^{1/2} \approx n + 1 + \tfrac{1}{2} - \tfrac{1}{8}(n+1)^{-1}, \tag{45}$$

where, as in (42) and (43), we have kept enough terms (Prob. 17-11) to avoid error in (40) when taking differences between large, nearly equal numbers. Substituting (42)–(45) into (40), we find

$$\dot{\rho}_{n,n+1} \approx -\Big\{[\mathscr{A} - \mathscr{B}(n+\tfrac{3}{2})]\Big[(n+\tfrac{3}{2}) - \Big(n + \frac{1}{8} - \frac{1}{8}\frac{1}{n+1}\Big)\Big] + \tfrac{1}{8}(\mathscr{B}+\varepsilon)$$

$$+ \frac{\nu}{Q}\Big[(n+\tfrac{1}{2}) - \Big(n + \frac{1}{2} - \frac{1}{8}\frac{1}{n}\Big)\Big]\Big\}\rho_{n,n+1}$$

$$= -\frac{1}{8}\Big(\frac{\mathscr{A}}{n+1} + \frac{\nu/Q}{n} - \frac{1}{2}\frac{\mathscr{B}}{n+1} + \varepsilon\Big)\rho_{n,\,n+1}, \tag{46}$$

for which we show that the quantity

$$\varepsilon = \Big[\frac{\mathscr{B}^2(n+\frac{3}{2})}{[\mathscr{A} - \mathscr{B}(n+\frac{3}{2})]^2}\Big]\Big\{\mathscr{A} - \mathscr{B}(n+\tfrac{3}{2}) - \frac{\nu}{Q}\Big(\frac{n+\frac{1}{2}}{n+\frac{3}{2}}\Big)\Big[\frac{\mathscr{A} - \mathscr{B}(n+\frac{3}{2})}{\mathscr{A} - \mathscr{B}(n+\frac{1}{2})}\Big]^2\Big\} \tag{47}$$

can be neglected as follows. The distribution (41) is strongly peaked about the steady-state average photon number $\bar{n}_{ss}$ of (33) with a width of $(\nu/Q\mathscr{B})^{1/2}$ (see Prob. 17-5). Accordingly, we take the magnitude of the curly braces in (47) to be approximately equal to $\mathscr{B}\,(\nu/Q\mathscr{B})^{1/2}$. This gives

$$|\varepsilon| \approx \frac{[(\nu/Q)\mathscr{B}]^{1/2}}{\bar{n}_{ss}}\Big(\frac{\mathscr{B}\bar{n}_{ss}}{\nu/Q}\Big)^2 < \frac{[(\nu/Q)\mathscr{B}]^{1/2}}{\bar{n}_{ss}} \ll \frac{\nu/Q}{\bar{n}_{ss}}.$$

Dropping both ε and the still smaller $\mathscr{B}$ term in (46) and setting the remaining n's equal to the average number $\bar{n}_{ss}$, we find

$$\dot{\rho}_{n,n+1}(t) = -\tfrac{1}{2}D\rho_{n,n+1}(t), \tag{48}$$

where the decay constant

$$D = \frac{1}{4}\frac{\mathscr{A} + (\nu/Q)}{\bar{n}_{ss}} \simeq \frac{\frac{1}{2}\mathscr{A}}{\bar{n}_{ss}}. \tag{49}$$

Substituting the integral of (48) into (38), we find the ensemble averaged electric field:

$$\langle E(t)\rangle = \langle E(0)\rangle \cos(\nu t)\exp(-\tfrac{1}{2}Dt). \tag{50}$$

The square of the Fourier transform of (50) gives the laser spectrum:

$$|\tilde{E}(\omega)|^2 = \Big|\int_0^\infty dt\,\exp(-i\omega t)\,\langle E(0)\rangle\cos\nu t\,\exp(-\tfrac{1}{2}Dt)\Big|^2$$

$$= \frac{\langle E(0)\rangle^2}{4}\Big[\frac{1}{(\omega-\nu)^2 + (\frac{1}{2}D)^2}\Big]. \tag{51}$$

This is a Lorentzian with the linewidth (full width at half maximum) of D given by (49), as depicted in Fig. 17-6.

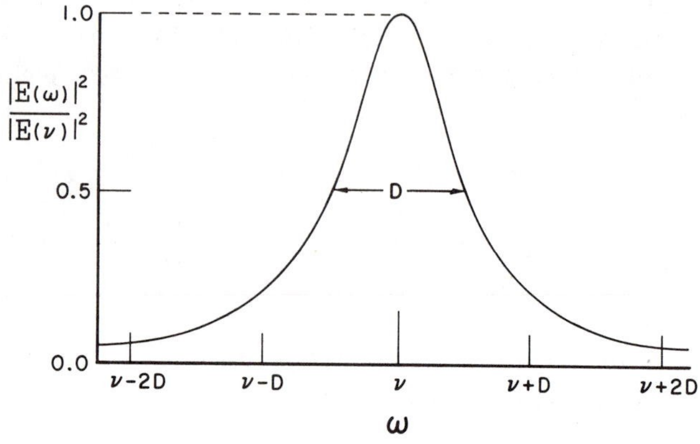

Figure 17-6. Normalized frequency spectrum given by (51).

As is true of all sustained oscillators, the field amplitude is constrained to fluctuate about its steady-state value ($\mathscr{E}\sqrt{\bar{n}_{ss}}$), but the phase can fluctuate freely. Hence the decay of the ensemble average is due to a diffusion in phase† (with diffusion coefficient D) resulting in a diminishing vector sum. Alternatively the field becomes uncorrelated with its value at an earlier time. It is an essential feature of the model that even for an initial pure case the "openness" of the system (due to, for example, spontaneous emission) leads to a mixture in the course of time. This mixture ultimately yields a zero ensemble field average. An analogous situation is found in ferromagnetism, where the magnetization of an open system (magnet) experiences the same kind of decay, albeit on a geological time scale. Measured in terms of the atomic and cavity lifetimes, the decay of $\langle E(t)\rangle$ is also very slow, typically estimated to take about 30 min. In any practical situation, of course, the linewidth of the laser is limited far more severely by mechanical vibrations and other considerations than by the quantum fluctuations leading to (49).

The linewidth can also be derived from the Fokker-Planck equation (25). In Prob. 17-16 the reader shows, using Eqs. (16.82) and (16.83), that Eq. (25) has the polar coordinate representation

$$\frac{d}{dt}P(r, \theta, t) = -\frac{1}{2}\frac{1}{r}\frac{\partial}{\partial r}\left[r^2\left(\mathscr{A} - \frac{\nu}{Q} - \mathscr{B}r^2\right)P(r, \theta, t)\right]$$

† The linewidth is interpreted in terms of Brownian motion in Sec. 20-3.

$$+\tfrac{1}{4}\mathscr{A}\frac{\partial^2}{\partial\theta^2}P(r,\theta,t). \tag{52}$$

For steady-state operation sufficiently far above threshould (but not so far that fourth-order theory is invalid), the changes in P along the radial (field amplitude) coordinate are restricted by the steady-state operating condition. Hence these changes can be neglected compared to those along the unconstrained θ coordinate, and (52) reduces to

$$\frac{d}{dt}P(\theta,t)=\frac{1}{4}\frac{\mathscr{A}}{\bar{n}_{ss}}\frac{\partial^2}{\partial\theta^2}P(\theta,t). \tag{53}$$

Here we have set $r^2=\bar{n}_{ss}$, its steady-state average value. Comparison of Eq. (53) with (16.66) and (16.74) shows that (53) gives the diffusion coefficient:

$$D=\frac{1}{2}\frac{\mathscr{A}}{\bar{n}_{ss}} \tag{54}$$

in agreement with (49), since $\mathscr{A}\approx\nu/Q$ near threshold. We discuss the philosophy of this line width in Secs. 18-3 and 18-4.

Problems

17-1. Derive the density operator equation of motion (24) by adding the linear loss and gain contributions of (16.51) with $\mathscr{A}=\mathscr{R}_a$ and $\nu/Q=\mathscr{R}_b$ to the fourth-order gain saturation term. This term is found most easily by iteration of (16.48), written in the form

$$\overbrace{[\mathscr{V},[\mathscr{V},\cdots,[\mathscr{V}}^{2n},\rho^0]\cdots]]=\begin{pmatrix}\rho_{aa}{}^{(2n)} & 0\\ 0 & \rho_{bb}{}^{(2n)}\end{pmatrix}$$

$$=\hbar^2g^2\begin{pmatrix}aa^\dagger\rho_{aa}{}^{(2n-2)}-a\rho_{bb}{}^{(2n-2)}a^\dagger & 0\\ 0 & a^\dagger a\rho_{bb}{}^{(2n-2)}-a^\dagger\rho_{aa}{}^{(2n-2)}a\end{pmatrix}. \tag{55}$$

17-2. Calculate the equation of motion (18) for the photon number probability ρ_{nn} by taking appropriate matrix elements of the density operator equation (24). *Hint*: The adjoint merely gives a multiplicative factor of 2. Why?

17-3. Calculate the equation of motion (20) for the average photon number $\langle n(t)\rangle$, using the photon rate equation (18) along the lines of (16.33).

17-4. Derive the Fokker-Planck equation (25) from the density operator equation (24). *Hint*: Let the adjoint turn into a complex conjugate when appropriate matrix elements have been taken. Furthermore, drop terms smaller than $\mathscr{B}|a|^2$ and also terms $\mathscr{B}|a|^2$ when compared only to $\mathscr{A}$, that is, not to $\mathscr{A}-(\nu/Q)$.

17-5. Show that the mean-square deviation $\langle n^2\rangle - \langle n\rangle^2 = \nu/Q\mathscr{B}$ for the near-threshold distribution (32). *Hint*: Set

$$\langle n\rangle = \frac{\mathscr{A} - (\nu/Q)}{\mathscr{B}} - 1,$$

In differences between large, similar numbers, the 1 can (and here does) matter. Furthermore, write n^2 in terms of $[\mathscr{A} - \mathscr{B}(n+1)]\,Q/\nu$ and $[\mathscr{A} - (n+2)\mathscr{B}]\,Q/\nu$ to calculate $\langle n^2\rangle$.

17-6. Show that the steady-state solution of the Fokker-Planck equation (25) is

$$P(a) = \mathscr{N}_P \exp\left\{\frac{1}{\mathscr{A}}|a|^2\left[\mathscr{A} - \frac{\nu}{Q} - \tfrac{1}{2}\mathscr{B}|a|^2\right]\right\}, \tag{56}$$

where $\mathscr{N}_P$ is a normalization constant. The distribution $P(a)$ represents the probability density for finding the electric field $\mathscr{E}a$.

17-7. Prove by mathematical induction that the steady-state solution of Eq. (17) is the strong-signal photon number distribution (34).

17-8. Show that the average photon number for the strong signal distribution (34) is (35).

17-9. Using the binomial theorem in conjunction with Eq. (18), show that the equation of motion for the kth moment $\langle n^k\rangle$ is given by

$$\begin{aligned}\frac{d}{dt}\langle n^k\rangle &= \sum_{n=1}^{\infty}[\mathscr{A} - \mathscr{B}(n+1)]\sum_{i=0}^{k-1}\binom{k}{i}(n^{i+1} + n^i)\rho_{nn} \\ &\quad - \frac{\nu}{Q}\sum_{n=0}^{\infty}\sum_{i=0}^{k-1}\binom{k}{i}(-1)^{k-1-i}n^{i+1}\rho_{nn} \\ &= \sum_{i=0}^{k-1}\binom{k}{i}\left\{\langle n^{i+1}\rangle\left[\mathscr{A} - (-1)^{k-1-i}\frac{\nu}{Q}\right] + \langle n^i\rangle\mathscr{A}\right. \\ &\qquad \left. - \mathscr{B}(\langle n^{i+2}\rangle + 2\langle n^{i+1}\rangle + \langle n^i\rangle)\right\}.\end{aligned}$$

17-10. Using (24), show that the off-diagonal element $\rho_{n,n+1}$ of the density operator obeys the equation of motion (40).

17-11. Show that inclusion of the next term in approximations (42) and (43) leads to an additive factor with the negligible magnitude of $\frac{1}{4}\varepsilon(\mathscr{B}Q/\nu)^2$, where the small quantity ε is given by (47).

17-12. Include z dependence in (16) by averaging over z as in the semiclassical theory of Chap. 8.

17-13. Show that gain atom excitation to the lower level at rate r_b yields a contribution to the density matrix equation of motion

$$\begin{aligned}\dot{\rho}_{nm}\Big|_{|b\rangle\text{ atoms}} &= -\left(\frac{\mathscr{N}_{n-1,\,m-1}\mathscr{A}_b}{1 + \mathscr{N}_{n-1,\,m-1}\mathscr{B}/\mathscr{A}}\right)\rho_{nm} \\ &\quad + \left\{\frac{[(n+1)(m+1)]^{1/2}\mathscr{A}_b}{1 + \mathscr{N}_{nm}\mathscr{B}/\mathscr{A}}\right\}\rho_{n+1,\,m+1},\end{aligned} \tag{57}$$

where the linear absorption coefficient

$$\mathscr{A}_b = 2r_b\left(\frac{g}{\gamma}\right)^2. \tag{58}$$

In particular, show that the diagonal element ρ_{nn} with active excitation to the lower level has the equation of motion

$$\dot{\rho}_{nn} = -\left[\frac{(n+1)\mathscr{A}}{1+(n+1)\mathscr{B}/\mathscr{A}}\right]\rho_{nn} + \left[\frac{\mathscr{A}_b}{1+(n+1)\mathscr{B}/\mathscr{A}} + \frac{\nu}{Q}\right](n+1)\rho_{n+1,\, n+1}$$
$$+ \left(\frac{n\mathscr{A}}{1+n\mathscr{B}/\mathscr{A}}\right)\rho_{n-1,\, n-1} - \left(\frac{\mathscr{A}_b}{1+n\mathscr{B}/\mathscr{A}} + \frac{\nu}{Q}\right) n\, \rho_{nn}. \tag{59}$$

17-14. Using the principle of detailed balance, show that the steady-state ($\dot{\rho}_{nn} = 0$) ρ_{nn} is given by

$$\rho_{nn} = Z^{-1}\frac{(\mathscr{A}^2Q/\nu\mathscr{B})^{n+(\mathscr{A}/\mathscr{B})\,(1+A_bQ/\nu)}}{[n + (\mathscr{A}/\mathscr{B})(1 + \mathscr{A}_bQ/\nu)]!} \tag{60}$$

and hence that the average photon number is given by

$$\langle n\rangle = \frac{\mathscr{A}}{\nu/Q}\left[\frac{\mathscr{A} - \mathscr{A}_b - (\nu/Q)}{\mathscr{B}}\right] \tag{61}$$

Near threshold ($\mathscr{A} \approx \nu/Q$), this result is very similar to the semiclassical intensity given by Eq. (8.55), although the latter includes effects due to a sinusoidal z dependence in the electric field.

17-15. The semiclassical equations of motion (2.25) and (2.26) for the probability amplitudes C_a and C_b are formally the same as the quantum versions (14.70) and (14.71), in which $g\sqrt{n+1}$ corresponds to $\wp E_0/\hbar$. Therefore the solutions of the quantum equations are the same as those of the semiclassical versions with this substitution. With this in mind, solve the general case ($\gamma_a \neq \gamma_b$, $\omega \neq \nu$) for the equation of motion for ρ_{nm}.

Ans.:

$$\dot{\rho}_{nm} = -[(n+1)R_{nm} + (m+1)R_{mn}^*]\,\rho_{nm}$$
$$+ (R_{n-1,\, m-1} + R_{m-1,\, n-1}^*)\sqrt{nm}\,\rho_{n-1,\, m-1}$$
$$- \frac{1}{2}\frac{\nu}{Q}(n+m)\,\rho_{nm} + \frac{\nu}{Q}[(n+1)\,(m+1)]^{1/2}\,\rho_{n+1,\, m+1}, \tag{62}$$

where the rate coefficients

$$R_{nm} = \frac{r_a g^2[\gamma_b(\gamma_{ab} + i\Delta) + g^2(n-m)]}{\gamma_a\gamma_b(\gamma_{ab}^2 + \Delta^2) + 2\gamma_{ab}^2g^2(n+1+m+1) + g^2(m-n)[g^2(m-n) + i\Delta(\gamma_a - \gamma_b)]} \tag{63}$$

in which $\Delta \equiv \omega - \nu$.

17-16. Show that the Fokker-Planck equation (25) has the polar coordinate representation (52).

References and Notes

The material in this chapter is based, in part, on:

M. O. Scully, D. M. Kim and W. E. Lamb, Jr., 1970, *Phys. Rev.* **A2**, 2529, 2534.

M. O. Scully and W. E. Lamb, Jr., 1967, *Phys. Rev.* **159**, 208.

M. O. Scully, W. E. Lamb, Jr., and M. J. Stephen, 1965, in: *Proceedings of the International Conference on Quantum Electronics, Puerto Rico, 1965,* ed., P. L. Kelley, B. Lax, and P. Tannenwald, McGraw-Hill Book Co., New York.

The time evolution of the photon statistical distribution for a laser as it builds up from vacuum is given in *Appl. Opt.* **9**, 2423 (1970) by M. Sargent III, M. O. Scully, and W. E. Lamb, Jr. The upper right-hand corner of that issue may be "thumbed" to see the results of a computer movie depicting the laser buildup. Experiments corroborating the theoretical predictions are reported in F. T. Arecchi, V. Degiorgio and B. Querzola, 1967, *Phys. Rev. Letters* **19**, 1168.

See also the following:

V. Ernst and P. Stehle, 1968, *Phys. Rev.* **176**, 1456.

J. A. Fleck, Jr., 1966, *Phys. Rev.* **149**, 309.

J. P. Gordon, 1967, *Phys. Rev.* **161**, 367.

M. Lax and W. H. Louisell, 1967, *IEEE J. Quantum Electronics* **3**, 47.

R. Loudon, 1974, *The Quantum Theory of Light,* Oxford University Press, New York.

H. Risken, 1965, *Z. Phys.* **186**, 85; 1966, *Z. Phys.* **191**, 302.

K. Shimoda, H. Takahasi and C. H. Townes, 1957, J. Phys. Soc. Japan **12,** 686.

Y. R. Wang and W. E. Lamb, Jr., 1973, *Phys. Rev.* **A 8,** 866

C. R. Willis, 1964, *J. Math. Phys.* **5**, 1241.

XVIII

QUANTUM LASER THEORY AND MEASUREMENT

18. Quantum Laser Theory and Measurement

In Chap. 17 we developed the quantum theory of the laser obtaining in particular expressions for the laser photon statistics and linewidth. In this chapter we discuss and amplify those results in terms of measurement theory. Section 18-1 provides an overview of measurement theory as it applies to the quantum electric field. The discussion leads naturally into a specific discussion of the measurement of photon statistics and the laser linewidth. In Sec. 18-2 a heuristic model of a photodetector is presented with reveals how the *photoelectron* statistics resulting from incident radiation is related to the *photon* statistics of that radiation.† In Sec. 18-3 we discuss the laser linewidth with the use of a simple two level atom beam much like that of Sec. 16-1. Our present discussion utilizes the density matrix technique throughout rather than the density matrix-state vector combination of Sec. 16-1. The laser spectrum "seen" by the atomic beam has, in fact, the same Lorentzian profile given in Sec. 17-3. In Sec. 18-4 the laser linewidth is discussed in terms of a quantum version of the classical correlation function $\langle E(t + \tau)E(t)\rangle$. In spite of the fact that *two* time correlations are apparently required, we show that a single time and solutions to the density operator equation of motion are sufficient. This reduction is an example of the so-called Onsager regression hypothesis, which shows how to proceed from single-time to multitime correlation functions.

18-1. On the Measurement of the Laser Field

Despite the analysis given by Bohr and Rosenfeld (1933) of the problems

† The results given are derived from a fundamental point of view by Scully and Lamb(1969).

with measurement in nonrelativistic quantum mechanics, it is probably safe to say that the theory of measurement is even less well developed for quantum electrodynamics. Some of the difficulty, no doubt, arises from the infinite number of degrees of freedom of quantum electrodynamics. In the particular case of a single high-Q cavity mode, however, it seems possible to regard the analogy (Sec. 14-1) between a radiation oscillator and a mechanical oscillator as so close that the measurement problems become equivalent. Our discussion of laser field measurement is based on this assumption.

The most that can possibly be known about the radiation oscillator at $t = 0$ is its wave function, say $\psi(E, 0)$, in the electric field "coordinate" representation. We assume that this state has been "prepared" somehow. Under the guidance of a definite Hamiltonian, this wave function evolves into $\psi(E, t)$ at time t. Any Hermitian operator $\mathcal{O}(E, -i\hbar\partial/\partial E)$ can (or so the theory contends) be "measured." Each measurement gives as a result one of the eigenvalues $\mathcal{O}_n$ of the operator $\mathcal{O}$. The probability of finding a particular value $\mathcal{O}_n$ when a series of measurements is made on an ensemble of similarly prepared systems is

$$W_n = \left| \int_{-\infty}^{\infty} dE\, \phi_n(E)^* \psi(E, t) \right|^2 ,$$

where $\phi_n(E)$ is the eigenfunction belonging to the eigenvalue $\mathcal{O}_n$. The measurements under discussion here are the best permitted. If carried out well, a measurement so disturbs the system of interest that it is pointless even to think of any subsequent measurement of any other operator. The pure case will become a hopeless mixture, even if the system is not physically destroyed.

In most physical research, measurement in this extreme form is not a concern. Certain scattering experiments are sometimes called measurements; however, they do not represent measurement of a Hermitian operator in the strict sense, and so we prefer to call them observations or "bad" measurements. Sometimes, especially when a nearly classical system is under study, an attempt is made to follow the time development between $t = 0$ and $t = t$ by making a series of observations. In our opinion, there is currently no satisfactory theory of such "bad" measurements. We do recognize the possibility of "watching" the bob of a pendulum clock swing back and forth. In a similar manner, at least in principle, the temporal oscillations of the intense and highly classical electric field in a laser could be followed by recording on a moving film the deflection of a stream of high-velocity electrons sent across a narrow laser beam.

Equation (17.50) reveals that the ensemble average of $E(t)$ is a damped oscillating function of time. This damping comes from phase diffusion of the fields for an ensemble of lasers which represents various possible histories of any one laser. An electron-beam probe of any one continuous wave laser would, of course, not show such a damping, but only a very slight amount of phase irregularity. The average of many similar film records, however, would reveal the damping phenomenon. In Sec. 18-3 we return to a specific discussion of the laser line width implied by this damping, now we turn to the measurement of photon statistics.

18-2. Photoelectron Statistics for the Laser Field

In principle, according to the assumed quantum theory of measurement, the total amount of energy in the single-mode optical cavity could be measured, since this is represented by the Hermitian-Hamiltonian operator. The result of such a measurement would be an integer multiple n of $\hbar\nu$, apart from the zero-point energy $\frac{1}{2}\hbar\nu$. Each time the measurement was repeated on a similarly prepared system, the n value could change. The statistical distribution of n values after many measurements would be given by the diagonal elements ρ_{nn} of the density matrix.

In practice, however, it would not be easy to determine ρ_{nn} in this manner. As a partial substitute, the number of photoelectrons emitted in a certain time interval might be counted. In the usual observations, the photoelectron counting is done with the detector located outside the laser cavity, and the relationship of the results to ρ_{nn} is further complicated by diffraction of the radiation escaping from the laser cavity. It would not be very practical, but would be simpler in principle, to place the photoelectric surface inside the laser cavity. Even here, the photoelectron counting statistics would give a somewhat blurred image of the photon statistical distribution ρ_{nn}. In this section, we consider this blurring effect in terms of P_m, the probability of ejection of m photoelectrons for a detector with quantum efficiency η (a unit quantum efficiency gives $P_m = \rho_{mm}$).

The usual procedure for obtaining the photocount distribution P_m breaks the time τ of observation into many small time intervals. In each of these, a quantum-mechanical calculation is carried out in low-order perturbation theory to determine the probability of obtaining a count. The calculation is then completed by classical probabilistic arguments for the number of counts observed in the larger time interval T. This type of analysis "looks" at the system at the end of each of the small time intervals, that is, since the usual experiments are so complicated and microscopically disruptive, it seems reasonable to assume that each count is equivalent to "looking" at the system. As is well known, upon looking at a system we destroy its wave function; for example, when we trace over the laser coordinates, we produce a statistical mixture. Hence, the question might well be asked, "Does the procedure of looking after every count give the same counting distribution as would be observed if the system were not interrupted until a large number of potential counts had accumulated?" The latter problem has been treated by Scully and Lamb (1969) in a fully quantum-mechanical fashion, whereas the former requires the use of classical probability theory. The principal result of the quantum treatment is the photoelectron counting distribution

$$P_m = \sum_{n=m}^{\infty} \binom{n}{m} \eta^m (1-\eta)^{n-m} \rho_{nn}. \tag{1}$$

We can understand this equation as follows. Consider a state of the field having just one photon $|1\rangle$. Let the probability of having a photoelectron ejected from a detector interacting with this field for a certain time be given by η. Now, if the state of the radiation field is $|n\rangle$, the probability of observing m photoelectrons should be proportional to

$$P_m{}^{(n)} \propto \eta^m, \tag{2}$$

which is to be multiplied by the probability that $n - m$ quanta were not absorbed, that is, $(1 - \eta)^{n-m}$:

$$P_m{}^{(n)} \propto \eta^m (1-\eta)^{n-m}, \tag{3}$$

but we, of course, do not know which m photons of the original number n were absorbed, so we must include a combinational factor

$$P_m{}^{(n)} = \binom{n}{m} \eta^m (1-\eta)^{n-m}. \tag{4}$$

This is Bernoulli's distribution for m successful events (counts) and $n - m$ failures, each event having a probability η. Now, if we have a distribution of n values, we must multiply (4) by ρ_{nn} and sum on n;

$$P_m = \sum_n P_m{}^{(n)} \rho_{nn} \tag{5}$$

which yields (1).

As a direct consequence of our model, Eq. (1) not only contains the small, η limit ($\eta \ll 1$), but also is valid for all $\eta (0 \leq \eta \leq 1)$. Clearly, if we wish to obtain the *photon* statistics by counting photoelectrons, we must require $\eta = 1$, for then, as we see from (1),

$$P_m = \rho_{mm}. \tag{6}$$

In all other cases ($\eta < 1$) we are measuring the *photoelectron* statistics, which in general can be very different.

Alternatively, we can write P_m in the $P(\alpha)$ representation by noting from (15.2) that

$$\rho_{nn} = \int d^2\alpha \, P(\alpha) \left[\frac{(\alpha^*\alpha)^n}{n!}\right] \exp\{-\alpha^*\alpha\}. \tag{7}$$

In Prob. 18-1, the reader shows that, with (7), Eq. (1) becomes

$$P_m = \int d^2\alpha \, P(\alpha) \left[\frac{(\alpha^*\alpha\eta)^m}{m!}\right] \exp\{-\alpha^*\alpha\eta\}. \tag{8}$$

Both (1) and (8) agree with the results of the usual treatment (see, for example, papers in Mandel and Wolf, 1970). Equation (1) is a more convenient formulation for our purpose, however, as we most easily find the photon statistical distribution in the n representation. We obtain $P(I)$, the probability of intensity I, only after an auxiliary calculation of some complexity.

We may now calculate the photocount distribution for a fully quantized laser by inserting ρ_{nn} given by (17.34) into (1). We find

$$P_m = \sum_n \binom{n}{m} \eta^m (1-\eta)^{n-m} \left[Z^{-1} \frac{(\mathscr{A}^2/\mathscr{B}\mathscr{C})^{n+\mathscr{A}/\mathscr{B}}}{(n+\mathscr{A}/\mathscr{B})!} \right]. \tag{9}$$

Summing the series, we find the basic relation

$$P_m = Z^{-1} \eta^m \frac{(\mathscr{A}/\mathscr{B}\mathscr{C})^{m+\mathscr{A}/\mathscr{B}}}{(m+\mathscr{A}/\mathscr{B})!} \times {}_1F_1\left(m+1, m+\frac{\mathscr{A}}{\mathscr{B}}+1, (1-\eta)\frac{\mathscr{A}^2}{\mathscr{B}\mathscr{C}}\right), \tag{10}$$

where ${}_1F_1$ is the confluent hypergeometric function. We summarize the notation as follows: $\mathscr{A}$ = linear gain, $\mathscr{B}$ = saturation parameter, $\mathscr{C} = \nu/Q$, η = detector parameter, and Z^{-1} = normalization factor, which is given by

$$Z = \sum_n \frac{(\mathscr{A}^2/\mathscr{B}\mathscr{C})^{n+\mathscr{A}/\mathscr{B}}}{(n+\mathscr{A}/\mathscr{B})!} = \left[\frac{(\mathscr{A}^2/\mathscr{B}\mathscr{C})}{(\mathscr{A}/\mathscr{B})!}\right] {}_1F_1\left(1;1+\frac{\mathscr{A}}{\mathscr{B}}; \frac{\mathscr{A}^2}{\mathscr{B}\mathscr{C}}\right).$$

18-3. Spectrum Analyzer

In Eq. (17.51) we calculated the spectrum associated with the damped oscillating electric field (17.50) and found a Lorentzian of full width at half maximum just equal to the phase diffusion constant D. For a laser described by a purely diagonal density matrix $\rho_{nn'}$, the average $\langle E(t) \rangle$ would be zero, and it might seem that a spectrum could not be defined. It is clear, however, that there is no real difficulty here. Any reasonable operational procedure for determining a spectrum would give the desired result. For example, a Fourier analysis could be made of a very long stretch of the film record mentioned in Sec. 18-1. The phase information available on the early part of the tracing would, in effect, represent preparation of an ensemble with a nonvanishing off-diagonal density matrix.

In this section, we determine the spectrum without making a "bad" measurement as defined in Sec. 18-1. We consider the following simple spectrometer. Imagine that we have a beam of two-level atoms, upper state $|2\rangle$ and lower state $|1\rangle$, separated by an energy $\hbar\omega_2 - \hbar\omega_1 = \hbar\omega$. The beam is prepared with each atom initially in the lower state $|1\rangle$, that is, its initial 2×2 density matrix has only one nonvanishing element:

$$\rho_{11}(t_0) = 1. \tag{11}$$

These atoms pass through the laser cavity and interact weakly with the laser radiation, as shown in Fig. 18-1. The time of flight τ is much greater than $1/D$,

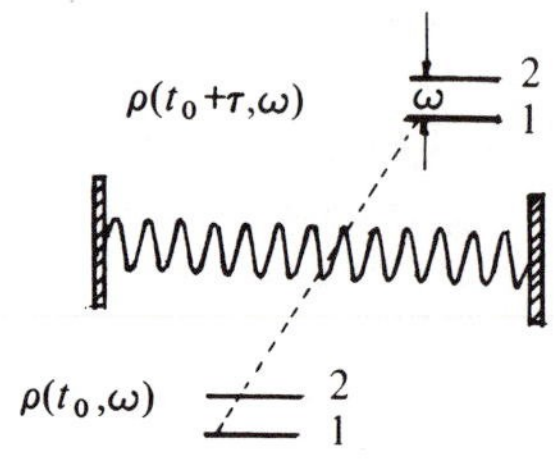

Figure 18-1. Schematic illustration of spectrum analyzer. Atom enters laser cavity in ground state $|1\rangle$ interacts with laser radiation, and emerges in linear superposition of states $|1\rangle$ and $|2\rangle$.

so that the effective atomic linewidth is much narrower than that of the laser radiation. The fraction of excited atoms ρ_{22} emerging from the cavity is determined by a suitable measurement. We then prepare a new beam of slightly different atomic frequency and repeat the experiment. Ultimately we obtain a distribution of relative excitations of different beams as a function of frequency. This plot of the relative effectiveness of the laser radiation in exciting atoms with different frequencies provides us with an operational definition of the spectral profile for the laser.

We now calculate the probability that an atom of frequency ω makes a transition to the state $|2\rangle$. To do this, we consider an atom injected at $t = t_0$ in state $|1\rangle$, determine the density matrix for the combined spectrometer atom-field system at time $t_0 + \tau$, and then trace over the radiation field. The spectrometer atom-field density matrix evolves from one in which all the atoms in the beam are initially in their ground states:

$$\rho_{\text{a-f}}(t_0) = \begin{pmatrix} 0 & 0 \\ 0 & 1 \end{pmatrix} \otimes \rho(t_0) \tag{12}$$

to one with possibly nonzero values of the general matrix element

$$\rho_{r,n;s,m}(t_0 + \tau), \quad r, s = 1, 2. \tag{13}$$

The probability of finding an atom of frequency ω in the upper state at $t_0 + \tau$ is

$$\rho_{22}(\omega, t_0 + \tau) = \sum_n \rho_{2,n;2,n}(\omega, t_0 + \tau). \tag{14}$$

It is this quantity which we wish to calculate.

During the time that the spectrometer atom is in the cavity, to a very good approximation, the pumping and damping of the laser field are going on as if the spectrometer atom were absent, that is, the presence of the spectrometer atom weakly coupled to the "massive" laser field hardly affects the optical oscillator. Hence the time rate of change of the density matrix for the spectrometer atom-laser system is given by the sum of the time derivatives in the absence of the atom plus the time derivative produced by the spectrometer atom interacting with the field:

$$\begin{aligned}\frac{d\rho}{dt} &= \left(\frac{d\rho}{dt}\right)_{\text{laser}} + \left(\frac{d\rho}{dt}\right)_{\text{spectrometer interaction}} \\ &= \left(\frac{d\rho}{dt}\right)_{\text{laser}} - \frac{i}{\hbar}[\mathscr{H}_0^{\text{atom}} + \mathscr{V}, \rho].\end{aligned} \tag{15}$$

Here $\mathscr{H}_0^{\text{atom}}$ is the unperturbed atomic Hamiltonian:

$$\mathscr{H}_0^{\text{atom}} = \hbar\begin{pmatrix}\omega_2 & 0 \\ 0 & \omega_1\end{pmatrix}, \tag{16}$$

and the interaction energy:

$$\mathscr{V} = g\sigma_{21}a + \text{adjoint}, \tag{17}$$

is the Schrödinger picture version of (14.66) in which σ_{21} is the Pauli spin-flip operator (1.42) for levels 1 and 2.

The density matrix elements that we need are those coupling $\rho_{1,n+1;1,n+1}(t_0)$ to $\rho_{2,n;2,n}(t_0 + \tau)$, that is, we want the equations of motion for the evolution

$$\rho_{1,n+1;1,n+1} \begin{matrix} \nearrow \rho_{2,n;1,n+1} \searrow \\ \searrow \rho_{1,n+1;2,n} \nearrow \end{matrix} \rho_{2,n;2,n}.$$

From (15), the laser contribution to the diagonal element $\dot{\rho}_{2,n;2,n}$ vanishes (steady state), yielding the equation of motion

$$\dot{\rho}_{2,n;2,n} = -\frac{i}{\hbar}\mathscr{V}_{2,n;1,n+1}\rho_{1,n+1;2,n}(t) + \text{c.c.} \tag{18}$$

The off-diagonal element $\rho_{1,n+1;2,n}$ has a laser contribution, namely, the decay term of Eq. (17.48). Including this and working in the Schrödinger picture, we find from (15) the equations of motion

$$\begin{aligned}\dot{\rho}_{1,n+1;2,n} = &-(i\nu - i\omega - \tfrac{1}{2}D)\rho_{1,n+1;2,n} \\ &- i\mathscr{V}_{1,n+1;2,n}\rho_{2,n;2,n} - i\rho_{1,n+1;1,n+1}\mathscr{V}_{1,n+1;2,n}\end{aligned} \tag{19}$$

$$\rho_{2,n;1,n+1} = \rho_{1,n+1;2,n}^{*}. \tag{20}$$

We now solve these equation to second order in the perturbation energy $\mathscr{V}$. Noting that at $t = t_0$ the atom-field density matrix factors and that the atom is in the lower state at that time, we have the initial conditions given by Eq. (12). The first nonvanishing contribution to $\rho_{2,n;2,n}$ is given by

$$\rho_{2,n;2,n}(\omega) \simeq i \int_{t_0}^{t_0+\tau} dt' \, \mathscr{V}_{2,n;1,n+1}(t') \rho^{(1)}{}_{1,n+1,2,n}(t') + \text{c.c.} \tag{21}$$

The argument ω in Eq. (21) indicates that we are considering an atom having that atomic frequency. The off-diagonal elements $\rho^{(1)}{}_{1,n+1;2,n}(t')$ are calculated from Eq. (19) as

$$\rho^{(1)}{}_{1,n+1;2,n}(t') = -i \int_{t_0}^{t'} dt'' \exp[-(i\nu - i\omega + \tfrac{1}{2}D)(t' - t'')] \times \mathscr{V}_{1,n+1;2,n} \rho_{nn}(t_0). \tag{22}$$

Inserting this into (21), we find

$$\rho_{2,n;2,n}(\omega) \simeq \int_{t_0}^{t_0+\tau} dt' \int_{t_0}^{t'} dt'' |\mathscr{V}_{2,n;1,n+1}|^2 \times \exp[-(i\nu - i\omega + \tfrac{1}{2}D)\,(t' - t'')]\rho_{nn}(t_0) + \text{c.c.} \tag{23}$$

In turn inserting this into (14), we find the probability that an atom of frequency ω has absorbed a laser photon in a time τ, that is,

$$\begin{aligned}\rho_{22}(\omega) &= -[g^2 \sum_{n=0}^{\infty} n\rho_{nn}(t_0)] \int_{t_0}^{t_0+\tau} dt' \int_{t_0}^{t'} dt'' \exp[-(i\nu - i\omega + \tfrac{1}{2}D)\,(t - t'')] \\ &\quad + \text{c.c.} \\ &\simeq \frac{g^2 D\tau \langle n \rangle}{(\omega - \nu)^2 + (D/2)^2}.\end{aligned} \tag{24}$$

This is in exact agreement with (17.51); Fig. 17-6 gives the ratio $\rho(\omega)/\rho(\nu)$, which equals $|E(\omega)/E(\nu)|^2$.

18-4. An Onsager Regression Hypothesis

The spectral profile for a system may be defined formally as the Fourier transform of the correlation function

$$G(t) = \tfrac{1}{2} Tr\{[a^\dagger(t)a(0) + a(t)a^\dagger(0)]\rho(0)_{\text{total}} + [a^\dagger(0)a(t) + a(0)a^\dagger(t)]\rho(0)_{\text{total}}\}, \tag{25}$$

where ρ_{total} is the density matrix for the total system of field plus reservoir. This section is intended to show that the linewidth implied by Eq. (25) agrees with that of (24) [or (17.51)].

In order to use (25), we must determine the time dependence of $a(t)$. Clearly it is not that of a free field, for then there would be no linewidth (delta-function spectrum). To obtain the temporal evolution of the correlation function $G(t)$, we must recall how "noise" entered the problem in the first place. We considered the radiation field to be acted upon by the pumping and damping atoms (reservoirs) and then traced the density matrix over the reservoir states. After contraction, the radiation field cannot be described by a state vector but is in a mixture requiring a density matrix for its specification. Thus we have extended definition (25) to include the reservoir states over which we will later trace. The time dependence of the operator $a(t)$ is now given by

$$a(t) = U^\dagger(t)a(0)U(t), \tag{26}$$

where $U(t)$ is the time development operator (see Prob. 6–7) for the combined laser-reservoir system. The correlation function with the atomic reservoirs included, but traced over, can be written in terms of these operators. In the following discussion, we consider only the quantity

$$g(t) = \mathrm{Tr}[a^\dagger(t)a(0)\rho_{\text{total}}(0)], \tag{27}$$

since the other terms of (25) are similarly obtained. Writing (27) in terms of (26), we have

$$\begin{aligned} g(t) &= \mathrm{Tr}_\rho \mathrm{Tr}_R \{[U^\dagger(t)a^\dagger(0)U(t)a(0)]R(0)\rho(0)\} \\ &= \mathrm{Tr}_\rho \{\mathrm{Tr}_R [U^\dagger(t)a(0)U(t)R(0)]a^\dagger(0)\rho(0)\}, \end{aligned} \tag{28}$$

where $R(0)$ is the reservoir density matrix which is uncoupled from the radiation field $\rho(0)$ at $t = 0$. Equation (28) is seen to have a simple form in terms of the new operator

$$\mathscr{A}(t) = \mathrm{Tr}_R[U^\dagger(t)a(0)U(t)R(0)], \tag{29}$$

namely,

$$g(t) = \mathrm{Tr}_\rho[\mathscr{A}^\dagger(t)\mathscr{A}(0)\rho(0)]. \tag{30}$$

We obtain an explicit form for the time dependence of the operators $\mathscr{A}(t)$ and $\mathscr{A}^\dagger(t)$ by recalling that the density matrix is given by

$$\rho_{n,n'}(t) = \{\mathrm{Tr}_R[U(t)R(0)\rho(0)U^\dagger(t)]\}_{n,n'}. \tag{31}$$

The time dependence of $\rho_{nn'}$, (t) is determined by the generalization of (17.48) derived in Prob. 18-2:

$$\dot\rho_{nn'}(t) = -[i\nu(n-n') + \tfrac{1}{2}D(n-n')^2]\rho_{nn'}(t). \tag{32}$$

The formal integral of this gives

$$\rho_{nn'}(t) = \rho_{nn'}(0)\exp\{-[i\nu(n-n') + \tfrac{1}{2}D(n-n')^2]t\}. \tag{33}$$

Comparing (31) with (33), we see that $U_{nn'}(t)$ is, in fact, diagonal. Hence Eq. (31) can be written as

$$\rho_{n,n'}(t) = \rho_{n,n'}(0)\,\mathrm{Tr}_R[U(t)_{n,n}R(0)U^\dagger(t)_{n',n'}]. \tag{34}$$

A comparison of (33) and (34) then shows that

$$\mathrm{Tr}_R[U(t)_{n,n}R(0)U^\dagger(t)_{n',n'}] = \exp[-\tfrac{1}{2}D(n-n')^2t - i\nu(n-n')t], \quad (35)$$

and we can write the time dependence of the matrix elements of the $\mathscr{A}$ operator (29) as

$$\begin{aligned}[\mathscr{A}(t)]_{n,n+1} &= a(0)_{n,n+1}\,\mathrm{Tr}_R[U(t)_{n,n}R(0)U^\dagger(t)_{n+1,n+1}] \\ &= a(0)_{n,n+1}\exp[-(\tfrac{1}{2}D - i\nu)t]. \quad (36)\end{aligned}$$

We can now write Eq. (28) as

$$G(t) = \mathrm{Tr}_\rho[a^\dagger(0)a(0)\rho(0)]\cos(\nu t)\exp(-\tfrac{1}{2}Dt). \quad (37)$$

It is clear that the Fourier transform of (37) and the other terms in Eq. (25) gives the spectral profile of Eq. (24). Accordingly, the decay of the two-time correlation function for the laser radiation (37) is identical to the single-time-averaged electric field as given by Eq. (17.50).

This section can be regarded as a demonstration of the Onsager regression hypothesis (Onsager, 1931). Lax (1967) has shown that Onsager's original statement for an equilibrium system is valid even for nonequilibrium situations, provided the system is Markoffian. Louisell and Marburger (1967) have extended the proof to show "that this result is always true, but that only under the conditions stated by Lax are the equations of motion the macroscopic ones."

Problems

18-1. Show with the coherent-state expansion for ρ_{nn} (7) that the photoelectron counting distribution (1) is given by (8).

18-2. Show along the lines of Sec. 17-3 that the general off diagonal density matrix element $\rho_{nn'}$, has the equation of motion (32) in the Schrödinger picture.

References

The material in this chapter is taken from the following:

M. O. Scully and W. E. Lamb, Jr., 1968, *Phys. Rev.* **166**, 246; 1969, *Phys. Rev.* **179**, 368.

Results of photon counting experiments are reviewed in these publications:

F. T. Arecchi, 1969, in: *Quantum Optics,* Proceedings of the International School of Physics, "Enrico Fermi," Course XLII, ed. by R. Glauber, Academic Press, New York.

E. R. Pike, 1970, in: *Quantum Optics,* Proceedings of the 10th Session of Scottish Universities' Summer School in Physics 1969, ed. by S. M. Kay and A. Maitland, Academic Press, London.

G. J. Troup, 1972, in: *Progress in Quantum Electronics,* Vol. 2, Part 1, ed. by J. H. Sanders and S. Stenholm, Pergamon Press, Oxford.

See also the following:

N. Bohr and L. Rosenfeld, 1933, Kgl. Danske Videnskab. Selskab, Mat. -fys. Medd. **12**; 1950, *Phys. Rev.* **78**, 194.

R. J. Glauber, 1965, in: *Quantum Optics and Electronics* (1964), Les Houches Summer Lectures, ed. by C. DeWitt, A. Blandin and C. Cohen-Tannoudji, Gordon and Breach, New York.

P. L. Kelley and W. H. Kleiner, 1964, *Phys. Rev.* **136**, A316.

M. Lax, *Phys. Rev.* **157**, 213, and his earlier works cited therein.

W. H. Louisell and J. H. Marburger, 1967, *J. Quantum Electronics* **QE-3**, 348.

L. Mandel, E. C. G. Sudarshan, and E. Wolf, 1964, *Proc. Phys. Soc.* **84**, 345.

L. Mandel and E. Wolf, eds., 1970, *Selected Papers on Coherence and Fluctuations of Light*, Dover, New York.

L. Onsager, 1931, *Phys. Rev.* **37**, 405.

XIX

RESERVOIR THEORY—NOISE OPERATOR METHOD

19. Reservoir Theory—Noise Operator Method

In Chap. 16 we developed the equation of motion for a system as it evolved under the influence of an unobserved (reservoir) system. We used the reduced density matrix concept and worked in an interaction picture. In this chapter we also consider the system-reservoir problem, but work with quantum operator equations of motion (see Sec. 7-4). Here too we eliminate the reservoir variables and find that the operator equations of motion for the system acquire damping terms and "noise" operators which produce fluctuations. The resulting equations resemble the classical Langevin equations which describe, for example, the Brownian motion of a particle suspended in a liquid or the behavior of a current as it passes through a resistor.

In Sec. 19-1 we review the Langevin equation theory of Brownian motion and show how the damping and diffusion terms enter the Fokker-Planck equation, such as that discussed in Sec. 16-2. Our review includes a brief derivation of the fluctuation-dissipation theorem and contains the chief mathematical techniques used in the rest of the chapter for quantum operators. In Sec. 19-2 we derive the (quantum) Langevin equations from the Heisenberg equations of motion for the uncomplicated but illuminating problem of a simple harmonic oscillator damped by a reservoir of other simple harmonic oscillators. We treated this problem earlier (Sec. 16-5) using the reduced density operator techniques. Unfortunately the method of Sec. 19-2 is limited to reservoirs of simple harmonic oscillators and hence is not sufficiently general for treating laser problems. In Sec. 19-3 we use perturbation theory to calculate the drifts of general operators from the Heisenberg equations of motion. The diffusion coefficients are then given in terms of these drifts by generalized Einstein relations. The laser linewidth is directly expressable in terms of a diffusion coefficient for the electric field operators. In our discussion we assume that the noise sources are correlated only over times much shorter than the drift times. This assumption is known as the Markoffian approximation and has the simple mathematical consequence of delta-function noise correlations [e.g., Eq. (6)]. More general discussions of noise operator theory are available in the references cited. In Chap. 20 we use the theory developed here to treat the laser again in a fully quantum-mechanical fashion.

19-1. Langevin Treatment of Brownian Motion

A particle suspended in a liquid obeys Newton's second law of motion (for each Cartesian component)

$$m\dot{v} = F_{\text{ext}}(t) + F_{\text{res}}(t), \tag{1}$$

where m is the mass of the particle, v is one Cartesian component of velocity, $F_{\text{ext}}(t)$ represents external forces, and $F_{\text{res}}(t)$ is the force exerted by the liquid reservoir. This second force damps the movement of the particle, as well as causing it to jitter about in Brownian motion. The damping component must clearly be proportional to some combination of powers of v since it vanishes for zero velocity. Experimentally it has been found that the first power suffices for many problems of interest.† Hence we set

$$F_{\text{res}}(t) = -m\Gamma\langle v\rangle + F_v(t), \tag{2}$$

where $F_v(t)$ is a random, rapidly fluctuating force with zero ensemble average:

$$\langle F_v(t)\rangle = 0 \tag{3}$$

(see Fig. 19-1 for an illustration of this average). In (2), we replace the ensemble average $\langle v\rangle$ by v inasmuch as the small fluctuations in v can be neglected in comparison to those of the force $F_v(t)$. Substituting (2) into (1) and ignoring the external force $F_{\text{ext}}(t)$, we find the Langevin equation:

$$m\dot{v}(t) = -m\Gamma v(t) + F_v(t). \tag{4}$$

The mean motion or "drift" of $v(t)$ is given immediately by (4) with (3):

$$m\langle\dot{v}(t)\rangle = -m\Gamma\langle v(t)\rangle. \tag{5}$$

The average $\langle[F_v(t)]^2\rangle$ does not vanish since negative swings of $F_v(t)$ yield positive squared values. Suppose that the minimum time in which $F_v(t)$ changes appreciably is called the correlation time τ_c. The average of the product $F_v(t)\,F_v(t')$ vanishes for $|t - t'| > \tau_c$, since then the product is as likely to be negative as positive. Hence $\langle F_v(t)\,F_v(t')\rangle$ is peaked about $t = t'$ and falls off to zero in a time difference $|t - t'| = \tau_c$. If τ_c is much less than

†See Reif (1965), Chap. 15.

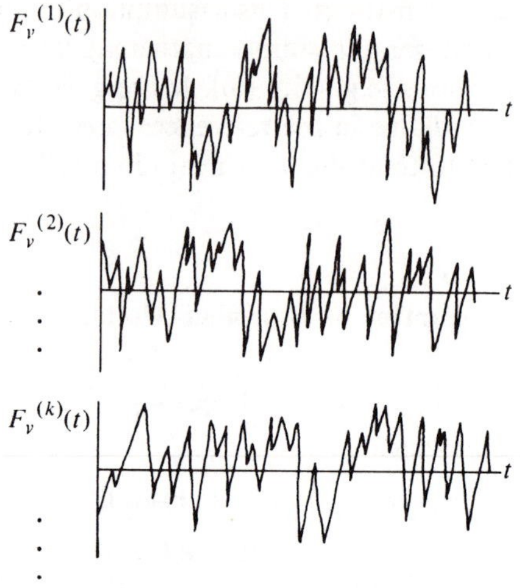

Figure 19-1. Possible time dependences of random force $F_v(t)$ for various members of the reservoir ensemble. The average value of $F_v(t)$ is zero, for members at a given point in time are as often positive as negative. The average of $F_v^2(t)$ is nonzero, since all members contribute positively. The average of $F_v(t)$ $F_v(t')$ is zero when t differs from t' by much more than the correlation time τ_c, since the product is then as often positive as negative for various members of the ensemble. When τ_c is short compared to other times, the average of $F_v(t)$ $F_v(t')$ acts as a delta function, Eq. (6) in the Markoffian approximation.

all other times of interest, for example, $1/\Gamma$ in (4), we can write

$$\langle F_v(t)\, F_v(t')\rangle = 2D_{vv}\, \delta(t - t'), \tag{6}$$

where D_{vv} is some constant expressing the magnitude of the fluctuating forces. When (6) is a good approximation, the system is said to be Markoffian. Our discussion will be limited primarily to such systems.

To relate the parameters Γ and D_{vv} to our work in Chap. 16, we consider the Fokker-Planck equation of motion (Sec. 16-3) for the probability density $P(v, t)$ for finding a particle with velocity component v. This equation is (16.74)

$$\frac{\partial}{\partial t} P(v, t) = -\frac{\partial}{\partial v}(M_1 P) + \frac{1}{2}\frac{\partial^2}{\partial v^2}(M_2 P). \tag{7}$$

Here M_n is the nth moment, defined by (16.73) as

$$M_n = \frac{\langle(\Delta v)^n\rangle}{\Delta t}$$

with the velocity increment

$$\Delta v(t) = v(t + \Delta t) - v(t), \tag{8}$$

where Δt is smaller than times characterizing the system (like $1/\Gamma$), but large compared to τ_c (which vanishes in the Markoffian approximation). We recall from Eqs. (16.75) and (16.77) that M_1 determines the "drift" [motion of the center of "mass" of the distribution $P(v, t)$], and M_2 the diffusion (spreading of P). We can calculate the moments from the Langevin equation (4) as follows. From (4), we find the formal integral

$$\Delta v(t) = -\Gamma \int_t^{t+\Delta t} dt'\, v(t') + \frac{1}{m} \int_t^{t+\Delta t} dt'\, F_v(t'). \tag{9}$$

Using the fact that $v\,(t)$ varies little in time Δt, we find

$$\begin{aligned} M_1 = \frac{\langle \Delta v \rangle}{\Delta t} &= -\Gamma\langle v\rangle + (m\,\Delta t)^{-1} \int_t^{t+\Delta t} dt' \langle F_v(t')\rangle \\ &= -\Gamma\langle v\rangle. \end{aligned} \tag{10}$$

Similarly, the second moment M_2 is given by

$$\begin{aligned} M_2 = \frac{\langle(\Delta v)^2\rangle}{\Delta t} = \frac{1}{\Delta t} \Big\{ (\Gamma\langle v^2\rangle)\,(\Delta t)^2 - \frac{2\Gamma}{m} \int_t^{t+\Delta t} dt' \int_t^{t+\Delta t} dt''\, \langle v(t')F_v(t'')\rangle \\ + m^{-2} \int_t^{t+\Delta t} dt' \int_t^{t+\Delta t} dt''\, \langle F_v(t')F_v(t'')\rangle \Big\}. \end{aligned} \tag{11}$$

The first term in braces is proportional to $(\Delta t)^2$ and yields a negligible value as $\Delta t \to 0$. To evaluate the second term, we derive first a very useful result for the correlation $\langle v(t)F_v(t)\rangle$. For this we note the identity

$$v(t) = v(t - \Delta t) + \int_{t-\Delta t}^{t} dt'\, \dot{v}(t'), \tag{12}$$

which implies that

$$\langle v(t)F_v(t)\rangle = \langle v(t - \Delta t)F_v(t)\rangle + \int_{t-\Delta t}^{t} dt' \langle[-\Gamma v(t') + \frac{1}{m} F_v(t')]F_v(t)\rangle. \tag{13}$$

The first term of the right-hand side of (13) vanishes because the velocity $v(t')$ at an earlier time $t' < t$ has no dependence on the future fluctuating force $F_v(t)$. Similarly the correlation $\langle v(t')\, F_v(t)\rangle$ vanishes except for the single point $t' = t$, which is multiplied by a zero time interval in the integral in (13). We are left with

$$\begin{aligned} \langle v(t)F_v(t)\rangle &= \frac{1}{m} \int_{t-\Delta t}^{t} dt' \langle F_v(t')F_v(t)\rangle \\ &= \frac{1}{m} \int_{-\infty}^{t} dt' \langle F_v(t')F_v(t)\rangle \\ &= \frac{1}{2m} \int_{-\infty}^{\infty} ds \langle F_v(0)F_v(s)\rangle, \end{aligned} \tag{14}$$

where in the last step we have taken the random force to be stationary in time, that is, the value depends only on the difference $t - t'$, not on t and t' individually. With the Markoffian, delta-correlated force of (6), this reduces to

$$\langle v(t)F_v(t)\rangle = \frac{D_{vv}}{m}. \tag{15}$$

Returning now to the second term in (11), we note that the maximum correlation between $v(t')$ and $F_v(t'')$ occurs for $t' \simeq t''$, since there exists no correlation for $t' < t''$ and decay sets in for $t' > t''$ Hence

$$\int_t^{t+\Delta t} dt' \int_t^{t+\Delta t} dt'' \langle v(t')\, F_v(t'')\rangle < \int_t^{t+\Delta t} dt' \int_t^{t+\Delta t} dt'' \langle v(t')F_v(t')\rangle$$

$$= \frac{D_{vv}}{m} (\Delta t)^2. \tag{16}$$

Thus the second term on the right-hand side of (11) is proportional to $(\Delta t)^2$, as was the first, and goes to zero as $\Delta t \to 0$. All that remains of (11) is the important result

$$\frac{\langle(\Delta v)^2\rangle}{\Delta t} = (\Delta t\, m^2)^{-1} \int_t^{t+\Delta t} dt' \int_t^{t+\Delta t} dt'' \langle F_v(t')F_v(t'')\rangle. \tag{17}$$

In terms of (6), this is

$$M_2 = \frac{\langle(\Delta v)^2\rangle}{\Delta t} = \frac{2D_{vv}}{m^2}. \tag{18}$$

Recognizing that it is this moment in the Fokker-Planck equation (7) that leads to diffusion [see Eq. (16.77)], we call D_{vv} the diffusion coefficient for the time development of $P(v, t)$. Problem 19-6 deals with the corresponding coefficient for diffusion in position x of a particle.

We remarked earlier that $F_{\text{res}}(t)$ was responsible for both the fluctuations and the dissipation (damping). Hence there ought to be a relation between the two. We can find this by considering the derivative

$$\frac{d}{dt}\langle v^2\rangle = 2\langle v\dot{v}\rangle = -2\Gamma\langle v^2\rangle + \frac{2}{m}\langle v(t)F_v(t)\rangle. \tag{19}$$

With (15), this gives

$$2D_{vv} = 2\Gamma m^2\langle v^2\rangle + m^2 \frac{d}{dt}\langle v^2\rangle. \tag{20}$$

In thermal equilibrium, the derivative vanishes and the law of equipartition of energy states

$$\tfrac{1}{2} m \langle v^2\rangle = \tfrac{1}{2}k_B T, \tag{21}$$

so that the diffusion coefficient becomes

$$D_{vv} = m\Gamma k_B T. \tag{22}$$

This relates the fluctuation and dissipation constants. A somewhat more general result (not requiring the Markoff approximation) is given by the substitution of (14) (derived with a bit more care) and (21) into (19), namely,

$$\Gamma = \frac{1}{2k_B T m}\int_{-\infty}^{\infty} ds \langle F_v(0)F_v(s)\rangle, \tag{23}$$

which is sometimes called the fluctuation-dissipation theorem. The same formula holds for the resistance R (in place of $m\Gamma$) encountered by a current passing through a resistor with the internal fluctuating voltage $V(t)$ [in place of $F_v(t)$]. The noise associated with this resistor is called "Johnson noise."

In the rest of this chapter, we develop analogs for quantum operators representing quantities of interest, for example, annihilation operators for simple harmonic oscillators (hence the radiation field) and spin-flip operators for two-level atoms. In Chap. 20 we then couple the various operators in a form appropriate for treating the laser.

19-2. Quantum Noise Treatment of Damped Harmonic Oscillator

Let us consider a system (A) consisting of a simple harmonic oscillator with frequency Ω and annihilation operator $a(t)$ which interacts with a reservoir (B) consisting of many other simple harmonic oscillators with closely spaced frequencies ω_k and annihilation operators $b_k(t)$ (see Fig. 19-2). The system-reservoir combination develops in time under the influence of the total Hamiltonian:

$$\begin{aligned}\mathscr{H} &= \mathscr{H}_A + \mathscr{H}_B + \mathscr{V} \\ &= \hbar\Omega(a^\dagger a + \tfrac{1}{2}) + \hbar\sum_k \omega_k(b_k{}^\dagger b_k + \tfrac{1}{2}) + \hbar\sum_k g_k(ab_k{}^\dagger + b_k a^\dagger).\end{aligned} \tag{24}$$

Since the system operators commute with the reservoir operators at a given time, the Heisenberg equations of motion (6.66) for the operators become

$$\begin{aligned}\dot{a}(t) &= \frac{i}{\hbar}[\mathscr{H}, a(t)] \\ &= \frac{i}{\hbar}[\mathscr{H}_A + \mathscr{V}, a(t)] \\ &= -i\Omega a(t) - i\sum_k g_k b_k(t),\end{aligned} \tag{25}$$

and similarly

$$\dot{b}_k = -i\omega_k b_k(t) - ig_k a(t). \tag{26}$$

We eliminate the reservoir operators $b_k(t)$ by substituting the formal integral of (26):

$$b_k(t) = b_k(t_0)\exp[-i\omega_k(t - t_0)] - ig_k\int_{t_0}^{t} dt'\, a(t')\exp[-i\omega_k(t - t')] \tag{27}$$

into (25). We find (changing the integration variable t' to $\tau = t - t'$)

$$\dot{a}(t) = -i\Omega a(t) - \sum_k g_k{}^2 \int_0^{t-t_0} d\tau\, a(t-\tau) \exp(-i\omega_k\tau)$$
$$- i \sum_k g_k b_k(t_0) \exp[-i\omega_k(t-t_0)]. \tag{28}$$

The time integration in this equation is similar to that encountered in the Weisskopf-Wigner theory of Sec. 14-4 [see Eq. (14.102)], and it is assumed here as it was there, that the interference time τ_c of $\sum_k g_k{}^2 \exp(-i\omega_k\tau)$ is much smaller than the time over which significant phase and amplitude modulations of $a(t)$ take place. Hence for times $t > t_0 + \tau_c$ the summation acts as a delta function [we ignore the principal part as in (14.104)], yielding

$$\int_0^{t-t_0} d\tau \sum_k g_k{}^2 \exp(-i\omega_k\tau)\, a(t-\tau) \simeq a(t) \int_0^{\infty} d\tau \sum_k g_k{}^2 \exp[i(\Omega - \omega_k)\tau]$$
$$= \tfrac{1}{2}\, \gamma a(t), \tag{29}$$

where the damping constant

$$\gamma = 2\pi[g(\Omega)]^2 \mathfrak{D}(\Omega) \tag{30}$$

as in Eq. (14.105).

We call the last term in (28),

$$f(t) = -i \sum_k g_k b_k(t_0) \exp[-i\omega_k(t-t_0)], \tag{31}$$

a "noise" operator because the "random" $b_k(t_0)$ introduce fluctuations or noise into the equations of motion for $a(t)$. The reservoir average $\langle f(t)\rangle_B$ vanishes in a fashion analogous to the classical counterpart $F_v(t)$ depicted in Fig. 19-1. Combining Eqs. (28), (30), and (31), we have

$$\dot{a}(t) = -(i\Omega + \tfrac{1}{2}\gamma)\, a(t) + f(t). \tag{32}$$

This equation has the same form as the Langevin equation (4) except for the rapidly varying time dependence of $a(t)$, caused by the $-i\Omega a(t)$ term. To apply the techniques of Sec. 19-1, we must eliminate this dependence, a feat easily accomplished through the use of the slowly varying (in times like $1/\gamma$) annihilation operator:

$$A(t) = \exp[i\Omega(t-t_0)]\, a(t) \tag{33}$$

and noise operator:

$$F(t) = \exp[i\Omega(t-t_0)]\, f(t)$$
$$= -i \sum_k g_k b_k(t_0) \exp[-i\,(\omega_k - \Omega)\,(t-t_0)]. \tag{34}$$

Equation (32) then becomes

$$\dot{A}(t) = -\tfrac{1}{2}\, \gamma A(t) + F(t), \tag{35}$$

which does have the form of the Langevin equation (4). Furthermore, since the $b_k(t_0)$ are random, their reservoir averages vanish, causing the reservoir averages of $F(t)$ and its adjoint $F^\dagger(t)$ to vanish as well:

$$\langle F(t)\rangle_B = \langle F^\dagger(t)\rangle_B = 0. \tag{36}$$

We can now calculate the mean motion (drift) of $A(t)$ and of the number operator $A^\dagger A$. In our discussion we relate the dissipation constant to the diffusion constants which arise. The drift is given immediately by the quantum Langevin equation (35) with the vanishing reservoir average (36):

$$\langle \dot{A}(t)\rangle_B = -\tfrac{1}{2}\gamma\langle A(t)\rangle_B. \tag{37}$$

Here we see that the mean value of the system operator goes to zero in time. Note that Eq. (37) is only averaged over the reservoir coordinates; it remains an operator in the system coordinates.

We suppose that the reservoir is in thermal equilibrium and hence that the average of the number operator $b_k^\dagger b_k$ is constant and is given by the Bose value (see Sec. 7-2 for discussion of thermal distributions):

$$\begin{aligned}\langle b_k^\dagger b_k\rangle_{\mathrm{B}} = \bar{n}(\omega_k) &= \frac{1}{Z}\sum_k \exp(-n_k\hbar\omega_k/k_{\mathrm{B}}T)\langle n_k|\,b_k^\dagger b_k\,|n_k\rangle\\ &= \frac{1}{\exp(\hbar\omega_k/k_{\mathrm{B}}T)-1}.\end{aligned} \tag{38}$$

Since the reservoir annihilation and creation operators are uncorrelated at any particular time, (38) can be generalized to,

$$\langle b_k^\dagger b_j\rangle_B = \bar{n}(\omega_k)\delta_{k,j}. \tag{39}$$

Using this relation with the noise operator value (34), we calculate the correlations $\langle F^\dagger(t)F(t)\rangle_B$ and $\langle F(t)F^\dagger(t)\rangle_B$ as follows:

$$\begin{aligned}\langle F^\dagger(t)F(t')\rangle_B &= \sum_k\sum_j g_k g_j\,\langle b_k^\dagger b_j\rangle_B \exp[i(\omega_k-\Omega)(t-t_0)\\ &\qquad - i(\omega_j-\Omega)(t'-t_0)]\\ &= \sum_k g_k^2\bar{n}(\omega_k)\;\exp[i(\omega_k-\Omega)\,(t-t')]\\ &= \pi[g(\Omega)]^2\,\mathfrak{D}(\Omega)\,\delta(t-t')\,\bar{n}(\Omega)\\ &= \gamma\bar{n}(\Omega)\,\delta(t-t').\end{aligned} \tag{40}$$

By analogy with the classical quantity D_{vv} in (6), we define the diffusion coefficient for $A^\dagger A$ through the equation

$$\langle F^\dagger(t)F(t')\rangle_B = 2\langle D_{A^\dagger A}\rangle_B\,\delta(t-t'). \tag{41}$$

Hence, from (40), the diffusion coefficient has the value

$$2\langle D_{A^\dagger A}\rangle_B = \gamma\bar{n}(\Omega). \tag{42}$$

Similarly, by noting that $b_k b_k^\dagger = b_k^\dagger b_k + 1$, we see from the derivation of (40) that

$$\langle F(t)F^\dagger(t')\rangle_B = 2\langle D_{AA^\dagger}\rangle_B\,\delta(t-t') = \gamma[\bar{n}(\Omega)+1]\,\delta(t-t'). \tag{43}$$

Using (40) and (43) in a derivation like that of Eqs. (12)–(15) for finding $\langle v(t)\,F_v(t)\rangle$, the reader can show (Prob. 19-1) that

$$\langle F^\dagger(t)A(t)\rangle_B = \langle A^\dagger(t)F(t)\rangle_B = \tfrac{1}{2}\gamma\bar{n}(\Omega) = \langle D_{A^\dagger A}\rangle_B. \tag{44}$$

This result with the Langevin equation (35) provides the answer for the mean time development of the system number operator $\langle A^\dagger(t)A(t)\rangle_B$:

$$\begin{aligned}\frac{d}{dt}\langle A^\dagger(t)A(t)\rangle_B &= -\gamma\langle A^\dagger(t)A(t)\rangle_B + \langle F^\dagger(t)A(t)\rangle_B + \langle A^\dagger(t)F(t)\rangle_B\\ &= -\gamma\langle A^\dagger(t)A(t)\rangle_B + \gamma\bar{n}(\Omega).\end{aligned} \tag{45}$$

Thus the steady-state value of the "intensity" operator $\langle A^\dagger(t)A(t)\rangle_B$ is $\bar{n}(\Omega)$ (times the system identity operator); this is nonzero in contrast to $\langle A^\dagger(t)\rangle_B$ and $\langle A(t)\rangle_B$, which decay to zero in time by (37). This fact is in agreement with the results of Sec. 16-3. Furthermore the commutator $[A(t), A^\dagger(t)]$ retains its unity reservoir average in time instead of decaying to zero. This it must do, of course, since the system-reservoir problem is treated as one large Hamiltonian system, but it is interesting to note that, if the noise sources were dropped, the averaged commutator would decay to zero along with $\langle A(t)\rangle_B$ (see Prob. 19-2).

Combination of Eqs. (45) and (42) yields the "Einstein relation"

$$2\langle D_{A^\dagger A}\rangle_B = \gamma\langle A^\dagger(t)A(t)\rangle_B + \frac{d}{dt}\langle A^\dagger(t)A(t)\rangle_B \tag{46}$$

for this damped harmonic oscillator problem. In Sec. 19-3 we derive more general relations valid for many system-reservoir problems, including the present one. These relations give the diffusion coefficients in terms of the drift coefficients and hence correspond to the classical fluctuation-dissipation theorem of Eq. (23) applied to Markoffian systems. In the present case, we already knew the value of the diffusion coefficient by virtue of the direct calculation in Eq. (40) together with (41). In the general case, calculation of a Langevin equation along the explicit lines of Eqs. (24)–(35) runs into insurmountable mathematical difficulties, and the drift and diffusion terms must be calculated from perturbation theory (as in Sec. 19-3). Hence the Einstein relations make it possible to avoid the effort of calculating the diffusion coefficients provided the drift terms are known.

To illustrate the nature of the difficulties of finding an explicit general Langevin equation, we attempt (unsuccessfully) to treat a simple harmonic oscillator (SHO) damped by a reservoir of two-level atoms (TLA) instead of other SHO's. The spin-flip operators σ_k and $\sigma_k{}^\dagger$ for this reservoir have an operator rather than a c-number for a commutator, and therein lies the rub. The Hamiltonian is

$$\mathscr{H} = \mathscr{H}_{\rm SHO} + \mathscr{H}_{\rm TLA} + \mathscr{V}$$

$$= \hbar\Omega a^\dagger a + \tfrac{1}{2}\hbar \sum_k \omega_k \sigma_{z,k} + \hbar \sum_k g_k(\sigma_k a^\dagger + \text{adjoint}). \tag{47}$$

The equations of motion for $a(t)$ and $\sigma(t)$ are

$$\dot{a}(t) = -i\Omega a(t) - i\sum_k g_k \sigma_k(t) \tag{48}$$

and

$$\dot{\sigma}_k(t) = -i\omega_k \sigma_k(t) + ia(t) g_k [\sigma_k^\dagger, \sigma_k]. \tag{49}$$

Solving formally for $\sigma_k(t)$ and noting (1.52) that $[\sigma_k^\dagger, \sigma_k] = \sigma_{z,k}$, we have

$$\sigma_k(t) = \sigma_k(t_0) \exp[-i\omega_k(t - t_0)] + ig_k \int_{t_0}^{t} dt'\, a(t')\sigma_{z,k}(t') \exp[-i\omega_k(t - t')]. \tag{50}$$

Unlike (27) for $b_k(t)$, this integral contains the time-dependent reservoir operator $\sigma_{z,k}(t)$. Substitution of (50) into (48) gives

$$\dot{a}(t) = -i\Omega a(t) - i\sum_k g_k \sigma_k(t_0) \exp[-i\omega_k (t - t_0)] - \sum_k g_k^2 \int_{t_0}^{t} dt'\, a(t')\sigma_{z,k}(t') \exp[-i\omega_k(t - t')]. \tag{51}$$

Here the "damping" term (last term of the right-hand side) still contains reservoir operators, which are just what we want to eliminate. In general, this kind of complication requires us to determine the drift and diffusion parameters by other means. To these we now turn.

19-3. General Calculation of Drift and Diffusion Coefficients

Let us assume that the system of interest, A, is described by a set of operators

$$\{a\} \equiv \{a_1, a_2, \ldots, a_\mu, \ldots\}, \tag{52}$$

where, for example, in the single oscillator problem

$$\{a\} = \{a, a^\dagger\}, \tag{53}$$

while in the two-level atomic problem

$$\{a\} = \{\sigma, \sigma^\dagger, \sigma_z\}. \tag{54}$$

For the class of problems concerning us, the commutation relations† have the form

$$[\mathscr{H}_A, a_\mu] = -\hbar\Omega_\mu a_\mu. \tag{55}$$

Here Ω_μ is a positive frequency ($\Omega_\mu = |\Omega_\mu|$) for positive frequency operators

†The commutation laws for the a_μ's, depending on the problem, may be those for Bose-Einstein or Fermi-Dirac statistics.

like the field annihilation operator $a(t)$, and is negative for the corresponding adjoint operators. The Heisenberg equation of motion (6.66) for these operators reduces to

$$\dot{a}_\mu(t) = \frac{i}{\hbar}[\mathscr{H}, a_\mu] = \frac{i}{\hbar}[\mathscr{H}_A + \mathscr{V}, a_\mu]$$

$$= -i\Omega_\mu a_\mu + \frac{i}{\hbar}[\mathscr{V}, a_\mu]. \tag{56}$$

We assume that the interaction energy $\mathscr{V}(t)$ consists of a sum of bilinear products, each containing one system operator and one reservoir operator, as, for example, in Eq. (24). The latter's reservoir average (36) vanishes, causing the average of $\mathscr{V}(t)$ itself to vanish:

$$\langle \mathscr{V}(t) \rangle_B = 0. \tag{57}$$

As such, $\mathscr{V}(t)$ acts very much as a noise operator.

We associate with each operator a_μ the slowly varying operator $A_\mu(t)$, defined by

$$A_\mu(t) = \exp(i\Omega_\mu t)\, a_\mu(t), \tag{58}$$

which, from (56), obeys the equation of motion

$$\dot{A}_\mu(t) = \frac{i}{\hbar}[\mathscr{V}(t), A_\mu(t)]. \tag{59}$$

We would like to obtain an equation of motion for A_μ in the absence of the reservoir operators contained in $\mathscr{V}(t)$. This equation would be the quantum Langevin equation

$$\dot{A}_\mu(t) = D_\mu(t) + F_\mu(t), \tag{60}$$

where $D_\mu(t)$ is the drift operator for $A_\mu(t)$, and $F_\mu(t)$ is the corresponding noise operator whose reservoir average vanishes:

$$\langle F_\mu(t) \rangle_B = 0. \tag{61}$$

In general, it is not possible to obtain the complete Langevin equation (60), but fortunately we need only certain averages to treat many problems of interest. Specifically, we need the drift term, given by the reservoir average of (60) as

$$\frac{d}{dt}\langle A_\mu(t) \rangle_B = \langle D_\mu(t) \rangle_B \tag{62}$$

and the diffusion coefficients, defined in terms of the "two-time" averages of noise operators:

$$\langle F_\mu(t) F_\nu(t') \rangle_B = 2\langle D_{\mu\nu} \rangle_B\, \delta(t - t'). \tag{63}$$

To find formulas for these quantities, we employ the mathematical techniques developed in the discussion of Brownian motion (Sec. 19-1), together with some perturbation theory. The difference

$$\Delta A_\mu(t) \equiv A_\mu(t + \Delta t) - A_\mu(t) \tag{64}$$

(where Δt is, as usual, a time smaller than those characterizing the system, like $1/\gamma$) can be calculated from the Heisenberg equations of motion (59) through the perturbative expansion:

$$\Delta A_\mu(t) = \frac{i}{\hbar}\int_t^{t+\Delta t} dt'\,[\mathscr{V}(t'), A_\mu(t)] + \left(\frac{i}{\hbar}\right)^2 \int_t^{t+\Delta t} dt' \int_t^{t'} dt''\,[\mathscr{V}(t'), [\mathscr{V}(t''), A_\mu(t)]] + \cdots. \tag{65}$$

We proceed to write the drift and diffusion coefficients in terms of the $A_\mu(t)$. From a formal integral of the Langevin equation (60), we have

$$\Delta A_\mu(t) = \int_t^{t+\Delta t} dt'\, D_\mu(t') + \int_t^{t+\Delta t} dt'\, F_\mu'(t). \tag{66}$$

Using the fact that $D_\mu(t)$ varies little in the time Δt and the zero reservoir average (61) of the noise operator $F_\mu(t)$, we find

$$\langle \Delta A_\mu(t)\rangle_B = \langle D_\mu(t)\rangle_B\, \Delta t,$$

that is,

$$\langle D_\mu(t)\rangle_B = \frac{\langle \Delta A_\mu(t)\rangle_B}{\Delta t}. \tag{67}$$

This is the quantum noise operator generalization of the drift (10) for Brownian motion.

To calculate the drift (67), we insert the perturbation expansion (65). The first-order integral contains the average $\langle[\mathscr{V}(t'), A_\mu(t)]\rangle_B$, which vanishes for $t' > t$ since $A_\mu(t)$ cannot acquire correlations with the future, random interaction energy $\mathscr{V}(t')$. Any nonzero correlation for $t \simeq t'$ can also be neglected inasmuch as it is multiplied by a time interval of approximately zero duration. We are left with the second-order term:

$$\langle D_\mu(t)\rangle_B \simeq -(\Delta t\, \hbar^2)^{-1} \int_t^{t+\Delta t} dt' \int_t^{t'} dt''\, \langle[\mathscr{V}(t'), [\mathscr{V}(t''), A_\mu(t)]]\rangle_B. \tag{68}$$

We can write this integral more conveniently by making changes in the variables as follows:

$$\int_t^{t+\Delta t} dt' \int_t^{t'} dt''\, \mathscr{I}(t', t'') = \int_t^{t+\Delta t} dt'' \int_{t''}^{t+\Delta t} dt'\, \mathscr{I}(t', t''), \tag{69}$$

for both integrations are carried out over the same area in the $t' - t''$ plane (see Fig. 19-2).

We further set

$$\tau' = t'' - t; \qquad \tau'' = t' - t'' = t' - t - \tau',$$

in terms of which (69) becomes

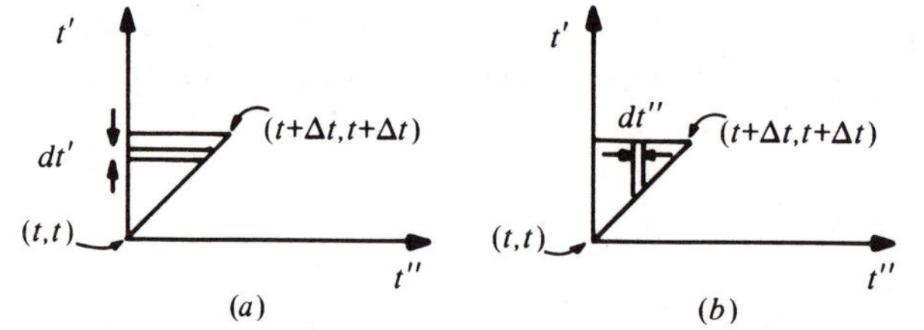

Figure 19-2. (a) Area in t'-t'' plane covered by integrals in left-hand side of (69). (b) Same area covered by integrals in right-hand side of (69).

$$\int_0^{\Delta t} d\tau' \int_0^{\Delta t-\tau'} d\tau'' \,\mathscr{I}(t+\tau'+\tau'', t+\tau'). \tag{70}$$

The drift term (68) becomes

$$\langle D_\mu(t)\rangle_B = -(\Delta t\hbar^2)^{-1} \int_0^{\Delta t} d\tau' \int_0^{\Delta t-\tau'} d\tau'' \langle \mathscr{V}[(t+\tau'+\tau''), [\mathscr{V}(t+\tau'), A_\mu(t)]]\rangle_B. \tag{71}$$

The $\mathscr{V}(t+\tau'+\tau'')$ and $\mathscr{V}(t+\tau')$ have nonzero correlation only for small τ''. Hence the integral over τ'' can be extended to ∞ without change in value. Furthermore the stationarity of the random processes makes it possible to let $\Delta t \to 0$ in the limit for τ'. Hence (with τ'' replaced by τ) we have the simpler formula

$$\langle D_\mu(t)\rangle_B = -\hbar^{-2} \int_0^\infty d\tau \,\langle[\mathscr{V}(t+\tau), [\mathscr{V}(t), A_\mu(t)]]\rangle_B. \tag{72}$$

To illustrate the formalism, let us consider again the simple harmonic oscillator with the system-reservoir Hamiltonian (24), which we write in terms of the slowly varying system annihilation operators $A(t) = \exp(i\Omega t)\, a(t)$ and reservoir annihilaton operators

$$B_k(t) = \exp(-i\omega_k t) b_k(t). \tag{73}$$

The interaction energy operator $\mathscr{V}(t)$ is given by

$$\begin{aligned}\mathscr{V}(t) &= \hbar \sum_k g_k a(t)\, b_k^\dagger(t) + \text{adjoint}\\ &= \hbar \sum_k g_k \exp[-i(\Omega - \omega_k)t]\, A(t) B_k^\dagger(t) + \text{adjoint}.\end{aligned} \tag{74}$$

We choose to find the drift for the annihilaton operator $A(t)$. Inserting $\mathscr{V}(t)$ of (74) first for the inside commutator of (72), we have

$$\langle D_A\rangle_B = -\frac{1}{\hbar}\int_0^\infty d\tau \sum_k g_k \langle[\mathscr{V}(t+\tau), -B_k(t)\exp[i(\Omega - \omega_k)t]\,]\rangle_B$$

$$= -\int_0^\infty d\tau \sum_j \sum_k g_j g_k \{\langle B_k(t) A(t+\tau) B_j^\dagger(t+\tau)\rangle_B$$

$$-\langle A(t+\tau) B_j^\dagger(t+\tau) B_k(t)\rangle_B\} \exp[i(\Omega-\omega_k)t - i(\Omega-\omega_j)(t+\tau)]$$

$$= \int_0^\infty d\tau \sum_k g_k^2 \langle [B_k(t), B_k^\dagger(t+\tau)]\rangle_B \langle A(t+\tau)\rangle_B \exp[-i(\Omega-\omega_k)\tau]$$

$$= -\tfrac{1}{2}\gamma \langle A(t)\rangle_B. \tag{75}$$

In reaching the final equality, we have used the fact that $\langle [B_k(t), B_j^\dagger(t+\tau)]\rangle_B = \delta_{j,k}\,\delta(\tau)$, which follows from the short correlation times of the $B_k(t)$ and the values of the relation $[B_k(t), B_j^\dagger(t)] = \delta_{jk}$. We then recover result (35) of Sec. 19-2.

To determine the diffusion coefficient in (63) we need the value of the correlation $\langle A_\mu(t)F_\nu(t)\rangle_B$. In analogy with the classical case in Eqs. (12)–(15), we find this by exploiting the identity

$$A_\mu(t) = A_\mu(t-\Delta t) + \int_{t-\Delta t}^{t} dt'\, \dot{A}_\mu(t') \tag{76}$$

in conjuction with the Langevin equation (60) for $\dot{A}_\mu$. We find

$$\langle A_\mu(t)F_\nu(t)\rangle_B = \langle A_\mu(t-\Delta t)F_\nu(t)\rangle_B$$

$$+ \int_{t-\Delta t}^{t} dt' \langle [D_\mu(t') + F_\mu(t')]F_\nu(t)\rangle_B. \tag{77}$$

The first term on the right-hand side goes to zero as $\Delta t \to 0$ since $A_\mu(t')$ with $t' < t$ has no dependence on the future force $F_\mu(t)$. Since $D_\mu(t')$ is a function of the $A_\mu(t)$, it also has no dependence on $F_\mu(t)$ except for $t' = t$, which is weighted by zero measure in the integral. Hence (77) reduces to [with (63)]

$$\langle A_\mu(t)F_\nu(t)\rangle_B = \int_{t-\Delta t}^{t} dt' \langle F_\mu(t')F_\nu(t)\rangle_B$$

$$= \langle D_{\mu\nu}\rangle_B. \tag{78}$$

The same method shows that

$$\langle F_\mu(t)A_\nu(t)\rangle_B = \langle D_{\mu\nu}\rangle_B. \tag{79}$$

With these relations, we can prove two useful results about the diffusion coefficients $\langle D_{\mu\nu}\rangle_B$, namely, their relations to ΔA_μ and ΔA_ν and the Einstein relation. For the first, we note from (66) that

$$\frac{\langle \Delta A_\mu(t)\,\Delta A_\nu(t)\rangle_B}{\Delta t} = \frac{1}{\Delta t}\int_t^{t+\Delta t} dt' \int_t^{t+\Delta t} dt'' \langle D_\mu(t')D_\nu(t'')$$

$$+ D_\mu(t')F_\nu(t'') + F_\mu(t')D_\nu(t'') + F_\nu(t')F_\nu(t'')\rangle_B. \tag{80}$$

The product $D_\mu(t')\,D_\nu(t'')$ varies little in Δt and can be factored out at time t, yielding a term proportional to Δt. This vanishes as Δt is taken sufficiently small. The maximum correlation between $D_\mu(t')$ and $F_\nu(t'')$ occurs for $t' \approx t''$ and is essentially a constant since D_μ is a function of the A_μ, which have the constant correlations (78) and (79). Hence

$$\frac{1}{\Delta t}\int_t^{t+\Delta t} dt' \int_t^{t+\Delta t} dt'' \langle D_\mu(t')F_\nu(t'') + F_\mu(t')D_\nu(t'')\rangle_B$$

$$< \int_t^{t+\Delta t} dt' \langle D_\mu(t')F_\nu(t') + F_\mu(t')D_\nu(t')\rangle_B$$

$$= 2\langle D_\mu(t)F_\nu(t)\rangle_B\, \Delta t,$$

which also vanishes as $\Delta t \to 0$. The last integral in (80) is just the diffusion coefficient $2\langle D_{\mu\nu}\rangle_B$. Hence we are left with

$$2\langle D_{\mu\nu}\rangle_B = \frac{\langle \Delta A_\mu(t)\, \Delta A_\nu(t)\rangle_B}{\Delta t}, \tag{81}$$

which is the quantum operator generalization of the second classical moment (18). This equation can be used with (65) to calculate $\langle D_{\mu\nu}\rangle_B$ directly by perturbation theory. It is easier, however, to find the drift D_μ first and then to use the generalized Einstein relations, which we now derive.

From the Langevin equation (60) and the correlations (78) and (79), we have

$$\begin{aligned}\frac{d}{dt}\langle A_\mu(t)\, A_\nu(t)\rangle_B &= \langle \dot{A}_\mu(t)A_\nu(t)\rangle_B + \langle A_\mu(t)\, \dot{A}_\nu(t)\rangle_B \\ &= \langle D_\mu(t)A_\nu(t)\rangle_B + \langle F_\mu(t)A_\nu(t)\rangle_B \\ &\quad + \langle A_\mu(t)D_\nu(t)\rangle_B + \langle A_\mu(t)\, F_\nu(t)\rangle_B \\ &= \langle D_\mu(t)A_\nu(t)\rangle_B + \langle A_\mu(t)D_\nu(t)\rangle_B + 2\langle D_{\mu\nu}\rangle_B.\end{aligned}$$

This gives the generalized Einstein relation

$$2\langle D_{\mu\nu}\rangle_B = -\langle D_\mu(t)\, A_\nu(t)\rangle_B - \langle A_\mu(t)\, D_\nu(t)\rangle_B + \frac{d}{dt}\langle A_\mu(t)\, A_\nu(t)\rangle_B, \tag{82}$$

which is our main result. This is used in Chap. 20 on the Langevin theory of the laser.

Problems

19-1. Prove Eq. (44), that is,

$$\langle F^\dagger(t)A(t)\rangle_B = \langle A^\dagger(t)F(t)\rangle_B = \tfrac{1}{2}\gamma\, \bar{n}(\Omega),$$

where $\bar{n}$ is the blackbody distribution given by (38). *Hint*: As for Eq. (38), assume that the reservoir average is thermal, as defined in terms of the photon number states as

$$\langle \mathcal{O}\rangle_B = \sum_n P_n\, \mathcal{O}_{nn} = \frac{1}{z}\sum_n \exp(-n\hbar\Omega/k_B T)\langle n|\mathcal{O}|n\rangle. \tag{83}$$

19-2. Show that the commutator

$$[A(t), A^\dagger(t)] = [a(t), a^\dagger(t)] = 1$$

for all time, provided that the noise operators in (34) and (31) are present
19-3. Solve for the equations of motion for the atom operators σ_a, σ_b [Eqs. (20.1 and 20.2)] and $\sum$ for a two-level atom coupled to a reservoir of simple harmonic oscillators by the interaction Hamiltonian

$$\mathscr{V}(t) = \hbar \sum_k g_k B_k \Sigma^\dagger \exp[-i(\omega_k - \omega)t] + \text{adjoint.} \tag{84}$$

Here $\Sigma(t)$ is the slowly varying operator corresponding to the spin-flip operator $\sigma(t)$. *Hint*: Use the perturbation theory expressions of Sec. 19-3; the method of Sec. 19-2 appears to work but presents some conceptual problems. The answers are given as Eqs. (20.7) and (20.9). Note that the expectation value of σ_a, $\langle\sigma_a\rangle = \text{Tr}(\rho\sigma_a) = \rho_{aa}$. Hence, since ρ_{aa} decays with the constant γ_a in the Schrödinger picture, that time dependence is shifted to σ_a in the Heisenberg picture.
19-4. Calculate the diffusion coefficient (81), using the perturbation expansion (65).
19-5. Calculate the drift for a simple harmonic oscillator damped by a two-level atom reservoir, using the perturbation theory formula (72).
19-6. Show, using an expansion for $x(t)$ like (12) for $v(t)$, that the average

$$\langle x(t)F_v(t)\rangle = 0. \tag{85}$$

Then show that the diffusion of a particle in Brownian motion is described by

$$\langle x^2\rangle = D_{xx}t = \frac{2k_B T}{m\Gamma} \tag{86}$$

by multiplying the Langevin equation (4) by $x(t)$ and manipulating. Note that the corresponding equation for three-dimensional Brownian motion includes a factor of 3. This is due to the facts that

$$\langle x^2\rangle = \langle y^2\rangle = \langle z^2\rangle = \tfrac{1}{3}\langle \mathbf{r}^2\rangle. \tag{87}$$

References

The material in this chapter is covered further in:

M. Lax, 1968, in: *Brandeis University Summer Institute Lectures (1966)*, Vol. II, ed. by M. Chretien, E. P. Gross and S. Deser, Gordon and Breach, New York.

W. H. Louisell, 1973, *Quantum Statistical Properties of Radiation in Pure and Applied Optics*, John Wiley & Sons, New York.

See also the following:

R. P. Feynman, R. B. Leighton, and M. Sands, 1965, *The Feynman Lectures on Physics*, Vol. 1, Addison-Wesley Publishing Co., Reading, Mass., Chap. 41.

H. Haken, 1969, *Handbuch der Physik*, Vol. XXV/2c, ed. by L. Genzel, Springer-Verlag, Berlin.

F. Reif, 1965, *Fundamentals of Statistical and Thermal Physics,* McGraw-Hill Book Co., New York, esp. Secs. 15-5 to 15-13.

I. R. Senitzky, 1970, *Phys. Rev.* **A3**, 421. Contains references to earlier works.

XX

LANGEVIN THEORY OF LASER FLUCTUATIONS

20. Langevin Theory of Laser Fluctuations

In this chapter we use the Langevin (Brownian motion) method of Chap. 19 to once again treat quantum mechanically both the field and the active atoms in a laser. The present treatment comprises a Heisenberg picture approach to the problem, as compared to the approach in Chap. 17, which represents the Schrödinger picture. The two methods are in complete agreement (as they should be!), a point discussed in some detail by Lax (1968). As before, the atoms are considered to comprise systems which interact with independent pumping and dissipation reservoirs, and the laser field is regarded as another system with reservoirs. The atom and field systems are then coupled by an electric-dipole perturbation energy, as depicted schematically in Fig. 20-1.

In Sec. 20-1 the atomic equations of motion with pumping and damping terms and associated noise operators are developed in the absence of coupling with the laser electric field. Diffusion coefficients which ultimately influence the laser's linewidth are calculated. In Sec. 20-2 the atomic equations are coupled to the single-mode laser field equations, and the former are then adiabatically eliminated (the atoms are assumed to follow the laser field instantly, i.e., γ, γ_a, $\gamma_b \gg \nu/Q$). The resulting Langevin equation for the field contains noise operators arising from both the atomic and field reservoirs. The relevant diffusion coefficients are then calculated. In Sec. 20-3 the steady-state solutions for the laser intensity and laser linewidth are obtained. These are compared to previous developments in the semiclassical (Chap. 8), density operator (Chap. 17), and purely classical (Chap. 4) approaches. We consider here the particularly simple laser model consisting of homogeneously broadened atoms and a unidirectional running-wave field (the z dependence cancels out). For greater generality, the reader is referred to the work of Lax (1968) and of Louisell (1973).

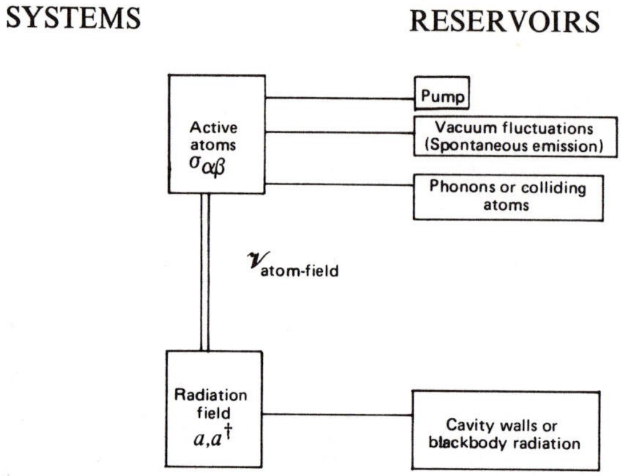

Figure 20-1. Diagram depicting the coupled atomic and field systems and their associated reservoirs encountered in the Langevin treatment of the laser. The reservoirs are taken to be independent of one another. For simplicity, we assume that the active atoms are located at z values for which sin Kz = 1. Hence z dependence is eliminated from the problem. Alternatively, we could consider a unidirectional ring. See Prob. 17-12 for a more general treatment with use of density matrix methods, including z dependence.

20-1. Atomic Drift and Diffusion

We represent the atoms by a three-level system as shown in Fig. 20-2. The operator representing the upper level a for the ith atom is the projection operator

$$\sigma_a{}^i(t) = (|a\rangle\langle a|)^i. \tag{1}$$

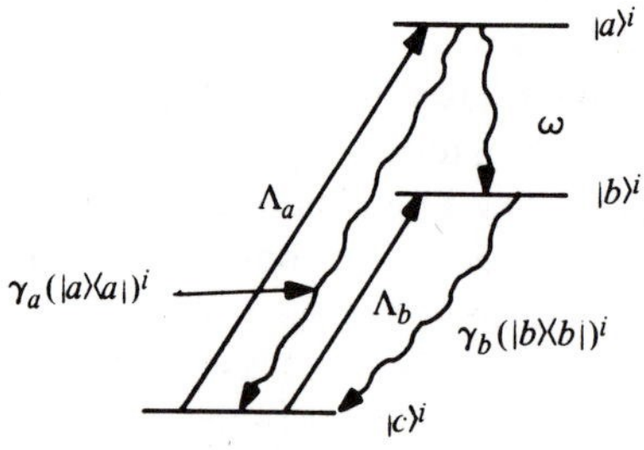

Figure 20-2. Three-level energy diagram showing pump operators Λ_a and Λ_b for the laser levels (a and b) and decay constants γ_a and γ_b. The superscript i indicates that the ith atom is being considered. Real transitions would involve some extra intermediate states. We assume they have been eliminated in a fashion akin to that in which the $|c\rangle$ state is eliminated here.

Similarly the operators for b and c are

$$\sigma_b^i(t) = (|b\rangle\langle b|)^i \tag{2}$$

and

$$\sigma_c^i(t) = (|c\rangle\langle c|)^i. \tag{3}$$

Normalization requires that

$$\sigma_a^i + \sigma_b^i + \sigma_c^i = I^i,$$

where I^i is the identity operator for the ith atom.

The excited atoms decay to the ground state (c) because of interaction with reservoirs (collisions or field fluctuations—spontaneous emission). The mathematical development is much like that of Sec. 19-2 and is the content of Prob. 19-3. Here we merely give the answer:

$$\frac{d}{dt}\langle[\sigma_\alpha^i(t)]\rangle_{\text{decay}} = -\gamma_\alpha\langle\sigma_\alpha^i(t)\rangle, \quad \alpha = a \text{ or } b. \tag{4}$$

The pump from the ground level adds terms proportional to σ_c^i, namely, $\lambda_\alpha'\sigma_c^i$. We assume that the ground state is sufficiently populated so that the pumping process affects it little (as in our semiclassical and density operator approaches). Hence we use pump operators

$$\Lambda_a^i = \lambda_\alpha'\sigma_c^i = \lambda_\alpha'(|c\rangle\langle c|)^i, \tag{5}$$

which are considered constant in time. We suppose furthermore that the atoms are sufficiently far apart that their states are independent of one another, that is, we take inner products

$$(\langle\alpha|)^i(|\beta\rangle)^j = \delta_{\alpha\beta}\delta_{ij}, \tag{6}$$

where α and β can label any level. Adding the pumping and dissipation contributions and including the corresponding noise operators, we have the formal Langevin equations for the level operators:

$$\frac{d}{dt}[\sigma_\alpha^i(t)] = \Lambda_\alpha^i - \gamma_\alpha\sigma_\alpha^i + F_\alpha^i(t). \tag{7}$$

Here for simplicity we have neglected spontaneous decay from level a to b (along with other transfer mechanisms).

Similarly we represent the complex atomic polarization by the spin-flip operator:

$$\sigma^i(t) = (|b\rangle\langle a|)^i. \tag{8}$$

This also decays because of reservoir interactions such as collisions and spontaneous emission, as discussed further in Prob. 19–3. Its equation of motion was found to be

$$\frac{d}{dt}[\sigma^i(t)] = -[\gamma + i\omega]\,\sigma^i(t) + f_\sigma^i(t), \tag{9}$$

in which we assumed that the pump is incoherent, that is, does not produce a superposition of states.

It is advantageous to work with an ensemble of many (N) atoms, just as it was convenient to use population matrices in the semiclassical theory. The corresponding atomic operators are defined by the expression

$$\mathcal{O} = \frac{1}{N}\sum_{i=1}^{N}\mathcal{O}^i. \tag{10}$$

Thus the average level operator

$$\sigma_a(t) = \frac{1}{N}\sum_{i=1}^{N}\sigma_a{}^i(t) = \frac{1}{N}\sum_{i=1}^{N}(|a\rangle\langle a|)^i. \tag{11}$$

We also find it convenient to define the slowly varying atomic flip operator:

$$\Sigma(t) = \exp(i\nu t)\,\sigma(t) = \frac{1}{N}\exp(i\nu t)\sum_{i=1}^{N}\sigma^i(t), \tag{12}$$

where ν is the laser oscillation frequency determined later. The equations of motion for these average operators are [from (7) and (9)]

$$\dot{\sigma}_a(t) = \Lambda_a - \gamma_a\sigma_a(t) + F_a(t) = D_a(t) + F_a(t), \tag{13}$$

$$\dot{\Sigma}(t) = -[\gamma + i(\omega - \nu)]\,\Sigma(t) + F_\Sigma(t) = D_\Sigma(t) + F_\Sigma(t), \tag{14}$$

$$\dot{\Sigma}^\dagger(t) = -[\gamma - i(\omega - \nu)]\,\Sigma^\dagger(t) + F_{\Sigma^\dagger}(t) = D_{\Sigma^\dagger} + F_{\Sigma^\dagger}, \tag{15}$$

where D_a, D_Σ, and $D_{\Sigma^\dagger}$ are drift terms for σ_a, Σ, and $\Sigma^\dagger$ (Sec. 19-3).

In our discussion of the laser line width, we need various atomic diffusion coefficients. These can be calculated by use of an "N-atom" extension of the Einstein relation (19.82).

For atomic operators only, we have the correlation†

$$\begin{aligned}\langle F_\mu(t)F_\nu(t')\rangle &= N^{-2}\sum_i\sum_j\langle F_\mu{}^i(t)F_\nu{}^j(t')\rangle \\ &= N^{-2}\sum_i\langle F_\mu{}^i(t)\,F_\nu{}^i(t')\rangle,\end{aligned}$$

where we have assumed that the various atomic reservoirs are uncorrelated. With the average atomic diffusion coefficient

$$\langle D_{\mu\nu}\rangle = \frac{1}{N}\sum_{i=1}^{N}\langle D_{\mu\nu}{}^i\rangle, \tag{16}$$

the noise correlation reads

$$\langle F_\mu(t)\,F_\nu(t')\rangle = \frac{1}{N}2\langle D_{\mu\nu}\rangle\,\delta(t - t'). \tag{17}$$

Using this with the Langevin equation $\dot{A}_\mu = D_\mu + F_\mu$, we have the N-atom relationship:

†In this chapter, the $\langle\,\rangle$ brackets denote a reservoir average. The result is, in general, a system operator, that is, *not* a c-number.

$$\frac{d}{dt}\langle A_\mu A_\nu\rangle = \langle D_\mu A_\nu\rangle + \langle A_\mu D_\nu\rangle + \langle F_\mu A_\nu\rangle + \langle A_\mu F_\nu\rangle$$

$$= \langle D_\mu A_\nu\rangle + \langle A_\mu D_\nu\rangle + \frac{1}{N} 2\langle D_{\mu\nu}\rangle,$$

that is,

$$2\langle D_{\mu\nu}\rangle = -N\langle D_\mu A_\nu\rangle - N\langle A_\mu D_\nu\rangle + N\frac{d}{dt}\langle A_\mu A_\nu\rangle. \tag{18}$$

In particular, a diffusion coefficient for the polarization operator is

$$2\langle D_{\Sigma^\dagger\Sigma}\rangle = -N\langle D_{\Sigma^\dagger}\Sigma\rangle - N\langle \Sigma^\dagger D_\Sigma\rangle + N\frac{d}{dt}\langle \Sigma^\dagger\Sigma\rangle. \tag{19}$$

To evaluate this quantity, we note from (6) that

$$\begin{aligned}\Sigma^\dagger\Sigma &= N^{-2}\sum_i\sum_j (|a\rangle\langle b|)^i(|b\rangle\langle a|)^j \\ &= N^{-2}\sum_i (|a\rangle\langle a|)^i \\ &= \frac{1}{N}\sigma_a.\end{aligned} \tag{20}$$

With this (19) becomes

$$\begin{aligned}2\langle D_{\Sigma^\dagger\Sigma}\rangle &= -N[-\gamma + i(\omega-\nu)]\langle\Sigma^\dagger\Sigma\rangle - N[-\gamma - i(\omega-\nu)]\langle\Sigma^\dagger\Sigma\rangle \\ &\quad + \langle\Lambda_a - \gamma_a\sigma_a\rangle \\ &= \langle\Lambda_a\rangle + (2\gamma - \gamma_a)\langle\sigma_a\rangle.\end{aligned} \tag{21}$$

Similarly, we find

$$2\langle D_{\Sigma\Sigma^\dagger}\rangle = \langle\Lambda_b\rangle + (2\gamma - \gamma_b)\langle\sigma_b\rangle. \tag{22}$$

Calculation of other atomic diffusion coefficients is deferred to Prob. 20-1. For our purposes, we need only (21) and (22).

20-2. Laser Equations of Motion

Equations (13)–(15) give the time rates of change of the atomic operators σ_a, σ_b (level occupancy operators), as well as Σ and $\Sigma^\dagger$ (electric dipole polarization operators) when solely under the influence of pumping and pumping reservoirs. The laser field annihilation operator $a(t)$ obeys an equation similar to that (19.32) developed in Sec. 19-2, namely,

$$\dot{a}(t) = -\left[\frac{1}{2}\frac{\nu}{Q} + i\Omega\right]a(t) + f(t). \tag{23}$$

Here we have expressed the field damping rate in terms of the cavity quality factor Q. In terms of the slowly varying operators

$$A(t) = a(t)\exp(i\nu t), \qquad F(t) = f(t)\exp(i\nu t), \tag{24}$$

Eq. (23) implies

$$\dot{A}(t) = -\left[\frac{1}{2}\frac{\nu}{Q} + i(\Omega - \nu)\right]A(t) + F(t). \tag{25}$$

The correlation properties of the noise operators F and $F^\dagger$ are given by the diffusion coefficients [from (19.42) and (19.43)]:

$$2\langle D_{A^\dagger A}\rangle = \frac{\nu}{Q}\,\bar{n}, \tag{26}$$

$$2\langle D_{AA^\dagger}\rangle = \frac{\nu}{Q}(\bar{n} + 1), \tag{27}$$

where for optical frequencies the average thermal photon number $\bar{n}$ (7.81) is tiny.

We now couple the atomic system to the field system as indicated in Fig. 20-1 by adding the electric-dipole perturbation energy,

$$\mathscr{V}(t) = \hbar g N \Sigma A^\dagger + \text{adjoint}, \tag{28}$$

to the Hamiltonian $\mathscr{H}$. The N appears here since N atoms are involved. With addition of appropriate commutators containing (28), Eqs. (13)–(15) and (25) become the coupled laser equations:

$$\dot{\sigma}_a = \Lambda_a - \gamma_a\sigma_a + ig[A^\dagger \Sigma - \Sigma^\dagger A] + F_a(t). \tag{29}$$

Similarly,

$$\dot{\sigma}_b = \Lambda_b - \gamma_b\sigma_b - ig[A^\dagger \Sigma - \Sigma^\dagger A] + F_b(t), \tag{30}$$

$$\dot{\Sigma} = -[\gamma + i(\omega - \nu)]\,\Sigma + ig(\sigma_a - \sigma_b)A + F_\Sigma(t), \tag{31}$$

$$\dot{A} = -\left[\frac{1}{2}\frac{\nu}{Q} + i(\Omega - \nu)\right]A - igN\Sigma + F(t). \tag{32}$$

We now suppose that the atoms relax rapidly to fluctuations about the reservoir averaged steady state of Eqs. (29)–(31), appropriate to the instantaneous laser field amplitude, that is,

$$\gamma, \gamma_a, \gamma_b \gg \frac{\nu}{Q}. \tag{33}$$

For this, $A(t)$ acts essentially as a constant in the atomic equations of motions, (29)–(31). We can then neglect the atomic time rates of change

$$\dot{\sigma}_a \simeq \dot{\sigma}_b \simeq \dot{\Sigma} \simeq 0, \tag{34}$$

if we simultaneously neglect frequency components in the noise operators with values greater than the atomic decay constants. This second neglect gives accurate results for the field since there is still enough bandwidth to treat the abridged atomic noise operators in a Markoffian fashion with respect to the more slowly varying field $A(t)$. With these approximations, Eq. (31) yields the spin-flip operator:

$$\Sigma(t) \cong \mathscr{D}(\omega - \nu)\,[ig(\sigma_a - \sigma_b)A(t) + F_\Sigma(t)], \tag{35}$$

where the complex denominator (9.6)

$$\mathscr{D}(\omega - \nu) = \frac{1}{\gamma + i(\omega - \nu)}. \tag{36}$$

In setting $\dot{\Sigma} = 0$, we neglect the more rapid noise fluctuations. Therefore one might ask why we keep the original noise operator $F_\Sigma(t)$ in (35) instead of one with truncated spectrum. This is done for convenience; the field $A(t)$ is too sluggish to tell the difference.

Substitution of (35) into (32) yields the field equation of motion:

$$\dot{A}(t) = -\left[\frac{1}{2}\frac{\nu}{Q} + i(\Omega - \nu)\right]A(t) + g^2 N\mathscr{D}(\omega - \nu)A(t)\,(\sigma_a - \sigma_b) + G(t), \tag{37}$$

where the new noise operator

$$G(t) = F(t) - igN\mathscr{D}(\omega - \nu)F_\Sigma(t). \tag{38}$$

From (37) with steady state solutions of (29) and (30), we see that noise contributed to the field by the level operators σ_a is of higher order in the atom-field interaction matrix element (g^2). In (37) we keep only noise terms to first order in g, that is, only the $F_\Sigma(t)$ term in (38), and therefore neglect the noise contributed by the level operators σ_a. Hence for (37) all we need is the reservoir average of the difference operator $N(\sigma_a - \sigma_b)$, namely,

$$N_a - N_b = N[\langle\sigma_a\rangle - \langle\sigma_b\rangle]; \tag{39}$$

in terms of this, Eq. (37) reads

$$\dot{A}(t) = -\left[\frac{1}{2}\frac{\nu}{Q} + i(\Omega - \omega)\right]A(t) + g^2\mathscr{D}(\omega - \nu)(N_a - N_b)A(t) + G(t). \tag{40}$$

In the adiabatic limit (33), we find the average equations

$$0 = \dot{N}_a = N\langle\Lambda_a\rangle - \gamma_a N_a - \mathscr{R}\,(N_a - N_b), \tag{41}$$

$$0 = \dot{N}_b = N\langle\Lambda_b\rangle - \gamma_b N_b + \mathscr{R}\,(N_a - N_b), \tag{42}$$

with the rate operator

$$\mathscr{R} = \frac{2g^2}{\gamma}\mathscr{L}(\omega - \nu)\,\langle A^\dagger A\rangle \tag{43}$$

and the dimensionless Lorentzian

$$\mathscr{L}(\omega - \nu) = \frac{\gamma^2}{\gamma^2 + (\omega - \nu)^2}. \tag{44}$$

To determine the average $N_a - N_b$ for (37), we divide (41) by γ_a and (42) by γ_b, and subtract to find

$$N_a - N_b = \mathscr{N} - \frac{\mathscr{R}}{R_s}(N_a - N_b),$$

with the unsaturated population difference

$$\mathscr{N} = N\left\langle\frac{\Lambda_a}{\gamma_a} - \frac{\Lambda_b}{\gamma_b}\right\rangle, \tag{45}$$

and the rate saturation parameter R_s of (8.38):

$$R_s = (\gamma_a^{-1} + \gamma_b^{-1})^{-1}.$$

Solving for the difference $N_a - N_b$, we have

$$N_a - N_b = \frac{\mathscr{N}}{1 + (\mathscr{R}/R_s)}. \tag{46}$$

Substituting (46) into (40), we find

$$\dot{A}(t) = -\left[\frac{1}{2}\frac{\nu}{Q} + i(\Omega - \nu)\right]A(t) + \left[\frac{g^2\mathscr{D}(\omega-\nu)\mathscr{N}}{1 + \mathscr{R}/R_s}\right]A(t) + G(t). \tag{47}$$

In keeping with our assumption that g^2 is small, we expand the denominator $1/(1 + \mathscr{R}/R_s)$ to first order in $\mathscr{R}/R_s$ (fourth order in g) to find

$$\dot{A}(t) = -\left[\frac{1}{2}\frac{\nu}{Q} + i(\Omega - \nu)\right]A(t) + \mathscr{A}_c A(t) - \mathscr{B}_c A\langle A^\dagger A\rangle + G(t), \tag{48}$$

with the complex linear gain coefficient

$$\mathscr{A}_c = g^2\mathscr{D}(\omega - \nu)\mathscr{N} \tag{49}$$

and the complex self-saturation coefficient

$$\mathscr{B}_c = 2\mathscr{A}_c\frac{g^2}{\gamma R_s}\mathscr{L}(\omega - \nu). \tag{50}$$

Here $\nu/2Q$ is the field decay constant; ν, Ω, and ω are the field oscillation, passive cavity, and atomic center frequencies, respectively; g is the atom-field coupling constant (without z dependence); $\mathscr{D}(\omega - \nu)$ is the complex Lorentzian denominator (36); $\mathscr{N}$ is the excitation difference (45); γ is the atomic polarization decay constant defined in (31); R_s is the rate saturation parameter (8.38); $\mathscr{L}(\omega - \nu)$ is the dimensionless Lorentzian (44); and $G(t)$ is the noise operator (38) containing contributions from both field and atomic reservoirs.

The noise correlations which ultimately determine the laser linewidth are $\langle G^\dagger(t)G(t')\rangle$ and $\langle G(t)G^\dagger(t')\rangle$. For (38) these are

$$\begin{aligned}\langle G^\dagger(t)G(t')\rangle &= \langle F^\dagger(t)F(t')\rangle + \left(\frac{gN}{\gamma}\right)^2\mathscr{L}(\omega - \nu)\langle F_\Sigma{}^\dagger(t)F_\Sigma(t')\rangle \\ &= \left[\frac{\nu}{Q}\bar{n} + N\left(\frac{g}{\gamma}\right)^2\mathscr{L}(\omega - \nu)\,2\langle D_\Sigma{}^\dagger{}_\Sigma\rangle\right]\delta(t - t')\end{aligned} \tag{51}$$

and

$$\langle G(t)G^\dagger(t')\rangle = \left[\frac{\nu}{Q}(\bar{n}+1) + N\left(\frac{g}{\gamma}\right)^2 \mathscr{L}(\omega-\nu)\, 2\langle D_{\Sigma\Sigma}\dagger\rangle\right]\delta(t-t'). \quad (52)$$

The diffusion coefficients $\langle D_\Sigma\dagger_\Sigma\rangle$ of (21) and $\langle D_{\Sigma\Sigma}\dagger\rangle$ of (22) can be expressed conveniently in terms of the reservoir averages of the steady-state solutions for the population operators in (41) and (42). For example,

$$N\langle \Lambda_a\rangle - \gamma_a N_a = \mathscr{R}(N_a - N_b), \quad (53)$$

which gives

$$2N\langle D_\Sigma\dagger_\Sigma\rangle = 2\gamma N_a + \mathscr{R}(N_a - N_a). \quad (54)$$

Similarly,

$$2N\langle D_{\Sigma\Sigma}\dagger\rangle = 2\gamma N_b - \mathscr{R}(N_a - N_b). \quad (55)$$

In particular, the rate term cancels in the sum of the correlations:

$$\langle G^\dagger(t)G(t')\rangle + \langle G(t)G^\dagger(t')\rangle$$
$$= 2\left[\frac{\nu}{Q}(\bar{n}+\tfrac{1}{2}) + \frac{g^2}{\gamma}\mathscr{L}(\omega-\nu)(N_a+N_b)\right]\delta(t-t'). \quad (56)$$

20-3. Classical Limit and the Laser Linewidth

With the Langevin theory it is possible to calculate the various correlations and averages determined earlier by density matrix methods, although the algebra is often more difficult and *aficionados* of the method frequently switch to the Fokker-Planck approach more or less in midstream. In this section we avoid this tempting latter course (but see Prob. 20-5) and give the classical limit and laser linewidth directly from the Langevin treatment.

For the classical limit, we define a dimensionless complex electric field

$$\alpha(t) = \mathrm{Tr}[\rho_f(t_0)\langle A(t)\rangle] \equiv \langle\langle A(t)\rangle\rangle = \frac{E(t)}{\mathscr{E}}\exp[-i\phi(t)], \quad (57)$$

in which $\rho_f(t_0)$ is a (constant) field density operator whose trace with the reservoir average of $A(t)$ yields a complex field expectation value. We further ignore the ordering between $A^\dagger(t)$ and $A(t)$ so that

$$\langle\langle A(t)\rangle\rangle\langle\langle A^\dagger(t)\rangle\rangle\langle\langle A(t)\rangle\rangle = \alpha(t)|\alpha(t)|^2. \quad (58)$$

With these approximations, we obtain from (48) the semiclassical expressions

$$\dot{E}(t) = -\frac{1}{2}\frac{\nu}{Q}E(t) + \tfrac{1}{2}E(t)\left\{\mathscr{A} - \mathscr{B}\left[\frac{E(t)}{\mathscr{E}}\right]^2\right\}, \quad (59)$$

$$\nu + \dot{\phi} = \Omega + \left(\frac{\omega-\nu}{\gamma}\right)\left[\mathscr{A} - \mathscr{B}\left[\frac{E}{\mathscr{E}}\right]^2\right], \quad (60)$$

where the real linear gain coefficient:

$$\mathscr{A} = 2\frac{g^2}{\gamma}\mathscr{N}\mathscr{L}(\omega - \nu) \tag{61}$$

and the (real) self-saturation coefficient:

$$\mathscr{B} = 2\mathscr{A}\frac{g^2}{\gamma R_s}\mathscr{L}(\omega - \nu) \tag{62}$$

reduce for $\omega = \nu$, $\gamma_a = \gamma_b = \gamma$ to those determined earlier in the density operator theory [(17.13) and (17.14)]. As discussed in Chap. 17, the semiclassical net gain coefficient of Chap. 8 $a_n = \frac{1}{2}(\mathscr{A} - \nu/Q)$, and the semiclassical self-saturation coefficient $\beta_n = (8\hbar\nu)^{-1}\,\varepsilon_0 V\mathscr{B}$, where V is the volume of the cavity, Equation (59) yields the steady-state average photon number:

$$\bar{n}_{ss} = \frac{\mathscr{A} - \nu/Q}{\mathscr{B}}. \tag{63}$$

The laser linewidth is given by the width of the spectral distribution:

$$I(\omega) = \int_{-\infty}^{\infty} dt\, \exp(-i\omega t)\,\langle A^\dagger(t)A(0)\rangle. \tag{64}$$

To calculate this quantity, we consider operation sufficiently far above threshold that the annihilation operator $A(t)$ can be written as the essentially classical expression

$$A(t) = \sqrt{\bar{n}_{ss}}\,\exp(-i\phi). \tag{65}$$

Here we ignore fluctuations in the amplitude inasmuch as this quantity is constrained to oscillate about its steady-state value. In contrast, the phase can change freely, allowing the "tip of the field (65) to diffuse out around a circle in the complex plane as depicted in Fig. 20-3. Hence we consider only the time dependence in $\phi(t)$. For operation very near threshold, the straight Langevin theory encounters severe difficulties and an alternative approach such as numerical integration of the Fokker-Planck or density matrix treatments is preferable if not actually required. In terms of (65), the correlation (64) is

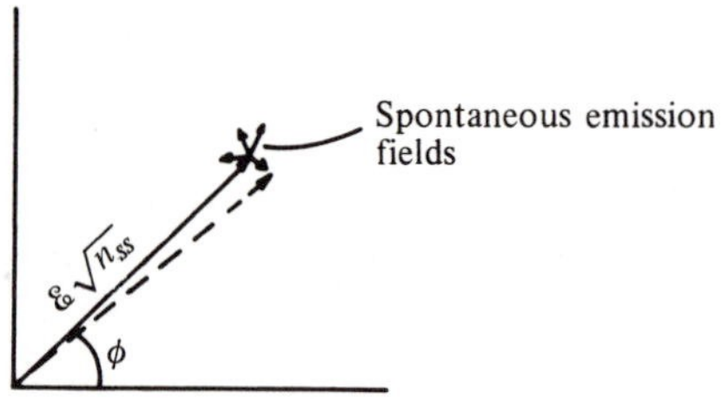

Figure 20-3. Electric field phasor with magnitude $\mathscr{E}\sqrt{\bar{n}_{ss}}$ buffeted about by small, spontaneously emitted fields.

$$\langle A^\dagger(t)A(0)\rangle = \bar{n}_{ss}\langle \exp\{i[\phi(t)-\phi(0)]\}\rangle$$
$$= \bar{n}_{ss}\exp\{-\tfrac{1}{2}\langle[\phi(t)-\phi(0)]^2\rangle\}, \tag{66}$$

where the second equality results from an analysis just like that for $\gamma = 1/T_2$ in Sec. 7-3. The argument of the exponential in (66) can be written in terms of the phase diffusion coefficient given by the general expression (19.81) as

$$2\langle D(\phi)\rangle = \lim_{\Delta t\to 0}\frac{\langle[\Delta\phi(t)]^2\rangle}{\Delta t}, \tag{67}$$

in which we take the limit $\Delta t \to 0$ since our problem is essentially Markoffian. The correlation (66) now reads

$$\langle A^\dagger(t)A(0)\rangle = \bar{n}_{ss}\exp[-\langle D(\phi)\rangle t]. \tag{68}$$

The Fourier transform of this (64) is a Lorentzian with full width at half maximum of $2\langle D(\phi)\rangle$. Hence this diffusion coefficient yields the laser linewidth that we seek.

To calculate its value, we solve for the phase difference

$$\Delta\phi(t) = \phi(t+\Delta t) - \phi(t) = \int_t^{t+\Delta t} dt'\,\dot{\phi}(t'). \tag{69}$$

Since from (65) $2i\phi = \ln A^\dagger - \ln A$, the equation of motion for $\phi(t)$ is seen from (65) to be

$$2i\frac{d\phi}{dt} = \frac{1}{A^\dagger}\frac{dA^\dagger}{dt} - \frac{1}{A}\frac{dA}{dt}, \tag{70}$$

which with (47) becomes

$$2i\frac{d\phi}{dt} = \frac{1}{\sqrt{\bar{n}_{ss}}}[\exp(-i\phi)\,G^\dagger(t) - \exp(i\phi)\,G(t)] + \text{constant}. \tag{71}$$

Integrating this equation from t to $t+\Delta t$ [i.e., evaluating (69)] and substituting the result into (67), we have

$$2\langle D(\phi)\rangle = -\tfrac{1}{4}\lim_{\Delta t\to 0}\frac{1}{(\sqrt{\bar{n}_{ss}})^2}\frac{1}{\Delta t}\int_t^{t+\Delta t} dt' \int_t^{t+\Delta t} dt''$$
$$\times \langle\{\exp[-i\phi(t')]\,G^\dagger(t') - \exp[i\phi(t')]\,G(t')\}$$
$$\times \{\exp[-i\phi(t'')]\,G^\dagger(t'') - \exp[i\phi(t'')]\,G(t'')\}\rangle$$
$$= \frac{1}{4}\frac{1}{\bar{n}_{ss}}\lim_{\Delta t\to 0}\frac{1}{\Delta t}\int_t^{t+\Delta t} dt' \int_t^{t+\Delta t} dt''$$
$$\times \langle G^\dagger(t')\,G(t'')\exp\{-i[\phi(t')-\phi(t'')]\} + \text{adjoint}\rangle.$$

Substituting (56) and taking the limit, we have

$$2\langle D(\phi)\rangle = \frac{1}{2}\frac{1}{\bar{n}_{ss}}\left[\frac{\nu}{Q}(\bar{n}+\tfrac{1}{2}) + \frac{g^2}{\gamma}\mathscr{L}(\omega-\nu)(N_a+N_b)\right]. \tag{72}$$

This expression can be written without the atomic coupling parameter g^2/γ

by use of the steady-state result (63). In fact, expansion of the population difference operator (46) corresponding to the third-order approximation gives the average saturated population difference:

$$N_a - N_b = \left[1 - 2\frac{g^2}{\gamma R_s}\mathscr{L}(\omega - \nu)\,\bar{n}_{ss}\right]\mathscr{N}$$

$$= \frac{1}{2}\frac{\nu}{Q}\left(\frac{g^2\mathscr{L}}{\gamma}\right)^{-1}. \tag{73}$$

Insertion of this value into (72) gives the laser linewidth:

$$2\langle D(\phi)\rangle = \frac{1}{2}\frac{\nu/Q}{\bar{n}_{ss}}\left(\bar{n} + \frac{1}{2} + \frac{1}{2}\frac{N_a + N_b}{N_a - N_b}\right)$$

$$= \frac{1}{2}\frac{\nu/Q}{\bar{n}_{ss}}\left(\bar{n} + \frac{N_a}{N_a - N_b}\right)$$

$$= \frac{1}{2}\frac{\nu/Q}{\bar{n}_{ss}}[\bar{n} + |\bar{n}_m|], \tag{74}$$

where the average number for the medium, $\bar{n}_m$, is defined by the expression

$$\bar{n}_m = \frac{1}{\exp(\hbar\omega/k_B T_m) - 1} \tag{75}$$

with a *negative* temperature T_m. The physical origin of the absolute value in (74) is the second-order correlation function for diffusion, in which the sign is squared, unlike the first-order correlation yielding loss or gain depending on the sign. Inasmuch as in steady-state operation the (saturated) gain equals the loss, the coefficient for both processes is $\nu/2Q$. For the temperatures $T = 0$, $T_m = -0$ ($N_b = 0$), Eq. (74) reduces to the linewidth calculated earlier (17.49) with use of density matrix techniques.

To gain some insight into (74), consider the thermal fluctuation contribution $(\nu/2Q)\bar{n}\hbar\omega/(\bar{n}_{ss}\hbar\omega)$. For temperatures much higher than $\hbar\omega/k_B$, the exponential in (16.32) defining $\bar{n}$ can be expanded, yielding

$$2\langle D(\phi)\rangle \cong \frac{\frac{1}{2}(\nu/Q)\,\bar{n}\hbar\omega}{\bar{n}_{ss}\,\hbar\omega} \simeq \frac{\frac{1}{2}(\nu/Q)\,k_B T}{\bar{n}_{ss}\hbar\omega}. \tag{76}$$

This is analogous to the diffusion coefficient (19.22) for a particle undergoing Brownian motion, here along an arc of a circle with radius $\sqrt{\bar{n}_{ss}\hbar\omega}$. Since the diffiusion coefficient of interest is that for the phase ϕ, we convert from diffusion on the arc by dividing by the radius squared [$(\Delta\phi)^2$ is involved].

For optical frequencies, $\hbar\omega \simeq 2\text{eV}$, which is much greater than room-temperature thermal energy, $k_B T \simeq \frac{1}{40}$ eV, so that $\bar{n} \simeq 0$. Thus for lasers the thermal contribution can be neglected. The contribution of the medium remains, however, and is due to spontaneous emission from the upper level a to the lower level b. To see this, we note that $\mathscr{A} \simeq \nu/Q$ and write (74) as (dropping the thermal contribution)

$$2\langle D(\phi)\rangle \simeq \frac{\frac{1}{2}\mathscr{A}\bar{n}_m}{\bar{n}_{ss}} = \frac{(g^2/\gamma)(N_a - N_b)}{\bar{n}_{ss}}\frac{N_a}{N_a - N_b}$$

$$= \frac{N_a(g^2/\gamma)\,\mathscr{E}^2}{(\sqrt{\bar{n}_{ss}}\,\mathscr{E})^2}, \tag{77}$$

where $\mathscr{E}$ is the electric field "per photon." Figure 20-3 depicts the electric field expectation value having the magnitude $\mathscr{E}\sqrt{\bar{n}_{ss}}$ and phase ϕ. Spontaneous emission adds a small fluctuating field with the mean-square rate $N_a(g^2/\gamma)\mathscr{E}^2$. Schematically this value reads

$$N_a\left(\frac{g^2}{\gamma}\right)\mathscr{E}^2 = \text{(number of upper-level atoms)}$$
$$\times\text{(probability/second of spontaneous emission into laser mode by a single atom)}$$
$$\times\text{(electric field/photon)}^2.$$

Division by the square of the electric field amplitude $E\sqrt{\bar{n}_{ss}}$ then plausibly† gives the phase fluctuation rate $2\langle D(\phi)\rangle$, a value understood entirely in terms of spontaneous emission. Other effects (e.g., shot noise) contribute small corrections which are neglected in our present treatment.

Problems

20-1. Using the Einstein relation (18) and derivations like the one for Eq. (20), show that the various atomic diffusion coefficients are given by

$$2\langle D_{aa}\rangle = \gamma_a\langle\sigma_a\rangle + \langle\Lambda_a\rangle$$
$$= \text{atomic rate out} + \text{atomic rate in},$$
$$2\langle D_{b\Sigma}\rangle = \gamma_b\langle\Sigma\rangle,$$
$$2\langle D_{\Sigma^\dagger b}\rangle = \gamma_b\langle\Sigma^\dagger\rangle,$$
$$2\langle D_{\Sigma a}\rangle = \gamma_a\langle\Sigma\rangle,$$
$$2\langle D_{a\Sigma^\dagger}\rangle = \gamma_a\langle\Sigma^\dagger\rangle,$$
$$\langle D_{\Sigma b}\rangle = \langle D_{a\Sigma}\rangle = \langle D_{b\Sigma^\dagger}\rangle = \langle D_{\Sigma^\dagger a}\rangle = 0,$$
$$2\langle D_{\Sigma\Sigma^\dagger}\rangle = \langle\Lambda_b\rangle + (2\gamma - \gamma_b)\langle\sigma_b\rangle.$$

20-2. If we had not neglected noise operators of $\mathcal{O}(g^2)$, we would have had to consider the expression

$$B = ig[A^\dagger\Sigma - \Sigma^\dagger A]$$

in the level operator equations (29) and (30). Substitution of (35) for the spin-flip operator $\Sigma(t)$ yields "noisy" terms which, however, do not have zero res-

†That is, $\Delta\phi \propto E_{\text{-spont}}/E_{\text{-stim}}$, so that $(\Delta\phi)^2 \propto 1/E_{\text{stim}}{}^2 \propto 1/\bar{n}_{ss}$.

ervoir average. Using the techniques which led to the Einstein relation (19.82), show that, for $\gamma \gg \omega - \nu$, B can be written as

$$B = -\mathscr{R}(\sigma_a - \sigma_b) - \frac{2g^2}{\gamma}\mathscr{L}(\omega - \nu)\sigma_a + F_B,$$

where

$$F_B = ig\mathscr{D}(\omega - \nu)\, A_c{}^\dagger F_\Sigma + \text{adjoint}.$$

is a pure noise operator, that is, $\langle F_B \rangle = 0$, and where $A_c{}^\dagger = A^\dagger(t_c)$ with $t_c \lesssim t$, which is uncorrelated with $F_\Sigma(t)$. The drift term so separated consists of the 1 that might be expected from σ_a due to spontaneous emission. *Hint:* Show first along the lines of Eq. s. (19.76)–(19.79) that $\langle A^\dagger(t)F_\Sigma(t)\rangle = ig\mathscr{D}(\nu - \omega)$ $\langle D_\Sigma{}^\dagger{}_\Sigma\rangle$ and that $\langle F_\Sigma{}^\dagger(t)A(t)\rangle = ig\mathscr{D}(\omega - \nu)\langle D_\Sigma{}^\dagger{}_\Sigma\rangle$. Then set $\langle D_\Sigma{}^\dagger{}_\Sigma\rangle/\gamma \approx \langle\sigma_b\rangle$, since γ is taken sufficiently large. Finally remove the reservoir averages, inserting the A_c. Note that $\langle\sigma_b\rangle$ can be replaced simply by σ_b since the additional noise introduced is negligible compared to that in F_B.

20-3. Calculate the correlation of the noise operator F_B of Prob. 20-2. This noise operator corresponds to the induced transitions. Specifically show that

$$N^2\langle F_B(t)F_B(t')\rangle = 2\left(\frac{g}{\gamma}\right)^2 \mathscr{L}(\omega - \nu)A^\dagger A(N_a + N_b)$$

$$= \text{photon contribution to shot noise.}$$

20-4. Show that, to order g^2/γ and $(g^2/\gamma)^2 A^\dagger A\rangle$,

$$\frac{d}{dt}\langle A^\dagger A\rangle = \left[\mathscr{A} - \frac{\nu}{Q}\right]\langle A^\dagger A\rangle + -\mathscr{B}\langle A^\dagger A A^\dagger A\rangle.$$

This agrees with the density operator results (17.20) in this approximation.

20-5. Calculate the Fokker-Planck equation corresponding to the Langevin equation (48).

References

See References of Chap 19.

For experimental evidence on laser linewidth and many references, see H. Gerhardt, H. Welling and A. Güttner, 1972, *Z. Phys.* **253**, 113.

XXI

OUTLOOK

21. Outlook

In this chapter we discuss several ways in which laser physics makes contact with other fields and contributes to them. In Sec. 21-1 we mention a few subjects that have been laser dominated, for example, nonlinear optics and light scattering. In Secs. 21-2, 21-3, and 21-4 we outline recent research efforts which, in our opinion, provide good examples of how the techniques applied extensively in the laser analysis may be profitably applied to other problems in nonequilibrium quantum statistical mechanics. These problems also involve open systems and are influenced by nonlinear coherent and incoherent events. It is reasonable, then, to expect that the techniques which have been extensively studied and applied to problems such as the laser will have wide application in other areas. Specifically, three problems have been chosen for presentation here:

a. A philosophical interpretation of laser "quasimodes" (Appendix B) in terms of the "true" modes of the universe.
b. The analogy between the laser and a second-order phase transition.
c. An outline of the quantum theory of Josephson radiation along the lines of the quantum theory of the laser (Sec. 16-2 and Chap. 17).

21-1. Impact of the Laser

The laser has had a profound influence on modern optical physics. For example, the current interest in nonlinear optics is a result of the high-intensity optical fields produced by the laser. This study is the subject of several textbooks, and we do not discuss it in any detail here. However, it should be noted that the essential ingredients for the understanding of nonlinear optical phenomena are contained in the expressions for the macroscopic polarization

$$P_\nu = \varepsilon_0 \chi_{\nu\mu}{}^{(1)} E_\mu + \varepsilon_0 \chi_{\nu\mu\rho}{}^{(2)} E_\mu E_\rho + \varepsilon_0 \chi_{\nu\mu\rho\sigma}{}^{(3)} E_\mu E_\rho E_\sigma, \tag{1}$$

as derived, for example, for the (nonlinear) laser problem in Sec. 9-1. Determination of the χ coefficients leads to an enhanced understanding of matter in general and its interaction with radiation in particular. Experimentally we are interested in observing radiation at a harmonic of the incident laser frequency. This radiation is produced, of course, by the nonlinear dipole moment, which is driven by the incident laser field.

An aspect of solid-state spectroscopy which has been stimulated by the laser is the scattering of radiation by, for example, density fluctuations produced by fluctuations in the entropy S and pressure P:

$$\delta\rho(\mathbf{r}, t) = \left(\frac{\partial\rho(\mathbf{r}, t)}{\partial S}\right)_P \delta S + \left(\frac{\partial\rho(\mathbf{r}, t)}{\partial P}\right)_S \delta P. \tag{2}$$

The first term in Eq. (2) leads to Rayleigh scattering, whereas the second term leads to Brillouin scattering. Again this is the subject of several review articles and summer school lectures (see the references at the end of the chapter) and are not presented in detail here. The availability of *tunable* dye and spin-flip Raman lasers has made considerable impact on this area of spectroscopy and on related ones. These systems provide very useful spectroscopic probes and will be utilized, no doubt, in many new ways in times to come.

In addition two new forms of spectroscopy have been created because of the laser. The first developed from an understanding of gas lasers, namely, the so-called Lamb dip or saturation spectroscopy (see Sec. 10-1 for background). This technique enables us to measure the line centers and natural lifetimes of atomic systems to an accuracy determined by the radiative decay of the atomic constituents, rather than the Doppler width of the corresponding inhomogeneous profile. Second, the availability of mode-locked lasers in the picosecond regime has led to a new kind of time-resolved spectroscopy which permits the measurement of temporal events on the subnanosecond time scale This has been applied to the measurement of lifetimes in liquids and solids and is presently being applied to the measurement of decay times in the electronic fluid of an excited semiconductor.

Most of the applications utilize the high irradiance of the laser field but do not require the associated coherence properties. Examples of phenomena requiring a coherent field for their observation are the optical echo and self-induced transparency processes discussed in Chap. 13. More recently the optical analog of nuclear magnetic resonance spin nutation has been observed. The application of the laser to so-called coherent or heterodyne detection provides a method for the elimination of detector noise in radiation systems. Furthermore the laser's high intensity has made the technique of self-beat spectroscopy practicable.

In most laser spectroscopy, the measurement of the spectral profile yields all available information since the scattered radiation or the absorption processes are Gaussian random processes. For these, correlations of an order higher than first are determined by the first-order correlations as discussed in Sec. 7-3. For more general processes, however, the photon-counting techniques discussed in Chaps. 16–18 provide nonredundant information concerning higher-order correlations, that is, information not contained in the spectral density (Fourier transform of the second-order correlation). In addition to the use of photon-counting techniques in the investigation of laser radiation emitted near threshold, these techniques hold promise for other areas of scientific investigations. For example, photon-counting techniques have recently been applied in studying macromolecular systems.

In yet another application of the laser, the utilization of picosecond neodymium-glass lasers to pump a potential X-ray laser has indicated that it may be possible to generate coherent X-rays in this fashion. An X-ray laser would open a whole new realm of physical investigations and consequently is being considered theoretically and experimentally by several workers. The device is quite different from those usually considered in laser technology since, for example, the pumping mechanism must overcome very short atomic lifetimes. This fact requires a more careful theoretical treatment of the pump than that given in Sec. 8-1. Furthermore the dipole approximation (Sec. 2-1) may not be valid since the lasing atoms occupy dimensions not necessarily small compared with the wavelength of the emitted radiation.

The extension of laser theories to include high-energy laser work involving, for example, the gas dynamic and TEA (transversely excited atmospheric) CO_2 systems is now being actively pursued. For example, the way in which an unstable resonator changes its optical properties when filled with an active driving medium is of current research interest.

21-2. Theory of a Laser "Quasimode"

One recent extension of laser theory considered in this text is the philosophical discussion by Lang, Scully, and Lamb (1973) of the question, "Why is the laser line so narrow?" An elementary response is that the laser gain compensates for the cavity losses, yielding a *constant* (i.e., undecaying) electric field amplitude which therefore has a zero width spectrum. The work of Lang et al, is concerned with the underlying (microscopic) laser physics of this process. They describe the laser in terms of the (true) modes of the laser universe system sketched in Fig. 21-1, rather than the discrete set of quasi modes (Appendix B). Normal modes of the open Fabry-Perot cavity cannot be rigorously defined because of diffraction losses and imperfect mirrors.

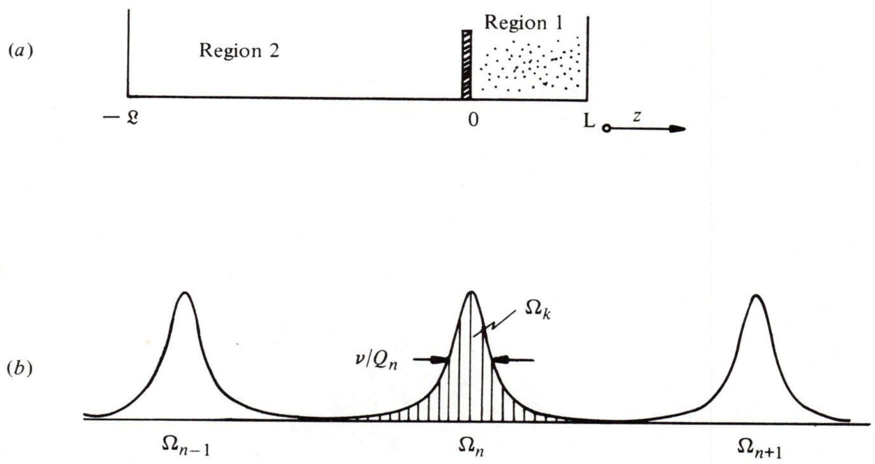

Figure 21-1. *(a) Small cavity (from $z = 0$ to $z = L$) representing laser inside large cavity ($z = -\mathfrak{L}$ to $z = 0$) approximating free space. (b) Pseudomode spectral profiles with representative lines for modes of larger cavity.*

This resonator problem was first investigated numerically by Fox and Li (1961) and analytically by Boyd and Gordon (1961), who showed that a high-Q Fabry-Perot type of cavity has a discrete set of quasi modes for which the diffraction leakage from the cavity is small. The usual laser theory (Chap. 8) begins with these Fox-Li modes simulating the cavity losses by, for example, ohmic currents smeared throughout the cavity.

From the present point of view, we consider the theory of an optical maser based on a model for the cavity which is coupled to the outside world by means of a dielectric mirror. In this model there are a large (infinite) number of modes of the universe corresponding to each of the Fox-Li modes of the lasing cavity. In this way, the coupling to the outside world replaces the leakage of radiation from our lasing cavity. The motives for this approach are as follows:

1. The sense in which the present multimode approach reduces to the usual quasimode treatment and therefore provides a rigorous foundation for the latter is a problem of current interest. For example, in the quantum theory of the laser, one assumes that the modes of the Fox-Li cavity may be quantized and later introduces a loss mechanism to simulate radiation leakage. This approach is good for most laser calculations but is not the most fundamental procedure imaginable, that is, the atoms in the mirrors were ignored in the calculation previously outlined. They are taken into account by means of an explicit boundary value problem, and in this sense we are not treating the radiation in free space but rather are treating the radiation as it is bounded by the atoms and electrons in the mirrors, as in Fig. 21-1*a*. A more rigorous calculation would involve the quantized field in vacuum and subsequent detailed consideration of these free-field modes and their interaction with the atoms and electrons in the mirrors. Since this is a very complicated problem, we have previously been satisfied with a less ambitious approach involving only the Fox-Li quasi modes and their subsequent quantization.

2. By determining how the multimode approach reduces to the single Fox-Li mode treatment, we hope to understand the mechanism leading to the extreme monochromaticity of the laser radiation. The argument that laser radiation is monochromatic as a consequency of gain narrowing is often advanced. We regard this argument as incomplete at best, and a more fundamental approach involving many modes (of the universe) is desired. The present calculation indicates that the narrowness of the laser linewidth should be regarded as a consequence of a locking phenomenon between these modes.

3. It is of interest to investigate the sense in which the present calculation, which does not treat cavity dissipation phenomenologically, still implies a fluctuation-dissipation theorem and indeed leads to it. That is to say, if we no longer consider the equation of motion for a single Fox-Li mode as being damped phenomenologically (in addition to which there must be a Langevin

noise source), in what fashion do we recover the corresponding fluctuation-dissipation theorem?

4. The fully quantized theory of laser oscillation of Chap. 17 could be reasonably and readily carried out by quantizing a radiation field in the entire space.

5. Finally, there is current interest in understanding the extent to which the laser threshold is related to the physics of phase transition phenomena. It has been asserted by other workers that the Bose condensation of liquid helium below its lambda point is analogous to the "condensation" of laser radiation into a single mode. It is clear that such a question can be analyzed only in the context of a many-mode laser analysis of the present variety.

Details of the calculations for the normal modes of the laser coupled to the rest of the world have been given by Lang, Scully, and Lamb (1973). For the present analysis we need only mention that the normal-mode functions $U_k(z)$ are given by

$$U_k(z) = \begin{cases} M_k \sin [k(z - L)]. & z > 0, \\ \xi_k \sin [k(z - \mathfrak{L})], & z < 0, \end{cases} \tag{3}$$

where k is a wave number, the lengths $\mathfrak{L}$ and L are defined in Fig. 21-1a, and the ξ_k factors alternate between plus and minus unity when we go from one mode to the next. The coefficients M_k are proportional to

$$\frac{\nu}{Q}\left[(\Omega_k - \Omega)^2 + \frac{\nu}{Q}\right]^{-1/2}, \tag{4}$$

where the finite cavity Q results from nonzero mirror transmission. A field (8.8) for the nth Fox-Li mode at time $t = 0$ is

$$E(z, 0) = \tfrac{1}{2} E_n(0) \exp(-i\phi_n)U_n(z) + \text{c.c.} \tag{5}$$

As indicated in Fig. 21-1b, this is given by the sum over modes of the universe as

$$E(z, 0) = \tfrac{1}{2} \sum_k \mathscr{E}_k(0)U_k(z) + \text{c.c.}, \tag{6}$$

where $\mathscr{E}_k(t)$ is the complex, slowly varying amplitude for the kth mode of the universe. In time, this field develops into

$$E(z, t) = \tfrac{1}{2} E_n(0) \exp[-i(\nu_n t + \phi_N)] \left\{\sum_k \frac{\mathfrak{L}}{L} M_k U_k(z) \exp[-i(\Omega_k - \nu_n)t]\right\} + \text{c.c.} \tag{7}$$

Evaluaation of the sum over wave number k gives

$$E(z, t) = \tfrac{1}{2} E_n(0) \exp[-i(\nu_n t + \phi_n)] \exp[-(\nu/2Q_n)t]U_n(z) + \text{c.c.}, \tag{8}$$

which indicates that the field is passing from the laser cavity region into the external world and the electric field thereby decays exponentially because of this leakage. Hence we see that our system does, in fact, mimic the decay associated with the Fox-Li quasi modes.

Inclusion of excited lasing atoms in the calculation yields a multimode amplitude-determining equation for $\mathscr{E}_k(t)$ which has the form of a complex combination of the real equations (9.18) and (9.19). This is

$$\dot{\mathscr{E}}_k(t) + i(\Omega_k - \nu_n)\mathscr{E}_k(t) + \frac{1}{2}\frac{\nu}{Q_n}\mathscr{E}_k(t) = a_g M_k \sum_\kappa M_\kappa \mathscr{E}_\kappa(t)$$
$$- \beta M_k \sum_\kappa \sum_{\kappa'} \sum_{\kappa''} M_\kappa M_{\kappa'} M_{\kappa''} \mathscr{E}_\kappa \mathscr{E}_{\kappa'}^* \mathscr{E}_{\kappa''}, \tag{9}$$

where a_g and β are gain and saturation constants, respectively, and κ, κ', κ'' are wave numbers. With knowledge of Sec. 9-2, one might expect from (9) that the cross saturation (terms like $\theta_{\kappa\kappa'}$) would be greater than the self-saturation (terms like β_κ) and would lead to mode suppression. However, the numerous terms with relative phase angles implicit in the complex product $\mathscr{E}_\kappa \mathscr{E}^*_{\kappa'} \mathscr{E}_{\kappa''}$ tend to wash out this strongly coupled interaction. With considerable analysis, it can be shown that (9) reduces to the single-mode amplitude- and frequency-determining equations (8.50) and (8.51).

Physically, the calculation reveals that the modes of the universe corresponding to the single quasi mode oscillate at one common, self-consistent frequency with specific relative phases. The modes are labeled by their wavelengths and thereby are, in principle, distinguishable entities, whereas their frequencies are *locked* together. The response of a given mode is analogous to that for van der Pol's triode oscillator circuit (Sec. 4-2), which locks to the frequency of an external injected signal. In the present problem, each universe mode locks to a self-consistent, internally generated field. We discuss this phenomenon further in the laser-phase transition analogy of Sec. 21-3. For now, we emphasize that the monochromaticity of a single-mode laser radiation is, in fact, a consequence of mode locking.

This result contrasts with that often heard in which gain narrowing is invoked to explain the monochromaticity of the laser. According to the latter explanation, the universe modes nearest the central frequency of the Fox-Li mode experience greater *net* gain (gain minus loss with Lorentzian line shape) than those in the wings of the loss distribution. Hence in time the central modes are amplified more than the wing modes and compete successfully for the available gain. Ultimately the radiation narrows into a delta function. This explanation is based more on mode competition (and suppression) than on mode locking and is consequently an incorrect interpretation of the physics.

It can also be shown that, when other additive losses augment the mirror transmission losses, a fluctuation-dissipation theorem (Sec. 19-1) very analogous to that expected by a lossy cavity description is obtained. A fully quantized version of the analysis makes close contact with the Bose-Einstein condensation alluded to in the motives for this research. Furthermore, it yields the small but nonzero linewidth of Sec. 17-3 caused by spontaneous emission fluctuations neglected in the semiclassical approach.

21-3. Phase Transition Analogy

We now turn to a comparison between the laser near threshold and matter near a phase transition, as developed by DeGiorgio and Scully (1970). A basic feature of the laser theory is that the problem essentially involves an open system, that is, atoms are injected and removed at random times so that the total problem is not necessarily well treated by Hamiltonian dynamics. The description of the laser radiation is itself a problem in nonequilibrium statistical mechanics. The finite linewidth of the laser is a consequence of spontaneous emission, which can be thought of as contributing a small electric field of random phase to the existing field. This leads to a phase diffusion of the total electromagnetic field in the fashion discussed in Sec. 20-3. The situation is analogous to the problem found in ferromagnetism, where the magnetization experiences a buffeting from external reservoir influences. This buffeting leads to a decay of the magnetization, albeit on a geological time scale, and this is very similar to the problem that leads to the nonzero linewidth of the laser. Measured in terms of reciprocal frequency, the decay time of the ensemble average electric field is also very long.

In many other ways the physics of the laser is analogous to that of a cooperative system near a phase transition. The root of the analogy is the fact that both the laser and the ferromagnet (or superconductor) are, to a conceptual idealization, mean field or self-consistent field theories. In the many-body ferromagnet or superconductor problem any given element of the ensemble sees all the other elements through the self-consistent mean field. For example, in the magnetic problem each spin communicates with all other spins through the average magnetization. Similarly the atoms in the laser contribute to the total electric field by means of their induced dipole. This dipole is in turn induced by the mean electric field as contributed by all of the other atoms in the ensemble.

In order to establish a completely satisfactory analogy it is necessary to extend the laser analysis to include the effects of a classical injected signal, which corresponds to the externally applied symmetry-breaking field in the magnetic problem, as in Fig. 21-2. Since this exercise requires an extension of the previous calculations, we give the result below.

Near the threshold region, the laser Fokker-Planck equation given by (17.25) is changed by the presence of the external signal to read as follows:

$$\frac{\partial P}{\partial t} = -\frac{1}{2}\frac{\partial}{\partial a}\left[\left(\mathscr{A} - \frac{\nu}{Q} - \mathscr{B}|a|^2\right)aP + \mathscr{A}\frac{\partial}{\partial a^*}P + 2\mathscr{S}P\right] + \text{c.c.} \quad (10)$$

Figure 21-2. External Symmetry Breaking field.

The steady-state solution ($\partial P/\partial t = 0$) to this equation is given by

$$P(\alpha) = \mathcal{N}_P \exp\left\{\frac{4}{\mathscr{A}}\left[\frac{1}{4}\left(\mathscr{A} - \frac{\nu}{Q} - \tfrac{1}{2}\mathscr{B}|\alpha|^2\right)|\alpha|^2 + \tfrac{1}{2}\mathscr{S}(\alpha + \alpha^*)\right]\right\}, \quad (11)$$

where $\mathscr{S}$ is taken to be real and $\mathcal{N}_P$ is a normalization constant.

Writing $P(\alpha)$ in Cartesian coordinate form $P(x, y)$ as in Prob. 16–10, we can calculate $\langle x\rangle$. The ensemble averaged electric field $\langle E\rangle$ is proportional to $\langle x\rangle$ since we have chosen $\mathscr{S}$ to be real. The effect of the injected signal on $\langle E\rangle$ is illustrated in Fig. 21-3 for both quantum and semiclassical treatments.

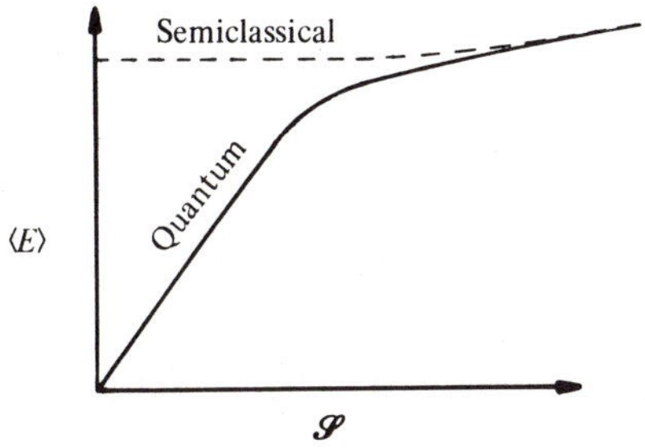

Figure 21-3. Electric field expectation value ⟨E⟩ (solid line) versus injected signal $\mathscr{S}$ as given by the quantum theory of radiation. Corresponding classical theory value (dashed line).

There we see that quantum mechanically the ensemble averaged field vanishes for $\mathscr{S} = 0$, a fact due to the phase diffusion of the field, as discussed in the linewidth calculations of Secs. 17-3 and 20-3. In contrast, the classical field does *not* vanish. For $\mathscr{S} > 0$, the ensemble average acquires nonzero values since the field or phasor is then constrained by the injected signal from diffusing out around a circle in the complex plane.

Let us now consider the corresponding expression for the probability density $P(M)$ for a ferromagnetic system of magnetization M near a phase transition. In thermodynamic equilibrium, this density is given by

$$P(M) \propto \exp[-F(M)/k_B T], \tag{12}$$

where the free energy $F(M)$ is given by (in the Landau approximation to the second-order phase transition)

$$F(M) = C(T - T_C)M^2 + DTM^4 + HM. \tag{13}$$

Here C and D are constants, T_C is the Curie temperature, and H is an externally injected signal which serves to constrain the magnetization M in a fashion analogous to $\mathscr{S}$ for the electric field E.

From (13), we see that the laser distribution $P(x, y)$ is given by

$$P(x, y) = \mathscr{N}_P \exp[-G(x, y)/k_L\sigma], \tag{14}$$

where $4k_L$ is the spontaneous emission rate of one atom, and $G(x, y)$ has a form quite analogous to that obtained in the Landau problem as given in Table 21-1 and plotted in Fig. 21-4. It should be emphasized that the

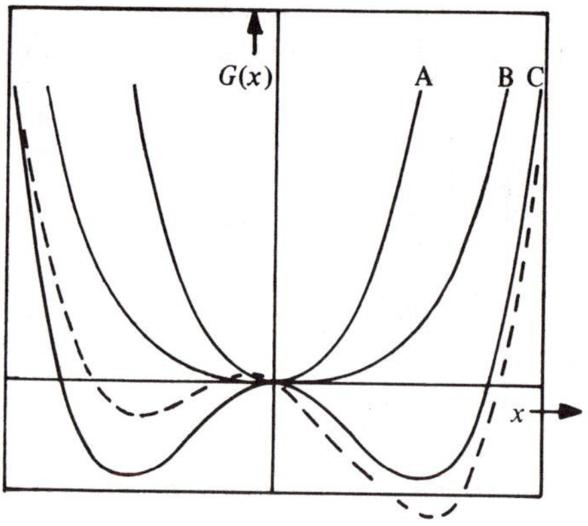

Figure 21-4. Plot of the energy function $G(x, y)$ for $y = 0$ and zero injected signal. Curves A, B, and C correspond to a laser approximately 7% below, at, and 7% above threshold. With $\mathscr{S} \neq 0$, the dip for $x > 0$ above threshold is greater than that for $x < 0$, that is, the free energy is less. The lowest value gives the solution realized in practice.

ferromagnetic order-disorder transition required three thermodynamic variables for its description, namely, M, H, and T. In order to establish a complete analogy the ferromagnetic problem must be viewed as involving the three variables E, $\mathscr{S}$, and σ. Hence the correspondence between the external magnetic field and the symmetry-breaking mechanism in the laser analysis is clearly established. The analogy is summarized in Table 21-1.

Table 21-1. Summary of Comparison Between the Laser and Ferromagnet in Mean-Field Approximation.

Laser variables not appearing elsewhere are defined as follows: σ is the zero-field excess number of atoms in the excited state over the lower state ($\sigma = r_a/\gamma$ for the simple case of Sec. 17-1); A and B are linear gain and saturation coefficients given by $\mathscr{A}/\sigma$ and $\mathscr{B}/\sigma$ respectively [Eqs. (17.13) and (17.14) give simple values for $\mathscr{A}$ and $\mathscr{B}$], and the complex scalar electric field $\mathscr{E}a = E \exp(i\phi) = x + iy$.

QUANTITY	LASER	FERROMAGNET
Order parameter	E	M
Reservoir variable	σ (population inversion)	T (temperature)
	σ_t (threshold inversion)	T_c (critical temperature)
Coexistence curve	$E = \begin{cases} 0, & \sigma < \sigma_t \\ \left[\dfrac{A}{B}\left(\dfrac{\sigma - \sigma_t}{\sigma}\right)\right]^{1/2}, & \sigma > \sigma_t \end{cases}$	$M = \begin{cases} 0, & T > T_c \\ \left[\dfrac{C}{D}\left(\dfrac{T - T_c}{T}\right)\right]^{1/2}, & T < T_c \end{cases}$
Symmetry breaking mechanism	$\mathscr{S}$ (classical injected signal)	H (external field)
Critical isotherm†	$E = (\mathscr{S}/\sigma_t B)^{1/3}$	$M = (H/T_c D)^{1/3}$
Zero field susceptibility	$\chi = \dfrac{\partial E}{\partial \mathscr{S}}\Big\vert_{\mathscr{S}=0} = \begin{cases} \{A(\sigma_t - \sigma)\}^{-1}, & \sigma < \sigma_t \\ \{2A(\sigma - \sigma_t)\}^{-1}, & \sigma > \sigma_t \end{cases}$	$\chi = \dfrac{\partial M}{\partial H}\Big\vert_{H=0} = \begin{cases} \{C(T - T_c)\}^{-1}, & T > T_c \\ \{2C(T_c - T)\}^{-1}, & T < T_c \end{cases}$
Thermodynamic potential	$G(E) = -\dfrac{A}{2}(\sigma - \sigma_t)E^2 + \dfrac{B}{4}\sigma E^4 - \mathscr{S}E\cos\phi + G_0$	$F(M) = -\dfrac{C}{2}(T - T_c)M^2 + \dfrac{D}{4}TM^4 - HM + F_0$
Statistical distribution	$P(E) = \exp\{-G(E)/k_L\sigma\}$	$P(M) = \exp\{-F(M)\, k_B T\}$

The paper by Lang et al. (1973) described in Sec. 21-2 offers an added dimension to the phase transition analogy based on macroscopic quantities, for it shows that the laser quasimode is the locked superposition of many modes of free space. We see that, below threshold, the free-space modes have a spread of frequencies and unrelated phases, whereas above threshold they assume the same phase and identical frequencies. Similarly in superconductivity, Cooper pairs of electrons of momenta by **k** and $-$ **k** have a spread of energies and unrelated phases above the critical temperature and have the same energy and phase below it.

The similarities are the result of the presence in all cases of a competition between an ordering or cooperative influence and a disordering force. In the

†value of order parameter at critical point.

locking problems (multimode laser operation of Chap. 9 included), the injected signals, internally generated or externally imposed, act in concert to order the oscillation either into a single frequency or into a periodic array. These ordering forces compete with the disordering sustaining forces that attempt to drive the various modes at unrelated frequencies. Similarly, in second-order phase transitions, such as spins aligning in a ferromagnet or Cooper pairs "phase locking" in a superconductor, there exists an ordering influence yielding lower energy that competes against the randomizing effects of thermal fluctuations. At low enough temperatures ($<T_C$), these fluctuations become small enough to allow the ordering forces to win.

A principal difference between the locking problems and the ordinary phase transitions is that the former are not in thermal equilibrium: the locking systems continuously dissipate energy, in distinct contrast to typical second-order phase transitions. The "dissipative phase transition" is not restricted to mode-locking phenomena. It may occur, for example, in a liquid layer heated from below (Benard instability; see Chandresekhar, 1961). For further discussion of this transition and the analogies in general, the reader is referred to recent review papers by Graham (1973) and Scully (1973).

21-4. Josephson Radiation

Recently, superconducting tunnel jutions have been used to measure the ratio $2e/h$ with sufficient precision to challenge the earlier results of atomic physics. This work is based on Josephson's demonstration that coherent radiation is emitted when two pieces of superconductor are separated by a thin film and maintained at a potential difference V. In Josephson's thesis it was suggested that determining the frequency of the emitted radiation ω and the voltage V would provide an excellent measurement of the ratio of the electron charge to Planck's constant, since $\omega/V = 2e/h$. Many other fascinating experiments have been reported, and no doubt others will continue to appear in the literature. In the next few pages we briefly discuss the Josephson device and the theory of Josephson radiation along the lines of the quantum laser theory. This analysis provides another application of the reduced density matrix (Lee and Scully, 1971) technique.

A breakthrough in the understanding of superconductivity came with Cooper's observation that, when a pair of electrons [with momenta $\mathbf{k}$ and $-\mathbf{k}$ and opposite spin $(\mathbf{k}\uparrow, -\mathbf{k}\downarrow)$] having an attractive interaction between them is added to a full Fermi sea, the pair condenses below the Fermi surface. This has the effect of "smearing" out the Fermi surface as in Fig. 21-5b. The Bardeen-Cooper-Schrieffer theory is based on these electron pairs.

Suppose that, following Josephson, we place two superconductors in close proximity (separate them by a thin oxide layer). Is the system still super-

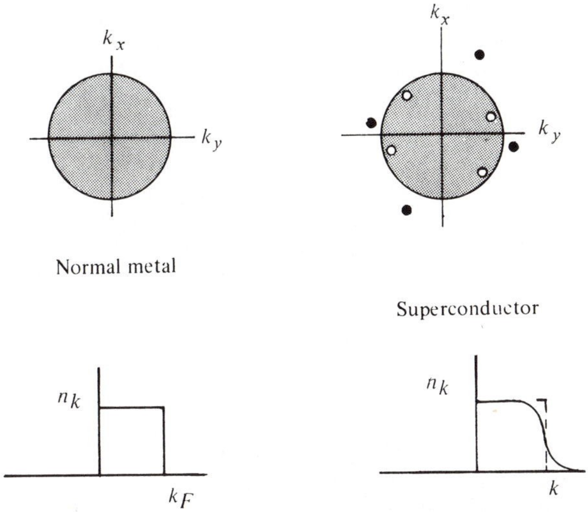

Figure 21-5. Illustration that electron pairs exist above the Fermi energy ε_F for a superconductor at $T = 0$.

conducting? The answer is yes; the pairs tunnel coherently across the thin barrier, and a current still flows without resistance. Now, suppose that a potential V_0 is set up between the superconductors. The effect of the voltage is to displace the Fermi sea of the left superconductor compared to the right, and (in order to conserve energy) light is emitted when a pair tunnels from left to right. Tunneling in the presence of a potential difference and the subsequent emission are depicted in Figs. 21-6*a* and 21-6*b*.

In the preceding paragraph we described the tunneling between two isolated superconductors. Of course, this is not the complete picture, and a more detailed analysis should include the effects of the normal wires connecting the superconductors, as well as the Q of the microwave cavity into which radiation is emitted. The junction area forms a resonance cavity, with a low $Q \approx$ 1–100, and under most circumstances only one mode will be excited. Let Ω denote the cavity eigenfrequency, and $a^\dagger$ and a the creation and annihilation operators, respectively, of that particular mode. We can write the coupling between the superconductors and the radiation as

$$\mathscr{V} = \hbar g(Sa^\dagger + aS^\dagger), \tag{15}$$

where S ($S^\dagger$) is the operator which transfers a pair from the left- to the right- (right- to left-) hand superconductor, and g is a c-number coupling constant.

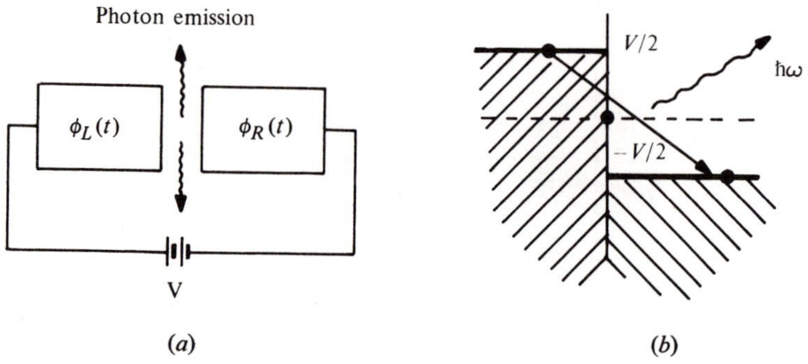

Figure 21-6. (a) Schematic representation of the AC Josephson Effect. A battery drives current through the circuit and sets up a voltage, V, across the junctions. The superconductors on the L.H.S. and the R.H.S. have phases $\phi_L = eVt/\hbar$ and $\phi_R = -eVt/\hbar$ respectively. Photons are emitted as pairs tunnel from the higher to lower potential. (b) Energy levels shift due to applied voltage and conservation of energy implies $\hbar\omega = (2e)V$.

This describes the tunneling of a coherent superposition of pairs across the junction with the emission (absorption) of a single photon. The effect of the finite cavity Q and the normal metal wires is to give an extra incoherent contribution to the time evolution of the density matrix (as discussed in Sec. 16-2), given by

$$\left(\frac{\partial \rho}{\partial t}\right)_{\text{cavity } Q} = -\frac{1}{2}\frac{\nu}{Q}(a^\dagger a\rho + \rho a^\dagger a - 2a\rho a^\dagger), \tag{16}$$

$$\left(\frac{\partial \rho}{\partial t}\right)_{\text{wire}} = -A(SS^\dagger\rho + \rho SS^\dagger - 2S^\dagger\rho S), \tag{17}$$

where A is a constant determined by the current flowing in the junction. The total equation of motion for the density operator is then

$$\dot{\rho} = -\frac{i}{\hbar}[\mathscr{H}_0 + \mathscr{V}, \rho] + \left(\frac{\partial \rho}{\partial t}\right)_{\text{cavity}} + \left(\frac{\partial \rho}{\partial t}\right)_{\text{wire}}. \tag{18}$$

Here $\mathscr{H}_0$ refers to the coupled superconductors described earlier, together with the frec-field Hamiltonian $\Omega a^\dagger a$ for the radiation field. The contribution of each term in Eq. (18) is illustrated in Fig. 21-7. With these expressions in hand we now proceed to a calculation of the linewidth of the radiation from a Josephson junction. The superconductor-oxide sandwich behaves like a capacitor of capacitance $\kappa \simeq 1\mu$F. The voltage V_0 is then a function of the

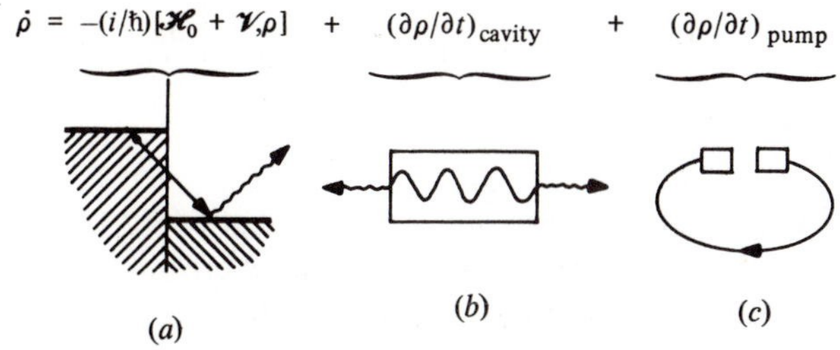

Figure 21-7. Illustration of the density matrix equation of motion. (a) illustrates two superconductors separated by the barrier and maintained at different chemical potentials. An electron pair drops from the higher potential on the left to the lower potential on the right, and emits a single photon. (b) depicts an imperfect resonance cavity having a finite quality factor "Q." (c) illustrates the role of the battery as electron pump, bringing the electrons from the right hand superconductor back to the left via wires of normal metal.

number of excess charges on the left- over the right-hand superconductor:

$$V_0 = \frac{1}{\kappa}(N_L - N_R) \equiv \frac{1}{\kappa} S_z, \tag{19}$$

where the notation S_z is motivated by analogy with a system having $N_L - N_R$ "spins" up. Hence a voltage term appears in $\mathscr{H}_0$, given by

$$\mathscr{H}_v = \frac{e}{\kappa}(N_L - N_R)^2 = \frac{e}{\kappa} S_z^2. \tag{20}$$

To obtain a linewidth, then, we would like to have the voltage correlation function, $\langle V_0(t)\, V_0(0)\rangle$, which is simply related to the number correlation, $\langle S_z(t)\, S_z(0)\rangle$, by Eq. (19). To get this correlation function we should look at the equation of motion for the reduced density matrix for the junction:

$$\sigma_{k,k} = \sum_n \rho_{k,n;k,n}, \tag{21}$$

where k is the pair wave vector:

$$(N_L - N_R)|k\rangle \equiv S_k|k\rangle = k|k\rangle.$$

This is obtained by adiabatic elimination of the photon coordinates implied in Eq. (18). We find

$$\dot{\sigma}_{k,k} = -\left[\frac{g^2\left(\frac{\nu}{Q}\right)^2}{\left(\Omega - \frac{2e^2k}{\hbar\kappa}\right)^2 + \left(\frac{\nu}{Q}\right)^2}\right]\sigma_{k,k}$$

$$+\left\{\frac{g^2\left(\frac{\nu}{Q}\right)}{\left(\Omega-\frac{2e^2(k+1)}{\hbar\kappa}\right)^2+\left(\frac{\nu}{2Q}\right)^2}\right\}\sigma_{k+1,k+1} \tag{22}$$

$$-A(\sigma_{k,k}-\sigma_{k-1,k-1}).$$

The form of Eq. (22) is clearly very analogous to that of (17.18), encountered for ρ_{nn} in the quantum theory of the laser. This expression may be used to calculate $\langle S_z(t)\, S_z(0)\rangle$ with the resultant linewidth

$$\Delta\nu=\frac{16e^3J_{DC}}{h^2\,\partial J/\partial V}\,. \tag{23}$$

This linewidth is of order 10^3 Hz. A more detailed discussion of the results of this section is given by Lee and Scully (1971).

References

Nonlinear Optics

G. C. Baldwin, 1969, *An Introduction to Nonlinear Optics*, Plenum Publishing Corp., New York.

N. Bloembergen, *Nonlinear Optics*, W. A. Benjamin, Inc., Reading, Mass. (1965).

P. A. Franken and J. Ward, 1963, *Rev. Mod. Phys.* **35**, 23.

J. A. Giordmaine, *Sci. Am.* **210** (4), 38.

Y. R. Shen, 1976, *Rev. Mod. Phys.* **48**, 1.

Light Scattering

G. B. Benedek, 1968, in: *Brandeis University Summer Institute Lectures* (1966), ed. by M. Cretien, E. P. Gross, and S. Deser, Gordon and Breach, New York.

H. Z. Cummins and P. E. Schoen, 1972, in: *Laser Handbook*, Vol. II, ed. by F. T. Arecchi and E. O. Schulz-Dubois, North-Holland Publishing Co., Amsterdam.

B. Chu, 1975, *Laser Light Scattering*, Academic Press, New York.

Tunable Dye Lasers

M. Hercher, B. B. Snavely, and C. K. N. Patel, 1973, in: *Coherence and Quantum Optics*, ed. by L. Mandel and E. Wolf, Plenum Publishing Corp., New York.

C. V. Shank, 1975, *Rev. Mod. Phys.* **47**, 649.

Spin-Flip Laser

A. Mooradian, S. R. J., Brueck, and F. A., Blum, 1970, *Appl. Phys. Letters* **17**, 481.

C. K. N. Patel and E. D. Shaw, 1970, Phys. Rev. Letters **24**, 451.

Saturation Spectroscopy

R. G., Brewer, 1972, *Science* **178**, 247. This review article contains a number of references.

K. Shimoda, Ed., 1976, *High Resolution Laser Spectroscopy*, Springer-Verlag, Heidelberg.

Pico-Second Laser Pulses

A. J. DeMaria, 1971, in: *Progress in Optics*, Vol. IX, ed. by E. Wolf, North-Holland Publishing Co., Amsterdam.

P. Rentzepis, Ed., 1976, *Picosecond Pulses*, Springer-Verlag, Heidelberg.

X-Ray Lasers

B. Lax and A. H. Guenther, 1972, *Appl. Phys. Letters* **21**, 361.

R. A. McCorkle, 1972, *Phys. Rev. Letters* **29**, 982.

M. O. Scully, W. H. Louisell, and W. B. McKnight, 1973, Opt. Comm. **9**, 246.

Self-Beat Spectroscopy

H. Z. Cummins and H. L. Swinney, 1970 in: *Progress in Optics*, Vol. VIII, ed. by E. Wolf, North Holland, Publishing Co., Amsterdam.

L. Mandel and E. Wolf, eds., 1973, *Coherence and Noise in Quantum Optics*, Plenum Publishing Corp., New York.

D. W. Schaefer and P. N. Pusey, 1972, *Phys. Rev. Letters* **29**, 843.

Phase Transition Analogy

S. Chandrasekhar, 1961, *Hydrodynamic and Hydromagnetic Stability*, Clarendon Press, Oxford.

V. DeGiorgio and M. O. Scully, 1970, *Phys. Rev.* **A2**, 1170.

R. Graham, 1973, in: *Quantum Statistics in Optics and Solid-State Physics*, Springer Verlag, Berlin.

R. Graham and H. Haken, 1970, *Z, Phys.* **237**, 31.

H. Haken, 1975, *Rev. Mod. Phys.* **47**, 67.

J. F. Scott, M. Sargent III, C. Cantrell, 1975, Opt. Comm. **15**, 13.

M. O. Scully, 1973, in: *Coherence and Quantum Optics*, ed. by L. Mandel and E. Wolf, Plenum Publishing Corp., New York.

A good treatment of the Bardeen-Cooper-Schrieffer theory of superconductivity in the Anderson pseudospin notation is given in C. Kittel, 1963, *Quantum Theory of Solids*, John Wiley & Sons, New York.

The quantum theory of Josephson radiation as presented here is reviewed in:

D. N. Rogovin, M. O. Scully, and P. A. Lee, 1973, "Quantum Theory of Josephson Radiation," in *Progress in Quantum Electronics*, Ed. by J. H. Sanders and S. Stenholm, Pergamon, New York.

P. A. Lee and M. O. Scully, 1971, *Phys. Rev. B* **3**, 769.

For a quantum noise treatment of this problem see

M. J. Stephen, 1969, *Phys. Rev. Letters* **21**, 1629.

Laser Cavity Modes

G. D. Boyd and J. P. Gordon, 1961, *Bell System Tech. J.* **40**, 489.

A. G. Fox and T. Li, 1961, *Bell System Tech. J.* **40**, 453.

R. Lang, M. O. Scully, and W. E. Lamb, Jr., 1973, *Phys. Rev. A.* **7**, 1788.

M. B. Spencer and W. E. Lamb, Jr., 1972, Phys, Rev. **A5**, 884.

These subjects are typically written up tutorially in the *Physics of Quantum Electronics* Summer School Proceedings, S. F. Jacobs, M. Sargent III, and M. O. Scully, Eds., Addison-Wesley Pub., Reading, Mass.

A

FIELD FROM DIPOLE SHEET

Appendix A. Field from an Infinite Sheet of Oscillating Dipoles

In this appendix we compute the field on the z axis contributed by an infinite sheet of induced dipoles oscillating in phase in the x-y plane (see Fig. 3-5). We find that the field is proportional to a dipole on the z axis but is shifted in phase by 90°. This result is valid for distances large compared to the size and spacing of the individual dipoles.

For computational convenience, we write the dipole moment in complex form:

$$\mathbf{p} = \hat{x} p_0 \exp(-i\nu t) \tag{1}$$

and obtain the single-dipole field in complex form (from Jackson, 1962, Sec. 9.2):

$$\mathbf{E}_{\text{dip}}(\mathbf{R}, t) = \frac{1}{4\pi\varepsilon_0 R} p_0 \exp[-i(\nu t - KR)] \left\{ \frac{1 - iKR}{R^2} [-\hat{x} + 3(\hat{x} \cdot \hat{n})\hat{n}] + K^2(\hat{n} \times \hat{x}) \times \hat{n} \right\}. \tag{2}$$

The unit vectors are shown in Fig. 3-5.

In terms of the Cartesian coordinates x, y, and z,

$$R^2 = x^2 + y^2 + z^2, \tag{3}$$

and the unit vector $\hat{n}$ is given by

$$\hat{n} = \frac{-x\hat{x} - y\hat{y} + z\hat{z}}{R}. \tag{4}$$

This gives

$$(\hat{n} \times \hat{x}) \times \hat{n} = (y\hat{z} + z\hat{y}) \times \frac{\hat{n}}{R} = [x(z\hat{z} - y\hat{y}) + (y^2 + z^2)\hat{x}] \frac{1}{R^2}, \tag{5}$$

and $\hat{x} \cdot \hat{n} = -x/R$. Hence the electric field (2) from a single dipole becomes

$$\mathbf{E}_{\text{dip}}(\mathbf{R}, t) = \hat{x}\,\frac{p_0 \exp[-i(\nu t - KR)]}{4\pi\varepsilon_0 R^3}\left[(1 - iKR)\left(-1 + \frac{3x^2}{R^2}\right) + K^2(y^2 + z^2)\right]$$
$$+ \text{ terms odd in } x. \tag{6}$$

To calculate the field produced by all dipoles in the x-y plane at the observation point $(0, 0, z)$, we integrate (6) over the entire x-y plane. Dropping the terms odd in x (which cancel in the x integration) and transforming to polar coordinates, we have the complex field:

$$\mathbf{E}(z, t) = \hat{x}\eta\,\frac{p_0 \exp(-i\nu t)}{4\pi\varepsilon_0}\int_0^\infty \rho\, d\rho \int_0^{2\pi} d\phi\, \frac{\exp(iKR)}{R^3}\left[(1 - iKR)\right.$$
$$\left.\times\left(-1 + \frac{3\rho^2\cos^2\phi}{R^2}\right) + K^2(\rho^2\sin^2\phi + z^2)\right]$$
$$= \hat{x}\eta\,\frac{p_0 \exp(-i\nu t)}{4\varepsilon_0}\int_0^\infty \rho\, d\rho\, \frac{\exp(iKR)}{R^3}\left[(1 - iKR)\right.$$
$$\left.\times\left(-2 + \frac{3\rho^2}{R^2}\right) + K^2(\rho^2 + 2z^2)\right],$$

where η is the number of dipoles per unit area

Now, changing the variable of integration ρ to R, using the relations

$$R^2 = \rho^2 + z^2 \tag{7}$$

and $2\rho\, d\rho = 2R\, dR$, we have

$$\mathbf{E}(z, t) = \hat{x}\eta\,\frac{p_0 \exp(-i\nu t)}{4\varepsilon_0}\int_z^\infty dR\, \exp(iKR)\left\{\frac{1 - iKR}{R^2}\left[1 - 3\left(\frac{z}{R}\right)^2\right]\right.$$
$$\left.+ K^2\left(1 + \frac{z^2}{R^2}\right)\right\}. \tag{8}$$

The integral over R in (8) can be conveniently written in terms of the functions

$$F_n = \int_z^\infty dR\, \frac{\exp(iKR)}{R^n}, \qquad n = 0, 1, 2, 3, 4. \tag{9}$$

Specifically, the integral has the value

$$J = F_2 - iKF_1 - 3z^2F_4 + 3iKz^2F_3 + K^2F_0 + K^2z^2F_2. \tag{10}$$

The zeroth-order term can be integrated immediately:

$$K^2F_0 = -iK\exp(iKR)\Big|_z^\infty \rightarrow iK\exp(iKz), \tag{11}$$

where the term with "$\exp(i\infty)$" can be neglected if we assume that the contributions far from the origin are diminished by an absorption factor $\exp(-R/a)$ due to, for example, the medium through which propagation takes place. The sum of the remaining terms is also $iK\exp(iKz)$, which we show by using the recursion relation obtained by integration by parts:

$$F_n = \frac{1}{iK}\frac{\exp(iKR)}{R^n}\bigg|_z^\infty + \frac{n}{iK}\int_z^\infty dR\,\frac{\exp(iKR)}{R^{n+1}}$$
$$= (iK)^{-1}\left[nF_{n+1} - \frac{\exp(iKz)}{z^n}\right]. \tag{12}$$

We write, successively, F_1 in terms of F_2, F_2 in terms of F_3, and F_3 in terms of F_4. Equation (10) becomes

$$J = F_2 - iK(iK)^{-1}\left[F_2 - \frac{\exp(iKz)}{z}\right] - 3z^2F_4 + 3iKz^2F_3 + iK\exp(iKz) + K^2z^2F_2$$
$$= \frac{\exp(iKz)}{z} - 3z^2F_4 + 3iKz^2F_3 + iK\exp(iKz) - (iK)^2z^2F_2$$
$$= \frac{\exp(iKz)}{z} - 3z^2F_4 + iKz^2F_3 + 2iK\exp(iKz)$$
$$= 2iK\exp(iKz).$$

Substituting this value for the integral in (8), we have the complex field from the dipole sheet:

$$\mathbf{E}(z, t) = \hat{x}\frac{1}{2}\frac{K\eta}{\varepsilon_0}ip_0\exp[-i(\nu t - Kz)]. \tag{13}$$

Hence the field resulting from the dipole sheet is shifted in phase from that of a dipole on the z axis by 90° and has the space dependence of the incident plane wave. In connection with the material in Sec. 3-2, the in-quadrature part of the dipoles ($\sin \nu t$) contributes to the absorption or amplification of the incident field, and the in-phase ($\cos \nu t$) part contributes to the index of refraction (see R. Feynman, R. Leighton, and M. Sands, *The Feynman Lectures on Physics*, Vol. I, Addison-Wesley, 1965, Chap. 31-1, for discussion of the latter contribution).

B

PASSIVE CAVITY MODES

Appendix B. Passive Cavity Modes

The maser (Chap. 5) operating at wavelengths of the order of a centimeter can oscillate in a cavity whose length is equal to the wavelength. It is difficult to make similar cavities much smaller than a millimeter and virtually impossible to make them as small as an optical wavelength. Thus for optical masers, lasers, a different approach is indicated. One solution is to use a cavity related to the Fabry-Perot etalon or interferometer, whose length is many times the wavelength of resonance. The cavity consists of two parallel reflectors separated by a distance L, between which electromagnetic waves are reflected and amplified by the laser medium as depicted in Fig. 8-2. Resonance can occur whenever the phase shift per pass, Φ, is an integral multiple of π, that is, whenever,

$$\Phi_n = \frac{\Omega_n L}{c} = n\pi, \tag{1}$$

where Ω_n is the resonant frequency corresponding to the integer n, and c is the speed of light in the host medium (for gases, in a vacuum). In this appendix we discuss the nature of "modes" in this cavity in the absence of a gain medium. The discussion provides a basis for the derivation including an active medium in Chap. 8. Section B-1 develops the passive cavity eigenvalue equation, Sec. B-2 gives numerical solutions of this eigenvalue equation by the method of successive approximations, and Sec. B-3 develops the analytic eigenfunctions for spherical cavity resonators.

B-1. The Passive Cavity Eigenvalue Equation

It is commonly thought that the Fabry-Perot interferometer is simultaneously resonant for uniform plane waves traveling along the longitudinal axis and at a number of discrete angles to this axis. Although this is what appears to happen, the true resonant nature of the interferometer is concealed by the continuous addition of new radiation. To appreciate this fact, consider a uniform plane wave distribution across one reflector. In traveling to the opposite reflector, the wave is diffracted, the outermost parts "spilling over" the opposite reflector to a greater extent than the central parts. In successive passes, the process is repeated, leaving the peripheral areas depleted in energy with respect to the central parts. We see below that after many passes the initial plane wave distribution reduces to one which remains essentially constant from pass to pass, apart from a complex phase factor. This reduction suggests that we define the ultimate distribution to be a "normal mode." Such a mode differs from the ordinary normal (undecaying) radiation modes and might more precisely be called a "quasimode." This fact is discussed further in Sec. 21-2. The reason that the quasimodes could not be resolved by the (passive) Fabry-Perot interferometer is that available light sources have not been sufficiently monochromatic for this purpose (see Prob. 8-10 for a discussion of resolving power). In a laser, however, the active medium very nearly compensates for the losses, thereby leading to a spectral line sufficiently narrow for resolution.

In this section we describe the passive cavity in terms of the Kirchhoff-Huygen principle. In essence we consider one reflector as an infinite number of elementary Huygen sources—sources of spherical wavelets, sum the contribution of each at corresponding points on the opposite reflector, and finally require the resulting distribution to be the same as that on the first reflector within a constant amplitude and phase factor. Solution of the resulting eigenvalue equation gives the electromagnetic distribution in the cavity and the diffraction losses for various cavity configurations (mirror sizes, shapes and separations, etc.). Our development is partly heuristic; a complete development from Maxwell's equations via one of Green's identities is given by Silver (1949).

Suppose that each point on a surface (Fig. B-1) is a source radiating a spherical wave. The potential u_P at the point P a distance r away from a unit source $\exp(-iKr)$ is $\exp(-iKr)/r$, in which K is the radiation wave number. Multiplying this by a source distribution factor Cu_S and including a direction correction to prevent propagation backward (we use $1 + \cos\theta$, where θ is the angle between the outward-drawn normal and the radius vector from the source to P), we find

$$u_P|_{dS} = C\frac{\exp(-iKr)}{r}(1 + \cos\theta)u_S\,dS.$$

Here dS is a differential surface area. Summation over the entire surface gives the total potential at P as

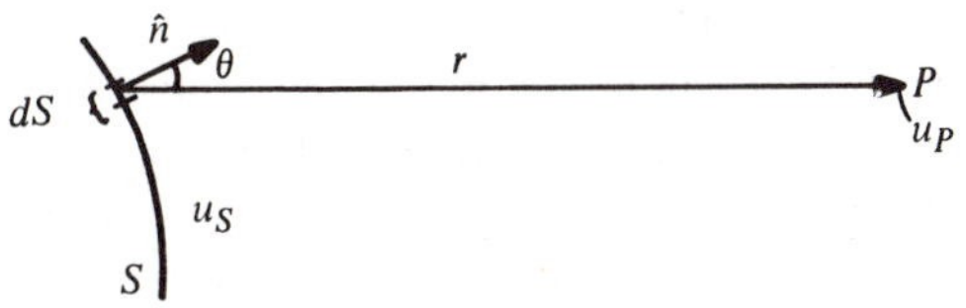

Figure B-1. Diagram defining parameters used in calculation of the potential u_P due to sources on the surface S with potential distribution u_S. The point P is a distance r from the differential surface area dS; $\hat{n}$ is the outward-drawn normal.

$$u_P = C \int_S dS\, u_S\, (1 + \cos\theta) \frac{\exp(-iKr)}{r}. \tag{2}$$

Silver's derivation yields $C = iK/4\pi$. Thus we see that the potential at P is approximately equal to an infinite sum of the amplitudes of spherical wavelets originating from differential surface elements of the surface with strength proportional to iKu_S times a directionality factor.

We now set up an eigenvalue equation describing the successive reflections between two reflectors by requiring the potential distribution on one reflector to be the same as that on the other (apart from a complex factor σ), as explained above. In symbols, we have from (2)

$$u_S' = \sigma u_S = \frac{iK}{4\pi} \int_S dS\, u_S\, (1 + \cos\theta) \frac{\exp(-iKr)}{r}, \tag{3}$$

where u_S and u_S' are the distributions on the two reflectors, respectively. This integral can be solved both numerically by the method of successive approximations and analytically for special cavity configurations. The numerical approach was demonstrated by Fox and Li (1961), who used an IBM 704 digital computer to facilitate the solution. In this method, one iterates Eq. (3), demanding that after some number of passes, say q, the amplitude u_{q+1} of the next pass reproduce the amplitude u_q withing a constant amplitude and phase factor σ, that is,

$$u_{q+1} = \sigma u_q. \tag{4}$$

The analytic method of Boyd and Gordon (1961) applies to spherical resonators for which (3) reduces approximately to a finite, two-dimensional Fourier transform. The prolate spheroidal functions have themselves as their own finite transforms and are hence solutions of the integral equation (3). In Sec. B-2 we discuss the results of Fox and Li for the plane parallel resonator; in Sec. B-3 we discuss spherical resonators along the lines of Boyd and Gordon.

B-2. Plane Parallel Resonators

Fox and Li (1961) found that an initial plane wave distribution changed under successive approximation into a dominant mode which appeared recurrently, only diminishing uniformly in amplitude. Intuitively, the method of successive approximations corresponds to bouncing a wave back and forth between the reflectors, ultimately yielding a "normal" mode under the influence of recurring diffraction losses since most modes have large losses in comparison to one dominant mode. Figure B-2 shows a plot of the relative amplitude of the field intensity for the 1st and 300th transits as plotted against the x coordinate, which is normalized in units of a, the mirror half width. Inasmuch as the distribution is even-symmetric about the center, only the positive x is given. The ripples in the amplitude correspond to the Fresnel zones, as would be expected in a classical Fabry-Perot interferometer. In particular, there exist $a^2/L\lambda = 6.25$ Fresnel zones, as seen from the center of the second reflector. In passing from the center to the edge, 3×6.25 zones appear, a number which agrees with the number of reversals in the amplitude distribution. In time the ripples become smaller, and after 300 approximations the variation in amplitude (within the constant amplitude and phase factor) is 0.03% of the final average value.

A question arises as to whether or not the same modes result for different initial conditions. If odd-symmetric initial conditions are used, the result does in fact turn out to be odd-symmetric. In particular, an initial condition was chosen for which the distribution from 0 to a was 180° out of phase with that from $-a$ to 0. As would be expected because the concentration of power is out further radially in this case than in the even-symmetric solution, the power losses of the odd-symmetry are greater. In either case, for Fresnel numbers larger than 1, the losses are nearly linearly decreasing functions of the Fresnel number N_F, as shown in Fig. B-6. In a variety of odd- and even-symmetry initial conditions, the only difference in the final dominant modes was the time required to reach them. The fact that the computer solution is incapable of giving more than the two lowest modes consistently is strong evidence of modal structure. Because the axial contributions of the electric and magnetic vectors are small, the waves may be termed transverse electromagnetic or expressed notationally (for the rectangular or circular reflectors) as TEM_{mn}, where m corresponds to the x axis or angular coordinate and the n to the y axis or radial coordinate. Figure B-3 illustrates some of the possible mode configurations.

B-3. Confocal Resonator Configurations

Among the most useful (because of ease in alignment and low diffraction

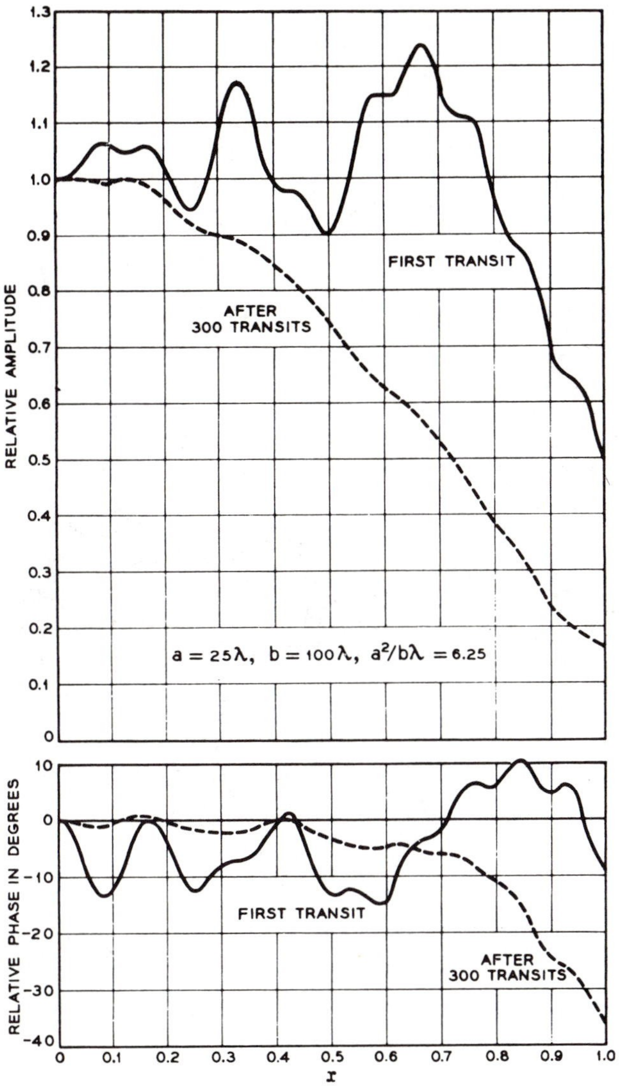

Figure B-2. Relative amplitude and phase distributions of field intensity for infinite strip mirrors. An initial plane wave distribution was used. From Fox and Li (1961). The "confocal parameter" b here equals the mirror spacing L.

loss) resonator configurations are the confocal spherical or parabolic resonators. The simplest confocal case consists of two identical spherical mirrors separated by their common radius of curvature. Inasmuch as, to good approximation at small angles, the focal point of the spherical mirror equals

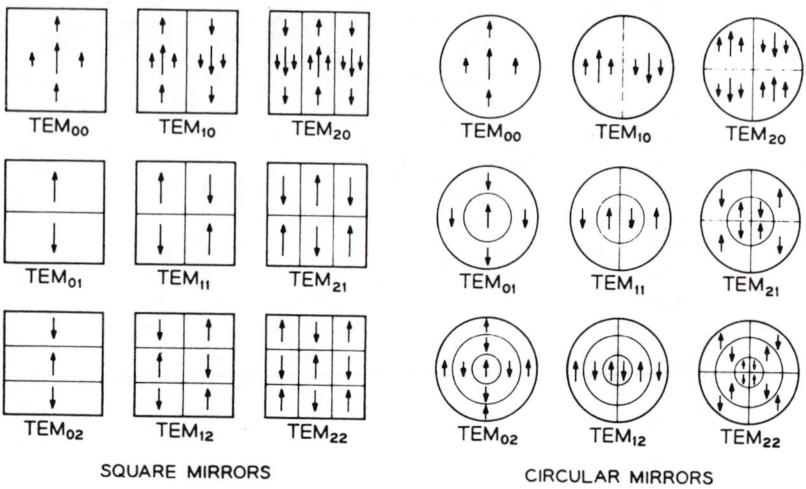

Figure B-3. Possible mode configurations for square and circular mirrors. Arrows represent either the electric field **E** *or the magnetic field* **H**. *Thus there is considerable degeneracy for varying polarization arrangements. Note that, by superimposing various modes, "doughnut"-shaped modes and others can result. From Fox and Li (1961).*

one half of the radius of curvature, the foci of this arrangement coincide and the configuration is termed confocal. Other spherical resonators are easily related to this one. We solve Eq. (3) for this case analytically as discussed by Boyd and Gordon (1961).

We see in Fig. B-4 that a line through z' perpendicular to the cavity axis (z) intercepts the left reflector at height y' (the discussion for the x axis follows easily, and we consider only variations in y). Since the mirror separation for a confocal cavity is the mirror radius of curvature R_c, we have

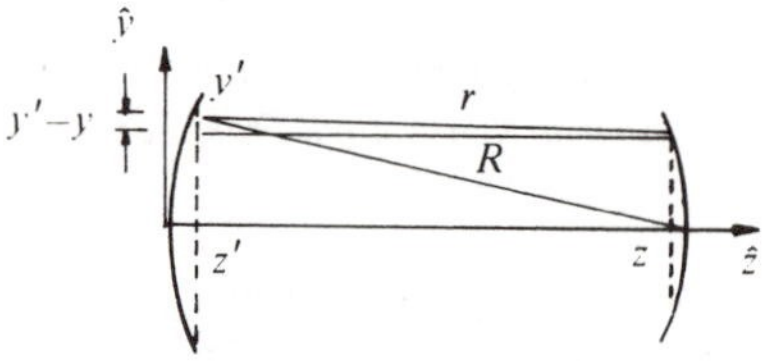

Figure B-4. Spherical resonator defining parameters used in Eqs. (5)–(8).

$$z' = R_c - (R_c^2 - y'^2)^{1/2} = R_c - R_c(1 - y'^2/R_c^2)^{1/2}$$
$$\simeq R_c - R_c + \frac{y'^2}{2R_c}, \tag{5}$$

where the approximation is valid for $y' \ll R_c$. Similarly position z for the right mirror is given by

$$z = R_c - [R_c - (R_c^2 - y^2)^{1/2}]$$
$$\simeq R_c - \frac{y^2}{2R_c}, \tag{6}$$

which with (5) yields the axial difference:

$$z - z' \simeq R_c - \frac{1}{2}\frac{y'^2 + y^2}{R_c}.$$

The desired length r from source to observation point then has the square

$$r^2 = (z - z')^2 + (y - y')^2$$
$$\simeq R_c^2 - y'^2 - y^2 + y^2 - 2yy' + y'^2$$
$$= R_c^2\left(1 - \frac{2yy'}{R_c^2}\right). \tag{7}$$

Neglect terms with factors of y^2/R_c^2.

The contribution to r^2 due to x-axis variations is easily seen to be $2xx'$. With this factor, Eq. (7) yields the value

$$r \simeq R_c - x\left(\frac{x'}{R_c}\right) - y\left(\frac{y'}{R_c}\right), \tag{8}$$

which may be substituted into integral equation (3).

Noting that the x and y dependences separate in the factor $\exp(-iKr)$, we try separable solutions $u_S(x, y)$, defined by

$$u_S(x, y) = f_m(x)g_n(y). \tag{9}$$

In terms of (8) and (9), Eq. (3) reads as

Limits on the integrals are for a square section of a spherical mirror.

$$\sigma_m\sigma_n f_m(x)g_n(y) = \frac{iK\exp(-iKR_c)}{2\pi R_c}\int_{-a}^{a} dx' \exp[-i(Kx/R_c)x']\, f_m(x')$$
$$\times \int_{-a}^{a} dy' \exp[-i(Ky/R_c)y']\, g_n(y'). \tag{10}$$

Considering x alone (y is obviously the same), we define the dimensionless variables

$$X = \sqrt{K/R_c}\, x, \qquad A = \sqrt{K/R_c}\, a, \tag{11}$$

in terms of which we have from (10)

$$\chi_m F_m(X) = \frac{1}{\sqrt{2\pi}}\int_{-A}^{A} dX' \exp(iXX')\, F_m(X'). \tag{12}$$

The integral in (12) is a finite Fourier transform, the ordinary transform being given by $A = \infty$. Slepian and Pollack (1961) have shown that the solutions of (12) are prolate spheroidal wave functions. We are primarily interested in large Fresnel numbers ($N_F = a^2/R_c\lambda \gg 1$), for which $A^2 = 2\pi N_F \gg 1$, and distributions with relatively small beam width. For these, we can extend the limit of integration in (12) to ∞, obtaining the ordinary Fourier transform. A complete set of functions with rapid radial decrease which have, themselves (up to a constant) as their own Fourier transforms consists of the Hermite-Gaussian functions of Eq. (1.22). Specifically, we have the integral

$$\frac{1}{\sqrt{2\pi}}\int_{-\infty}^{\infty} dX' \exp(iXX')\, H_m(X') \exp(-X'^2/2) = i^m H_m(X) \exp(-X^2/2), \qquad (13)$$

so that

$$F_m(X) = F_m(0) H_m(X) \exp(-X^2/2) \qquad (14)$$

is a solution of (12) for $A = \infty$ and $\chi_m = i^m$. The lowest three Hermite-Gaussian functions are plotted in Fig. 1-3.

We can determine the approximate field characteristic *within* the reflectors by evaluating the Fourier transforms, using (13) in (10). After some algebra, we find

$$E(z, y, z) = E_0 h(z) H_m(Xh) H_n(Yh) \exp\left[-\frac{K(x^2+y^2)}{R_c + 4z^2/R_c}\right]$$

$$\times \exp\left\{-iK\left[\tfrac{1}{2}R_c + z + \frac{2z(x^2+y^2)}{R_c^2 + 4z^2}\right] + i(1+m+n)\left(\frac{\pi}{2} - \phi\right)\right\}, \qquad (15)$$

where $h(z) = [2R_c^2/(R_c^2 + 4z^2)]^{1/2}$ and $\tan\phi = (R_c - 2z)/(R_c + 2z)$. We see that the lowest-order eigenfunction ($m = n = 0$) is just the Gaussian beam discussed in Probs. 8-10 and 8-11 and depicted in Fig. B-5. This beam is characterized by the "Rayleigh length" z_0, which is the distance from the focal point ($z = 0$) to the point at which the beam width w_s (spot size) is $\sqrt{2}$ larger than the value w_0 for $z = 0$. For (15), this length is $R_c/2$. Furthermore, the radius of curvature of a constant phase front is given by (Prob. 8-11)

$$R = z + \frac{z_0^2}{z}. \qquad (16)$$

Hence, for $z = z_0 = R_c/2$, this radius is the same as the mirror radius, so that the mirrors coincide with equiphase surfaces.

The eigenfrequencies of the modes are simply calculated from (15) by the requirement that the phase shift between the mirrors ($z = -R_c/2$ to $+R_c/2$) be an integral multiple (q) of π, that is,

$$q\pi = \left[K_{mnq}R_c - (1+m+n)\left(\frac{\pi}{2} - 0\right)\right] - [0].$$

This gives the eigenfrequency (writing $R_c = \mathrm{L}$)

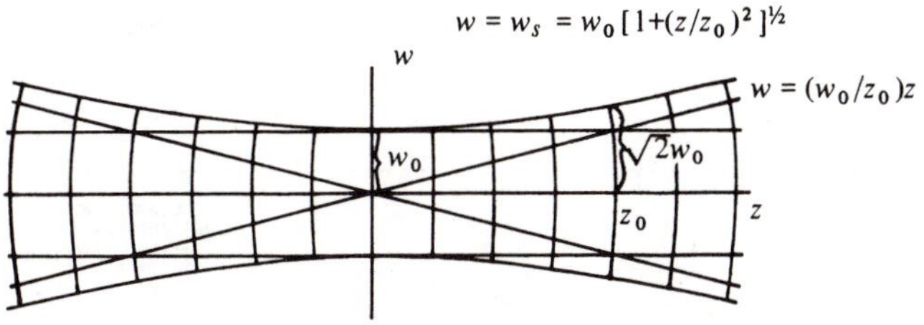

Figure B-5. Diagram of Gaussian beam with Rayleigh length parameter z_0, focal-point waist size $w_0 = \sqrt{2z_0/K}$. For ease of reading, we have drawn the figure with $w_0 = 0.5$, $z_0 = 2$, implying a wavelength of $\lambda = \pi w_0^2/z_0 \simeq 0.393$. In laser resonators, there are very different values, such as $z_0 = 50$ cm, $\lambda = 10^{-4}$ cm, which give $w_0 \simeq 0.5$ mm, that is, a beam diameter of about 1 mm. This beam is extremely skinny compared to that shown. Hence only atoms near the z axis contribute to amplification in a corresponding laser. Unstable resonators (see Fig. B-7) lead to much fatter beams, but feature high losses.

$$\Omega_{mnq} = K_{mnq}c = \pi(2q + 1 + m + n)\frac{c}{2L}. \tag{17}$$

In particular, the TEM_{00q} mode has $\Omega_{00q} = \pi(2q + 1)(c/2L)$, which differs from definition (1) for Ω_q by $\pi(c/2L)$. The additional phase shift of π results from the shift that a Gaussian beam encounters in going from $z = -z_0$ to $+ z_0$.

The fractional loss per pass a_l is given by

$$a_l = 1 - |\sigma_m\sigma_n|^2 = 1 - |\chi_m\chi_n|^2. \tag{18}$$

For the Hermite-Gaussian approximation, this loss vanishes since $\chi_m = i^m$. When the prolate spheroidal functions are used, the representative values given in Fig. B-6 are obtained.

In general, a resonator described by this analysis can be formed by placing reflectors of curvature (16) at any pair of phase fronts in Fig. B-5. It can be shown (see Siegman (1971), p. 322) that the corresponding beam widths at the reflectors are both finite only if the condition

$$0 \leq \left(1 - \frac{L}{R_1}\right)\left(1 - \frac{L}{R_2}\right) \leq 1 \tag{19}$$

is satisfied. Other configurations (see Fig. B-7) have considerably higher loss and are termed "unstable." Nevertheless their larger beam size can be a decided advantage in high-power laser applications since more of the active medium can contribute to amplification. Although the stable cavity laser

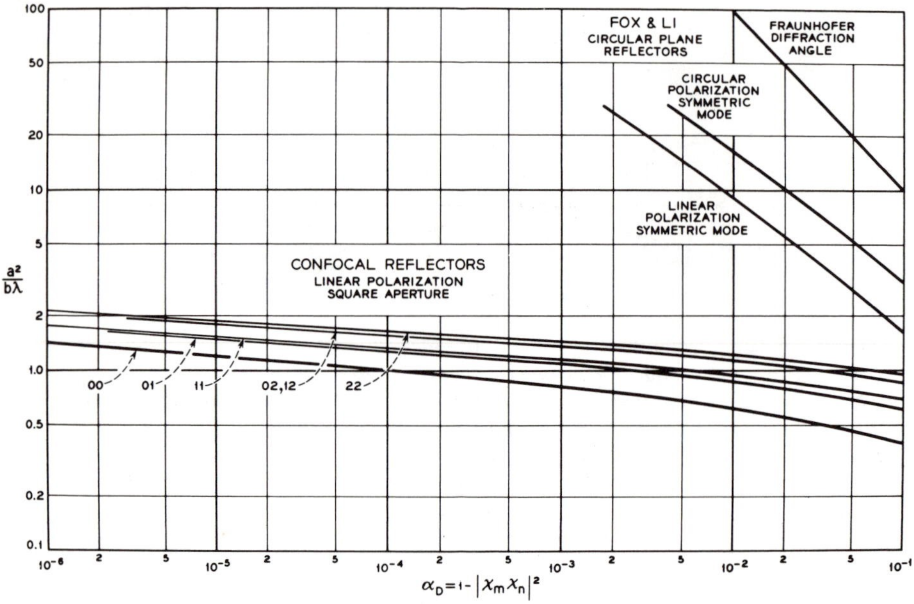

Figure B-6. Fresnel number versus fractional loss per pass for confocal cavities and plane parallel. From Boyd and Gordon (1961) and Fox and Li (1961), respectively.

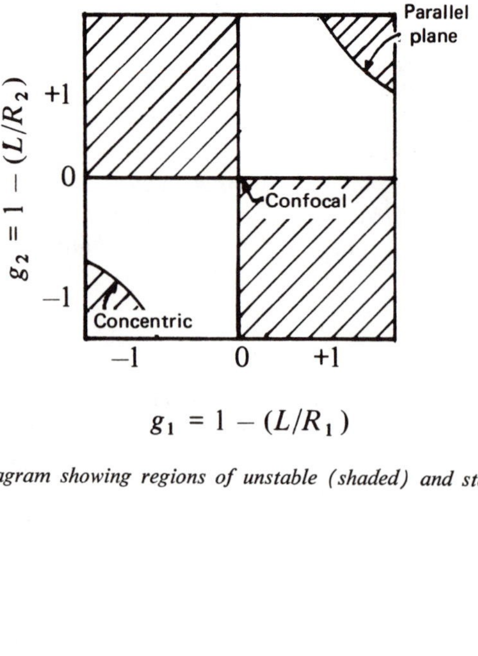

Figure B-7. Diagram showing regions of unstable (shaded) and stable resonator configurations.

radiation distribution is described reasonable well by the passive cavity results of this appendix, the unstable cavity distributions depend critically on the active medium. Hence the unstable cavity laser is still an area of active research.

The multimode laser theories of this book effectively assume that all the modes have the same transverse variations, that is, transverse variations are ignored. This is not accurate, however, for the description of interaction between, for example. TEM_{00q} and TEM_{01q} modes since the larger beam size of the latter leads to contributions from atoms having little effect on the TEM_{00q} mode. In the language of Sec. 8-2, the two modes "burn holes" with only partial overlap. This phenomenon reduces the mode interaction below that predicted by the theories Chaps. 8–10.

References

See the references for Chap. 8 and the following:

S. Silver, 1949, *Microwave Antenna Theory and Design,* M.I.T. Radiation Laboratory Series, Vol. 12, McGraw-Hill Book Co., New York (1949).

D. Slepian and H. O. Pollack, 1961, *Bell Syst. Tech. J.* **40,** 43.

C

PLASMA DISPERSION FUNCTION

Appendix C. The Plasma Dispersion Function

In this appendix we describe the plasma dispersion function $Z(\upsilon)$, defined equivalently by

$$Z(\upsilon) = iKu \int_0^\infty d\tau \exp[-\upsilon\tau - (Ku/2)^2\tau^2] \tag{1}$$

and

$$Z(\upsilon) = \frac{iK}{\sqrt{\pi}} \int_{-\infty}^{\infty} dv \frac{\exp[-(v/u)^2]}{\upsilon + iKv}. \tag{2}$$

Our treatment is oriented toward the gas laser theory of Chap. 10, but the function, as its name suggests, occurs also in plasma physics (see Fried and Conte, 1961, for a brief description of this use along with much other information). In general, the function describes the response of an ensemble of exponentially decaying systems with Gaussian frequency distribution. As developed in Prob. 10-4, $Z(\upsilon)$ is the Fourier transform of an exponential times a Gaussian, that is, Eq. (1), or equivalently the convolution of the Fourier transform of an exponential (complex Lorentzian) times the Fourier transform of the Gaussian (a Gaussian), that is, (2). Both forms follow from the double integral definition

$$Z(\upsilon) = \frac{iK}{\sqrt{\pi}} \int_0^\infty d\tau \int_{-\infty}^{\infty} dv \exp[-(v/u)^2] \exp[-(\upsilon + iKv)\tau]. \tag{3}$$

In this appendix we develop some properties of the function, summarize the results in Table C-1, illustrate the behavior of $Z(\upsilon)$ for typical gas laser parameters in Fig. C-1, and give a formula for numerical evaluation.

A useful relationship for $Z(\upsilon)$ is expressed in terms of the error function (erf) of complex argument. For this, we define the complex variable

TABLE C-1. *Properties of the Plasma Dispersion Function $Z(\upsilon)$*
Here the complex number $\upsilon = \gamma + i(\omega - \nu)$.

Definitions

$$Z(\upsilon) = iKu \int_0^\infty d\tau \exp[-\upsilon\tau - (Ku/2)^2\tau^2]$$

$$= \frac{iK}{\sqrt{\pi}} \int_{-\infty}^{\infty} dv \frac{\exp[-(v/u)^2]}{\upsilon + iKv}$$

$$= i\sqrt{\pi} \exp[(\upsilon/Ku)^2] [1 + \text{erf}(-\upsilon/Ku)]$$

$$= i \sum_{n=0}^{\infty} z_n \left(\frac{\upsilon}{Ku}\right)^n, \qquad z_{2n} = \frac{2^n\sqrt{\pi}}{(2n)!!}, \qquad z_{2n+1} = -\frac{2^{2n+1}n!}{(2n+1)!}$$

Limits

$$Z(\upsilon) \simeq i\sqrt{\pi}, \textit{ for } |\upsilon/Ku| \ll 1$$

$$Z(\upsilon) \simeq \exp(-\xi^2) [i\sqrt{\pi} - 2\int_0^\xi dx \exp(-x^2)], \qquad \xi = \frac{\omega - \nu}{Ku}, \text{ for } |\gamma/Ku| \ll 1$$

$$Z(\upsilon) \simeq \frac{iKu}{\upsilon}, \text{ for } |\upsilon/Ku| \gg 1$$

Properties

$$Z(\upsilon^*) = -Z^*(\upsilon)$$

$$\frac{dZ(\upsilon)}{d\upsilon} = \frac{2}{Ku}\left[\frac{\upsilon}{Ku} Z(\upsilon) - i\right]$$

$$i\zeta = -\frac{\upsilon}{Ku} \tag{4}$$

and change the variable of integration in (1) from τ to $x = Ku\tau/2$. Then (1) becomes

$$Z(\zeta) = 2i \int_0^\infty dx \exp(2i\zeta x - x^2)$$

$$= 2i \int_0^\infty dx \exp[-(x - i\zeta)^2 - \zeta^2]$$

$$= 2i \exp(-\zeta^2) \int_{-\infty}^{i\zeta} dx' \exp(-x'^2)$$

$$= i\sqrt{\pi} \exp(-\zeta^2)[1 + \text{erf}(i\zeta)], \tag{5}$$

where $\text{erf}(i\zeta)$ is the error function of a complex argument.

A convenient form for numerical evaluation when $|\upsilon/Ku| < 1$ is given by (1) with the exponential $\exp(-\upsilon\tau)$ expanded in a Taylor series. We find

$$Z(\upsilon) = 2i \sum_{n=0}^{\infty} \frac{1}{n!}\left(-\frac{2\upsilon}{Ku}\right)^n \int_0^\infty dx\, x^n \exp(-x^2)$$

$$= 2i \sum_{n=0}^{\infty} \left[\frac{1}{(2n)!} \left(\frac{2v}{Ku} \right)^{2n} \int_0^{\infty} dx\, x^{2n} \exp(-x^2) \right.$$

$$\left. + \frac{1}{(2n+1)!} \left(-\frac{2v}{Ku} \right)^{2n+1} \int_0^{\infty} x\, dx\, x^{2n} \exp(-x^2) \right]$$

$$= i \sum_{n=0}^{\infty} \left[\frac{2^n \sqrt{\pi}}{(2n)!!} \left(\frac{v}{Ku} \right)^{2n} - \frac{2^{2n+1} n!}{(2n+1)!} \left(\frac{v}{Ku} \right)^{2n+1} \right]$$

$$= i \sum_{n=0}^{\infty} z_n \left(\frac{v}{Ku} \right)^n. \tag{6}$$

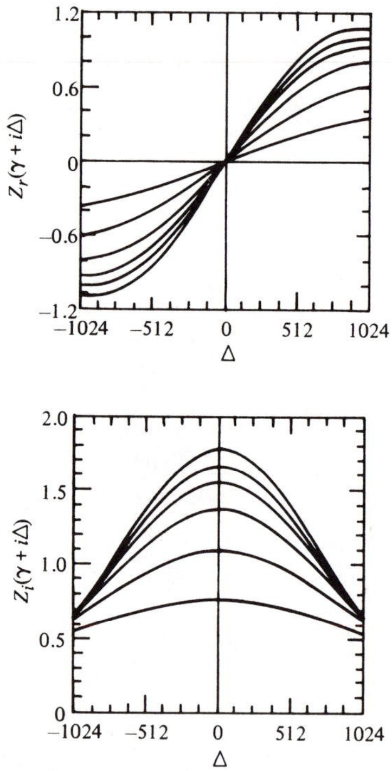

Figure C-1. Plasma dispersion function versus normalized detuning. Curves in order of decreasing magnitudes are for $\gamma = 0, 50, 100, 250, 500, 1024$, $Ku = 1024$.

The first few values of the z_n coefficients are $z_0 = \sqrt{\pi}$, $z_1 = -2$, $z_2 = \sqrt{\pi}$, $z_3 = -4/3$, $z_4 = \sqrt{\pi}/2$, etc. For values of $|v/Ku| > 1$, a continued fraction is generally more convenient (see Fried and Conte, 1961). In particular, we see from (6) that in the extreme Doppler limit ($Ku \gg |v|$)

$$Z(v)\big|_{|v/Ku| \ll 1} \simeq i\sqrt{\pi}. \tag{7}$$

A less extreme limit, in which only $\gamma/Ku \ll 1$, is given in Table C-1. This result follows from completion of the square in (1) with $\upsilon = i(\omega - \nu)$. In the other limit for which $Ku \rightarrow 0$, the Gaussian in (1) can be dropped, resulting in the formula

$$Z(\upsilon)|_{Ku\rightarrow 0} \rightarrow \frac{iKu}{\upsilon}. \qquad (8)$$

Thus, in particular, out in the wings where $\omega - \nu \gg Ku$, the function reduces to a complex Lorentzian. These properties and others given in Probs. 10-2 through 10-4 are summarized in Table C-1.

In Fig. C-1 graphs of the real and imaginary parts of $Z(\upsilon)$ versus normalized detuning $[(\omega - \nu)/Ku]$ are given for several values of γ/Ku. The imaginary part appears to be a combination of a Lorentzian and a Gaussian, which is not surprising from a convolution of the two functions. For reference, in Fig. C-2 a Lorentzian is plotted with a Gaussian having the same width and area. It can be seem that the Lorentzian has relatively substantial wing areas.

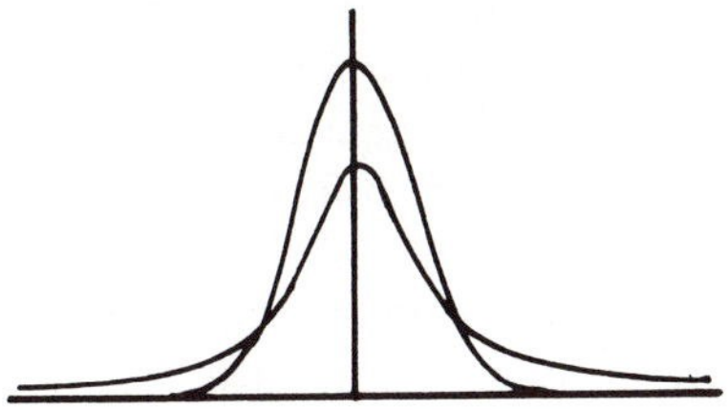

Figure C-2. Graphs of a Lorentzian and a Gaussian having the same area and full width at half maximum. Note the relatively large wing area of the Lorentzian.

Reference

B. D. Fried and S. D. Conte, 1961, *The Plasma Dispersion Function (Hilbert Transform of the Gaussian)*, Academic Press, New York.

D

GAS LASER PERTURBATION THEORY

Appendix D. Gas Laser Perturbation Theory

In this appendix we evaluate the first- (10.50), second- (10.52), and third- (10.53) order perturbation integrals, using the multimode perturbation energy (10.8). The results are valid for an arbitrary amount of Doppler broadening and therefore follow from the more general Zeeman theory by Sargent, Lamb, and Fork (1967). Unlike the treatment in that work, the time integrations for the complex polarization $\mathscr{P}_n(t)$ of Eq. (10.13) are performed before the velocity integral, for the calculation is then both simpler and more generally applicable. Some discussion of the Doppler limit relative to the exact result is given for single-mode operation. The Doppler limit coefficients for the amplitude- and frequency-determining equations (9.18) and (9.19) are summarized in Tables 10-1 and 10-2. Both the general and Doppler limit versions of the coefficients have been programmed for the computer.

The zeroth-order contribution to the population matrix is given by (10.49). The first-order contribution (10.50) to the off-diagonal element ρ_{ab} is given by the rate equation result (10.19) with the population difference $\rho_{aa} - \rho_{bb}$ replaced by its zero-order value, $N(z, v, t)$. The reasoning involved in the calculation is the same as that for the rate equation solution itself. The first-order contribution to the complex polarization is then given by the rate equation result (10.28).

We are primarily interested in calculating the third-order contribution to ρ_{ab} and to $\mathscr{P}_n(t)$. Before doing this, however, it is interesting to calculate the second-order contribution to the population difference, $\rho_{aa}{}^{(2)} - \rho_{bb}{}^{(2)}$. In doing so, it is advantageous to use the summation index ρ for $\mathscr{V}_{ba}$ and σ for $\mathscr{V}_{ab}$, for then the phase factors have the same form. After using trigonometric addition formulas for the two products with form $U_\rho(z' - v\tau'')U_\sigma(z' - v\tau'' - v\tau''')$ and discarding odd functions of v, we find

$$(\rho_{aa}{}^{(2)} - \rho_{bb}{}^{(2)})(z', v, t') = -\frac{1}{4}\left(\frac{\wp}{\hbar}\right)^2 N \sum_\rho \sum_\sigma E_\rho E_\sigma \exp[i(\nu_\rho - \nu_\sigma)\, t' + \phi_\rho - \phi_\sigma]$$

$$\times \int_{-\infty}^{t'} dt'' \int_{-\infty}^{t''} dt''' \, \{\exp[-(i\nu_\rho - i\nu_\sigma + \gamma_a)\tau''] + \exp[-(i\nu_\rho - i\nu_\sigma + \gamma_b)\tau'']\}$$

$$\times \{\exp[-(i\omega - i\nu_\sigma + \gamma)\tau'''] + \exp[-(i\nu_\rho - i\omega + \gamma)\tau''']\}$$

$$\times \tfrac{1}{2}\{\cos[(K_\rho - K_\sigma)z'] \cos(Kv\tau''') - \cos[(K_\rho + K_\sigma)z'] \cos(2Kv\tau'' Kv\tau''')\}. \quad (1)$$

Here we note that the rapidly varying spatial dependence of the rate constant R, which we neglected in (10.21) is multiplied by $\cos(2Kv\tau'' + Kv\tau''')$ in the second-order contribution. Hence, for appreciable values of v, this term tends to average out in the time integration over τ''. Note that the last line in (1) reduces to $\sin K_\rho z \sin K_\sigma z$ for zero velocity, as it should according to (9.10).

We now solve for the third-order contribution to the complex polarization, given by (13) as

$$\mathscr{P}_n{}^{(3)} = 2\wp \exp(i\nu_n t + i\phi_n) \int_{-\infty}^{\infty} dv \frac{1}{\mathscr{N}} \int_0^L dz\, U_n{}^*(z)\, \rho_{ab}{}^{(3)}(z, v, t).$$

Substituting the perturbation energy (8) into (53) for $\rho_{ab}{}^{(3)}$ with subscripts as indicated in the perturbation tree of Fig. 10-4, we find

$$\mathscr{P}_n{}^{(3)} = \tfrac{1}{4}\, i\wp^4\hbar^{-3} \sum_\mu \sum_\rho \sum_\sigma E_\mu E_\rho E_\sigma \exp(i\Psi_{n\mu\rho\sigma}) \frac{1}{\mathscr{N}} \int_0^L dz\, N(z) U_n{}^*(z)$$

$$\times \int_{-\infty}^{\infty} dv\, W(v) \int_0^{\infty} d\tau' \int_0^{\infty} d\tau'' \int_0^{\infty} d\tau''' \exp[-i(\omega - \nu_\mu + \nu_\rho - \nu_\sigma)\tau' - \gamma\tau']$$

$$\times \{\exp[-i(\nu_\rho - \nu_\sigma)\tau'' - \gamma_a\tau''] + \exp[-i(\nu_\rho - \nu_\sigma)\tau'' - \gamma_b\tau'']\}$$

$$\times U_\mu(z - v\tau')$$

$$\times \{\exp[-i(\omega - \nu_\sigma)\tau''' - \gamma\tau'''] U_\rho{}^*(z - v\tau' - v\tau'')$$

$$\times U_\sigma(z - v\tau' - v\tau'' - v\tau''')$$

$$+ \exp[-i(\nu_\rho - \omega)\tau''' - \gamma\tau'''] U_\sigma(z - v\tau' - v\tau'')$$

$$\times U_\rho{}^*(z - v\tau' - v\tau'' - v\tau''')\}. \quad (2)$$

The first product of four U's can be reduced, via trigonometric identities [see, for example, (9.10) and the following discussion] and neglect of functions odd in v and rapidly varying in z to

$$\tfrac{1}{8}\{\cos[(K_n - K_\mu - K_\rho + K_\sigma)z] \cos[Kv(\tau''' - \tau')]$$

$$+ \cos[(K_n - K_\mu + K_\rho - K_\sigma)z] \cos[Kv(\tau''' + \tau')]$$

$$+ \cos[(K_n + K_\mu - K_\rho - K_\sigma)z] \cos[Kv(\tau''' + 2\tau'' + \tau')]\}. \quad (3)$$

The second product is given by (3) with the indices ρ and σ interchanged. In substituting (3) into (2), we replace expressions like $\cos[Kv(\tau''' \pm \tau')]$ by $\exp[-iKv(\tau''' \pm \tau')]$, as shown in Prob. 10-9. Equation (2) then reduces to

$$\mathscr{P}_n^{(3)}(t) = \tfrac{1}{16}\,\wp^4(\hbar^3 Ku)^{-1} \sum_\mu\sum_\rho\sum_\sigma E_\mu E_\rho E_\sigma \exp(i\Psi_{n\mu\rho\sigma}) \sum_{l=1}^{4}\sum_{w=1}^{3} T_{lw}, \quad (4)$$

where the third-order integrals

$$T_{lw} = iN_{lw}K\pi^{-1/2}\int_{-\infty}^{\infty} dv\, \exp[-(v/u)^2] \int_0^\infty d\tau' \int_0^\infty d\tau'' \int_0^\infty d\tau'''$$
$$\times \exp[-(\upsilon_{l1} + is_{w1}Kv)\tau' - (\upsilon_{l2} + is_{w2}Kv)\tau'' - (\upsilon_{l3} + is_{w3}Kv)\tau''']. \quad (5)$$

Here the factors s_{wk} are defined by the matrix

$$s = \begin{pmatrix} -1 & 0 & 1 \\ 1 & 0 & 1 \\ 1 & 2 & 1 \end{pmatrix}, \quad (6)$$

the excitation parameters N_{lw} are given in Table D-1, and the complex frequencies υ_{lk} are listed in Table D-2. With practice, one can actually write expression (5) defining the matrix (6) and Tables D-1 and D-2 by *inspection* of the perturbation tree in Fig. 10-3. This technique is particularly valuable in

TABLE D-1. *Definitions of the N_{lw} which Appear in the Third-Order Integrals (5), in Terms of N_{2l} (9.16)*

The relation $n = \mu - \rho + \sigma$, satisfied by significant terms in the amplitude- and frequency-determining equations (9.18) and (9.19), has been used to simplify the complicated subscript dependence in (3).

N_{lw}	$w =$ 1	2	3
$l = 1$	$N_{2(\rho-\sigma)}$	$\bar{N}$	$N_{2(\rho-\mu)}$
2	$\bar{N}$	$N_{2(\rho-\sigma)}$	$N_{2(\rho-\mu)}$
3	$\bar{N}$	$N_{2(\rho-\sigma)}$	$N_{2(\rho-\mu)}$
4	$N_{2(\rho-\sigma)}$	$\bar{N}$	$N_{2(\nu-\mu)}$

TABLE D-2. *Definitions of Complex Frequencies υ_{lk} Appearing in the Third-Order Integrals (5)*

υ_{lk}	$k =$ 1	2	3
$l = 1$	$\gamma + i(\omega - \nu_\mu + \nu_\rho - \nu_\sigma)$	$\gamma_a + i(\nu_\rho - \nu_\sigma)$	$\gamma + i(\omega - \nu_\sigma)$
2	$\gamma + i(\omega - \nu_\mu + \nu_\rho - \nu_\sigma)$	$\gamma_a + i(\nu_\rho - \nu_\sigma)$	$\gamma + i(\nu_\rho - \omega)$
3	$\gamma + i(\omega - \nu_\mu + \nu_\rho - \nu_\sigma)$	$\gamma_b + i(\nu_\rho - \nu_\sigma)$	$\gamma + i(\nu_\rho - \omega)$
4	$\gamma + i(\omega - \nu_\mu + \nu_\rho - \nu_\sigma)$	$\gamma_b + i(\nu_\rho - \nu_\sigma)$	$\gamma + i(\omega - \nu_\sigma)$

dealing with more complicated problems, such as the multimode ring laser (O'Bryan and Sargent, 1973), or the still more complex Zeeman ring laser (Hanson and Sargent, 1974).

The integrals (5) can be evaluated in terms of the plasma dispersion function (10.29) in two ways. As for the dispersion function itself, [see (10.31)] one can complete the square in v and perform the velocity integration. The resulting time integrals can be reduced to linear combinations of the dispersion function by suitable changes of integration. It is both simpler and more general, however, to perform the time integrations immediately, obtaining

$$T_{lw} = iN_{lw}K\pi^{-1/2}\int_{-\infty}^{\infty} dv \exp[-(v/u)^2]$$
$$\times \{(\upsilon_{l1} + is_{w1}Kv)(\upsilon_{l2} + is_{w2}Kv)(\upsilon_{l3} + is_{w3}Kv)\}^{-1} \tag{7}$$

and then to separate the frequency factor into partial fractions. The result is the linear combination of plasma dispersion functions given by (10.29). In fact, the second line of (7) can be expanded as

$$\left(\frac{1}{s_1\upsilon_1 - s_2\upsilon_1}\right)\left(\frac{s_1}{\upsilon_1 + is_1Kv} - \frac{s_2}{\upsilon_2 + is_2Kv}\right)\left(\frac{1}{\upsilon_3 + is_3Kv}\right)$$
$$= \left(\frac{1}{s_1\upsilon_2 - s_2\upsilon_1}\right)\left[\left(\frac{s_1}{s_1\upsilon_3 - s_3\upsilon_1}\right)\left(\frac{s_1}{\upsilon_1 + is_1Kv} - \frac{s_3}{\upsilon_3 + is_3Kv}\right)\right.$$
$$\left. - \left(\frac{s_2}{s_2\upsilon_3 - s_3\upsilon_2}\right)\left(\frac{s_2}{\upsilon_2 + is_2Kv} - \frac{s_3}{\upsilon_3 + is_3Kv}\right)\right], \tag{8}$$

in which we have suppressed the l and w subscripts for typographical simplicity (υ_k stands for υ_{lk} and s_k for s_{wk}). Using definition (10.29) with (7) and (8), we find

$$T_{lw} = N_{lw}\left[\frac{1}{s_2\upsilon_1 - s_1\upsilon_2}\right]\left\{s_1\left[\frac{Z(\upsilon_3/s_3) - Z(\upsilon_1/s_1)}{s_1\upsilon_3 - s_3\upsilon_1}\right]\right.$$
$$\left. - s_2\left[\frac{Z(\upsilon_3/s_3) - Z(\upsilon_2/s_2)}{s_2\upsilon_3 - s_3\upsilon_2}\right]\right\}. \tag{9}$$

In particular, for $w = 1$ ($s_3 = -s_1 = 1$, $s_2 = 0$), we have

$$T_{l1} = \left(\frac{N_{l1}}{\upsilon_{l2}}\right)\left[\frac{Z(\upsilon_{l1}) + Z(\upsilon_{l3})}{\upsilon_{l1} + \upsilon_{l3}}\right], \tag{10}$$

for which we have used the relation [which follows immediately from (10.29)]

$$Z(-\upsilon) = -Z(\upsilon).$$

Similarly, for $w = 2$ ($s_3 = s_1 = 1$, $s_2 = 0$), we have

$$T_{l2} = -\left(\frac{N_{l2}}{\upsilon_{l2}}\right)\left[\frac{Z(\upsilon_{l1}) - Z(\upsilon_{l3})}{\upsilon_{l1} - \upsilon_{l3}}\right] \tag{11}$$

and for $w = 3$ ($s_3 = s_1 = 1$, $s_2 = 2$),

$$T_{l3} = \frac{1}{2}\left(\frac{N_{l3}}{\upsilon_{l1} - \upsilon_{l2}/2}\right)\left[\frac{Z(\upsilon_{l3}) - Z(\upsilon_{l1})}{\upsilon_{l3} - \upsilon_{l1}} - \frac{Z(\upsilon_{l3}) - Z(\upsilon_{l2}/2)}{\upsilon_{l3} - \upsilon_{l2}/2}\right]. \tag{12}$$

The complex frequency υ_{l3} sometimes equals υ_{l1}, for which

$$\lim_{\upsilon_{l3}\to\upsilon_{l1}} \frac{Z(\upsilon_{l3}) - Z(\upsilon_{l1})}{\upsilon_{l3} - \upsilon_{l1}} = \frac{dZ(\upsilon_{l1})}{d\upsilon_{l1}}. \tag{13}$$

As the atomic speed $u \to 0$, $Z(\upsilon) \to iKu/\upsilon$, and (9), (11), and (12) reduce to

$$T_{lw} \underset{u\to 0}{\to} \frac{iKuN_{lw}}{\upsilon_{l1}\upsilon_{l2}\upsilon_{l3}}, \tag{14}$$

as they should according to (7). This limit gives the same result for $\mathscr{P}_n(t)$ as that derived earlier (9.14) for the stationary atom case.

For large Doppler broadening ($Ku \gg$ atomic decay rates and various beat frequencies),

$$Z(\upsilon) \simeq i\sqrt{\pi}, \tag{15}$$

an approximation which will be referred to as the "strong Doppler limit." The third-order integrals reduce to

$$T_{l1} \simeq 2i\sqrt{\pi}N_{l1}\{\upsilon_{l2}(\upsilon_{l1} + \upsilon_{l3})\}^{-1}, \qquad T_{l2} \simeq T_{l3} \simeq 0. \tag{16}$$

Values for which $\gamma \ll Ku$ but which are accurate for any detuning ("Doppler limit") can be obtained by use of (10.32):

$$T_{l1} \cong i\sqrt{\pi}\, N_{l1}\{\upsilon_{l2}(\upsilon_{l1} + \upsilon_{l3})\}^{-1}\,[\exp\{-[\mathrm{Im}(\upsilon_{l1})/Ku]^2\} + \exp\{-[\mathrm{Im}(\upsilon_{l3})/Ku]^2\}]. \tag{17}$$

In the strong Doppler limit (15), the third-order contribution (4) becomes

$$\mathscr{P}_n^{(3)}(t) = \tfrac{1}{16}\, i\pi^{1/2}\wp^4(\hbar^3 Ku)^{-1} \sum_\mu\sum_\rho\sum_\sigma E_\mu E_\rho E_\sigma \exp(i\Psi_{n\mu\rho\sigma})$$
$$\times [\mathscr{D}_a(\nu_\rho - \nu_\sigma) + \mathscr{D}_b(\nu_\rho - \nu_\sigma)]\,[N_{2(\rho-\sigma)}\,\mathscr{D}(\omega - \tfrac{1}{2}\nu_\mu + \tfrac{1}{2}\nu_\rho - \nu_\sigma)$$
$$+ \bar{N}\mathscr{D}(-\tfrac{1}{2}\nu_\mu - \tfrac{1}{2}\nu_\sigma + \nu_\rho)]. \tag{18}$$

The multimode amplitude and frequency equations are (9.18) and (9.19), with the linear net-gain coefficient α_n and the linear mode pulling coefficient σ_n given in Table 10-1 and the third-order saturation coefficients given by

$$\vartheta_{n\mu\rho\sigma} = \tfrac{1}{32}\,\nu\wp^4(\hbar^3 Ku\varepsilon_0)^{-1} \sum_{l=1}^{4}\sum_{w=1}^{3} T_{lw}. \tag{19}$$

In the strong Doppler limit of (15), Eq. (19) reduces to the value in Table 10-2. One can also identify that value directly from Eq. (18) by including a multiplicative factor of $\nu/2\varepsilon_0$.

References

See the references for Chap. 10 and the following
D. R. Hanson and M. Sargent III, 1974, Phys. Rev. **A9,** 466
C. L. O. Bryan III and M. Sargent III, 1973, Phys. Rev. **A8,** 3071. 466.
M. Sargent III, W. E. Lamb, Jr., R. L. Fork, 1967, Phys. Rev. **164**, 436.

E

GAS LASER STRONG – SIGNAL THEORY

Appendix E. Gas Laser Strong-Signal Theory

This appendix gives the details of the gas laser strong-signal theory motivated in Sec. 10-3. The complex polarization $\mathscr{P}(t)$ of (10.55) is obtained and combined with the self-consistency relations (8.11) and (8.12) to yield amplitude- and frequency-determining equations. It is shown that the lowest-order approximation to the theory yields the rate equation results of Sec. 10-1.

It is left as an exercise for the reader (Prob. 10-10) to show from the equations of motion for the population matrix elements (10.10)–(10.12) that the equations for the in-phase and in-quadrature polarization components C_n and S_n of (10.56) and those for the population difference D (10.58) and sum M (10.59) are as follows:

$$\dot{S}_n = -\gamma S_n - (\omega - \nu)\, C_n - \frac{\wp^2 E_n}{\hbar} \sin(K_n z)\, D, \tag{1}$$

$$\dot{C}_n = -\gamma C_n + (\omega - \nu_n)\, S_n, \tag{2}$$

$$\dot{D} = \lambda_a - \lambda_b - \gamma_{ab} D - \tfrac{1}{2}(\gamma_a - \gamma_b)\, M + \frac{E_n}{\hbar} \sin(K_n z)\, S_n, \tag{3}$$

$$\dot{M} = \lambda_a + \lambda_b - \gamma_{ab} M - \tfrac{1}{2}(\gamma_a - \gamma_b)\, D. \tag{4}$$

We can reduce this set of four differential equations into a pair of two integrodifferential equations by substituting the formal integral of (2) for C_n:

$$C_n = (\omega - \nu_n) \int_{-\infty}^{t} dt'\, S_n(z', v, t') \exp[-\gamma(t - t')] \tag{5}$$

into (1) for S_n, and the integral of (4) for M:

$$M = -\tfrac{1}{2}(\gamma_a - \gamma_b) \int_{-\infty}^{t} dt'\, D(z', v, t') \exp[-\gamma_{ab}(t - t')] + \frac{\lambda_a + \lambda_b}{\gamma_{ab}}$$

into (3) for D, thus obtaining

$$\dot{S}_n = -\gamma S_n - (\omega - \nu_n)^2 \int_{-\infty}^{t} dt' \, S_n(z', v, t') \exp[-\gamma(t - t')] - \frac{\wp^2 E_n}{\hbar} \sin(K_n z) \, D, \tag{6}$$

$$\dot{D} = -\gamma_{ab} D + \tfrac{1}{2}(\gamma_a - \gamma_b)^2 \int_{-\infty}^{t} dt' \, D(z', v, t') \exp[-\gamma_{ab}(t - t')] + \frac{E_n}{\hbar} \sin(K_n z) \, S_n + \gamma_a \gamma_b \gamma_{ab}^{-1} N(z, v, t). \tag{7}$$

We now substitute the Fourier expansions (10.60) and (10.61) for S_n and D into these equations of motion, remembering that $N(z, v, t)$ is slowly varying in both z and t and that $d/dt = \partial/\partial t + v\partial/\partial z$. Equating coefficients of $\exp[(2j + 1)iK_n z]$ resulting from (10.60), we have

$$\{(2j + 1)\, iKv + \gamma + (\omega - \nu_n)^2 \, [(2j + 1)\, iKv + \gamma]^{-1}\} \, q_{2j+1} = \frac{1}{2} \frac{\wp E_n}{\hbar} (q_{2j+2} - q_{2j}). \tag{8}$$

Similarly equating the coefficients of $\exp(2ijKz)$ resulting from (10.61), we find

$$\{2ijKv + \gamma_{ab} - \tfrac{1}{4}(\gamma_a - \gamma_b)^2 \, (2ijKv + \gamma_{ab})^{-1}\} = \frac{1}{2} \frac{\wp E_n}{\hbar} (q_{2j+1} - q_{2j-1}) + \gamma_a \gamma_b \gamma_{ab}^{-1} \delta_{j0}. \tag{9}$$

Dividing by the factors in curly braces, we find the single equation for the Fourier coefficients:

$$q_j = \bar{E}_n \mathscr{D}_j (q_{j+1} - q_{j-1}) + \delta_{j0}, \tag{10}$$

where the dimensionless amplitude

$$\bar{E}_n \equiv \sqrt{I_n} \equiv \wp E_n / \sqrt{2\gamma_a \gamma_b \hbar^2}, \tag{11}$$

and the sums of complex denominators

$$\mathscr{D}_{2j+1} = \tfrac{1}{2}(\tfrac{1}{2}\gamma_a \gamma_b)^{1/2} \, \{\mathscr{D}[(2j + 1)\, Kv - (\omega - \nu_n)] + \mathscr{D}[(2j + 1)\, Kv + (\omega - \nu_n)]\}, \tag{12}$$

$$\mathscr{D}_{2j} = \tfrac{1}{2}(\tfrac{1}{2}\gamma_a \gamma_b)^{1/2} \, \{\mathscr{D}_a[2jKv] + \mathscr{D}_b[2jKv]\}.$$

Equation (10) can be solved for q_{j+1} in terms of q_j and q_{j-1}:

$$q_{j+1} = q_{j-1} + (\bar{E}_n \mathscr{D}_j)^{-1}(q_j - \delta_{j0}). \tag{13}$$

A similar equation yields q_{j-1} in terms of q_j and q_{j+1}. Hence, once any two coefficients in sequence [or, by (10), one out of sequence] are known, all the rest can be determined. This fact will be useful in determining the z and v dependences of the population difference D. To calculate the polarization, however, we need only q_1 and q_{-1}, for all other coefficients multiply exponentials which vanish in the Fourier projection of (10.60). We now proceed to calculate these coefficients in terms of a continued fraction.

Dividing (10) by q_j, we have

$$1 = \bar{E}_n \mathscr{D}_j \left(\frac{q_{j+1}}{q_j} - \frac{q_{j-1}}{q_j}\right), \quad j \neq 0.$$

Introducing the ratio

$$r_j = -\left(\frac{q_{j+1}}{q_j}\right) \text{ for } j \geq 0, \tag{14}$$

we can write (13) as

$$r_{j-1} = \bar{E}_n \mathscr{D}_j \, (1 + \bar{E}_n \mathscr{D}_j r_j)^{-1}. \tag{15}$$

We have thus reduced the linear second-order difference equation (equation containing three consecutive indices) to a nonlinear first-order difference equation. Iteration of the latter produces a continued fraction. In fact, for $j = 1$, we have

$$r_0 = \frac{\bar{E}_n \mathscr{D}_1}{1 + \bar{E}_n \mathscr{D}_1 r_1}.$$

Furthermore,

$$r_1 = \frac{\bar{E}_n \mathscr{D}_2}{1 + \bar{E}_n \mathscr{D}_2 r_2},$$

which gives

$$r_0 = \frac{\bar{E}_n \mathscr{D}_1}{[1 + I_n \mathscr{D}_1 \mathscr{D}_2/(1 + \bar{E}_n \mathscr{D}_2 r_2)]}.$$

In general, this iteration procedure gives the continued fraction

$$r_0 = \frac{\bar{E}_n \mathscr{D}_1}{[1 + I_n \mathscr{D}_1 \mathscr{D}_2/[1 + I_n \mathscr{D}_2 \mathscr{D}_3/[1 + \cdots}. \tag{16}$$

For negative j, we use the ratio

$$r'_j = \frac{q_j}{q_{j+1}}, \tag{17}$$

which with (10) becomes

$$r'_j = \bar{E}_n \mathscr{D}_j [1 + \bar{E}_n \mathscr{D}_j r_{j-1}]^{-1}.$$

For $j = -1$, this equation can be iterated as in (16) yielding the continued fraction

$$r'_{-1} = \bar{E}_n \mathscr{D}_{-1}/[1 + I_n \mathscr{D}_{-1} \mathscr{D}_{-2}/[1 + I_n \mathscr{D}_{-2} \mathscr{D}_{-3}/[1 + \cdots. \tag{18}$$

Now the sums of complex denominators $\mathscr{D}_{-j} = \mathscr{D}_j^*$ due to the fact that S_n and D are real. Hence $r'_{-1} = r_0^*$. Further, by the definitions of r_j and r'_j,

$$q_1 = -\, r_0 q_0 \tag{19}$$

$$q_{-1} = r'_{-1} q_0 = r_0^* q_0. \tag{20}$$

Combining these relations with (10) for q_0, one has

$$q_0 = \bar{E}_n \mathscr{D}_0 (q_1 - q_{-1}) + 1 = - \bar{E}_n \mathscr{D}_0 (r_0 + r_0^*) q_0 + 1$$

which gives

$$q_0 = [1 + 2\bar{E}_n \mathscr{D}_0 \mathrm{Re}(r_0)]^{-1}. \tag{21}$$

Inasmuch as the results of this theory are all in terms of the real part of the continued fraction for r_0, we introduce the real function

$$\mathfrak{F} = 2\gamma_{ab}(\sqrt{2\gamma_a\gamma_b}\,\bar{E}_n)^{-1}\mathrm{Re}(r_0) = 2\gamma_{ab}\hbar\,(\wp E_n)^{-1}\,\mathrm{Re}\,(r_0). \tag{22}$$

With this, q_0 is given by

$$q_0 = (1 + I_n \mathfrak{F})^{-1}, \tag{23}$$

and with (19) and (20)

$$q_1 = - r_0(1 + I_n \mathfrak{F})^{-1}, \tag{24}$$

$$q_{-1} = r_0^*(1 + I_n \mathfrak{F})^{-1}. \tag{25}$$

Substituting the Fourier series (10.60) for $S_n(z, v, t)$ into the integral (5) for $C_n(z, v, t)$, we find

$$C_n(z, v, t) = -i(\omega - \nu_n)N(z, v, t)\wp \sum_{j=-\infty}^{\infty} q_{2j+1}(v) \exp[(2j + 1)iK_n z]$$
$$\times \int_{-\infty}^{t} dt' \exp\{[-\gamma - i(2j + 1)Kv](t - t')\}$$
$$= -i(\omega - \nu_n)N\wp \sum_{j=-\infty}^{\infty} q_{2j+1}\, \mathscr{D}[(2j + 1)Kv] \exp[(2j + 1)iK_n z]. \tag{26}$$

Combining this and (10.60) in (10.56), we have

$$\mathscr{P}_n(z, v, t) = -\wp N \sum_{j=-\infty}^{\infty} q_{2j+1} \exp[(2j + 1)iK_n z]$$
$$\times \{i + (\omega - \nu_n)\mathscr{D}[(2j + 1)Kv]\}.$$

By (10.57), we have

$$\mathscr{P}_n(t) = -\wp \sum_{j=-\infty}^{\infty} \frac{2}{L}\int_0^L dz\, N(z, t) \sin(K_n z) \exp[(2j + 1)iK_n z]$$
$$\times \int_{-\infty}^{\infty} dv\, W(v)\{q_{2j+1}(v)\{i + (\omega - \nu_n)\mathscr{D}[(2j + 1)Kv]\}$$
$$= -\wp\bar{N}\int_{-\infty}^{\infty} dv\, W(v)\{q_{-1}[i + (\omega - \nu_n)\mathscr{D}(-Kv)]$$
$$- q_1[i + (\omega - \nu_n)\mathscr{D}(Kv)]\}$$
$$= -2\wp\bar{N}\int_{-\infty}^{\infty} dv\, W(v)\, \{i\,\mathrm{Re}(r_0) + (\omega - \nu_n)\,\mathrm{Re}[r_0\mathscr{D}(Kv)]\}\,(1 + I_n\mathfrak{F})^{-1}. \tag{27}$$

Combining this polarization with the self-consistency equation (8.11) and choosing the steady state ($\dot{E}_n = 0$), we have the amplitude-determining equation:

$$\frac{\nu}{2Q_n} E_n = \frac{\nu}{2\varepsilon_0} 2\wp \bar{N} \int_{-\infty}^{\infty} dv\, W(v) \operatorname{Re}(r_0)(1 + I_n \mathfrak{F})^{-1}.$$

With relation (22) between $\operatorname{Re}(r_0)$ and $\mathfrak{F}$, this becomes

$$\frac{\hbar\varepsilon_0\gamma_{ab}}{\wp^2\bar{N}Q_n} = 2\int_0^{\infty} dv\, W(v)\, \mathfrak{F}(v, \omega - \nu_n, I_n)\, [1 + I_n\, \mathfrak{F}(v, \omega - \nu_n, I_n]^{-1}, \quad (28)$$

where we have used the evenness of the integrand in the variable v to change the lower limit of integration to 0 by inserting a factor of 2 on the right-hand side. This equation determines the intensity I_n of oscillation only implicitly, as contrasted with the third-order perturbational treatment, which results in the explicit equation (10.39). Nevertheless, Eq. (28) is quite amenable to numerical analysis, as illustrated in Sec. 10-3. Using the self-consistency equation (8.12), we have the frequency-determining equation:

$$\frac{\nu_n - Q_n}{\omega - \nu_n} = 2\bar{N}(\nu\wp/\varepsilon_0 E_n) \int_0^{\infty} dv\, W(v) \operatorname{Re}[r_0(v, \omega - \nu_n, I_n)\mathscr{D}(Kv)]$$
$$\times [1 + I_n\, \mathfrak{F}(v, \omega - \nu_n, I_n)]^{-1}. \quad (29)$$

It is convenient to express the amplitude-determining equation (28) in terms of the threshold excitation $\bar{N}_T$, given by (10.40), or equivalently by use of the relation $I_n = 0$ with $\omega = \nu_n$ in (28) and (10.29). Further using the relative excitation $\mathfrak{N} = \bar{N}/\bar{N}_T$, we have

$$\mathfrak{N}^{-1} = 2Ku[\gamma_{ab}Z_i(\gamma)]^{-1} \int_0^{\infty} dv\, W(v)\, \mathfrak{F}(v, \omega - \nu_n, I_n)$$
$$\times [1 + I_n\, \mathfrak{F}(v, \omega - \nu_n, I_n)]^{-1}. \quad (30)$$

The lowest-order approximation to the complex polarization $\mathscr{P}_n(t)$ (27) is given by truncation of the continued fraction (16) for r_0 in the first term:

$$r_0 \simeq \bar{E}_n\mathscr{D}_1 = \frac{1}{4}\frac{\wp E_n}{\hbar}\{\mathscr{D}[Kv - (\omega - \nu_n)] + \mathscr{D}[Kv + (\omega - \nu_n)]\}, \quad (31)$$

giving

$$\begin{aligned} \mathfrak{F} &= 2\gamma_{ab}(2\gamma_a\gamma_b)^{-1/2}\bar{E}_n^{-1} \operatorname{Re}(r_0) \\ &\simeq 2\gamma_{ab}(2\gamma_a\gamma_b)^{-1/2} \operatorname{Re}(\mathscr{D}_1) \\ &= \frac{1}{2}\frac{\gamma_{ab}}{\gamma}[\mathscr{L}(\omega - \nu_n + Kv) + \mathscr{L}(\omega - \nu_n - Kv)]. \end{aligned} \quad (32)$$

This approximation is equivalent to the rate equation approximation discussed in Sec. 10-1, for it gives the same value for the population difference. In fact, from (10.61) and (23)

$$D(z, v, t) \simeq N(z, v, t)q_0 = N(z, v, t)(1 + I_n \mathfrak{F})^{-1}$$
$$\simeq N(z, v, t) \left\{1 + \tfrac{1}{2} I_n \frac{\gamma_{ab}}{\gamma} [\mathscr{L}(\omega - \nu_n + Kv) + \mathscr{L}(\omega - \nu_n - Kv)]\right\}^{-1}, \quad (33)$$

which is the same as the difference (10.22) Hence the polarization (27) must also reduce to the rate equation value (10.27). We can prove this by using (31) and a partial fraction expansion as follows:

$$(\omega - \nu_n)\mathrm{Re}[r_0 \mathscr{D}(Kv)] \simeq \tfrac{1}{4}(\omega - \nu_n) \frac{\wp E_n}{\hbar} \mathrm{Re}\{\mathscr{D}(Kv)\mathscr{D}[Kv - (\omega - \nu_n)] + \mathscr{D}(Kv)\, \mathscr{D}[Kv + (\omega - \nu_n)]\}$$
$$= \frac{1}{4} \frac{\wp E_n}{\hbar} \mathrm{Re}\{i\mathscr{D}(Kv) - i\mathscr{D}[Kv - (\omega - \nu_n)] - i\mathscr{D}(Kv) + i\mathscr{D}[Kv - (\omega - \nu_n)]\}$$
$$= \frac{1}{4} \frac{\wp E_n}{\hbar} \mathrm{Re}\{i\mathscr{D}[Kv + (\omega - \nu_n)] - i\mathscr{D}[Kv - (\omega - \nu_n)]\}. \quad (34)$$

Using the relation $\mathrm{Re}(z) = \frac{1}{2}(z + z^*)$ with (31) and (34), we find

$$i\,\mathrm{Re}(r_0) + (\omega - \nu_n)\,\mathrm{Re}[r_0 \mathscr{D}(Kv)] = \tfrac{1}{4} i \frac{\wp E_n}{\hbar} \{\mathscr{D}[Kv + (\omega - \nu_n)] + \mathscr{D}^*[Kv - (\omega - \nu_n)]\}.$$

Combining this with (32) and the integral (27), we find

$$\mathscr{P}_n(t) = -\, i \wp^2 \frac{E_n}{\hbar} \bar{N} \int_{-\infty}^{\infty} dv\, W(v)\, \mathscr{D}(\omega - \nu_n + Kv) \times \left\{1 + \tfrac{1}{2} I_n \frac{\gamma_{ab}}{\gamma} [\mathscr{L}(\omega - \nu_n + Kv) + \mathscr{L}(\omega - \nu_n - Kv)]\right\}^{-1}, \quad (35)$$

where we have replaced $\mathscr{D}^*[Kv - (\omega - \nu_n)]$ by $\mathscr{D}[Kv + (\omega - \nu_n)]$ under the integral there by introducing a multiplicative factor of 2. Equation (35) is the same as (10.27) obtained by use of the rate equation approximation and is discussed in some detail in Secs. 10-1 and 10-3.

References

The material in this appendix was adapted from:
S. Stenholm and W. E. Lamb, Jr., Phys. Rev. 181, 618.

See also the references of Chap. 10.

F

ZEEMAN LASER PERTURBATION THEORY

Appendix F. Zeeman Laser Perturbation Theory

In this appendix we calculate the polarization of a multilevel medium subject to the vector electric field of Eq. (12.1). Two derivations are given: one for a medium with angular momenta $J = 1 \longleftrightarrow J = 0$, and one for arbitrary J values. The results of the first calculation are contained, of course, in the second, but it is thought that the explicit calculation for a simple case reveals more structure than the general calculation. Ironically enough, the general calculation requires fewer equations, for several matrix elements can be represented by a single term having variable indices.

F-1. Calculation for $\mathbf{J = 1 \longleftrightarrow J = 0}$

This calculation is motivated in Sec 12-2, where we note that terms contributing to the complex polarization $\mathscr{P}_-$ are given by those for $\mathscr{P}_+$ with an exchange of minuses and pluses. Hence we write equations leading to $\mathscr{P}_+$ alone. This shortens the calculation without loss of generality.

In zeroth order, we drop all terms in the equations of motion (12.35)–(12.38) containing matrix elements of the perturbation energy and find non-vanishing contributions only to the diagonal elements. We have

$$\dot{\rho}_{aa} \simeq -\gamma_a \rho_{aa} + \lambda_a, \tag{1}$$

with the integral

$$\rho_{aa}^{(0)} = \lambda_a \int_{-\infty}^{t} dt' \exp[-\gamma_a(t - t')] = \frac{\lambda_a}{\gamma_a}. \tag{2}$$

In first order, we have non-vanishing contributions for ρ_{+b} and ρ_{-b}, for which we write only [from (12.35)]

$$\rho_{+b}^{(1)}(z, v, t) = \frac{i}{\hbar}\int_{-\infty}^{t} dt' \exp[-(i\omega_{+b} + \gamma)(t - t')]\,\mathscr{V}_{+b}(t)$$
$$\times [\rho_{++}^{(0)}(z', v, t') - \rho_{bb}^{(0)}(z', v, t'')]$$
$$= \frac{i}{\hbar} N(z, t)\, w(v) \int_0^{\infty} d\tau' \exp[-(i\omega_{+b} + \gamma)\,\tau']\,\mathscr{V}_{+b}(t - \tau'), \qquad (3)$$

where the population inversion density is taken to have the form

$$N(z, t)\, W(v) = \lambda_a(z, v, t)\gamma_a^{-1} - \lambda_b(z, v, t)\gamma_b^{-1}. \qquad (4)$$

As in the scalar theory, we assume a Maxwellian velocity distribution for $W(v)$ in our explicit calculations.

In second order, we obtain nonvanishing contributions both to the populations ρ_{++} and ρ_{--} and to the quadrupole term ρ_{+-}. We have from (12.36)

$$\rho_{++}^{(2)}(z, v, t) = \frac{i}{\hbar}\int_0^{\infty} d\tau' \exp(-\gamma_a\tau')\,\mathscr{V}_{+b}\rho_{b+} + \text{c.c.}$$
$$= -\hbar^{-2}N(z)\, W(v) \int_0^{\infty} d\tau' \int_0^{\infty} d\tau'' \exp(-\gamma_a\tau')$$
$$\times \{\mathscr{V}_{+b}(t - \tau') \exp[(i\omega_{b+} + \gamma)\tau'']\,\mathscr{V}_{b+}(t'')$$
$$+ \mathscr{V}_{b+}(t') \exp[-(i\omega_{+b} + \gamma)\tau'']\,\mathscr{V}_{+b}(t'')\} \qquad (5)$$

and a similar expression for $\rho_{--}^{(2)}$, given by (5) with the plus subscript replaced by minus. The lower-level population

$$\rho_{bb}^{(2)}(z, v, t) = -[\rho_{++}^{(2)} + \rho_{--}^{(2)}] \quad \text{with } \gamma_a \longleftrightarrow \gamma_b. \qquad (6)$$

The quadrupole term is given by a formal integral of (12.38) as

$$\rho_{+-}^{(2)}(z, v, t) = -\frac{i}{\hbar}\int_0^{\infty} d\tau' \exp[-(i\omega_{+-} + \gamma_a)\tau']\,[\mathscr{V}_{+b}\rho_{b-} - \mathscr{V}_{b-}\rho_{+b}]$$
$$= -\hbar^{-2}NW \int_0^{\infty} d\tau' \int_0^{\infty} d\tau'' \exp[-(i\omega_{+-} + \gamma_a)\tau']$$
$$\times \{\mathscr{V}_{+b}(t') \exp[-(i\omega_{b-} + \gamma)\tau'']\,\mathscr{V}_{b-}(t'')$$
$$+ \mathscr{V}_{b-}(t') \exp[-(i\omega_{+b} + \gamma)\tau'']\,\mathscr{V}_{+b}(t'')\}. \qquad (7)$$

Integrating (12.35) and using (5), (6), and (7), we have the third-order contribution:

$$\rho_{+b}^{(3)}(z, v, t) = \frac{i}{\hbar}\int_0^{\infty} d\tau' \exp[-(i\omega_{+b} + \gamma)\tau']\{\mathscr{V}_{+b}(t')[\rho_{++}^{(2)} - \rho_{bb}^{(2)}]$$
$$+ \mathscr{V}_{-b}(t')\rho_{+-}^{(2)}\}$$
$$= -i\hbar^{-3}NW \int_0^{\infty} d\tau' \int_0^{\infty} d\tau'' \int_0^{\infty} d\tau''' \exp[-(i\omega_{+b} + \gamma)\tau']$$
$$\times \Big[\mathscr{V}_{+b}(t')\,[\exp(-\gamma_a\tau'') + \exp(-\gamma_b\tau'')]$$

$$\times \{\mathscr{V}_{+b}(t'') \exp[-(i\omega_{b+} + \gamma] \mathscr{V}_{b+}(t''')$$
$$+ \mathscr{V}_{b+}(t'') \exp[-(i\omega_{b+} + \gamma)\tau'''] \mathscr{V}_{+b}(t''')\}$$
$$+ \mathscr{V}_{+b}(t') \exp(-\gamma_b\tau'')$$
$$\times \{\mathscr{V}_{-b}(t'') \exp[-i(\omega_{b-} + \gamma)\tau'''] \mathscr{V}_{b-}(t''')$$
$$+ \mathscr{V}_{b-}(t'') \exp[-(i\omega_{-b} + \gamma)\tau'''] \mathscr{V}_{-b}(t''')\}$$
$$+ \mathscr{V}_{-b}(t') \exp[-(i\omega_{+-} + \gamma_a)\tau'']$$
$$\times \{\mathscr{V}_{+b}(t'') \exp[-(i\omega_{b-} + \gamma)\tau'''] \mathscr{V}_{b-}(t''')$$
$$+ \mathscr{V}_{-b}(t'') \exp[-(i\omega_{+b} + \gamma)\tau'''] \mathscr{V}_{+b}(t''')\}\Big]. \quad (8)$$

Substituting the interaction energy (12.33) into the first-order integral (3), we have

$$\rho_{+b}^{(1)}(z, v, t) = -\tfrac{1}{2} i \frac{\wp}{\hbar} NWE_+ \exp[-i(\nu_+ t + \phi_+)]$$
$$\times \int_0^\infty d\tau' \exp[-(i\omega_{+b} - i\nu_+ + \gamma)\tau'] U(z').$$

Expanding $U(z') = \sin[K(z - v\tau')]$, using a trigonometric addition formula, and keeping only the terms even in v [the odd term vanishes in the integration over v provided $W(-v) = W(v)$], we have

$$\rho_{+b}^{(1)}(z, v, t) = -\tfrac{1}{2} i \frac{\wp}{\hbar} NWE_+ \exp[-(i\nu_+ t + \phi_+)] \sin Kz$$
$$\times \int_0^\infty d\tau' \exp[-(i\omega_{+b} - i\nu_+ + \gamma)\tau'] \sin[K(z - v\tau')]$$
$$= -\tfrac{1}{2} i \frac{\wp}{\hbar} NW(v)E_+ \exp[-i(\nu_+ t + \phi_+)] \sin Kz$$
$$\times [\mathscr{D}(\omega_{+b} - \nu_+ + Kv) + \mathscr{D}(\omega_{+b} - \nu_+ - Kv)]. \quad (9)$$

Performing the projections required by (12.34) and using definition (10.29) of the plasma dispersion function $Z(v)$, we have the complex polarization:

$$\mathscr{P}_+^{(1)} = -i \frac{\wp^2}{\hbar K u} \bar{N} E_+ Z[\gamma + i(\omega_{+b} - \nu_+)]. \quad (10)$$

This is the same as the scalar case (10.28) except that we have used ω_{+b} in place of ω and there is a companion equation for the other polarization. Further substituting the interaction energy (12.33) into the third-order integral (8), we find

$$\rho_{+b}^{(3)}(z, v, t) = i\left(\frac{\frac{1}{2}\wp}{\hbar}\right)^3 NWE_+ \exp[-i(\nu_+ t + \phi_+)] \int_0^\infty d\tau' \int_0^\infty d\tau'' \int_0^\infty d\tau'''$$
$$\times \exp[-(i\omega_{+b} - i\nu_+ + \gamma)\tau'] \, U(z')U^*(z'')U(z''')$$

$$\times \Big\{E_+^2[\exp(-\gamma_a\tau'') + \exp(-\gamma_b\tau'')]$$

$$\times \{\exp[-(i\omega_{b+} + i\nu_+ + \gamma)\tau''']$$

$$+ \exp[-(\omega_{+b} - i\nu_+ + \gamma)\tau''']\}$$

$$+ E_-^2 \exp(-\gamma_b\tau'') \{\exp[-i(\omega_{b-} + i\nu + \gamma)\tau''']$$

$$+ \exp[-(i\omega_{-b} - i\nu + \gamma)\tau''']\}$$

$$+ E_-^2 \exp[-(i\omega_{+-} - i\nu_+ + i\nu_- + \gamma_a)\tau'']$$

$$\times \{\exp[-(i\omega_{b-} + i\nu_- + \gamma)\tau''']$$

$$+ \exp[-(i\omega_{+b} - i\nu_+ + \gamma)\tau''']\}\Big\}. \quad (11)$$

We now substitute this contribution into Eq. (12.34) for the complex polarization. In so doing, the product of four sines (for the standing-wave case) reduces via trigonometric identities and neglect of functions odd in v to

$$\tfrac{1}{8}\{\cos[Kv(\tau''' - \tau')] + \cos[Kv(\tau''' + \tau')] + \cos[Kv(\tau''' + 2\tau'' + \tau')]\}. \quad (12)$$

Although it is possible to evaluate the resulting integrals exactly along the lines of Appendix D, we make the delta-function approximation, valid in the limit of large Doppler broadening. This procedure capitalizes on the fact that, for large velocities, only the term with $\cos[Kv(\tau''' - \tau')]$ attains slowly varying values in the time integrals, thus yielding an appreciable value. The other terms tend to average to zero. Performing the velocity integration first by completing the square, we have

$$(\sqrt{\pi}\,u)^{-1}\int_{-\infty}^{\infty} dv \exp[-(v/u)^2]\cos[Kvf(\tau', \tau'', \tau''')]$$

$$= \sqrt{\pi}\int_{-\infty}^{\infty} dx \exp[-(x^2 + iKuxf)] = \exp[-\tfrac{1}{4}(Kuf)^2].$$

Thus, for large Ku and $f = \tau''' - \tau'$, the exponential acts like $\delta(\tau''' - \tau')$:

$$\int_0^{\infty} d\tau''' \, G(\tau', \tau''') \exp[-\tfrac{1}{4}(Ku)^2(\tau''' - \tau')^2] = 2\sqrt{\pi}\,(Ku)^{-1}G(\tau', \tau'). \quad (13)$$

Carrying out the integrals in this approximation, we have the third-order contribution:

$$\mathscr{P}_+^{(3)}(t) = \tfrac{1}{8}i\sqrt{\pi}\,\frac{\wp^4}{\hbar^3 Ku}\bar{N}E_+\int_0^{\infty} d\tau'\int_0^{\infty} d\tau''\Big\{E_+^2[\exp(-\gamma_a\tau'') + \exp(-\gamma_b\tau'')]$$

$$\times \{\exp(-2\gamma\tau') - \exp[-2(i\omega_{+b} - i\nu_+ + \gamma)\tau']\}$$

$$+ E_-^2 \exp(-\gamma_b\tau'') \{\exp[-(i\omega_{+-} - i\nu_+ + i\nu_- + 2\gamma)\tau']$$

$$+ \exp[-(i\omega_{+b} + i\omega_{-b} - i\nu_+ - i\nu_- + 2\gamma)\tau']\}$$

$$+ E_-^2 \exp[-(i\omega_{+-} - i\nu_+ + i\nu_- + \gamma_a)\tau''] \{\exp[-(i\omega_{+-} - i\nu_+$$

$$+ i\nu_- + 2\gamma)\tau'] + \exp[-2(i\omega_{+b} - i\nu_+ + \gamma)\tau']\}\Big\}. \quad (14)$$

Performing the remaining time integrations, we have finally

$$\begin{aligned}\mathscr{P}_+^{(3)}(t) = \tfrac{1}{16}\, i\, \wp^4 \bar{N}(\hbar^3 Ku)^{-1} E_+ \{ & E_+^2 2\gamma_{ab}(\gamma\gamma_a\gamma_b)^{-1}[1 + \gamma\mathscr{D}(\omega_{+b} - \nu_+)] \\ & + E_-^2\gamma_b^{-1}[\mathscr{D}(\delta) + \mathscr{D}(\omega_0 - \nu_0)] \\ & + E_-^2\mathscr{D}_a(2\delta)[\mathscr{D}(\delta) + \mathscr{D}(\omega_{+b} - \nu_+)]\}, \end{aligned} \quad (15)$$

where the average frequencies

$$\omega_0 = \tfrac{1}{2}(\omega_{+b} + \omega_{-b}), \qquad \nu_0 = \tfrac{1}{2}(\nu_+ + \nu_-), \quad (16)$$

and the magnetic field splitting

$$\delta = \frac{\mu_B g H}{\hbar} - \tfrac{1}{2}(\nu_+ - \nu_-) \simeq \frac{\mu_B g H}{\hbar}.$$

Substituting the first- (10) and third- (15) order contributions to the complex polarization $\mathscr{P}_+(t)$ into the self-consistency equations (12.11) and (12.12), we obtain the intensity- and frequency-determining equations (12.43) and (12.44) with coefficients defined in Table 12-1. A physical interpretation in terms of hole burning of the populations ρ_{++}, ρ_{--}, and ρ_{bb}, as well as that of the quadrupole element ρ_{+-}, has been given by Fork and Sargent (1966).

F-2. Calculation for Arbitrary J Values

We now calculate the polarization (12.27) of a medium with arbitrary J values subject to the electric-dipole interaction energy (12.25). The equations of motion for the general density matrix are as follows:

$$\begin{aligned}\dot{\rho}_{a'b'} = & -(i\omega_{a'b'} + \gamma)\,\rho_{a'b'} \\ & + \frac{i}{\hbar}\sum_{a''} \mathscr{V}_{a''b'}\rho_{a'a''} - \frac{i}{\hbar}\sum_{b''} \mathscr{V}_{a'b''}\rho_{b''b'}, \end{aligned} \quad (17)$$

$$\begin{aligned}\dot{\rho}_{a'a''} = & -(i\omega_{a'a''} + \gamma_a)\rho_{a'a''} \\ & + \frac{i}{\hbar}\sum_{b''}(\mathscr{V}_{b''a''}\rho_{a'b''} - \mathscr{V}_{a'b''}\rho_{b''a''}) + \lambda_{a'}\delta_{a'a''}, \end{aligned} \quad (18)$$

$$\begin{aligned}\dot{\rho}_{b''b'} = & -(i\omega_{b''b'} + \gamma_b)\rho_{b''b'} \\ & + \frac{i}{\hbar}\sum_{a''}(\mathscr{V}_{a''b'}\rho_{b''a''} - \mathscr{V}_{b''a''}\rho_{a''b'}) + \lambda_{b'}\delta_{b'b''}, \end{aligned} \quad (19)$$

$$\dot{\rho}_{b'a'} = \dot{\rho}_{a'b'}^*, \quad (20)$$

where for simplicity we have taken $\gamma_{a'b'} = \gamma$, $\gamma_{a'} = \gamma_a$, and $\gamma_{b'} = \gamma_b$. The first-order contribution is given by

$$\rho_{a'b'}^{(1)}(z, v, t) = \frac{i}{\hbar}\int_{-\infty}^{t} dt' \exp[-(i\omega_{a'b'} + \gamma)(t - t')]$$
$$\times [\rho_{a'a'}^{(0)} - \rho_{b'b'}^{(0)}]\, \mathscr{V}_{a'b'}(t')$$
$$= \frac{i}{\hbar} W(v)N(z, t)\int_{0}^{\infty} d\tau'\, \mathscr{V}_{a'b'}(t') \exp[-(i\omega_{a'b'} + \gamma)\tau'], \qquad (21)$$

where we take $N_{a'b'} = N$. Using (18), (19), and (21), we have for the second-order components $\rho_{a'a''}^{(2)}$ and $\rho_{b''b'}^{(2)}$:

$$\rho_{a'a''}^{(2)}(z, v, t)$$
$$= \frac{i}{\hbar}\int_{0}^{\infty} d\tau' \exp[-(i\omega_{a'a''} + \gamma_a)\tau'] \sum_{b''} [\mathscr{V}_{b''a''}(t')\rho_{a'b''}^{(1)}(z, v, t)$$
$$- \mathscr{V}_{a'b''}(t')\rho_{b''a''}^{(1)}(z, v, t)]$$
$$= -\frac{1}{\hbar^2} W(v) \int_{0}^{\infty} d\tau' \int_{0}^{\infty} d\tau'' \exp[-(i\omega_{a'a''} + \gamma_a)\tau']$$
$$\sum_{b''} \{\mathscr{V}_{b''a''}(t')N(z, t)\, \mathscr{V}_{a'b''}(t'') \exp[-(i\omega_{a'b''} + \gamma)\tau'']$$
$$+ \mathscr{V}_{a'b''}(t')N(z, t)\, \mathscr{V}_{b''a''}(t'') \exp[-(i\omega_{b''a''} + \gamma)\tau'']\}, \qquad (22)$$

where $\tau'' = t' - t''$, and

$$\rho_{b''b'}^{(2)}(z, v, t) = -\frac{1}{\hbar^2} W(v) \int_{0}^{\infty} d\tau' \int_{0}^{\infty} d\tau'' \exp[-(i\omega_{b''b'} + \gamma_b)\tau']$$
$$\times \sum_{a''} \{\mathscr{V}_{a''b'}(t')N(z, t)\, \mathscr{V}_{b''a''}(t'') \exp[-(i\omega_{b''a''} + \gamma)\tau'']$$
$$+ \mathscr{V}_{b''a''}(t')N(z, t)\, \mathscr{V}_{a''b'}(t'') \exp[-(i\omega_{a''b'} + \gamma)\tau'']\}. \qquad (23)$$

Combining (17) with (22) and (23), we have the third-order components:

$$\rho_{a'b'}^{(3)}(z, v, t) = \frac{i}{\hbar}\int_{0}^{\infty} d\tau' \exp[-(i\omega_{a'b'} + \gamma)\tau'] [\sum_{a''} \mathscr{V}_{a''b'}(t')\rho_{a'a''}^{(2)}(z, v, t)$$
$$- \sum_{b''} \mathscr{V}_{a'b''}(t')\rho_{b''b'}^{(2)}(z, v, t)]$$
$$= -\frac{i}{\hbar^3} W(v) \int_{0}^{\infty} d\tau' \int_{0}^{\infty} d\tau'' \int_{0}^{\infty} d\tau''' \exp[-(i\omega_{a'b'} + \gamma)\tau']$$
$$\times \sum_{a''}\sum_{b''} \Big\{ \mathscr{V}_{a''b'}(t') \exp[-(i\omega_{a'a''} + \gamma_a)\tau'']$$
$$\times \{\mathscr{V}_{b''a''}(t'')N(z, t) \exp[-(i\omega_{a'b''} + \gamma)\tau''']\, \mathscr{V}_{a'b''}(t''')$$
$$- \mathscr{V}_{a'b''}(t'')N(z, t) \exp[-(i\omega_{b''a''} + \gamma)\tau''']\, \mathscr{V}_{b''a''}(t''')\}$$
$$+ \mathscr{V}_{a'b''}(t') \exp[-i(\omega_{b'b''} + \gamma_b)\tau'']$$
$$\times \{\mathscr{V}_{a''b'}(t'')N(z, t) \exp[-(i\omega_{b''a''} + \gamma)\tau''']\, \mathscr{V}_{b''a''}(t''')$$
$$+ \mathscr{V}_{b''a''}(t'')N(z, t) \exp[-(i\omega_{a''b'} + \gamma)\tau''']\, \mathscr{V}_{a''b'}(t''')\}\Big\}, \qquad (24)$$

where $\tau''' = t'' - t'''$.

In substituting the perturbation energy $\mathscr{V}_{a'b'}$ of (12.25) into these expressions, we can save some effort by using subscript products like $a'b'$ on E, ν, and ϕ with the value plus for $a' = b' + 1$ and minus for $a' = b' - 1$. we can then write (12.25) as

$$\mathscr{V}_{a'b'} = -\tfrac{1}{2}\sqrt{2}\ \wp_{a'b'}E_{a'b'}\exp\left[-i(\nu_{a'b'}t + \phi_{a'b'})\right]U(z). \qquad (25)$$

Substituting this into the first-order term (21), we have

$$\rho_{a'b'}{}^{(1)}(z, v, t) = -\tfrac{1}{2}\sqrt{2}\,iW(v)N(z)\ \wp_{a'b'}E_{a'b'}\exp[-i(\nu_{a'b'}t + \phi_{a'b'})]$$
$$\times \int_0^\infty d\tau' \exp[-(i\omega_{a'b'} - i\nu_{a'b'} + \gamma)\tau']\ U(z - v\tau').$$

Performing the time integration as in Sec. 10-1 [Eq. (10.16), standing-wave case], we have

$$\rho_{a'b'}{}^{(1)}(z, v, t) = -\tfrac{1}{4}\sqrt{2}\,i\wp_{a'b'}\hbar^{-1}N(z)W(v)E_{a'b'}\exp[-i(\nu_{a'b'}t + \phi_{a'b'})]$$
$$\times \sin Kz\,[\mathscr{D}(\omega_{a'b'} - \nu_{a'b'} + Kv) + \mathscr{D}(\omega_{a'b'} - \nu_{a'b'} - (Kv)]. \qquad (26)$$

Combining this with the complex polarization $\mathscr{P}_\pm(t)$ of (12.28), we have the first-order contribution:

$$\mathscr{P}_+{}^{(1)}(t) = -2\bar{N}(\hbar K u)^{-1}\sum_{a'}\sum_{b'}\delta_{a',\,b'+1}|\wp_{a'b'}|^2E_+Z[\gamma + i(\omega_{a'b'} - \nu_+)]. \qquad (27)$$

Hence the first-order coefficients are given by

$$\sigma_\pm + ia_\pm = \nu\bar{N}\wp^2(\hbar K u\varepsilon_0)^{-1}\sum_{a'}\sum_{b'}\delta_{a',\,b'+1}|\wp_{a'b'}/\wp^2|^2$$
$$\times Z[\gamma + i(\omega_{a'b'} - \nu_+)] - \tfrac{1}{2}i\frac{\nu}{Q_+}. \qquad (28)$$

To facilitate evaluation of the third-order contribution (24), we represent (24) by the perturbation "tree" in Fig. F-1. The total contribution is the sum of the products of terms along the "limbs" when the three time integrations are performed and summations over a'' and b'' are included. To further simplify the calculation, we temporarily introduce the mode phasor

$$\mathscr{M}_{a'b'} = \wp_{a'b'}\hbar^{-1}E_{a'b'}\exp[-i(\nu_{a'b'}t + \phi_{a'b'})]. \qquad (29)$$

The first limb of (24) is then

$$① = \tfrac{1}{4}\sqrt{2}\,iNW\sum_{a''}\sum_{b''}\mathscr{M}_{a''b'}\mathscr{M}_{b''a''}\mathscr{M}_{a'b''}$$
$$\times \int_0^\infty d\tau' \int_0^\infty d\tau'' \int_0^\infty d\tau'''\ U(z')U^*(z'')U(z''')$$
$$\times \exp[-i(\omega_{a'b'} - \nu_{a''b'} + \nu_{a''b''} - \nu_{a'b''})\tau' - \gamma\tau']$$
$$\times \exp[-i(\omega_{a'a''} + \nu_{a''b''} - \nu_{a'b''})\tau'' - \gamma_a\tau'']$$
$$\times \exp[-i(\omega_{a'b''} - \nu_{a'b''})\,\tau''' - \gamma\tau'''].$$

The other branches also contain the product of mode factors $\mathscr{M}_{a''b'}\mathscr{M}_{b''a''}\mathscr{M}_{a'b''}$

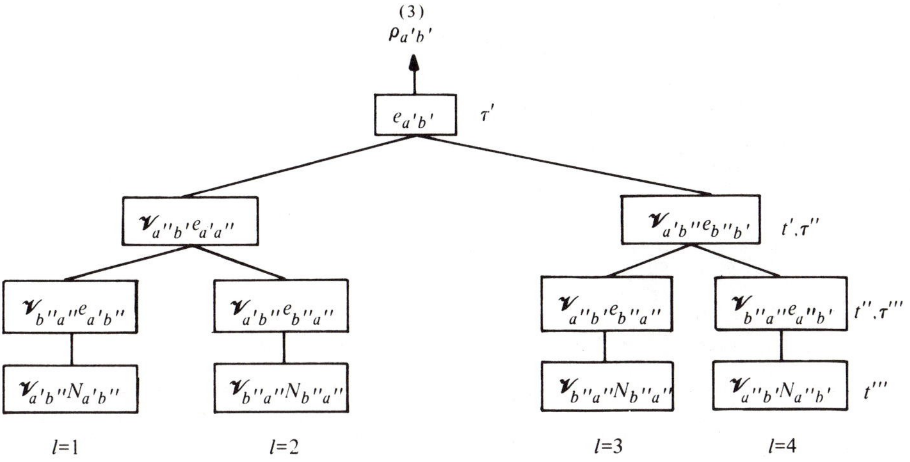

Figure F-1. Perturbation tree for third-order contribution $\rho_{ab}^{(3)}$ of Eq. (24). The $e_{\alpha\beta}$ are defined in the simpler tree of Fig. 10-3. Connecting lines with vertical ascent indicate multiplication; horizontal rows, addition. The earliest perturbation is represented by the first (bottom) row of boxes. Inasmuch as $\mathscr{V}_{a'b'}$ represents a general perturbation, the tree can be used with unidirectional and bidirectional ring lasers as well as with the two-mirror standing-wave laser.

and the same product of the U's. Furthermore, the exponential for τ' is the same. Hence the sum of the four branches is

$$\begin{aligned}
\rho_{a'b'}^{(3)} &= ① + ② + ③ + ④ = \tfrac{1}{4}\sqrt{2}\, iN(z,t)W(v) \sum_{a''}\sum_{b''} \mathscr{M}_{a''b'}\mathscr{M}_{b''a''}\mathscr{M}_{a'b''} \\
&\times \int_0^\infty d\tau' \int_0^\infty d\tau'' \int_0^\infty d\tau'''\, U(z')U^*(z'')U(z''') \\
&\times \exp[-i(\omega_{a'b'} + \nu_{a''b'} + \nu_{a''b''} - \nu_{a'b''})\tau' - \gamma\tau'] \\
&\times \Big\{\exp[-i(\omega_{a'a''} + \nu_{a''b''} - \nu_{a'b''})\tau'' - \gamma_a\tau''] \\
&\qquad \times \{\exp[-i(\omega_{a'b''} - \nu_{a'b''})\tau''' - \gamma\tau'''] \\
&\qquad\quad + \exp[-i(\nu_{a''b''} - \omega_{a''b''})\tau''' - \gamma\tau''']\} \\
&\quad + \exp[-i\omega_{b''b'} + \nu_{a''b''} - \nu_{a''b'})\tau'' - \gamma_b\tau''] \\
&\qquad \times \{\exp[-i(\nu_{a''b''} - \omega_{a''b''})\tau''' - \gamma\tau'''] \\
&\qquad\quad + \exp[-i(\omega_{a''b'} - \nu_{a''b})\tau''' - \gamma\tau''']\}\Big\}. \qquad (30)
\end{aligned}$$

In combining this with the-complex polarization $\mathscr{P}_\pm$ of (12.28), we encounter a product of four U's which can be reduced to expression (12). Using the

Doppler limit approach of Eq. (13), we perform the time integration and find for (12.28)

$$\mathscr{P}_{\pm}^{(3)}(t) = \int_{-\infty}^{\infty} dv\, W(v) \frac{1}{\mathscr{N}} \int_0^L dz\, U^*(z) \sum_{a'}\sum_{b'} \delta_{a',\, b'\pm 1}\, \wp_{b'a'}\, \rho_{a'b'}^{(3)}(z, v, t)$$

$$= \tfrac{1}{2} i \sqrt{\pi}\, \bar{N}(\hbar^3 Ku)^{-1} \sum_{a'}\sum_{b'} \delta_{a',\, b'\pm 1} \sum_{a''}\sum_{b''} \wp_{b'a'}\wp_{a''b'}\wp_{b''a''}\wp_{a'b''}$$

$$\times E_{a''b'}E_{b''a''}E_{a'b''} \exp[i(\nu_{\pm} - \nu_{a''b'} + \nu_{a''b''} - \nu_{a'b''})t + i(\phi_{\pm} - \phi_{a''b'} + \phi_{a''b''} - \phi_{a'b''})]$$

$$\times \Big\{ \mathscr{D}_a(\omega_{a'a''} + \nu_{a''b''} - \nu_{a'b''})$$

$$\times \tfrac{1}{2}\{\mathscr{D}(\tfrac{1}{2}\omega_{a'b'} + \tfrac{1}{2}\omega_{a'b''} - \tfrac{1}{2}\nu_{a''b'} + \tfrac{1}{2}\nu_{a''b''} - \nu_{a'b''})$$

$$+ \mathscr{D}(\tfrac{1}{2}\omega_{a'b'} + \tfrac{1}{2}\omega_{b''a''} - \tfrac{1}{2}\nu_{a''b'} - \tfrac{1}{2}\nu_{a'b''} + \nu_{a''b''})\}$$

$$+ \mathscr{D}_b(\omega_{b''b'} + \nu_{a''b''} - \nu_{a''b'})$$

$$\times \tfrac{1}{2}\{\mathscr{D}(\tfrac{1}{2}\omega_{a'b'} - \tfrac{1}{2}\omega_{a''b''} - \tfrac{1}{2}\nu_{a''b'} - \tfrac{1}{2}\nu_{a'b''} + \nu_{a''b''})$$

$$+ \mathscr{D}(\tfrac{1}{2}\omega_{a'b'} + \tfrac{1}{2}\omega_{a''b'} - \nu_{a''b'} + \tfrac{1}{2}\nu_{a''b''} - \tfrac{1}{2}\nu_{a'b''})\}\Big\}. \tag{31}$$

To determine the third-order coefficients, we consider the three sets of allowed transitions for $\mathscr{P}_+$ in Fig. F-2. The first case involves E_+ alone and hence leads to self-saturation terms. Specifically, multiplying (31) by $\nu/2\varepsilon_0$ for $a'' = a'$ and $b'' = b'$, we find the complex third-order coefficient of $E_+{}^3$:

$$\rho_+ + i\beta_+ = \frac{2\hbar^2\gamma_a\gamma_b}{\wp^2}\, \vartheta_{++++} = F_3 \sum_{a'}\sum_{b'} \delta_{a',\, b'\pm 1} |\wp_{a'b'}/\wp|^4$$

$$\times [\gamma\mathscr{D}(\omega_{a'b'} - \nu_+) + 1]. \tag{32}$$

Here $F_3 = \frac{1}{4}(\gamma_{ab}/\gamma)F_1$, and F_1 is defined in Table 12-1. The mode-pushing ρ_+ and self-saturation β_+ coefficient are given separately in Table 12-2.

Similarly, for case 2 of Fig. F-2, $b'' = b'$, $a'' = a' - 2$, $E_{a''b'} = E_{b''a''} = E_-$, and $E_{a'b''} = E_+$. This yields the complex cross-saturation coefficient:

$$\vartheta_{+--+} = \tfrac{1}{4} iF_1 \frac{1}{\wp^2\hbar^2} \sum_{a'}\sum_{b'} \delta_{a',\, b'+1} |\wp_{a'b'}|^2 |\wp_{a'-2,b'}|^2$$

$$\times \{\mathscr{D}_a(2\delta_a)[\mathscr{D}(\omega_{a'b'} - \nu_+) + \mathscr{D}(\delta_a)]$$

$$+ \gamma_b^{-1}[\mathscr{D}(\omega_{a'b'} - \delta_a - \nu_+) + \mathscr{D}(\delta_a)]\}, \tag{33}$$

where the frequency difference,

$$\delta_a = \frac{\mu_B g_a B}{\hbar} - \tfrac{1}{2}(\nu_+ - \nu_-) \simeq \frac{\mu_B g_a B}{\hbar}, \tag{34}$$

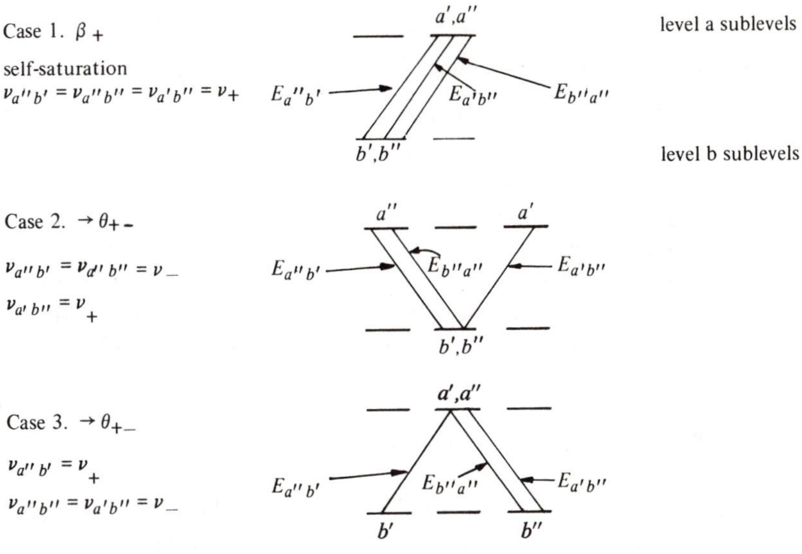

Figure F-2. Diagrams depicting the three allowed combinations of fields which contribute to $\rho_{a',b'=a'-1}$ in third order.

is approximately equal to the Zeeman splitting of the upper levels. Here we have calculated from definitions (12.18)

$$\tfrac{1}{2}(\omega_{a'b'} + \omega_{a''b'}) - \tfrac{1}{2}(\nu_+ + \nu_-) = \omega_{a'b'} - \frac{\mu_B B g_a}{\hbar} + \tfrac{1}{2}(\nu_+ - \nu_-) - \nu_+. \tag{35}$$

Equation (33) contains both a coherent "electric quadrupole" contribution (terms proportional to $\mathscr{D}_a(2\delta_a)$) and an ordinary incoherent cross saturation.

Finally, case 3 subscripts $a'' = a'$, $b'' = b' + 2$ and the mode amplitudes $E_{a''b'} = E_+$, $E_{b''a''} = E_{a'b''} = E_-$. We find the complex coefficient

$$\begin{aligned}\vartheta_{++--} = \tfrac{1}{4} F_1(\wp\hbar)^{-2} \sum_{a'}\sum_{b'} \delta_{a',\, b'+1} |\wp_{a'b'}|^2 |\wp_{a',\, b'+2}|^2 \\ \times \{\gamma_a^{-1} [\mathscr{D}(\omega_{a'b'} - \delta_b - \nu_+) + \mathscr{D}(\delta_b)] \\ + \mathscr{D}_b(2\delta_b) [\mathscr{D}(\omega_{a'b'} - \nu_+) + \mathscr{D}(\delta_b)]\}.\end{aligned} \tag{36}$$

The mode cross-pushing and cross-saturation terms are then given by

$$\begin{aligned}\tau_{+-} + i\theta_{+-} &= \frac{2\hbar^2 \gamma_a \gamma_b}{\wp^2} [\vartheta_{+--+} + \vartheta_{++--}] \\ &= \tfrac{1}{4}\, i\, \frac{\gamma_a \gamma_b}{\wp^4} F_1 \sum_{a'}\sum_{b'} \delta_{a',\, b'+1} |\wp_{a'b'}|^2 \{ |\wp_{a'-2,b'}|^2\end{aligned}$$

$$\times \{\mathscr{D}_a(2\delta_a)[\mathscr{D}(\omega_{a'b'} - \nu_+) + \mathscr{D}(\delta_a)] + \gamma_b^{-1}[\mathscr{D}(\omega_{a'b'} - \delta_a - \nu_+) + \mathscr{D}(\delta_a)]\} + |\wp_{a',b'+2}|^2 \{\mathscr{D}_b(2\delta_b)[\mathscr{D}(\omega_{a'b'} - \nu_+) + \mathscr{D}(\delta_b)] + \gamma_a^{-1}[\mathscr{D}(\omega_{a'b'} - \delta_b - \nu_+) + \mathscr{D}(\delta_b)]\}\}. \quad (37)$$

The corresponding formulas for $\rho_- + i\beta_-$ are given by (32) with $a' = b' - 1$. The cross-saturation coefficients $\tau_{-+} + i\theta_{-+}$ are given by (37) with ν_+ and ν_- interchanged (in the δ's too), $b' = a' + 1$, and $|\wp_{a'-2,b'}|^2$ replaced by $|\wp_{a'+2,b'}|^2$. We recommend that the reader interested in performing this kind of calculation draw the figure for E_- corresponding to Fig. F-2 and calculate the third-order coefficients from the complex polarization (31). Most of the coefficients calculated here are summarized in Table 12-2.

F-3. Sum Rules

As discussed in Sec. 12-3, the zero- (magnetic) field limit is particularly revealing with regard to the electric field polarization. For that discussion, the following sum rules are required:

$$\sum_{a'}\sum_{b'} \delta_{a',b'\pm1}|\wp_{a'b'}|^4$$

$$= \begin{cases} \wp^4 J(J+1)(2J+1)(2J^2+2J+1)/60, & \Delta J = 0, \quad (38) \\ \wp^4 (J+1)(2J+1)(2J+3)(6J^2+12J+5)/60, & \Delta J = \pm 1, J \longleftrightarrow J+1 \quad (39) \end{cases}$$

$$\sum_{a'}\sum_{b'} \delta_{a',b'\pm1}|\wp_{a'b'}|^2[|\wp_{a'-2,b'}|^2 + |\wp_{a',b'+2}|^2]$$

$$= \begin{cases} \wp^4 J(J+1)(2J+1)(2J-1)(2J+3)/60, & \Delta J = 0, \quad (40) \\ \wp^4 (J+1)(2J+3)(2J+1)(2J^2+4J+5)/60, & \Delta J = \pm 1, J \longleftrightarrow J+1. \quad (41) \end{cases}$$

We can derive these sum rules from the definitions of the matrix elements (12.23) and the simple sum formulas (see Gradshteyn and Ryzhik, 1965)

$$\sum_{j=1}^{N} j^2 = N(N+1)(2N+1)/6, \quad (42)$$

$$\sum_{j=1}^{N} j^4 = N(N+1)(2N+1)(3N^2+3N-1)/30. \quad (43)$$

Summation formulas for j and j^3 are not needed inasmuch as the lower limits of the summations in (38)–(41) are negatives of the upper limits, causing sums of odd powers to add to zero. We prove (38) for $\Delta J = 0$ explicitly and leave the others as problems. Using (12.23) and dropping odd powers, we have

$$\sum_{a'=-J}^{J} |\wp_{a',b'=a'+1}|^4 = \tfrac{1}{16}\wp^4 \sum_{a'=-J}^{J} (J - a')^2 (J + a' + 1)^2$$

$$= \tfrac{1}{16}\wp^4 \sum_{a'=-J}^{J} [J^2(J+1)^2 - a'^2(2J^2 + 2J - 1) + a'^4]$$

$$= \tfrac{1}{16}\,\wp^4[J^2(J+1)^2(2J+1) - (2J^2+2J-1)J(J+1)$$
$$\times (2J+1)/3 + J(J+1)(2J+1)(3J^2+3J-1)/15]$$
$$= \tfrac{1}{16}\tfrac{1}{15}\,\wp^4 J(J+1)(2J+1)[8J^2+8J-4]$$
$$= \tfrac{1}{60}\,\wp^4 J(J+1)\,(2J+1)\,(2J^2+2J+1). \qquad Q.E.D.$$

Note that by symmetry this result follows for $a' = b' + 1$ as well.

References

See references for Chap. 12 and the following:

R. L. Fork and M. Sargent III, 1966, in: *Proceedings of the International Conference on the Physics of Quantum Electronics,* ed. by P. L. Kelley, B. Lax, and P. E. Tannenwald, McGraw-Hill, Book Co., New York.

I. S. Gradshteyn and I. M. Ryzhik 1965, *Table of Integrals, Series and Products,* fourth edition, Academic Press, New York, p. 1.

G

SUPERRADIANCE

Appendix G. Collective Spontaneous Emission (Superradiance)

Superradiance is the collective spontaneous emission of a system of many atoms. We discuss ensembles confined to a volume small compared to a cubic wavelength, although extensions are available.† The atoms are strongly coupled by their common interaction with a resonant electric field. Hence it is not possible to treat the interaction between the field and an independent atom; the entire system must be treated at once. In our treatment, we assume that the atoms are affected only by the atom-field interaction energy; atom-atom interactions (collisions) and other energies are ignored. In practice this requires that the atomic separation d be large compared to atomic dimensions on the order of a Bohr radius a_0, that is, the wavelength $\lambda >> d >> a_0$. This collective spontaneous emission or superradiance is illustrated by photon echo. In our earlier treatment of these echoes (Sec. 13-3), we showed that, when the phases of the individual dipoles became equal again, the dipoles generated a large macroscopic polarization producing the echo and at other times yielded negligible polarization. In fact, randomly phased dipoles generally radiate at normal rates, yielding an energy proportional to N, the number of atoms. But when their phases become aligned, they radiate energy in a collective fashion at a rate initially proportional to N^2. This latter rate is characteristic of superradiance. A difference between the present treatment and photon echo lies in the method of preparation. Here spontaneous emission creates the phasing; two special pulses perform this operation in photon echo.

In this appendix we consider first (Sec. G-1) the collective radiation of a two-atom, single-mode field system, using the quantum techniques developed

†Superradiance was originally defined and discussed by Dicke (1954). For discussions of extensions, see Eberly (1972).

in Chap. 14. Even for this limited system, the phenomenon of radiation trapping appears. The theory is then extended (Sec. G-2) to three atoms, for which superradiance begins to show. The general case of N atoms is considered and is compared in Sec. G-3 to a single system with angular momentum $J = \frac{1}{2}N$. The analogy is particularly felicitous, for it makes it possible to dispense with somewhat awkward notation and to treat various extensions simply.

G-1. Two-Atom, Single-Mode Field Interaction

The interaction energy for two atoms and a single-mode field (see Fig. G-1) is given by

$$\mathscr{V} = -e\mathbf{r}_1 \cdot \mathbf{E}_1 - e\mathbf{r}_2 \cdot \mathbf{E}_2, \tag{1}$$

where $\mathbf{r}_1$ and $\mathbf{r}_2$ are the atomic $\mathbf{r}$ vectors, and $\mathbf{E}_1$ and $\mathbf{E}_2$ the applied electric fields at the centers of the atoms. Provided the atoms are close, the fields are the same, and we can write

$$\mathscr{V} = -e(\mathbf{r}_1 + \mathbf{r}_2) \cdot \mathbf{E}. \tag{2}$$

In quantum-mechanical form, this is given by the operator

$$\mathscr{V} = g_\lambda[(\sigma_1 + \sigma_1{}^\dagger) + (\sigma_2 + \sigma_2{}^\dagger)]\,(a_\lambda + a_\lambda{}^\dagger). \tag{3}$$

Making the rotating-wave approximation as before (14.65) and going into the interaction picture, we have

$$\mathscr{V}_I(t) = g_\lambda(\sigma_1 + \sigma_2)\exp[i(\nu_\lambda - \omega)t]\,a_\lambda{}^\dagger + \text{adjoint}. \tag{4}$$

To understand how this Hamiltonian affects the system, we consider the state vector

$$|\psi_{\text{a-f}}(t)\rangle = U(t, 0)|\psi_{\text{a-f}}(0)\rangle, \tag{5}$$

where the U matrix is a solution of the Schrödinger equation (6.63) given by the perturbation series

$$\begin{aligned} U(t, 0) &= 1 + \frac{-i}{\hbar}\int_0^t dt'\,\mathscr{V}(t') + \left(\frac{-i}{\hbar}\right)^2 \int_0^t dt' \int_0^{t'} dt''\mathscr{V}(t')\mathscr{V}(t'') + \cdots \\ &\approx 1 + \frac{-i}{\hbar}\,t\,\{g_\lambda(\sigma_1 + \sigma_2)a_\lambda{}^\dagger + g_\lambda(\sigma_1{}^\dagger + \sigma_2{}^\dagger)a_\lambda\} + \frac{1}{2}\left(\frac{-i}{\hbar}\right)^2 t^2\{\}\,\{\}. \end{aligned} \tag{6}$$

Suppose that the state vector (5) initially has the value

$$|\psi_{\text{a-f}}(0)\rangle = |b_1b_2\rangle|2_\lambda\rangle, \tag{7}$$

that is, the field has two photons with wavelength λ and both atoms are in the lower state (Fig. G-1). Then at time t the state vector

$$\begin{aligned} |\psi_{\text{a-f}}(t)\rangle = C_0|b_1b_2\rangle|2_\lambda\rangle &+ \sqrt{2}C_1[|a_1b_2\rangle + |b_1a_2\rangle]|1_\lambda\rangle \\ &+ \sqrt{2}C_2|a_1a_2\rangle|0_\lambda\rangle, \end{aligned} \tag{8}$$

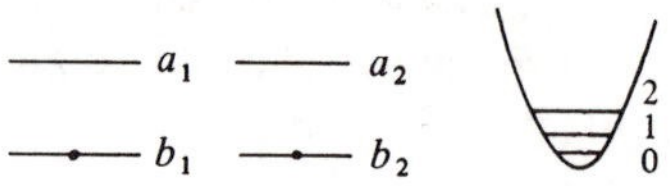

Figure G-1. Two-atom/field system shown with both atoms in their lower state and with two photons in the field. Because of electric-dipole interaction energy, transitions to possible excited atomic states can occur in time.

where the amplitudes C_0, C_1, and C_2 can be evaluated with the use of (6), for example, $C_2 = (-ig_\lambda t/\hbar)^2$ to second order.

The point of this little derivation is that the atom-field Hamiltonian (4) can cause transitions between $|b_1b_2\rangle|2_\lambda\rangle$ and the symmetric state $2^{-1/2}[|a_1b_2\rangle + |b_1a_2\rangle]\,|1_\lambda\rangle$ and between this state and $|a_1a_2\rangle|0_\lambda\rangle$. Now there are four obvious states of the system, for either atom can be in either of two states independently of the other:

$$(|a_1a_2\rangle|0_\lambda\rangle, \qquad |a_1b_2\rangle|1_\lambda\rangle, \qquad |b_1a_2\rangle|1_\lambda\rangle, \qquad |b_1b_2\rangle|2_\lambda\rangle).$$

For this problem, it is clearly advantageous to combine the one-photon states symmetrically. There must then be a fourth state orthogonal to those in (8). Finding that the antisymmetric state $2^{-1/2}[a_1b_2\rangle - |b_1a_2\rangle]\,|1_\lambda\rangle$ is the required state, we are now confronted with the fascinating fact that a system described by this antisymmetric state cannot decay to the lower state $|b_1b_2\rangle \times |2_\lambda\rangle$! The two atoms play "catch" with a photon and never miss. In short, radiation is trapped as depicted in Fig. G-2.

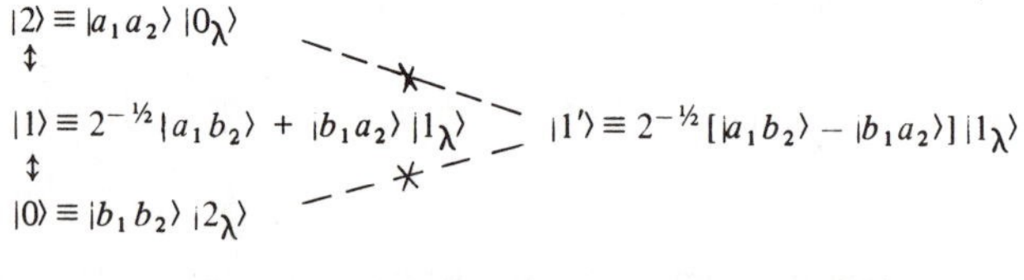

Figure G-2. Diagram of states used for two-atom/field system. The electric-dipole perturbation energy (4) can cause transitions between symmetric states (first column), but not between these states and the antisymmetric state $|1'\rangle$ (second column).

Our discussion can be summarized in the calculation of the matrix elements between these states. The writing can be simplified by naming the states according to their atomic energies and degrees of symmetry as follows:

$$|0\rangle = |b_1b_2\rangle|2_\lambda\rangle, \tag{9}$$

$$|1\rangle = 2^{-1/2}[|a_1b_2\rangle + |b_1a_2\rangle]|1_\lambda\rangle, \tag{10}$$

$$|1'\rangle = 2^{-1/2}[|a_1b_2\rangle - |b_1a_2\rangle]|1_\lambda\rangle, \tag{11}$$

$$|2\rangle = |a_1a_2\rangle|0_\lambda\rangle. \tag{12}$$

Here the totally symmetric states are unprimed; the antisymmetric state is primed. This notation is easily extended to the many-atom case, which we consider shortly.

We designate the set of vectors (9)–(12) and later extensions as the Dicke representation. The reader can easily verify the matrix elements:

$$\langle 1\,|(\sigma_1 + \sigma_2)a^\dagger|2\rangle = 2^{1/2}, \tag{13}$$

$$\langle 1'|(\sigma_1 + \sigma_2)a^\dagger|2\rangle = 0, \tag{14}$$

$$\langle 0\,|(\sigma_1 + \sigma_2)a^\dagger|1\rangle = 2, \tag{15}$$

$$\langle 0\,|(\sigma_1 + \sigma_2)a^\dagger|1'\rangle = 0. \tag{16}$$

This confirms our earlier conclusion that the perturbation energy (4) can cause transitions only between the symmetric states; the antisymmetric state is stable. This situation occurs elsewhere in varying degree; for example, the 2^3s state of helium is metastable and corresponds to an antisymmetric combination of the two electrons. It is possible to create a system in the antisymmetric state using Hamiltonians other than the electric dipole of (4). Specifically, a Stern-Gerlach apparatus can be used to obtain one atom in its upper state, and another apparatus to obtain a second atom in its ground state. The two atoms can be combined well within a wavelength of the resonant radiation in a bottle. Then the state vector for the system $|a_1b_2\rangle|1_\lambda\rangle$ can be written as a linear combination of the symmetric and antisymmetric states $|1\rangle$ and $|1'\rangle$. The symmetric state decays in time, leaving a 50% probability that the antisymmetric state exists at long times.

G-2. Many-Atom Field System

In this section, we consider first the three-atom system, which contains the essential formal characteristics of the many-atom case, and then the many-atom system itself. In so doing, it is convenient to denote each atom by its position in the eigenket rather than by a subscript; for example, the ket $|aab\rangle$ represents a set of three atoms the first two of which are in the upper state and the third in the lower. It is also convenient to leave the state of the field unspecified. This information can always be included when required.

For a three-atom system, there are 2^3 possible states, for each atom can be in the upper or lower level independently of the others. As for the two-atom system, the problem is simplest when various symmetric and antisymmetric combinations of the single eigenkets are used. Although there is only one set of symmetric combinations for the discrete energies, there are a number of possible antisymmetric combinations. The choice that Dicke made (and one corresponding to angular momentum theory—an important relationship) is given for the three-atom system in Fig. G-3. Similarly to the two-atom case (4), the atomic part of the three-atom interaction energy

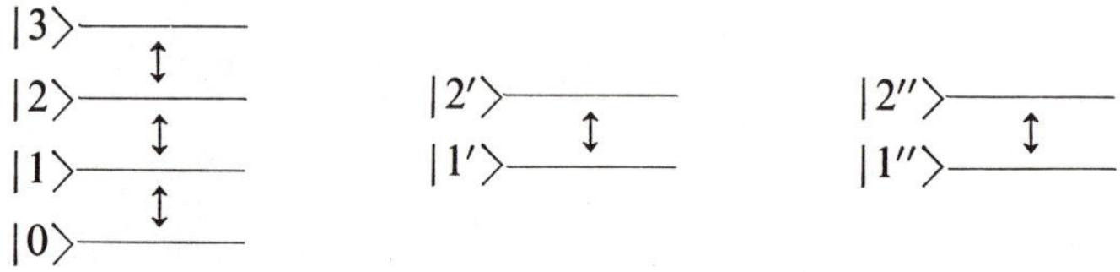

$$|3\rangle \equiv |aaa\rangle$$
$$|2\rangle \equiv 3^{-1/2}[|aab\rangle + |aba\rangle + |baa\rangle]$$
$$|1\rangle \equiv 3^{-1/2}[|bba\rangle + |bab\rangle + |abb\rangle]$$
$$|0\rangle \equiv |bbb\rangle$$

$$|2'\rangle \equiv 6^{-1/2}[|aab\rangle + |aba\rangle - 2|baa\rangle]$$
$$|1'\rangle \equiv 6^{-1/2}[|bba\rangle + |bab\rangle - 2|abb\rangle]$$

$$|2''\rangle \equiv 2^{-1/2}[|aab\rangle - |aba\rangle]$$
$$|1''\rangle \equiv 2^{-1/2}[|bba\rangle - |bab\rangle]$$

Figure G-3. Diagram of states used for three-atom/field system. The electric-dipole interaction energy can cause transitions only between states in the same column, that is, states having the same number of primes.

$$\mathscr{V}_{\text{atom}} = \sigma_1 + \sigma_2 + \sigma_3 \tag{17}$$

can cause transitions only between states in the same column (other matrix elements vanish). In particular, the matrix element

$$\begin{aligned}\langle 1|\,\mathscr{V}_{\text{atom}}|2\rangle &= 3^{-1/2}[\langle bba| + \langle bab| + \langle abb|]\,(\sigma_1 + \sigma_2 + \sigma_3) \\ &\qquad \times [|aab\rangle + |aba\rangle + |baa\rangle]\, 3^{-1/2} \\ &= \tfrac{1}{3}(2 \times 3) = 2.\end{aligned} \tag{18}$$

The rate at which a transition between $|2\rangle$ and $|1\rangle$ takes place is (to first order) proportional to the square of the matrix element (18), and hence the atomic contribution to this rate is four—greater than the number of atoms! At this point, we are beginning to see superradiance, for which the transition rates can initially be proportional to as much as N^2, where N is the number of atoms. Conversely, the matrix element between antisymmetric states $\langle 2''|\mathscr{V}|1''\rangle$ contains an atomic contribution of unity, whose square is less than the number of atoms.

Consider now N atoms which are in symmetric states. As for the three-atom case, the atomic part of the general interaction energy

$$\mathscr{V}_{\text{atom}} = \sum_{i=1}^{N} \sigma_i \tag{19}$$

causes transitions between the symmetric states, but not between these states and those with less symmetry. It is possible to prepare a state vector with only symmetric states by starting in the ground state $|0\rangle$(all atoms in lower state) and applying an electric field. The combinations to be used are, then, as follows:

$$|N\rangle \equiv |aa \ldots a\rangle,$$
$$|N-1\rangle \equiv N^{-1/2}\,[aa \ldots b\rangle + |aa \ldots ba\rangle + |ba \ldots a\rangle],$$
$$\vdots$$
$$|n_a\rangle \equiv \left[\frac{N!}{(n_a!n_b!)}\right]^{-1/2} \sum_{\text{perm}} |\underbrace{a \ldots a}_{n_a} \; \underbrace{b \ldots b}_{n_b}\rangle, \tag{20}$$
$$\vdots$$
$$|1\rangle \equiv N^{-1/2}[|bb \ldots a\rangle + |bb \ldots ab\rangle + \cdots + |ab \ldots b\rangle],$$
$$|0\rangle \equiv |bb \ldots b\rangle.$$

These vectors are the general set corresponding to the first column in Fig. G-3 for the three-atom case. The number of atoms in the upper level is n_a and that in the lower level n_b, so that the total number

$$N = n_a + n_b. \tag{21}$$

The summation over permutations in (20) indicates that all *distinct* arrangements of n_a atoms in the upper state and n_b atoms in the lower state are included. The number of these states is $N!/(n_a!\, n_b!)$.

Now consider the matrix element

$$\langle n_a - 1|\, \mathscr{V}_{\text{atom}}|n_a\rangle = \langle n_a - 1 \mid \sum_{i=1}^{N} \sigma_i \mid n_a\rangle \tag{22}$$

in an extension of our discussion for three atoms [see Eq. (18)]. The ith term for this matrix element is zero unless the ith atom is in the upper state. There are $(N-1)!\,[\,(n_a-1)!n_b!]^{-1}$ such eigenkets in the symmetric combination $|n_a\rangle$. Corresponding to each of these is a single eigenbra with identical states except for the ith atom, which is in the lower state. This eigenbra alone has a nonzero inner product with the corresponding eigenket rotated by σ_i. Hence the matrix element for the ith term

$$\langle n_a - 1|\sigma_i|n_a\rangle = \left[\frac{N!}{(n_a-1)!\,(n_b+1)!}\,\frac{N!}{n_a!n_b!}\right]^{-1/2} \left[\frac{(N-1)!}{(n_a-1)!\,n_b!}\right]. \tag{23}$$

There are N terms in the perturbation energy (19), so that the matrix element

(22) is given by (23) multiplied by N. Simplifying the resulting expression, we have

$$\langle n_a - 1 | \mathscr{V}_{\text{atom}} | n_a \rangle = [n_a(n_b + 1)]^{1/2}. \tag{24}$$

Now consider the state for which $n_a = n_b = \frac{1}{2}N$. The probability that emission occurs is initially proportional to the absolute value squared of this matrix element and hence to $N^2 I_0$, where I_0 gives the field strength. Because there are only N atoms, this response is termed superradiant. Semiclassically, the electric field amplitude is proportional to the polarization, which for N atoms in phase is, in turn, proportional to N. Hence the intensity is initially proportional to N^2. In the fully quantum-mechanical calculation, we have not explicitly referred to phase, but the choice of symmetric states implicitly requires that the atoms radiate in phase with one another. It is this cooperative radiation by a group of coupled atoms which constitutes superradiance. Because the initial decay of the atoms is greatly accelerated, the decay time of the coupled ensemble is correspondingly reduced.

G-3. Angular Momentum Treatment

At this point it is helpful to note the formal identity between states of similar symmetry [same column in Figs. G-2 and G-3, and in Eq. (20)] and those for a single system having a given angular momentum.† Specifically, we denote the states in the first column (totally symmetric states) by the ket $|rm\rangle$, where the total "angular-momentum"-like quantum number

$$r = \tfrac{1}{2}N, \tag{25}$$

and the magnetic sublevel quantum number

$$m = \tfrac{1}{2}(n_a - n_b), \tag{26}$$

which takes on the $N + 1$, $2r + 1$ values, i.e. $-\frac{1}{2}N$, $-\frac{1}{2}N + 1, \ldots, \frac{1}{2}N - 1$, $\frac{1}{2}N$. In particular, the symmetric states for two- and three-atom systems are related to the angular momentum states as depicted in Fig. G-4.

The antisymmetric (or less symmetric) states are given by angular momentum states with less total angular momentum. The two-atom antisymmetric state has zero angular momentum, for one "spin" points up, representing the state $|a\rangle$, and the other points down, representing $|b\rangle$. Hence the m value is also zero. Similarly the states in columns 2 and 3 of Fig. G–3 for the three-atom case have total angular momentum $\frac{1}{2}$ with m values $\pm\frac{1}{2}$, for one spin always subtracts from the sum of the other two.

Using the definitions of r (25) and m (26) along with (21) for n_a and n_b, we find the relationships

$$n_a = r + m, \qquad n_b = r - m \tag{27}$$

†See Feynman, Leighton, and Sands, *The Feynman Lectures on Physics,* Vol. III, Addison-Wesley, 1965, Chap. 18, for a discussion of how a system with angular momentum r is equivalent to N spin-$\frac{1}{2}$ systems.

for the symmetric case ($r = N/2$). Thus the matrix element (24) connecting the state $|n_a\rangle$ with $|n_a - 1\rangle$ becomes

$$\langle r, m - 1|R|rm\rangle = \tfrac{1}{2}[(r + m)(r - m + 1)]^{1/2}, \tag{28}$$

which is precisely the result found by using the lowering operator R in angular momentum theory.†

$r = \tfrac{1}{2}N = 1$

$m = 1$	———	$|aa\rangle = |r = 1, m = 1\rangle$
0	———	$2^{-1/2}[|ab\rangle + |ba\rangle] = |1\ 0\rangle$
-1	———	$|bb\rangle = |1, -1\rangle$

$r = \tfrac{1}{2}N = \tfrac{3}{2}$

$m = \tfrac{3}{2}$	———	$|aaa\rangle = |\tfrac{3}{2}\,\tfrac{3}{2}\rangle$
$\tfrac{1}{2}$	———	$3^{-1/2}[|aab\rangle + |aba\rangle + |baa\rangle] = |\tfrac{3}{2}\,\tfrac{1}{2}\rangle$
$-\tfrac{1}{2}$	———	$3^{-1/2}[|abb\rangle + |bab\rangle + |bba\rangle] = |\tfrac{3}{2}\,-\tfrac{1}{2}\rangle$
$-\tfrac{3}{2}$	———	$|bbb\rangle = |\tfrac{3}{2}\,\tfrac{3}{2}\rangle$

Figure G-4. Diagram giving relationships between symmetric states of N atoms and the angular momentum states for a single system having total angular momentum $\tfrac{1}{2}N$. The additions of spins representing state $|a\rangle$ (spin up) and $|b\rangle$ (spin down) are given.

The diagram for the complete N-atom system in angular momentum nomenclature is given in Fig. G-5. Here the degeneracy (number of discrete states) of the states with magnetic quantum number m and a particular angular momentum $r < N/2$ is given by the total number of states with $m = r$ minus the total number for $m + 1$:

$$\text{Degeneracy} = \frac{N!}{n_a!n_b!} - \frac{N!}{(n_a + 1)!\,(n_b - 1)!}$$

$$= \frac{N!}{(n_a + 1)!n_b!}[(n_a + 1) - n_b].$$

Now $n_a = \tfrac{1}{2}N + m = \tfrac{1}{2}N + r$. Similarly, $n_b = \tfrac{1}{2}N - r$, so that the degeneracy of $|r, m\rangle$ is

$$\frac{N!(2r + 1)}{(\tfrac{1}{2}N + r + 1)!\ (\tfrac{1}{2}N - r)!}. \tag{29}$$

For a single-atom state the intensity of spontaneous radiation

†See Dicke and Wittke (1960), Chap. 9.

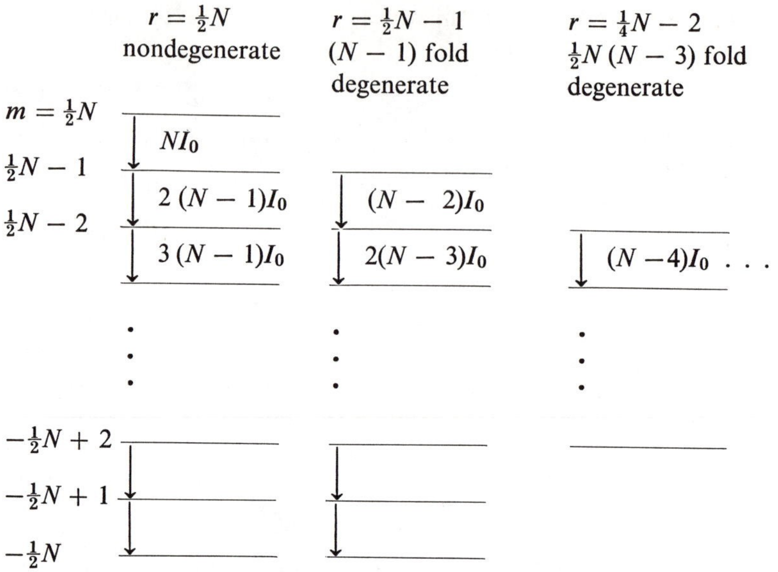

Figure G-5. Diagram for N-atom system in angular momentum terminology, showing matrix elements between certain levels.

$$I_0 = \frac{2\pi\sigma(\omega)\wp^2}{\hbar} \tag{30}$$

according to the Weisskopf-Wigner theory. For the states having $m = r = \frac{1}{2}N$, the intensity

$$I = I_0 (\tfrac{1}{2}N \times 2)(\tfrac{1}{2}N - \tfrac{1}{2}N + 1) = NI_0, \tag{31}$$

which would be predicted using the Weisskopf-Wigner theory. However, for states having $r = \frac{1}{2}N$, $m = 0$,

$$I = I_0 \tfrac{1}{2}N (\tfrac{1}{2}N + 1), \tag{32}$$

which is coherent spontaneous emission differing dramatically from that predicted by the Weisskopf-Wigner theory.

Another phenomenon not contained in the Weisskopf-Wigner theory is radiation trapping, introduced earlier in the discussion of the two-atom case. If there is one excited atom in a system of N atoms, then the probability for being in the totally symmetric state $r = \frac{1}{2}N$ is N^{-1}, for there are N possible such states. Thus the probability for being in a nonradiative state $1 - N^{-1} \to 1$ as $N \to \infty$, that is, the radiation is trapped. It is possible to consider this phenomenon and others using the formalism developed here. The reader is referred to the literature† for further discussion.

†See, for example, Bonifacio, Kim, and Scully (1969).

Problem

G-1. Show that the dipole moment vanishes for a Dicke state $|rm\rangle$. Note that this differs from the photon echo problem, in which there exists a macroscopic dipole moment. The situation is analogous to the relationship between a number state $|n\rangle$ and the coherent state $|\alpha\rangle$ [see Arecchi et al. (1972)].

References

See the references for Chaps. 1 and 14 and the following:

F. T. Arecchi, E. Courtens, R. Gilmore and H. Thomas, 1972, *Phys. Rev.* **A6**, 2211.

R. Bonifacio, D. Kim, and M. O. Scully, 1969, *Phys. Rev.* **187**, 441.

R. H. Dicke and J. P. Wittke, 1960, *Introduction to Quantum Mechanics*, Addison-Wesley Publishing Co., Reading, Mass.

J. H. Eberly, 1972, *American J. Phys.* **40**, 1374.

For a recent treatment and many references, see L. Allen and J. H. Eberly, 1975, *Optical Resonance and Two-Level Atoms*, John Wiley & Sons, New York.

H

THE COHERENT STATE

Appendix H. The Coherent State

In this appendix we describe the coherent state of Chap. 15 by requiring that, to be as classical as possible, the state yield minimum uncertainty for all time (the classical fields have zero uncertainties) when subject to a simple harmonic potential characteristic of the electromagnetic radiation. We find that the corresponding probability density $\psi^*\psi$ "coheres," that is, does not spread in time. In Sec. H-1 we derive the conditions for minimum uncertainty (4) and (8), and then use these to find a simple differential equation (11) for the wave functions. The solution is a Gaussian with an arbitrary real width and a complex center. In Sec. H-2 we derive the Green function, which specifies the time dependence of the wave function in a simple harmonic potential well. The width remains real (and the function of minimum uncertainty) only if given by $(\hbar/M\Omega)^{1/2}$. In Sec. H-3 we write the wave function with this width as a state vector expanded in terms of the photon number states. This last expression is Glauber's coherent state, discussed in Chap. 15.

H-1. The Minimum-Uncertainty Wave Function

To derive the conditions for minimum uncertainty, we define

$$\delta q = q - \langle q \rangle, \qquad \delta p = p - \langle p \rangle, \tag{1}$$

in terms of which the uncertainty product

$$(\Delta q)^2 (\Delta p)^2 = \int_{-\infty}^{\infty} dq \psi^*(q)\, (\delta q)^2 \psi(q) \int_{-\infty}^{\infty} dp \psi^*(q)\, (\delta p)^2 \psi(q)$$

$$= \int_{-\infty}^{\infty} dq\, (\delta q \psi)^* \, (\delta q \psi) \int_{-\infty}^{\infty} dq\, (\delta p \psi)^* \, (\delta p \psi). \tag{2}$$

We apply the Schwartz inequality,† which reads in the notation of (2) as

$$(\Delta q)^2 (\Delta p)^2 \geq \left| \int dq\, (\delta q\psi)^* (\delta p\psi) \right|^2 = \left| \int dq\psi^*\, \delta q\, \delta p\psi \right|^2, \tag{3}$$

where equality holds for

$$\delta p\psi = iC\, \delta q\psi \tag{4}$$

for some complex constant C. The last term in (3) can be written as

$$\begin{aligned} \left| \int dq\psi^*\, \delta q\, \delta p\psi \right|^2 &= \left| \int dq\psi^* \left[\tfrac{1}{2}(\delta q\, \delta p - \delta p\, \delta q) + \tfrac{1}{2}(\delta q\, \delta p + \delta p\, \delta q)\right] \psi \right|^2 \\ &= \left| \int dq\psi^* \left[\tfrac{1}{2} i\hbar + \tfrac{1}{2}(\delta q\, \delta p + \delta p\, \delta q)\right]\psi \right|^2 \\ &= \tfrac{1}{4} \left| i\hbar + \left[\int dq\psi^*\, \delta q\, \delta p\psi + \text{c.c.}\right] \right|^2 \\ &= \tfrac{1}{4}\hbar^2 + \tfrac{1}{4}\left[\int dq\psi^*\, (\delta q\, \delta p + \delta p\, \delta q)\psi\right]^2, \end{aligned} \tag{5}$$

where we have used the commutation relation

$$[\delta q, \delta p] = [q, p] = i\hbar. \tag{6}$$

Combining (3) and (6), we find

$$(\Delta q)^2 (\Delta p)^2 \geq \tfrac{1}{4}\hbar^2, \tag{7}$$

which yields the uncertainty relation (15.3).

The two conditions for equality in (7), that is, minimum uncertainty, are (4) and [from (5)]

$$\int dq\psi^*\, (\delta q\, \delta p + \delta p\, \delta q)\psi = 0. \tag{8}$$

Substituting (4) into (8), we find

$$(C - C^*) \int dq\psi^*\, (\delta q)^2 \psi = 0, \tag{9}$$

which can be true only if C is a real number which from (4) must have the dimensions of a mass times a frequency. Inserting (1) into (4), we find

$$[p - \langle p \rangle]\psi(q) = iC[q - \langle q \rangle]\psi(q).$$

This yields the differential equation

$$\frac{d\psi}{dq} = \left[\frac{-C}{\hbar} q + \left(\frac{2C}{\hbar}\right)^{1/2} \chi\right]\psi(q), \tag{10}$$

where the complex dimensionless constant

$$\chi = (2C\hbar)^{-1/2}(C\langle q \rangle + i\langle p \rangle). \tag{11}$$

The differential equation (10) can be integrated with the solution

---

†See Schiff (1969). Intuitively, one may understand the result by noting that in coordinate space the functions $\delta_p\psi$ and $\delta_q\psi$ overlap themselves more than they overlap one another unless (4) holds.

$$\psi(q) = \mathcal{N}_\psi \exp[-\tfrac{1}{2}(C/\hbar)q^2 + (2C/\hbar)^{1/2}\chi q]. \tag{12}$$

The normalization constant $\mathcal{N}_\psi$ is determined by the condition $\int dq\psi^*\psi = 1$. It is convenient to define the dimensionless parameter

$$\xi' = \left(\frac{C}{\hbar}\right)^{1/2} q \tag{13}$$

for this purpose. We find

$$\begin{aligned}
\int_{-\infty}^{\infty} dq\psi^*(q)\psi(q) &= \left(\frac{C}{\hbar}\right)^{-1/2} \mathcal{N}_\psi{}^2 \int_{-\infty}^{\infty} d\xi' \exp[-\xi'^2 + \sqrt{2}(\chi + \chi^*)\xi'] \\
&= \left(\frac{C}{\hbar}\right)^{-1/2} \mathcal{N}_\psi{}^2 \int_{-\infty}^{\infty} d\xi' \exp\{-[\xi' - \tfrac{1}{2}\sqrt{2}(\chi + \chi^*)]^2 \\
&\qquad + \tfrac{1}{2}(\chi + \chi^*)^2\} \\
&= \left(\frac{C}{\hbar}\right)^{-1/2} \mathcal{N}_\psi{}^2 \exp[\tfrac{1}{2}(\chi + \chi^*)^2] \int_{-\infty}^{\infty} d\xi'' \exp(-\xi''^2) \\
&= \left(\frac{C}{\pi\hbar}\right)^{-1/2} \mathcal{N}_\psi{}^2 \exp[\tfrac{1}{2}(\chi + \chi^*)^2].
\end{aligned}$$

Setting this to 1, we have the normalization constant

$$\mathcal{N}_\psi = \left(\frac{C}{\pi\hbar}\right)^{1/4} \exp[-\tfrac{1}{4}(\chi + \chi^*)^2]. \tag{14}$$

The wave function becomes

$$\begin{aligned}
\psi(q) &= \left(\frac{C}{\pi\hbar}\right)^{1/4} \exp[-\tfrac{1}{4}(\chi + \chi^*)^2] \exp(-\tfrac{1}{2}\xi'^2 + \sqrt{2}\chi\xi') \\
&= \left(\frac{C}{\pi\hbar}\right)^{1/4} \exp[\chi^2 - \tfrac{1}{4}(\chi + \chi^*)^2] \exp[-\tfrac{1}{2}(\xi' - \sqrt{2}\chi)^2].
\end{aligned}$$

Dropping the factor $\exp[\tfrac{1}{4}(\chi^2 - \chi^{*2})]$, which has unit modulus, we have

$$\psi(q) = \left(\frac{C}{\pi\hbar}\right)^{1/4} \exp[\tfrac{1}{2}(\chi^2 - \chi\chi^*)] \exp[-\tfrac{1}{2}(\xi' - \sqrt{2}\chi)^2]. \tag{15}$$

A distinguishing feature of this function is that the coefficient of q^2 [see (13)] in the exponential is $C/2\hbar$, a real number, for C is real because of the conditions (4) and (8) for minimum uncertainty. We now calculate the time development of (15) subject to a simple harmonic potential characterized by mass M and frequency Ω. We find that, unless $C = M\Omega$, the coefficient of q^2 becomes complex in time and hence that $\psi(q, t)$ fails to maintain minimum uncertainty for all times.

H-2. Time Development of Minimum-Uncertainty Wave Packet

To find the time dependence of a wave function, we can calculate the Green function $G(x, x_0, t)$, which is the solution of Schrödinger's equation with the initial condition

$$G(x, x_0, 0) = \delta(x - x_0). \tag{16}$$

The wave function at time t is then given by the superposition

$$\psi(x, t) = \int dx_0\, G(x, x_0, t)\psi(x_0, 0). \tag{17}$$

A formal integral of the Schrödinger equation for $G(x, x_0, t)$ gives the value

$$G(x, x_0, t) = \exp(-i\mathscr{H}t/\hbar)\, \delta(x - x_0), \tag{18}$$

where $\mathscr{H}$ is the Hamiltonian with derivatives taken with respect to x. Another convenient form for $G(x, x_0, t)$ can be derived by writing the wave function (17) as

$$\psi(x, t) = \sum_n C_n u_n(x) \exp(-in\Omega t), \tag{19}$$

where the expansion coefficients C_n are given by

$$C_n = \int dx_0\, u_n^*(x_0)\, \psi(x_0, 0). \tag{20}$$

Identifying the coeffiicients of $\psi(x_0, 0)$ in the integrands of (17) and (19) with (20), we find the bilinear form:

$$G(x, x_0, t) = \sum_n u_n^*(x_0)\, u_n(x) \exp(-in\Omega t), \tag{21}$$

which, of course, reduces to $\delta(x - x_0)$ at $t = 0$.

In principle it is possible to find the Green function for the simple harmonic oscillator by summing the series in (21) with use of the Hermite-Gaussian functions (1.23). It is easier however, to work with the Heisenberg picture operators $x(t)$ and $p(t)$ with the definition and equation of motion

$$p(t) = M\dot{x}(t), \tag{22}$$

$$\dot{p}(t) = -M\Omega^2 x(t). \tag{23}$$

Combining these, we have the equation for $x(t)$ alone:

$$\ddot{x}(t) + \Omega^2 x(t) = 0 \tag{24}$$

with the solution

$$x(t) = x(0)\cos \Omega t + p(0)\,(M\Omega)^{-1} \sin \Omega t. \tag{25}$$

It is convenient to take $x(0)$ and $p(0)$ to be the Schrödinger picture operators encountered in wave mechanics (Chap. 1). Then the operator $x(0)$ has the eigenvalue equation

$$x(0)\, \delta(x - x_0) = x_0\, \delta(x - x_0), \tag{26}$$

which with (18) gives

$$x(-t)\, G(x, x_0, t) = \exp(-i\mathscr{H}t/\hbar)\, x(0) \exp(i\mathscr{H}t/\hbar)\exp(-i\mathscr{H}t/\hbar)\delta(x - x_0)$$

$$= x_0 G(x, x_0, t). \tag{27}$$

This is intuitively understandable from (26) for, at some $-t$, $G(x, x_0, t)$ was, in fact, just the delta function $\delta(x - x_0)$. Using (25) for $x(-t)$, we find

$$[x(0) \cos \Omega t - p(0)(M\Omega)^{-1} \sin \Omega t - x_0]G(x, x_0, t) = 0. \tag{28}$$

Since $x(0)$ and $p(0)$ are the Schrödinger picture operators, we find that (28) yields the simple differential equation

$$-i\hbar (M\Omega)^{-1} \sin \Omega t \frac{\partial G}{\partial x} = (x \cos \Omega t - x_0)G.$$

This can be written as

$$\frac{1}{G}\frac{\partial G}{\partial x} = iM\Omega(\hbar \sin \Omega t)^{-1} (x \cos \Omega t - x_0),$$

which has the integral

$$\ln(G/G_0) = iM\Omega(\hbar \sin \Omega t)^{-1} (\tfrac{1}{2}x^2 \cos \Omega t - x_0 x) + \text{constant},$$

that is,

$$G = G_0 \exp [\tfrac{1}{2}iM\Omega(\hbar \sin \Omega t)^{-1} (x^2 \cos \Omega t - 2x_0 x) + \text{constant}].$$

We evaluate the constant by noting from the bilinear form (21) that, for the (real) Hermite-Gaussian functions, $G(x, x_0, t)$ is symmetric in x and x_0. Hence

$$G = \mathcal{N}_G' \exp \{\tfrac{1}{2}iM\Omega(\hbar \sin \Omega t)^{-1} [(x^2 + x_0^2) \cos \Omega t - 2x_0 x]\}. \tag{29}$$

Further using the transitivity relation

$$\int dx'\, G(x, x', t')\, G(x', x_0, t'') = G(x, x_0, t' + t''),$$

for $t' + t'' = 0$, we find the normalization constant:

$$\mathcal{N}_G' = \left(\frac{\tfrac{1}{2}M\Omega}{\hbar\pi |\sin \Omega t|}\right)^{1/2}. \tag{30}$$

For $t = 0$, this gives $\delta(x - x_0)$ as required. For $t = \pi/2\Omega$, we obtain a plane wave. This is truly a wave packet whose width is extraordinarily time dependent!

In the calculation for the wave function (17), it is convenient to introduce a normalized coordinate like (13):

$$\xi = \left(\frac{M\Omega}{\hbar}\right)^{1/2} x \tag{31}$$

and the ratio

$$R = \frac{C}{M\Omega}. \tag{32}$$

The Green function (29) is then written as

$$G(\xi, \xi_0, t) = \mathcal{N}_G \exp\{\tfrac{1}{2}i(\sin \Omega t)^{-1} [(\xi^2 + \xi_0^2) \cos \Omega t - 2\xi_0\xi]\}, \tag{33}$$

where the normalization constant (30) reduces to

$$\mathcal{N}_G = (2\pi |\sin \Omega t|)^{-1/2}. \tag{34}$$

The initial wave function (15) becomes

$$\psi(\xi_0, 0) = \mathcal{N}_\psi \exp(-\tfrac{1}{2}R\xi_0^2 + \chi'\xi_0), \tag{35}$$

where the complex displacement

$$\chi' = \sqrt{2R}\chi, \tag{36}$$

and the normalization constant

$$\mathcal{N}_\psi = \left(\frac{C}{\pi\hbar}\right)^{1/4} \exp(-\tfrac{1}{2}\chi^2 - \tfrac{1}{2}|\chi|^2). \tag{37}$$

The time-dependent wave function is given by

$$\begin{aligned}\psi(\xi, t) &= \int d\xi_0\, G(\xi, \xi_0, t)\, \psi(\xi_0, 0) \\ &= \mathcal{N}_\psi \mathcal{N}_G \int d\xi_0 \exp\{i(\sin \Omega t)^{-1} [\tfrac{1}{2}(\xi^2 + \xi_0^2) \cos \Omega t - \xi_0\xi] \\ &\qquad\qquad -\tfrac{1}{2}R\xi_0^2 + \chi'\xi_0\}.\end{aligned} \tag{38}$$

The argument of the exponential can be reduced by completion of the square in ξ_0 to

$$\begin{aligned}\{\ \} &= \tfrac{1}{2}\xi_0^2(i \cot \Omega t - R) + \xi_0(\chi' - i\xi/\sin \Omega t) + \tfrac{1}{2}i\xi^2 \cot \Omega t \\ &= -\tfrac{1}{2}(R - i \cot \Omega t)\left(\xi_0^2 - 2\xi_0 \frac{\chi' - i\xi/\sin \Omega t}{R - i \cot \Omega t}\right) + \tfrac{1}{2}i\xi^2 \cot \Omega t \\ &= -\tfrac{1}{2}(R - i \cot \Omega t)\left(\xi_0 - \frac{\chi' - i\xi/\sin \Omega t}{R - i \cot \Omega t}\right)^2 + \frac{1}{2}\frac{(\chi' - i\xi/\sin \Omega t)^2}{R - i \cot \Omega t} \\ &\qquad + \tfrac{1}{2}i\xi^2 \cot \Omega t.\end{aligned} \tag{39}$$

With this, the integral in (38) over ξ_0 yields the factor $[2\pi/(R - i \cot \Omega t)]^{1/2}$. We have

$$\begin{aligned}\psi(\xi, t) &= \mathcal{N}_\psi \mathcal{N}_G \left(\frac{2\pi}{R - i \cot \Omega t}\right)^{1/2} \exp\left[\frac{\tfrac{1}{2}(\chi' - i\xi/\sin \Omega t)^2}{R - i \cot \Omega t} + \tfrac{1}{2}i\xi^2 \cot \Omega t\right] \\ &= \mathcal{N}_\psi \mathcal{N}_G \left(\frac{2\pi}{R - i \cot \Omega t}\right)^{1/2} \\ &\qquad \times \exp\left\{\frac{\tfrac{1}{2}\chi'^2 - i\chi'\xi/(\sin \Omega t) - \tfrac{1}{2}\xi^2(1 - iR \cot \Omega t)}{R - i \cot \Omega t}\right\}\end{aligned} \tag{40}$$

The coefficient of ξ^2 in the exponent of the minimum-uncertainty wave function (16) is real. That in (40) is not real for all time unless the ratio R is unity, that is, unless

$$C = M\Omega. \tag{41}$$

In other respects, (40) satisfies the conditions for minimum uncertainty, and hence the choice (41) in (40) yields the desired wave function.

This conclusion is illustrated by the time development of the probability density $\psi^*\psi$ for (40). Some straightforward algebra gives

$$\psi^*\psi = \left(\frac{M\Omega}{\pi\hbar}\right)^{1/2} \frac{1}{\sigma} \exp\{-[\xi - \xi_0 \cos(\Omega t + \phi')]^2/\sigma^2\}, \tag{42}$$

where the displacement $\xi_0 = |\chi'|(\sin^2\phi + \cos^2\phi/R^2)^{1/2}$, the phase $\phi' = \tan^{-1}(R\tan\phi)$, $\chi' = |\chi'|\exp(-i\phi)$, and the width

$$\sigma = \left(\frac{\cos^2\Omega t + R^2\sin^2\Omega t}{R}\right)^{1/2}. \tag{43}$$

We see that the probability density is a Gaussian wave packet whose center executes simple harmonic motion and whose width (43) is, in general, variable in time. The packet coheres, that is, has constant width only if $R = 1$.

Thus choosing (41) in (40), we find the coherent wave function:

$$\psi(\xi, t) = \left(\frac{M\Omega}{\pi\hbar}\right)^{1/4} \exp[-\tfrac{1}{2}a^2 - \tfrac{1}{2}|a|^2]\exp(i\Omega t/2)\exp(ia^2e^{-i\Omega t}\sin\Omega t + \sqrt{2}\,a\xi e^{-i\Omega t} - \tfrac{1}{2}\xi^2), \tag{44}$$

with the complex displacement

$$a = (2M\Omega\hbar)^{-1/2}[M\Omega\langle q\rangle + i\langle p\rangle]. \tag{45}$$

Further simplifying (44) and dropping the irrelevant phase factor $\exp i\Omega t/2$, we obtain

$$\begin{aligned}\psi(\xi, t) &= \left(\frac{M\Omega}{\pi\hbar}\right)^{1/4} \exp[-\tfrac{1}{2}a^2(1 - 2i\sin\Omega t\cos\Omega t - 2\sin^2\Omega t) - \tfrac{1}{2}|a|^2 \\ &\qquad -\tfrac{1}{2}(\xi - \sqrt{2}\,ae^{-i\Omega t})^2 + (ae^{-i\Omega t})^2] \\ &= \exp[\tfrac{1}{2}(ae^{-i\Omega t})^2 - \tfrac{1}{2}|a|^2]\phi_0(\xi - \sqrt{2}\,ae^{-i\Omega t}),\end{aligned} \tag{46}$$

where ϕ_0 is the lowest-order Hermite-Gaussian function of (1.22). We see that the coherent $\psi(\xi, t)$ is the same as the initial state (15) with the displacement χ given by $ae^{-i\Omega t}$ and the width parameter C by $M\Omega$.

H-3. Expansion in Number States

To expand the coherent wave function (46) in the Hermite-Gaussian functions (1.23) without displacement $[\phi_n(\xi)]$, we note that the Taylor series expansion for a function $f(x - x_0)$ can be written as

$$\begin{aligned}f(x - x_0) &= f(x) - x_0 f'(x) + x_0^2 f''(x)/2! - \cdots \\ &= \exp[-x_0(d/dx)]f(x).\end{aligned} \tag{47}$$

Hence (46) can be written (for typographical simplicity we take $t = 0$) as

$$\psi(\xi, 0) = \exp[\tfrac{1}{2}(a^2 - aa^*)] \exp[-\sqrt{2}\, a(d/d\xi)]\, \phi_0(\xi). \tag{48}$$

We can find a formula for the derivative $d/d\xi$ in terms of the annihilation and creation operators a and $a^\dagger$ by noting that the momentum operator is given by

$$p = \tfrac{1}{2}i(2M\hbar\Omega)^{1/2}(a^\dagger - a) = -i\hbar\frac{d}{dq} = -\, i\hbar\left(\frac{M\Omega}{\hbar}\right)^{1/2}\frac{d}{d\xi} = -i(M\hbar\Omega)^{1/2}\frac{d}{d\xi},$$

that is,

$$\frac{d}{d\xi} = -\frac{1}{\sqrt{2}}(a^\dagger - a). \tag{49}$$

Inserting this into (48), we obtain

$$\psi(\xi) = \exp[\tfrac{1}{2}(a^2 - aa^*)] \exp[a(a^\dagger - a)]\phi_0(\xi). \tag{50}$$

Now the operator (see Messiah, 1961, p. 442)

$$\exp[a(a^\dagger - a)] = \exp\{\tfrac{1}{2}a^2[a^\dagger, a]\} \exp(aa^\dagger) \exp(-aa). \tag{51}$$

Furthermore, $\exp(-aa)\phi_0 = \phi_0$, and the wave function reduces to

$$\begin{aligned}\psi(\xi) &= \exp(-\tfrac{1}{2}aa^*) \exp(aa^\dagger)\phi_0(\xi)\\ &= \sum_{n=0}^{\infty} \frac{a^n}{\sqrt{n!}} \exp(-\tfrac{1}{2}|a|^2)\phi_n(\xi),\end{aligned} \tag{52}$$

where we have used (14.32). This wave function is the coordinate representative of the coherent state:

$$|a\rangle = \sum_{n=0}^{\infty} \frac{a^n}{\sqrt{n!}} \exp(-\tfrac{1}{2}|a|^2)|n\rangle. \tag{53}$$

Note that, if we include the time dependence in (52) in the standard way of Eq. (1.13), we find

$$\psi(\xi, t) = \sum_{n=0}^{\infty} \left[\frac{(ae^{-i\Omega t})^n}{\sqrt{n!}}\right] \exp(-\tfrac{1}{2}|a|^2)\phi_n(\xi), \tag{54}$$

which agrees with the replacement of a by $ae^{-i\Omega t}$ in (52), as indicated by Eqs. (46) and (48).

Finally, forming the probability density, we find

$$\begin{aligned}\psi^*\psi &= \left(\frac{M\Omega}{\pi\hbar}\right)^{1/2} \exp[-\tfrac{1}{2}(ae^{-i\Omega t})^2 - |a|^2 - \tfrac{1}{2}(a^*e^{i\Omega t})^2]\\ &\qquad \times \exp[-\xi^2 + 2\sqrt{2}\xi\, \mathrm{Re}\,(ae^{-i\Omega t})]\\ &= \left(\frac{M\Omega}{\pi\hbar}\right)^{1/2} \exp[-2\mathrm{Re}(ae^{-i\Omega t})^2] \exp\{-[\xi - \sqrt{2}\,\mathrm{Re}\,(ae^{-i\Omega t})]^2\\ &\qquad + 2\mathrm{Re}(ae^{-i\Omega t})^2\}\end{aligned}$$

$$= \left(\frac{M\Omega}{\pi\hbar}\right)^{1/2} \exp\{-[\xi - \sqrt{2}\,|a|\cos(\Omega t + \phi)]^2\}, \tag{55}$$

where the complex

$$a = |a|\exp(-i\phi).$$

Here we see that the wave packet oscillates back and forth with constant spread $[\hbar/(M\Omega)]^{1/2}$. This conclusion follows, of course, from the more general expression (42).

Now, using relation (14.6) between the electric field and position q, we find the corresponding wave function:

$$\psi(E, t) = (\sqrt{2}\pi\mathscr{E})^{-1/2} \exp[\tfrac{1}{2}(ae^{-i\Omega t})^2 - \tfrac{1}{2}|a|^2] \times \exp\{-\tfrac{1}{2}[(\sqrt{2}\mathscr{E})^{-1}E - \sqrt{2}\,ae^{-i\Omega t}]^2\} \tag{56}$$

and the probability distribution:

$$\psi(E,t)^*\psi(E,t) = (\sqrt{2}\pi\mathscr{E})^{-1} \exp\{-[(\sqrt{2}\mathscr{E})^{-1}E - \sqrt{2}\,|a|\cos(\Omega t + \phi)]^2\}. \tag{57}$$

References

See references of Chap. 15 and:

E. H. Kennard, 1927, Z. Physik **44,** 326.

A. Messiah, 1961, *Quantum Mechanics,* North-Holland Publishing Co., Amsterdam.

L. I. Schiff, 1968, *Quantum Mechanics,* 2nd ed., McGraw-Hill Book Co., New York.

E. Schrödinger, 1926, Naturwissenschaften **14,** 664.

I. R. Senitzky, 1954, Phys. Rev. **95,** 904.

I

GENERAL QUANTUM LASER EQUATIONS

Appendix I. General Quantum Laser Equations of Motion

In this appendix we calculate the equations of motion for the reduced density matrix describing the laser field. The calculation is similar to the atomic beam version in Chap. 17 but deals with atoms whose levels a and b decay by spontaneous emission to lower-lying levels c and d, respectively, as depicted in Fig. I-1. This model is typical of laser behavior and parallels the semiclassical discussion of Chap. 8 more closely than it corresponds to the atomic beam approach. Inasmuch as the beam transit times are taken to have the distribution $\gamma \exp(-\gamma\tau)$, which simulates spontaneous emission, the results are the same for equal decay constants ($\gamma_a = \gamma_b = \gamma$). A by-product of our discussion (Sec. I-2) is a derivation of the semiclassical equations of motion (7.34)–(7.36) for the reduced atomic density matrix. This development as-

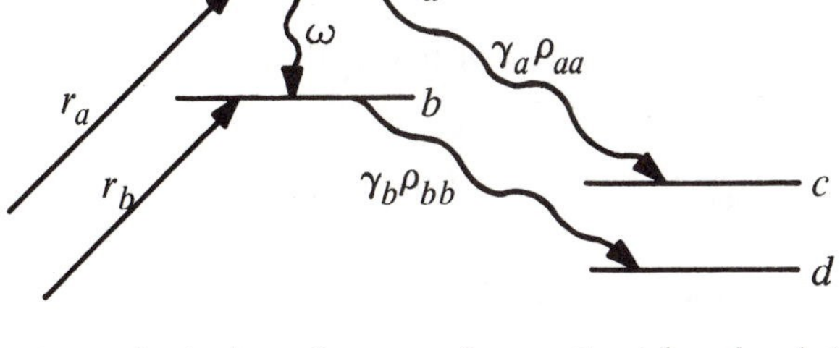

Figure I-1. Atomic level scheme for atoms. Laser action takes place between levels a b, which decay to levels c and d with decay constants γ_a and γ_b, respectively. The energy spacing between a and b is $\hbar\omega$. The excitation rates to levels a and b are given by r_a and r_b, respectively. In the text, we consider primarily excitation to the upper level alone ($r_b = 0$). The general model depicted here corresponds quite closely to the semiclassical treatment of Chap. 8 and is developed in the problems.

sumes that the field is given by a coherent state (15.11) and that off-diagonal correlations between the field and atomic coordinates can be neglected. In particular, this "factorization" of the atom-field density matrix discards the spontaneous emission into the laser field itself, emission producing the major part of the laser linewidth (apart, of course, from mechanical vibrations). This role of spontaneous emission is discussed intuitively in Sec. 20-3 and analytically in Secs. 17-3, 18-2, and 20-3.

I-1. Quantum Laser Equations of Motion

In the notation of Sec. 14-4, a representative state vector for the combined atom-laser-free-field system is given by

$$|\psi_{\text{a}-l-\text{f}}(t)\rangle = \sum_n \{[C_{an\{0\}}(t)|a\rangle + C_{bn\{0\}}(t)|b\rangle]|\{0\}\rangle$$

$$+ \sum_r [C_{cn\{1_r\}}(t)|c\rangle + C_{dn\{1_r\}}(t)|d\rangle]|1_r\}\rangle\}|n\rangle, \tag{1}$$

where, for example, the probability amplitude $C_{an\{0\}}(t)$ is that for the upper atomic level, an n-photon laser field, and no photons in all other modes, and $C_{cn\{1_r\}}(t)$ is the amplitude for level c, an n-photon laser field, and one photon in the rth mode with no photons in the remaining modes. The density operator composed of such state vectors is given by

$$\rho_{\text{a}-l-\text{f}}(t) = \sum_\psi P_\psi |\psi_{\text{a}-l-\text{f}}(t)\rangle\langle\psi_{\text{a}-l-\text{f}}(t)|$$

$$= \sum_n\sum_m \Big\{[\rho_{an\{0\};\, am\{0\}}|a\rangle\langle a| + \rho_{bn\{0\};\, bm\{0\}}|b\rangle\langle b|] \otimes |\{0\}\rangle\langle\{0\}|$$

$$+ \sum_r [\rho_{cn\{1_r\};\, cm\{1_r\}}|c\rangle\langle c|$$

$$+ \rho_{dn\{1_r\};\, dm\{1_r\}}|d\rangle\langle d|]\,|\{1_r\}\rangle\langle\{1_r\}|\Big\}|n\rangle\langle m|$$

$$+ \text{terms off atomic or free-field diagonals.} \tag{2}$$

Here the density matrix elements

$$\rho_{an\{0\};\, am\{0\}}(t) = \sum_\psi P_\psi\, C_{an\{0\}}(t) C_{am\{0\}}{}^*(t), \tag{3}$$

$$\rho_{bn\{0\};\, bm\{0\}}(t) = \sum_\psi P_\psi\, C_{bn\{0\}}(t) C_{bm\{0\}}{}^*(t), \tag{4}$$

with similar expressions for the remaining elements in (2).

We desire to find the reduced laser-field density matrix element $\rho_{nm}(t)$. This is given by the trace over atomic and free-field coordinates, that is, by

$$\rho_{nm}(t) = \rho_{an\{0\};\, am\{0\}} + \rho_{bn\{0\};\, bm\{0\}} + \rho_{cn;\, cm} + \rho_{dn;\, dm}, \tag{5}$$

in which the reduced matrix elements $\rho_{cn;\,cm}$ and $\rho_{dn,\,dm}$ are defined by

$$\rho_{cn;\,cm} = \sum_r \rho_{cn\{1_r\};\,cm\{1_r\}}, \tag{6}$$

$$\rho_{dn;\,dm} = \sum_r \rho_{dn\{1_r\};\,dm\{1_r\}}. \tag{7}$$

The Weisskopf-Wigner decay calculation leading to Eq. (14.104) proceeds here for decay from level a to level c independently of the laser interaction between levels a and b. Hence the equation of motion for $C_{an\{0\}}(t)$ is just the sum of the stimulated term like (14.70) and the spontaneous contribution of (14.104), namely,

$$\dot{C}_{an\{0\}}(t) = -\tfrac{1}{2}\gamma_a C_{an\{0\}}(t) - ig\sqrt{n+1}\exp[i(\omega-\nu)t]C_{b,n+1,\{0\}}(t). \tag{8}$$

Similarly, the damped equation for $\dot{C}_{b,n+1,\{0\}}(t)$ is given by

$$\dot{C}_{b,n+1,\{0\}}(t) = -\tfrac{1}{2}\gamma_b C_{b,n+1,\,\{0\}}(t) - ig\sqrt{n+1}\exp[-i(\omega-\nu)t]\,C_{an\{0\}}(t). \tag{9}$$

Use of these equations shows that the matrix elements (3) and (4) decay to zero in times $1/\gamma_a$ and $1/\gamma_b$, respectively. Hence, for an atom excited at time t, the field matrix element (5) at time $t+\tau$ for τ greater than the atomic lifetimes reduces to

$$\rho_{nm}(t+\tau) \simeq \rho_{cn;\,cm}(t+\tau) + \rho_{dn;\,dm}(t+\tau). \tag{10}$$

We can derive the equations of motion for these reduced elements (6) and (7) along the lines of the Weisskopf-Wigner theory of Sec. 14-4. Specifically, from (6) and the lower-level equation (14.98), we find

$$\begin{aligned}\dot{\rho}_{cn;\,cm}(t) &= \sum_\psi P_\psi \sum_r \dot{C}_{cn\{1_r\}} C_{cm\{1_r\}}^* + \text{c.c.}\\ &= -i\sum_\psi P_\psi \sum_r g_r \exp[i(\Omega_r-\omega)t]\,C_{an\{0\}} C_{cm\{1_r\}}^* + \text{c.c.}\\ &= -i\sum_r g_r \exp[i(\Omega_r-\omega)t]\,\rho_{an\{0\};\,cm\{1_r\}} + \text{c.c.}\end{aligned} \tag{11}$$

In turn, from (14.97) and (14.98), we find

$$\begin{aligned}\dot{\rho}_{an\{0\};\,cm\{1_r\}} &= \sum_\psi P_\psi\{-i\sum_s g_s \exp[-i(\Omega_s-\omega)t]C_{cn\{1_s\}}\,C_{cm\{1_r\}}^*\\ &\qquad + ig_r\exp[-i(\Omega_r-\omega)t]\,C_{an\{0\}}C_{am\{0\}}^*\}\\ &\simeq ig_r\exp[-i(\Omega_r-\omega)t]\rho_{an\{0\};\,am\{0\}}.\end{aligned} \tag{12}$$

Here the approximation follows since $|C_{cm\{1_s\}}|^2 \ll 1$. A formal integral of (12) gives

$$\rho_{an\{0\};\,cm\{1_r\}}(t) = ig_r\exp[-i(\Omega_r-\omega)t]\int_0^t dt'\exp[-i(\Omega_r-\omega)(t-t')] \times \rho_{an\{0\};\,am\{0\}}(t').$$

Combining this with (11), we find

$$\dot{\rho}_{cn;\,cm}(t) = \sum_r g_r^2 \int_0^t dt' \exp[-i(\Omega_r - \omega)(t - t')]\rho_{an\{0\};\,am\{0\}}(t') + \text{c.c.} \tag{13}$$

This is the same kind of integral as (14.99), here including a complex conjugate and a sign change. Solving it as for (14.99) and using γ_a to indicate explicitly decay from level a, we have

$$\dot{\rho}_{cn;\,cm}(t) = \gamma_a \rho_{an\{0\};\,am\{0\}}(t). \tag{14}$$

Similary, $\rho_{dn;\,dm}(t)$ has the equation of motion

$$\dot{\rho}_{dn;\,dm}(t) = \gamma_b \rho_{bn\{0\};\,bm\{0\}}(t). \tag{15}$$

These equations are plausible, for they say that the total spontaneous emission from $|a\rangle$ (or $|b\rangle$), regardless of wavelength, leaves the atom in state $|c\rangle$ (or $|d\rangle$). The laser field plays a negligible role in this process since its frequency is far from resonance. With the use of (14) and (15), Eq. (10) can be written as

$$\rho_{nm}(t + \tau) = \int_0^\tau d\tau' \,[\gamma_a \rho_{an\{0\};\,am\{0\}}(t + \tau') + \gamma_b \rho_{bn\{0\};\,bm\{0\}}(t + \tau')]. \tag{16}$$

Thus our problem reduces to finding $\rho_{an\{0\};\,am\{0\}}$ and $\rho_{bn\{0\};\,bm\{0\}}$ as functions of time. We obtain them using the element definitions (3) and (4) with the equations of motion (8) and (9).

Suppose that an atom is excited to the upper state $|a\rangle$ at time t. Then an atom-laser-free field state vector (1) factors as

$$|\psi_{a-l-f}(t)\rangle = |a\rangle|\{0\}\rangle|\psi(t)\rangle, \tag{17}$$

where $|\psi(t)\rangle$ is a representative field state vector (see Sec. 17-1). With reference to Eq. (1), we see that $C_{an\{0\}}(t) = C_n(t)$ and the other initial probability amplitudes vanish. According to the equations of motion (8) and (9), $C_{b,n+1,\{0\}}$ couples with $C_{an\{0\}}$ in time. Analogously to Eqs. (17.7) and (17.8) we write

$$C_{an\{0\}}(t + \tau') = C_n(t)\mathscr{C}_{an}(t + \tau'), \tag{18}$$

$$C_{b,n+1,\{0\}}(t + \tau') = C_n(t)\mathscr{C}_{b,n+1}(t + \tau'), \tag{19}$$

in which the amplitudes $\mathscr{C}_{an}(t)$ and $\mathscr{C}_{b,n+1}(t)$ are solutions of (8) and (9), respectively.

In terms of the $\mathscr{C}$'s, the density matrix elements of (3) and (4) are written as

$$\rho_{an\{0\};\,am\{0\}}(t + \tau') = \rho_{nm}(t)\,\mathscr{C}_{an}(t + \tau')\,\mathscr{C}_{am}^*(t + \tau'), \tag{20}$$

$$\rho_{bn\{0\};\,bm\{0\}}(t + \tau') = \rho_{n-1,\,m-1}(t)\,\mathscr{C}_{bn}(t + \tau')\,\mathscr{C}_{bm}^*(t + \tau'). \tag{21}$$

Thus Eq. (10) for $\rho_{nm}(t + \tau)$ reduces to

$$\rho_{nm}(t + \tau) = \rho_{nm}(t) \int_0^\infty d\tau' \,\gamma_a \,\mathscr{C}_{an}(t + \tau')\,\mathscr{C}_{am}^*(t + \tau')$$

$$+ \rho_{n-1,m-1}(t) \int_0^\infty d\tau' \, \gamma_b \mathscr{C}_{bn}(t + \tau') \mathscr{C}_{bm}{}^*(t + \tau'). \quad (22)$$

Here we have extended the limit of integration to ∞ (as in the semiclassical theory) for mathematical convenience. The coarse-grained time rate of change of ρ_{nm} is then given by the rate r_a of excitation to $|a\rangle$ multiplied by the change induced by a single atom, that is,

$$\begin{aligned}\dot{\rho}_{nm}(t) &= r_a[\rho_{nm}(t + \tau) - \rho_{nm}(t)] \\ &= -\rho_{nm}(t) r_a \left[1 - \int_0^\infty d\tau' \, \gamma_a \mathscr{C}_{an}(t + \tau') \mathscr{C}_{am}{}^*(t + \tau')\right] \\ &\quad + \rho_{n-1,m-1}(t) r_a \int_0^\infty d\tau' \, \gamma_b \mathscr{C}_{bn}(t + \tau') \mathscr{C}_{bm}{}^*(t + \tau'). \quad (23)\end{aligned}$$

Note that this is a generalization of Eq. (17.9), in which γ_a can be different from γ_b. Furthermore we have established that Eq. (17.9) does, in fact, account properly for spontaneous emission.

The integrals in this equation have a particularly simple meaning. Consider, for example, the integral for the upper level. Schematically it reads

$\int_0^\infty d\tau \, \gamma_a \times$ (Probability that atom is in $|a\rangle$ at time $t + \tau'$ in presence of *nm*-photon field)

= total probability of spontaneous emission from a to c in this field

= probability that *nm* field does *not* stimulate atom to emit.

Similarly, the b integral is the probability of stimulated emission by an $n - 1$, $m - 1 =$ photon field. Thus we can write the coarse-grained equation of motion (23) in the schematic form:

$\dot{\rho}_{nm}(t) = -\rho_{nm}(t) \times$ (rate of atomic excitation to upper level)
$\times$ (probability of stimulated emission by an *nm*-photon field)
$+ \rho_{n-1,m-1}(t) \times$ (rate of atomic excitation to $|a\rangle$)
$\times$ (probability of stimulated emission by an $(n - 1, m - 1)$-photon field).

In particular, the equation of motion for the diagonal element ρ_{nn}, the probability of n photons, has the intuitively reasonable interpretation

$\dot{\rho}_{nn}(t) = -$(probability of n photons) $\times$ (rate of excitation to $|a\rangle$)
$\times$ (probability of stimulated emission by n-photon field)
$+$ (probability of $n - 1$ photons) $\times$ (rate of excitation to $|a\rangle$)
$\times$ (probability of stimulated emission by $(n - 1)$-photon field).
(24)

This result is reminiscent of the first Fokker-Planck development in Sec. 16-3.

In view of the decomposition (18) and (19), the $\mathscr{C}$'s have the equations of motion (8) and (9). The solutions are particularly simple for the case of equal decay constants ($\gamma_a = \gamma_b = \gamma$) and resonance ($\nu = \omega$), for which Eq. (23) reduces to (17.9). More general solutions are discussed in Prob. 17-15.

I-2. Reduction to the Semiclassical Theory

In this section we derive the semiclassical density matrix equations of motion (7.34)–(7.36) for the electric-dipole perturbation energy. We suppose that the field is described by a coherent state (15.11) inasmuch as this yields the most nearly classical field. The essential approximation we make is that the atomic coordinates are uncorrelated with those of the field, that is, the atom-field density matrix factors as

$$\rho_{\text{a-f}}(t) = \rho_{\text{atom}}(t) \otimes \rho_{\text{field}}(t). \tag{25}$$

The reader will recall that initially the combined matrix does factor, but that correlations develop in time represented by the joint probability amplitudes C_{an} and $C_{b,n+1}$. The factorization (25) leads to a zero linewidth instead of (17.44), since the off-diagonal field elements ρ_{nm} no longer decay.

From the probability amplitude equations of motion (8) and (9), we find

$$\dot{\rho}_{an;\,an} = -\gamma_a \rho_{an;\,an} - \{ig\sqrt{n+1}\exp[i(\omega-\nu)t]\rho_{b,n+1;\,an} + \text{c.c.}\}, \tag{26}$$

$$\dot{\rho}_{bn;\,bn} = -\gamma_b \rho_{bn;\,bn} + \{ig\sqrt{n}\exp[i(\omega-\nu)t]\,\rho_{bn;\,a,n-1} + \text{c.c.}\}, \tag{27}$$

$$\dot{\rho}_{an;\,bn} = -\gamma_{ab}\rho_{an;\,bn} - ig\exp[i(\omega-\nu)t]\,[\sqrt{n+1}\,\rho_{b,n+1;\,bn} - \sqrt{n}\,\rho_{an;\,a,n-1}]. \tag{28}$$

The atom-only equations of motion are given by the trace over the combined matrix, that is, by $\rho_{\text{atom}}(t) = \text{Tr}_{\text{field}}\rho_{\text{a-f}}(t)$. Therefore the component equations of motion for $\rho_{\text{atom}}(t)$ are given by the field trace over (26)–(28). To perform these traces, we factor the density matrix, giving

$$\rho_{b,n+1;\,an} = \rho_{ba}\rho_{n+1,n}, \qquad \rho_{b,n+1;\,bn} = \rho_{bb}\rho_{n+1,n},$$
$$\rho_{an;\,a,n-1} = \rho_{aa}\rho_{n,n-1}. \tag{29}$$

The matrix element $\rho_{n+1,n}$ for a coherent state $\rho = |a\rangle\langle a|$ is given from (15.11) as

$$\rho_{n+1,n} = \frac{a^{n+1}}{\sqrt{(n+1)!}}\frac{(a^*)^n}{\sqrt{n!}} = \frac{a}{\sqrt{n+1}}\rho_{nn}. \tag{30}$$

Hence the field trace over (26) gives

$$\dot{\rho}_{aa} = \sum_n \dot{\rho}_{an;\,an} = -\gamma_a\rho_{aa} - \left\{ig\sum_n \exp[i(\omega-\nu))t\sqrt{n+1}\;\rho_{ba}\rho_{n+1,n} + \text{c.c.}\right\}$$
$$= -\gamma_a\rho_{aa} - \left\{ig\exp[i(\omega-\nu)t]\,\rho_{ba}\sum_n \sqrt{n+1}\frac{a}{\sqrt{n+1}}\rho_{nn} + \text{c.c.}\right\}$$

$$= -\gamma_a \rho_{aa} - \{iga \exp[i(\omega - \nu)t] \rho_{ba} + \text{c.c.}\}. \tag{31}$$

Here, from (14.60), $ga = -(\wp \mathscr{E} a/\hbar) \sin Kz$, so that (31) is just the expression (7.34) with the interaction energy $\mathscr{V}_{ab} = -\wp \mathscr{E}\, a \sin Kz$ as desired. A similar calculation yields electric-dipole versions for (7.35) and (7.36) from (27) and (28).

INDEX